SOTHIS PRESS

Sothis Press aims to completely reimagine the public domain book with an emphasis on universal access to captivatingly designed print editions. We believe that technology will enable every single book ever published to be perpetually accessible in beautiful print form. Sothis is beauty in ubiquity.

www.sothis.press
1 Waterhouse Square, London, ECIN 2ST

The Journal of Comparative Neurology and Psychology, 1908

Henry H. Donaldson

SOTHIS PRESS

ISBN 978-1-0365-0670-4

The Journal of Comparative Neurology and Psychology, 1908 is in the public domain.

This edition is published by Sothis Press © 2023. The company is registered in England and Wales. The Company Number is 14900389.

THE JOURNAL

OF

COMPARATIVE NEUROLOGY AND PSYCHOLOGY

VOLUME XVIII

1908

PHILADELPHIA

THE WISTAR INSTITUTE OF ANATOMY AND BIOLOGY

The Journal of
Comparative Neurology and Psychology

CONTENTS OF VOLUME XVIII, 1908

Number 1, January, 1908.

Number 2, April, 1908.

Number 3, June, 1908.

Number 4, October, 1908.

Number 5, November, 1908.

Number 6, December, 1908.

SUBJECT AND AUTHOR INDEX.

VOLUME XVIII.

Authors' names are in small caps. References to subjects and authors of original articles are marked with an asterisk.

The Journal of
Comparative Neurology and Psychology

VOLUME XVIII | JANUARY, 1908 | NUMBER 1

AN EXPERIMENTAL STUDY OF IMITATION IN CATS.[1]

BY

CHARLES SCOTT BERRY

(From the Harvard Psychological Laboratory.)

WITH TWO FIGURES.

I. PROBLEM AND METHOD.

This paper gives an account of some experiments which were made for the purpose of determining to what extent imitation is a factor in the development of the cat. It is a continuation of the study of imitation which was begun with the white rat.[2]

The experiments now to be described were made with four Manx cats, a mother cat and three kittens, which I have designated by the letters M, X, Y and Z.

NAME	COLOR	SEX	DATE OF BIRTH
M	gray	female	unknown
X	gray	female	August, 1906
Y	gray	male	August, 1906
Z	gray	female	August, 1906

My experiments were begun in October, 1906, and concluded in March, 1907. During this time I fed the cats bread and milk

[1] This investigation was carried on under the direction of Doctor ROBERT M. YERKES, to whom I am greatly indebted for the suggestion of the problem and general method.

[2] The Imitative Tendency of White Rats. *Journal of Comparative Neurology and Psychology*, vol. 16, pp. 333–361. 1906.

and raw meat, with the occasional addition of some vegetables. They were fed only once a day, and the feeding time was immediately after the experiment. Enough bread and milk were given to keep the animals in good condition, and the raw beef was used principally as a reward for the performance of the required act. In all the experiments the trained cat was fed each time it performed the act which the situation required. All of the cats were perfectly tame and very active.

The general method of testing the imitative tendency of cats was as follows: Either separately or together the cats were given opportunity to learn to perform a certain act or series of acts. In case all learned it of their own initiative no tests of imitation could be made, but if one, or more, of the individuals failed, after abundant opportunity, to discover the appropriate mode of reaction, it was given a chance to learn by watching another cat perform the act.

II. EXPERIMENTS.

Experiment 1. Jumping from Box to Table.

Method.—A box, open on one side, was placed upon a table 81 cm. high. A second table 10 cm. lower was placed 56 cm. from the first. The act to be performed was to jump from the box to the meat, which was placed in plain sight on the lower table.

Results.—X, who was put into the box alone, was afraid to jump. M was put in with her, and she saw M jump from the box to the table five times, but still she was afraid to follow. Y and Z were tested in like manner with the same result. I now drew the lower table 8 cm. nearer the first, and put the three kittens into the box together, but still they were afraid to jump. M was now put in with them. The second time she jumped to the meat, X followed her. When X was put back with Y and Z she jumped down at once, and in less than a minute was followed by Z. After Y saw Z jump to the meat five times he jumped to the floor. No further trials were made.

Experiment 2. Opening Door by Pulling Knot.

Method.—A wooden box (Box I) 73 cm. long, 50 cm. wide, and 56 cm. deep was closed on one side by a door of wire netting of half inch mesh. Near one end of the opposite side was a door (20 x 15 cm.) which was held shut by a wooden crossbar. From the end of this bar a string passed up over the box and entered it at the middle of the end farthest from the door. To open the door it was necessary for the cat to pull a knot at the end of the string. The knot was close against the inside of the box (see Fig. 1).

Results.—By fastening a piece of meat to the knot I taught M to open the door. At the end of the fourth day, the association between the pulling of the knot and the opening of the door seemed to be well formed.

X, Y and Z were put into the box together. In less than five minutes X was pulling the knot. With Z watching, X went to the

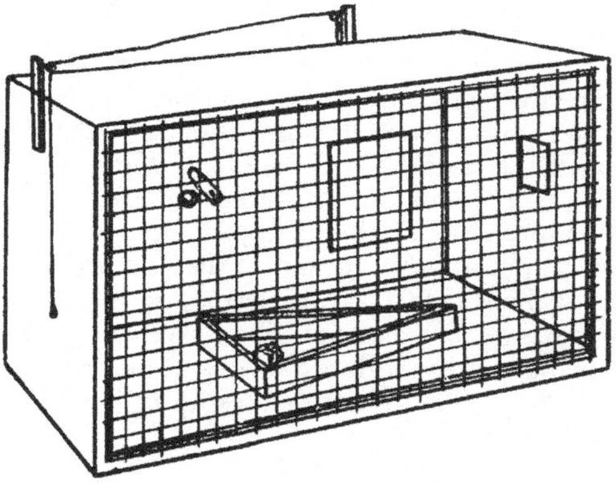

BOX I.

FIG. 1.

knot, seized it and pulled hard enough to open the door. After they had been fed and put into the box again, Z pulled the knot first. X then tried it, and finally Y seized it and succeeded in opening the door. When they were put back Z at once opened the door by pulling the knot. Left in the box alone for ten minutes, Z did not get out again.

The following is Z's record for the next two days when in the box alone. Z learned to get out in less time than it required to teach M.

Oct. 27 she got out,	Oct. 28,
1st time in 3′	1st time in 2′
2d time in 1′	2d time in 30″

X and Y were now tested separately. The following table gives the time each was in the box alone.

Date	X	Y	Date.	X	Y
Oct. 26	20′	20′	Oct. 31	30′	30′
Oct. 27	20′	20′	Nov. 1	25′	25′
Oct. 29	30′	30′	Nov. 3	30′	30′
Oct. 30	30′	30′			

Neither succeeded in getting out. On Nov. 2 I put them into the box together. Y found the knot and opened the door. As soon as X saw Y pulling the knot she took hold of it and pulled. After having been put back they pulled at the knot turn about for a few minutes, and then they gave up.

Tables 1, 2 and 3 show how quickly the cats learned to open the door when they were given an opportunity to imitate.

In the tables I have given the number of times that a certain act was performed or witnessed by the subjects of the experiment, and the time which the imitator consumed in performing the act. In Table I (Y imitating Z), for example, the second column gives the number of times that Z got out of the box; the third column, the number of times that Y saw Z escape; the fourth column, the number of times that Y got out when given an opportunity to imitate by being left in the box alone, and the fifth column, the time required by Y (the imitator) in escaping.

TABLE I.

Y imitating Z.

Date	Z gets out	Y sees	Y gets out alone	Time
Nov. 7	1	1		
Nov. 7			1	5′
Nov. 7			1	5′
Nov. 7			1	5′
Nov. 8	1	1		
Nov. 8			1	1′
Nov. 8			1	18′
Nov. 8			1	4′
Nov. 8			1	1′
Nov. 8			1	1′
Nov. 8			1	1′30″
Nov. 8			1	2′
Nov. 8			1	45″
Totals	2	2	11	

TABLE II.

X imitating Z.

DATE	Z GETS OUT	X SEES	X FOLLOWS Z OUT	X GETS OUT FIRST	X GETS OUT ALONE	TIME
Nov. 5	1	1	1			
Nov. 5					1	7′
Nov. 5					1	3′
Nov. 5	2	2	1	1		
Nov. 6					1	5′
Nov. 6	5	5	1	4		
Nov. 7	5	5	1	4		
Nov. 8	2	2		2		
Nov. 9					1	2′
Nov. 9	2	1		2		
Nov. 9					1	1′
Nov. 9	2	2	1	1		
Totals	19	18	5	14	5	

TABLE III.

X imitating Y.

DATE	Y GETS OUT	X SEES	X FOLLOWS Y OUT	X GETS OUT FIRST	X GETS OUT ALONE	TIME
Nov. 10	2	2	1	1		
Nov. 10					1	1′
Nov. 10					1	1′
Nov. 12	5	5		5		
Nov. 12					1	1′
Nov. 12	1	1		1		
Nov' 12	2	2		2		
Nov. 13	6	6	3	3		
Nov. 14	3	3	1	2		
Nov. 15	5	5	4	1		
Nov. 19	2	2	2			
Nov. 19					1	5′
Nov. 19					1	1′
Nov. 19	4	4	4			
Nov. 20	1	1	1			
Nov. 20					1	3′
Nov. 20					1	1′
Nov. 20					1	1′
Nov. 20					1	1′
Nov. 21					1	15′
Nov. 21					1	2′.
Nov. 21	2	2	2			
Nov. 22					1	6′
Nov. 22	1	1		1		
Totals	34	34	18	16	12	

Z was used a few times instead of Y. The conduct of X was just the same whichever cat was used. During the trials of the first two or three days X imitated Z very closely. Sometimes, even after Z had opened the door, she stayed behind long enough to pull the knot before following her out. Frequently X started for the door when Z or Y began to pull the knot. She looked back and forth from the knot to the door until the door opened then she dashed out ahead of the other cat. She seemed to understand that pulling the knot opened the door. At other times she quietly looked on while Y opened the door, and then followed him out. As she made but little effort to get out when in the box alone I tried to arouse her to greater activity by spreading a wet towel over the bottom of the box, but this expedient failed to produce the desired result.

Experiment 3. Opening Door by Turning Button and Pulling Loop.

Method.—A door of wire netting extended the full length of one side of a wooden box (Box II) 72 cm. long, 47 cm. wide, and 48 cm. deep. In one end of this box, 5 cm. above the floor, a small door (15 x 17 cm.) was made. This door was constructed of wire netting so that food placed on the outside of the box could easily be seen. The door was opened from the inside by turning a wooden button and pulling a loop which was concealed by the button. The button was 8 cm. to the right and a little above the door.

Results.—Z was in the box alone:

Nov. 1	30′	Nov. 5	30′
Nov. 2	30′	Nov. 6	30′
Nov. 3	30′	Nov. 7	30′

She turned the button several times during the first two days, but she did not pull the loop. During the last four days she did not touch the button. Nov. 7, after Z had been in the box alone for thirty minutes without touching the button, X was put in with her. They got out four times in less than fifteen minutes. The first two times X turned the button and Z pulled the loop, but the last two times Z both turned the button and pulled the loop. Nov. 8, Z, while alone in the box, did not even touch the button, but on Nov. 9 she succeeded in getting out in fifteen minutes. After being put back she did not get out again in thirty minutes. However, after X was put in with her they succeeded in getting out in twenty-five minutes. X turned the button and Z pulled the loop. Nov. 10, Y was in the box alone for thirty minutes and in with Z for fifteen minutes. Neither got out although X did turn the button. Nov. 12, Y got out in less than five minutes, but when put back he did not get out again in forty-five minutes. Although X was now put in with him they did not succeed in getting out. Nov. 13, Y was in the box alone for thirty minutes and in with Z for ten minutes, but neither got out. On Nov. 14, Y, although left in the box alone for thirty minutes, did not even touch the button. After X and Z were put in with him they got out:

1st time in 5′
2d time in 10′
3d time in 3′
4th time in 1′
5th time in 1′
6th time in 1′

The first time Z turned and closed the button; then X turned the button and Z pulled the loop. The second time Z turned the

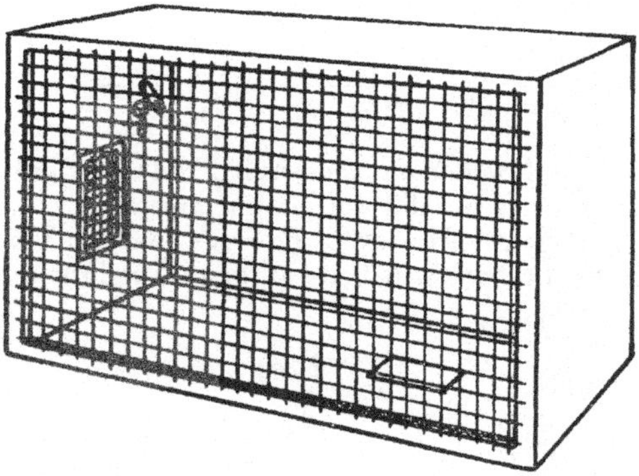

BOX II.

FIG. 2.

button and both X and Z pulled the loop. The third time Z turned the button and X pulled the loop. The last three times Z both turned the button and pulled the loop. In all these trials Y merely looked on. Z was taken out and X and Y were left in the box together for fifteen minutes, but they made no efforts to get out. Nov. 15, Y saw Z open the door four times. Each time Z opened the door both were fed, then Y was put into the box alone for five minutes. During these four trials Y touched the button but once, but after Z had turned the button the fifth time Y jumped up and pulled the loop, thus opening the door.

After being put back, he went directly to the button, turned it and pulled the loop. He now opened the door five times in less than five minutes. The association seemed to be perfectly formed.

TABLE IV.

X imitating Y.

Date	Y gets out	X sees	X follows Y out	X gets out first	X opens door	Time
Nov. 16	2	2	2			
Nov. 16					1	7′
Nov. 16					1	7′
Nov. 16	2	2	2			
Nov. 17	2	2		2		
Nov. 17					1	2′
Nov. 17	3	3	3			
Nov. 22					1	10′
Nov. 22					1	1′
Nov. 22					1	1′
Nov. 22					1	1′30″
Nov. 22					1	15″
Totals	9	9	7	2	8	

Each time X followed Y out she was fed and then put back alone for five minutes. If she did not get out during that time Y was put in with her again. However, the first two times X opened the door Y was in the box with her. Y first turned the button then X pulled the loop.

TABLE V.

M imitating Y.

Date	Y gets out	M sees	M follows Y out	M gets out alone	Time
Nov. 16	6	3	6		
Nov. 19	7	7	7		
Nov. 20	7	7	7		
Nov. 25	6	6	6		
Nov. 22	7	7	7		
Totals	33	30	33		

M did not watch Y very closely until he had opened the door several times, then she began to pay close attention, especially when he went to the button. In all the tests she had with Y she did not once attempt to turn the button. Generally she was inactive when in the box alone. During the first trial on Nov. 28 she scratched at the loop after X had turned the button. The

TABLE VI.

M imitating X.

DATE	X GETS OUT	M SEES	M FOLLOWS X OUT	M GETS OUT ALONE	TIME
Nov. 22..........	5	5	5		
Nov. 28..........	4	4	4		
Nov. 28..........				1	1′
Nov. 28..........				1	1′
Nov. 28..........				1	45″
Nov. 28..........				1	30″
Nov. 28..........				1	1′
Nov. 28..........				1	15″
Totals.......	9	9	9	6	

second time she was put back after following X out, she turned the button and scratched at the loop; but it was not until X had opened the door four times that she pulled the loop hard enough to open the door. She always pulled the loop with her claws, whereas X generally used her teeth.

Experiment 4. Getting Food by Turning Button.

Method.—A hole three-fourths of an inch in diameter was bored in the middle of Box I. This hole was covered both on the inside and outside of the box by wooden buttons. Meat was placed in the hole from the outside. To get the meat the cat had to turn the inside button (see Fig 1).

Results.—X when put into the box turned the button in less than five minutes.

TABLE VII.

M imitating X.

DATE	X GETS MEAT	M SEES	M GETS MEAT	TIME
Nov. 23.................	7	7		
Nov. 26.................	7	7		
Nov. 27.................	1	1		
Nov. 27.................			1	1′
Nov. 27.................			1	1′
Nov. 27.................			1	2′
Nov. 27.................			1	1′
Nov. 27.................			1	30″
Nov. 28.................			1	2′
Totals.............	15	15	6	

Each day M was put into the box alone for ten minutes before X was put in with her. After X had turned the button and eaten the meat she was taken out and M was left alone in the box for five minutes. She got no meat unless she turned the button. Occasionally she smelled of the button but she made no effort to turn it until the end of the fifteenth trial.

Experiment 5. Raising Small Trap-door.

Method.—A door (7 x 9 cm.) was made in the bottom of Box II. A narrow opening was left in the front end of the door. By inserting the claws in this opening the cat could easily raise the door and get the meat placed under it. The doorway was closed on the under side so that the cat could not get out of the box.

Results.—X learned unaided to open the door in less than twenty minutes and Z learned almost as soon. Y was left in the box alone:

Nov. 26	20′
Nov. 27	25′
Nov. 28	20′
Nov. 30	90′
Dec. 1	30′

Although he worked at the door more or less, he did not once succeed in getting it open, as he almost invariably scratched in the wrong place.

TABLE VIII.
Y imitating X and Z.

Date	X opens the door	Y sees	Y opens the door	Time
Dec. 3	5	5		
Dec. 5	2	2		
Dec. 5			1	4′
Dec. 5	2	1		
Dec. 7	5	5		
Dec. 8	7	6		
Dec. 10	7	7		
Dec. 11	5	5		
Dec. 12	2	2		
Dec. 13	3	3		
Dec. 14	18	18		
Dec. 14			1	1′
Dec. 14			1	1′
Dec. 14			1	10″
Dec. 14			1	5″
Dec. 14			1	5″
Dec. 14			1	5″
Dec. 14			1	5″
Totals	56	54	8	

In the tests of imitation which were now made Z instead of X was used part of the time to open the door. The behavior of Y was just the same whichever cat was used. The general method was to take X out after she had opened the door once, and let Y try the door alone for five minutes. If he did not get it open X was then put in with him again. However, on the last day of the experiment X was allowed to open the door six times in succession before she was taken out of the box.

During the first part of the experiment Y imitated X very closely. When X was taken out he frequently tried the trap-door; but during the latter part of the experiment he only looked on while X opened the door and ate the meat. *During the first series of six trials on the last day Y merely looked on, during the second series he smelled of the door each time X opened it, and during the third series he reached through the door after X had taken out the meat. After X had been taken out of the box upon the conclusion of the third series of trials Y went to the door and opened it at once. After that he opened the door as fast as I could put in the meat and close it.*

Experiment 6. Rolling Ball into Hole.

Method.—In Box I a hole large enough to admit a tennis ball was made in the middle of the bottom of the box, 12 cm. from one end. In the middle of the end of the box next to the hole and 25 cm. above the floor a small door (6 x 6 cm.) was placed. This door, which opened inward, was held shut by a wooden crossbar. The mechanical devise was of such a nature that when the ball rolled through the hole and fell into a box below, the pressure on the box raised the crossbar and permitted the door to fly open. The opening of the door exposed to view a small piece of meat which the cat easily could reach. In order to make it easier for the cat to roll the ball into the hole a wooden triangle (44 x 44 x 29 cm.) was fastened to the bottom of the box with the hole at its apex (see Fig. 1).

Results.—Below are given the periods during which the cats were given an opportunity to discover that meat could be obtained by rolling the ball into the hole.

Two or three times the ball was knocked into the hole accidentally while the kittens were playing together. Strange as it may seem, X was the only one of the kittens that showed any disposition to play with the ball. It is true that occasionally one of the

TABLE IX.

Date	Y	Z	M, Z	M, Y, Z
Dec. 1		30′		
Dec. 3		50′		
Dec. 7		30′		
Dec. 8		30′		
Dec. 10			30′	
Dec. 11			30′	
Dec. 15	20′			
Dec. 17				30′
Dec. 22				45′
Jan. 1				60′
Jan. 2				60′
Jan. 3				60′
Jan. 4				50′
Jan. 5				60′
Total	$\frac{1}{3}$ hr.	$2\frac{1}{3}$ hr.	1 hr.	$6\frac{1}{12}$ hr.

kittens struck it, but never twice in succession. In one week I taught X to roll the ball into the hole from any part of the triangle.

After X had rolled the ball into the hole two or three times in succession in the presence of Y; she was taken out and Y was left in the box alone for five minutes, then X was put in with him again. This was continued until Y learned to roll the ball into the hole. Y got no meat when X rolled the ball into the hole unless he got to the door first (see Table X).

During the first few trials Y merely looked on, but gradually he reached a point where he occasionally struck at the ball when X was rolling it. The next step was to strike at it when he was in the box alone. When X got the ball almost to the hole Y gave the closest attention, and when the ball went in he dashed to the door and tried to get the meat first. Not infrequently when X got the ball almost to the hole Y knocked it in. Soon after he had reached this stage he rolled the ball into the hole when he was in the box alone.

In the case of Z the method was the same as that employed with Y, except that Z was generally fed when X rolled the ball into the hole. Only twice in the forty-one trials of Table XI did Z touch the ball. As far as I could see the only thing Z learned was to associate the opening of the door with the hole, but not with the rolling of the ball. When in the box alone she devoted most of her time to the hole. Y, on the contrary, first formed the association between the rolling of the ball and the opening of the door.

TABLE X.

Y imitating X.

DATE	X ROLLS BALL IN HOLE	Y SEES	Y ROLLS BALL IN HOLE	TIME
Jan. 10	6	6		
Jan. 12	6	5		
Jan. 14	10	10		
Jan. 15	13	13		
Jan. 16	12	12		
Jan. 17	6	6		
Jan. 17	2	2		
Jan. 18	2	2		
Jan. 19	9	9		
Jan. 21	3	3		
Jan. 21			1	4
Jan. 21			1	1′
Jan. 21			1	12′
Jan. 21	2	[illegible]		
Jan. 22			1	2′
Jan. 22			1	1′
Jan. 22			1	15′″
Jan. 22			1	15″
Jan. 22			1	30″
Jan. 22			1	45″
Jan. 22			1	15″
Jan. 22			1	60″
Jan. 22			1	15″
Jan. 22			1	10″
Totals	71	70	13	

TABLE XI.

Z imitating X.

DATE	X ROLLS BALL IN HOLE	Z SEES	Z ROLLS BALL IN HOLE	TIME
Dec. 12	2	1		
Dec. 13	3	3		
Dec. 14	4	4		
Dec. 17	3	2		
Jan. 7	1	1		
Jan. 8	2	2		
Jan. 9	6	6		
Jan. 10	4	4		
Jan. 22	4	3		
Jan. 23	12	7		
Totals	41	33		

My next method was to roll the ball into the hole four times in succession myself, and then place it in the farthest corner of the triangle, and leave Z alone with it for five minutes. Z was given a small piece of meat each time the ball went into the hole. During the first ten or fifteen trials Z merely looked on. It was not long, however, before she began to strike the ball when it was rolling. Her interest gradually increased until finally she rolled the ball into the hole of her own accord.

TABLE XII.

Z imitating Me.

Date	I roll ball in hole	Z sees	Z rolls ball in hole	Time
Jan. 24	25	23		
Jan. 25	25	24		
Jan. 28	20	18		
Jan. 28			1	1′
Jan. 28			1	4′
Jan. 28			1	30″
Jan. 28			1	45″
Jan. 28			1	15″
Jan. 28			1	25″
Jan. 28			1	31″
Jan. 28			1	15″
Jan. 28			1	15″
Totals	70	65	9	

For M the method was the same as that used with Z except that I rolled the ball into the hole five times in succession instead of four. During the latter part of the experiment the method was varied somewhat by giving M a chance at the ball each time after

TABLE XIII.

M imitating Me.

Date	I roll ball in hole	M sees	M rolls ball in hole	Time
Jan. 25	15	14		
Jan. 26	30	26		
Jan. 28	25	23		
Jan. 29	25	23		
Jan. 30	25	25		
Jan. 31	25	23		
Feb. 1	25	21		
Feb. 4	25	24		
Feb. 5	26	24		
Feb. 6			1	1
Feb. 6	22	22		
Feb. 7	9	9		
Feb. 7			1	3
Feb. 7	6	6		
Feb. 7			6	
Feb. 8			20	8′30″
Totals	258	240	28	

I had rolled it into the hole. The second day she struck at it several times as it was rolling toward the hole. There seemed to be no method in her attempts, for several times she knocked the ball away from the hole when otherwise it would have gone in. On January 31 for the first time she struck the ball when it was not in motion. From this time on it was an easy matter to get her to strike it by tapping on the floor beside it. When she was left alone occasionally she smelled of the ball, but she spent most of her time watching me and washing herself. It was not until the last two days of the experiment that she deliberately rolled the ball into the hole.

Experiment 7. Learning to Catch Mice.

Method.—A cage 112 cm. long, 83 cm. wide, and 190 cm. high was inclosed on three sides with wire netting. A mouse put into this cage could neither escape nor conceal itself.

Results.—January 2. A large black mouse was placed in the cage with Z. At first Z very cautiously smelled of it. Then when the mouse ran she ran after it, striking it with her paw. Although she became rougher in her play during the last half hour, she did not once growl or strike the mouse with her claws. At the end of one hour the mouse was taken out of the cage uninjured.

January 3. Y was put into the cage with the same mouse for one hour. When the mouse ran he ran after it, but at first he did not touch it. After a few minutes, however, he began to strike it. When it climbed up the side of the cage he sat and watched it until it came down again. Unlike Z, he used his claws and switched his tail. During the last few minutes of the hour he did not seem to be much interested in his companion.

January 4. X was put into the cage with the same mouse for one hour. At first, like the other cats, she merely smelled of the mouse and followed it about the cage, but soon she began to strike it with her paws. A few times she seized it in her mouth. As far as I could see she never used her claws. She played with it much as she played with the tennis ball in Experiment 6. Her interest abated somewhat during the latter part of the hour. The mouse when taken out of the cage apparently was uninjured, and began to wash itself. However, two days later it was found dead in its cage, possibly from injuries received in the experiment.

February 14. Y was in the cage with a gray mouse for fifteen minutes. She followed it about striking it gently with her paws. When it ran up the side of the cage she ran up after it and brought it down in her mouth, but she did not injure it.

February 15. The same gray mouse was put into the cage with Z for twenty minutes. The mouse climbed up to the top of the cage. Z went up and smelled of it four times before she knocked it down with her paw. She did not pay very much attention to it during the last five minutes.

February 15. Y was put in with the mouse for twenty minutes. He soon discovered it up at the top of the cage. After he had gone up and smelled of it three times he seized it with his teeth and threw it down. He switched his tail and his claws rattled on the floor as he ran after it, but he never growled. In all of these trials the cats had not been fed meat for at least twenty-four hours.

February 16. A white mouse was put in with X. The cat played with it as usual. After a few minutes M (the mother cat) was put into the cage with X. She killed and ate the mouse while X looked on. X did not dare to approach as M growled very ominously whenever X moved. After M had finished eating the mouse I took her out and put another mouse in with X. She played with it just as she had played with the other one. I could not see that her behavior was influenced in the least by the tragedy she had just witnessed. When Z was put into the cage with her X seized the mouse in her mouth whenever Z approached, but as long as Z did not move she played with it as usual. When the mouse was given to Z she would not let X have it. After a few minutes Z was taken out and M was put in with X. M killed the mouse at once and began to play with it. She let X have it, but the latter would not eat it, until M had exposed the raw flesh; then she ate it at once. M was now removed and another mouse put in with X. She played with it as usual, but made no attempt to kill it.

February 19. A white mouse was put into the cage with X for ten minutes. She was no rougher with it than usual. But when Y was put in with her she seized the mouse and began to growl. When the mouse ran toward Y he did not attempt to seize it even when it was nearer to him than it was to X. After removing Y, I fed M a little meat in sight of X. She at once left the mouse went to the side of the cage next to M and began to mew. Apparently she did not realize that she had fresh meat at her disposal. A mouse was now given to M, who killed and ate it while X looked on. X was allowed to smell of the blood on the floor; then another mouse was given to her. She played with it as usual. Apparently she had not profited in the least by M's experience.

Next Y was tested with the same mouse. He was not as rough as usual in his play. I now put M in with him. She killed the mouse and began to eat it. After she exposed the raw flesh I gave it to Y who ate it at once. After he had finished eating it I gave him another mouse, but he did not attempt to injure it. Five minutes later Z and X were put in with Y. Growling fiercely, he seized the mouse. In fifteen minutes he had killed and eaten it.

February 20. A white mouse was put into the cage with Z. As the mouse tried to bite her she picked it up and tossed it about as if it were a rag ball. She did not seem to be angry in the least. When X was put in with her she growled and seized the mouse, but after a few minutes she let X have it. After they had played with it turn about for a few minutes, Y was put in with them. He killed and ate the mouse while they looked on. He was now taken out and a brown mouse was put in with X and Z. X seized it and killed it almost instantly. In less than five minutes she had eaten it. X was now removed and a brown mouse was put in with Z. She played with it but did not attempt to kill it.

February 21. Z played more gently with the mouse than usual. X was now allowed to kill and eat the mouse while Z looked on; then Z was given another mouse. She played with it a little while, then refused to take further notice of it. I put M, X and Y on the outside of the cage but they did not arouse Z in the least. She simply ignored the mouse.

March 6. Z played very gently with a mouse, until I put X in with her, then she seized it; but X soon succeeded in getting it away from her. After X had almost killed it, I gave it to Z again. She seized it savagely and held on to it, growling almost continuously. In less than a minute the mouse was dead and Z had begun to eat it.

March 7. A big brown mouse was given to Z. She played with it but did not attempt to injure it. Fifteen minutes later X was allowed to kill the mouse while Z looked on. After she had half eaten it I gave it to Z who soon finished it.

March 8. Z played with a mouse but made no attempt to kill it. But as soon as she saw X and M, who were now placed on the outside of the cage, she seized the mouse and began to growl fiercely. In seven minutes she had killed and eaten it.

March 13. Z was put into the cage alone with a mouse. In fifteen minutes she had killed and eaten it.

Experiment 8. Getting Meat out of Bottle.

Method.—In the opening occupied by the small trap-door in Box II a pint milk bottle was firmly fastened. It was partly filled with cloth so that the cat could easily reach meat which was placed on top of the filling.

Results.—M got the meat in four minutes. Y was successful in ten minutes, but Z failed completely, although she worked hard for twenty minutes. Her method was to stick her nose into the bottle and then reach for the meat with her paws on the outside. She also tried to get her nose and paw into the bottle at the same time.

X tried the same tactics as Z, except that she balanced herself with her nose in the bottle, and then reached for the meat simultaneously with both paws.

The next day Z again failed to get the meat in a trial of twenty minutes. After she had ceased trying X was put in with her. Although she went to work very energetically at the bottle she did not succeed in getting the meat, but her efforts did arouse Z to renewed efforts with the result that this time she was successful. She was now removed and X was left in the box alone for thirty-five minutes. She did not get the meat. On the following three days X was tested alone for the following periods:

Jan. 21	60′
Jan. 22	20′
Jan. 25	40′

She did not once succeed in getting the meat. After X had been in the box alone for forty minutes during the trial of January 23 I put Y in with her. *She watched* Y *very closely as he reached into the bottle and took out the six pieces of meat. After* Y *was removed* X *went to the bottle and got the meat in less than two minutes.* At first she used her old method, but finding that did not work she went at it as Y had done. In further trials she got the meat as skillfully as did X and Y.

Experiment 9. Getting down from Top of Cage

Method.—The kittens frequently climbed up on top of the cage which was used in Experiment 7, but they could not get down without help. I arranged a broad board (170 cm. long) in such a way that by jumping 40 cm. to this board, walking down it to the lower end, and then jumping 60 cm. they could reach the floor.

Results.—All three cats were placed on top of the cage, then meat was thrown on the floor in front of it. They were greatly excited. X got down by the way of the board in three minutes; Y doubled up to follow X, but his courage failed. Z who did not see X get down now jumped down as X did. Y looked on, and again doubled up to jump, but his courage was insufficient. After seeing X get down two more times he followed her down.

III. DISCUSSION OF RESULTS.

In the discussion of imitation it is essential that the term be defined objectively if it is to have much value for the comparative psychologist. That is, it must be so defined that the imitation is always from the standpoint of the observer. I think that MORGAN'S use of the term is satisfactory in this respect, for he says, "in the case of an imitative action the stimulus is afforded by the performance by another of an action similar in character to that which constitutes the response."[3] The acts of organisms are generally classified as instinctive, voluntary, and habitual. For each class there is a corresponding type of imitation.

As an illustration of instinctive imitation MORGAN cites the case of a hen pecking on the ground, and the chick imitating her action. It is the pecking of the mother hen that acts as a stimulus for the instinctive act of the chick.

Automatic or habitual imitation I use to designate those cases where the imitative act is simply an involuntary performance of an acquired, as opposed to an instinctive act. An example of this is the involuntary whistling of a tune that one hears another whistling. Here the act is involuntary, but not instinctive.

On the subjective side voluntary imitation is conscious purposive imitation. The act of another is imitated with a definite end in view. The test for this kind of imitation is refusal to imitate until

[3] MORGAN, C. L. Habit and Instinct, p. 168. *London.* 1896.

the benefits that would come from imitating have been perceived or experienced. For example, suppose two cats are put into a box together. One cat opens a door by turning a button, while the other cat merely looks on. Both pass out and are fed. If now, when the second cat is put back it goes to the button and turns it, thus opening the door, this would be an instance of voluntary imitation.

In the nine experiments with cats which have been described I have found instances of imitation. So the question is not, "do cats learn by imitation?" but instead, "what is the nature and extent of their imitation?"

In the first place, what evidence is there for voluntary imitation? In Experiment 4, M refused to turn the button until she had seen X turn it several times and get meat. Her failure was not due to lack of hunger, for after she turned the button once she continued to turn it as fast as I could put the meat in and close the hole.

I consider this a fair example of voluntary imitation, for M refused to turn the button until she had seen X repeatedly get meat by turning it. If it were merely instinctive imitation we should have expected M to scratch at the button while X was turning it, but this she did not do. She merely watched X, and when X was taken out of the box she went to the button and turned it. Of course it may be said that the act was purely accidental, but her manner seemed to indicate that such was not the case.

In Experiment 6, Y refused to roll the ball into the hole until he had experienced the results that came from performing the act. It was then, and not until then, that he began to roll the ball and watch the door. In Experiment 7, it was not until X had seen several mice killed and had eaten two that she seized and killed a mouse when it was put into the cage with her.

It seems to me the fairest way of interpreting these cases is to admit that they are instances of voluntary imitation of a low order. I say of a low order, because the imitation did not occur until the required act had been performed many times by the trained animal. In many cases I think it is not so much the association of the trained-animal-performing-the-act with the-getting-of-food, as it is an association of the-act-being-performed with the-getting-of-food. For example, in Experiment 6, Y, I think, first formed the association of X-rolling-ball with the-getting-of-food, but as the act was repeated by X the ball seemed more and more to attract the attention of Y until the association changed to rolling-of-ball with getting-of-food. The facility with which an animal imitates will depend, in large measure, upon how closely it attends to what the trained animal is doing. If it does not watch closely what is being done, the association is almost sure to be the-trained-cat with the-getting-of-food. And if this association is once stamped in, it is doubtful whether imitation can occur.

In voluntary imitation the act is performed not merely from impulse, but for the food or freedom that may result from its performance. In instinctive imitation, the performance of the act by the imitatee is sufficient stimulus to call out a similar response on the part of the imitator. In other words, the animal sees and then finds itself performing the act.

The subject of instinctive imitation has been passed over very hurriedly by most students of animal behavior. They seem to conclude that if a high type of voluntary imitation does not exist among the lower animals, imitation is of but little importance. Now I am convinced from my work with rats and cats that instinctive imitation is a factor of very great importance in the mental development of these animals. In nearly all my experiments instances of instinctive imitation were common. For example, in Experiment 2, Z seeing X pull at the knot, went to it, seized it and pulled hard enough to open the door. After they were fed and put back into the box, Z pulled the knot first, X then tried it, and after she had stopped, Y seized it and pulled hard enough to open the door. It was through instinctive imitation that the cats learned to get out of the box. X was the first cat to find the knot, yet it was Z imitating X who opened the door. The next time Y opened the door after Z had pulled the knot. When they were put back for the third time Z went directly to the knot and opened the door.

Z, being the most intelligent of the three cats, was the first to acquire the association between the pulling of the knot and the opening of the door. The other two cats subsequently learned to get out by imitating Z. I think this experiment well illustrates the importance of instinctive imitation.

Experiment 3 is also very illuminating in respect to instinctive imitation. After Z had been thoroughly tested without succeeding in getting out, X was put in with her. They got out four times in less than fifteen minutes. The first two times X turned the button and Z pulled the loop. The last two times Z both turned the button and pulled the loop. Here Z learned to get out of the box by imitating X, the less intelligent of the two. Y learned from Z, and X learned from Y.

Let us consider the nature of the associations formed in a case of instinctive imitation. X knows how to get out of the box. Y has been tested but has not succeeded in learning to get out. Y sees X pulling at the knot and he instinctively scratches at it a little, until X succeeds in pulling it hard enough to open the door. Both pass out and are fed. A few more times X opens the door, assisted in part by Y. Now if Y is put into the box alone and he opens the door by pulling the knot what associations have been formed? The first time he imitated X in scratching at the knot, the act was an instance of instinctive imitation, for he had no knowledge of an end to be attained beyond the mere performance of the act. But when simultaneously with Y's scratching, X opens the door, and they both secure food, the condition has been provided for the formation of an association between the scratching at that spot and the opening of the door. If upon being put back Y should scratch and thus open the door, the association formed would be quite independent of X, for the first time X opened the door Y did not associate it with the pulling of the knot by X, but with his own scratching at or near the knot. The first time Y scratched at the spot the stimulus was X scratching at that spot; the second time the stimulus was food to be obtained.

Not only is instinctive imitation of great importance in itself, but it is also important in that it leads up to voluntary imitation. It seldom happens that a cat learns by going through the act with the trained cat only once; generally it must see and help the imitatee perform the act many times before it is able to perform it alone.

Now in all these trials, after the first one, the imitator either looks on or participates in the act with a knowledge of the end to be attained. Here we have to some extent voluntary imitation, for the imitator is influenced not only by his own movements, but by seeing the other cat perform similar movements. The next step in the learning process is to form the association by observing the other cat perform the act and by sharing with him its benefits.

Let me point out more clearly the different steps involved in learning by imitation

1. Through instinctive imitation the cat performs the act once. As far as performing the act the second time is concerned the cat now is on the same basis as the animal that has accidentally performed the act once. But if the trained cat continues to perform the act, then the imitator has in addition to its first experience the experience of the trained cat to help in stamping in the association. Here it is that the transition to voluntary imitation occurs.

2. Voluntary imitation, where the imitator gets food each time the imitatee performs the required act (Experiment 2).

3. Voluntary imitation, where the imitator is not fed when the imitatee performs the required act, but is free to imitate (Experiment 4).

4. Voluntary imitation, where the imitator observes from another compartment the imitatee perform the required act. For reasons already stated[4] I do not think that imitation of this kind is to be found in rats and cats.

In the course of these experiments there were many instances of automatic imitation. In Experiment 6, Z formed the habit of looking into the hole in the bottom of the box. If another cat looked into the hole, she would almost invariably take a look. Again, when I changed the nature of the mechanism, yet used the same box, the trained cat went to the place where the string had been and scratched there. After doing this a few times she made no further efforts, but if later another cat went to that same spot and scratched the first went and did likewise.

Evidently automatic imitation enables an animal to retain what otherwise would soon be forgotten. Unlike human beings, they are very dependent upon external stimuli to enable them to utilize their past experience. For this reason automatic imitation plays an important part in enabling them to retain and utilize their past. If four or five kittens are taught to perform an act that results in the securing of food, the chances are that such an act will be performed by the individual members of that group much longer if they are kept together than it would if they were separated. For when one performs the act, the others automatically or voluntarily imitate him. In this way acts that have once been learned may be retained and made the basis of the performance of more complex acts.

[4] The Imitative Tendency of White Rats. *Journal of Comparative Neurology and Psychology,* vol. 16, p. 360. 1906.

It frequently happened during these experiments that the imitation was not exact. For example, in Experiment 3 M learned to pull the loop from imitating X, yet M always pulled the loop with her claws whereas X generally used her teeth. THORNDIKE would not call this a case of imitation, for in commenting on the results of his experiments with cats he says: "Good evidence that he did not imitate is the fact that, whereas 1 (whom he saw) pulled the loop with his teeth, 7 pulled it with his paw."[5]

To say that this is not a case of imitation is as absurd as to say that the small boy does not imitate his father because his father uses his right hand to drive a nail, whereas he, the small boy, being left-handed, uses his left hand. Just as the stimulus for the small boy was his "father driving a nail," not his "father driving a nail with his right hand," so in this experiment the stimulus for M was "X pulling the loop," not "X pulling the loop with her teeth."

In Experiment 6, Z and M learned to roll the ball into the hole from watching me do it. From the way they acted I have reason to think the association formed was, ball-rolling-into-hole with getting-of-meat. Here the attention was centered on the most striking element of the complex, the rolling of the ball. Their attention was focused, not so much upon what I was doing as upon what the ball was doing. As soon as the ball began to roll they lost all interest in me and watched it. This was especially noticeable after I had performed the act several times. This simply shows that certain elements of a given complex are likely to be singled out, and these enter into the association to the exclusion, in large measure, of other elements.

I am also led to believe that cats are credited with more instincts than they really possess. It is commonly reported that they have an instinctive liking for mice, and that mice have an instinctive fear of cats. It is supposed that the odor of a mouse will arouse a cat, and that the odor of a cat will frighten a mouse. My experiments tend to show that this belief is not in harmony with the facts. When cats over five months old were taken into the room where mice were kept they did not show the least sign of excitement. A cat would even allow a mouse to perch upon its back, without attempting to injure it. Nor did the mice show any fear of the cats. I have seen a mouse smell of the nose of a cat without showing any sign of fear.

[5] THORNDIKE. Animal Intelligence. *Psychol. Review, Monograph Supp.*, vol. 2, no. 4, p. 58. 1898.

It was not until the mouse began to run that the interest of the cat was aroused. The cat then ran after it, playfully striking it with her paw, becoming rougher the longer she played with it. The instinct seems to be for the cat to run after that which runs from it. I think it is evident from Experiment 7 that it is through imitation that the average cat learns to kill and eat mice. If this is true, it shows the extreme importance of imitation in the mental development of the cat. Furthermore it indicates that much that has commonly been attributed to instinct is, in reality, due to imitation.

However, a potent factor in this learning to kill mice is the mere presence of another cat. As a rule, when one of the cats was left with a mouse it merely played with it without showing any signs of anger; but the moment another cat approached its attitude changed at once. It now seized the mouse and began to growl. In this way one kitten may happen to kill a mouse in trying to keep another kitten from getting it.

My experiments have demonstrated furthermore, the fact that important individual differences appear in cats of the same litter. One individual has more intelligence than another, and there are marked variations in the learning ability of the same individual in different experiments.

To sum up, I think my experiments have shown: (1) that voluntary imitation of a certain type exists in cats; (2) that cats, to some extent, imitate human beings; (3) that instinctive imitation in cats is more important than students of animal behavior have supposed; and (4) that cats do not instinctively kill and eat mice, but learn to do so by imitation.

ORIENTATION IN THE WHITE RAT.

BY

HARVEY CARR AND JOHN B. WATSON

(From the Psychological Laboratory of the University of Chicago.)

WITH ONE FIGURE.

In a previous paper[1] the present writers advanced the conclusion that kinæsthetic and organic data play the fundamental rôle in the reactions of the white rat to the maze. This conclusion was reached by eliminating the other senses singly or in groups. It was not denied that the rat may occasionally use the data from these other senses or that it could use them if the occasion demanded. The present experiments attempt to supplement this conclusion. In them, conditions were imposed upon the rat which would tend to bring the kinæsthetic factor into strong relief if, as assumed, it does possess fundamental importance in the determination of conduct within the maze. Two experiments were made: (1) After learning the maze, starting always from *O*, the rats were placed in the positions[2] marked x_1, x_2, x_3, in the true pathway headed in either the right or the wrong direction and their method of obtaining orientation under these novel conditions was observed. The conclusion mentioned above was then theoretically discussed in the light of the new facts thus obtained, to see if difficulties and contradictions appear. (2) After the reactions to the maze became automatic, certain of the runways were either shortened or lengthened. The disturbing effect of these alterations upon the rats' conduct and their methods of learning to adjust themselves to the new conditions were observed. The two experiments will be discussed in order.

EXPERIMENT I.

THE EFFECT OF STARTING THE RAT AT DIFFERENT POSITIONS.

When the trained rat is put down in the maze at unfamiliar starting points, several possibilities of conduct are open to it:

[1] WATSON, J. B., *Psychological Review, Monograph Supplement*, vol. 8, no. 2, 1907.

[2] See cut of maze, p. 28. A similar but unsatisfactory test was reported in the previous paper. See p. 81, *loc. cit.*

(1) the animal may not have profited in the least by its previous experience in the maze; the situation may offer a problem *de novo*; (2) the rat may orient itself immediately as does a human

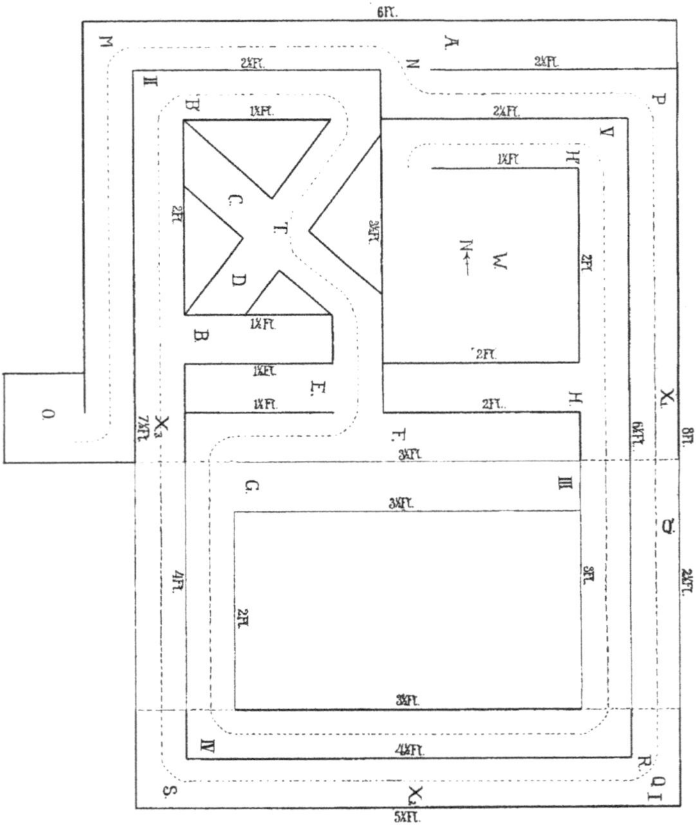

Fig. 1.

being, when, in a partially strange situation, he suddenly finds some thoroughly familiar landmark; (3) immediate orientation may not occur, and yet the situation may not be entirely new to the rat; it may exhibit some random movements before starting properly; but its conduct might be wholly different from an animal which had not previously learned the maze; (4) if the last condition obtains, can the rat learn in time to orient itself immediately when put down at random at any one of two, three or four such starting places?

On the basis of results obtained from our work during the past summer, which is presented in detail on page 33, combined with the previous work of WATSON[3] and of CARR,[4] we are ready to give more or less satisfactory data bearing upon the above questions. (1) The situation does not present a problem *de novo*. (2) Nor does immediate orientation occur. (3) There is a period of random effort; the rat may wander about, turn around in the alleys several times or run up and down the pathway for a variable distance, acting as though lost or in a new situation. In conscious terms, its behavior suggests uncertainty, perplexity, and lack of confidence. Finally, a change of behavior is observable. The suggestion of perplexity and uncertainty is gone, the rat starts off with a sudden burst of increased speed and every movement thereafter is characterized by the precision and regularity which mark the functioning of an automatic habit. The remaining part of the maze is run in the normal and habitual manner. This change of conduct has been termed "getting the cue." The "cue" may come suddenly while the animal is running backward in the maze with irregular speed; the rat may suddenly stop, turn quickly and start off at full speed toward the food-box. The change often comes gradually, especially when the animal starts off running in the right direction. After the cue has apparently been obtained, it may be lost for a time and again found after a short interval; however the cue once obtained is rarely lost. Furthermore, once the animal attains orientation, it traverses the rest of the maze without error. This change from random to controlled activity is striking and characteristic, but extremely difficult of description except in anthropomorphic terms. (4) The rat can learn with a sufficient number of trials to orient itself immediately, starting at random from any one, two, or three definite positions in the maze. The number of trials necessary to accomplish this feat has not been determined accurately. One set of rats learned to start from any one of six cul-de-sacs on the

[3] *Ibid.*, pp. 82 and 83.
[4] Heretofore unpublished.

basis of an average of eighteen trials for each animal. This would imply that under these conditions, three trials per rat were required by it in order to learn to start at random from any one of six cul-de-sacs. A greater number of trials, however, is necessary when the animal is forced to start at random from six such positions in the true pathway. In the latter case, orientation at these positions does not become immediate in less than five or six trials.[5]

With these facts bearing upon the behavior during the establishment of orientation before us, we may now well ask the question: how does the rat attain orientation? Can he do it in terms of kinæsthetic data alone? From our previous work upon the behavior of normal, blind, and anosmic rats in tests of this kind in the Hampton Court maze, it appeared, since no difference in conduct between the normal and defective animals could be found with respect to their ability to attain orientation when put down in the maze at unfamiliar starting points, that visual and olfactory data are at least not largely employed by them as a means of controlling their movements. This conclusion is based upon the assumption that the processes employed as control by the defective rats are the same as those which would have been employed by them had they been normal. Let us suppose, for example, that a normal rat does use visual data, or the data from some other "distance" sense, for controlling his movements when the automatic (kinæsthetic-motor) character of the act is interfered with, as is the case at first when the animal is started in the maze at some point other than the customary one. What would be the nature of the orienting process? Evidently the animal would have to move at random until distinctive familiar visual or other extraorganic stimulation appeared, at which time the automatic series would be restored and the animal might thereafter have no further need for distance sense data during the remainder of the trip around the maze.

If, however, we deny to the rat the possibility (or better, the probability) of its using distance sense data in the way described above, it becomes necessary for us to answer the question: how can a kinæsthetic-motor series, which has been thrown out of gear become readjusted without control data from some other sense avenue?

Summarized from CARR's unpublished results.

If we assume that each separate "unit" (possibly a runway) of the maze affords some characteristic set of kinæsthetic impulses which can be utilized as a stimulus to secure the proper adjustment to the succeeding unit, we might have a situation where a distinctive set of kinæsthetic impulses would serve the animal for control exactly like a set of distinctive visual cues, for example. There are four ways in which distinctive kinæsthetic groups of impulses might arise. (*a*) Two runways may be of unequal length. (*b*) They may be of equal length, but occur in different positions in the total series, *i. e.*, they are preceded by different conditions. (*c*) They may be alike in every respect with the exception that the one is entered by a turn to the right, while the other is entered by a turn to the left. In rounding a corner at a high rate of speed, the body sways over to the inside, the weight is thrown on one side, while the feet on the outer side are braced in order to maintain equilibrium. Such differences are so gross and fundamental that it is idle to deny that they possess functional influence upon subsequent behavior. (*d*) The alleys may be of the same length and be entered by the same direction of turn, but present possible differences in their stimulating effect because they extend in different directions. It is difficult to conceive why and how this can be so, and the possibility is suggested only because of certain observed facts. The successful functioning of an automatic habit depends upon the rat's orientation in relation to cardinal positions. Change the direction of the path and the automatic act is disturbed to some extent. The same act accomplished in two different directions is thus *different in some way to the animal.* Thus, it is theoretically possible for the rat to adapt its behavior successfully to a series of objective conditions wholly in terms of the differences in kinæsthetic stimulation, which it offers, without relying to any extent upon data contributed through any of the distance senses. We have no intention of maintaining that the rat discriminates these possible differences in kinæsthetic values in any overtly conscious or intellectual manner, viz., that they know "right" and "left" or cardinal directions, or that they consciously evaluate in any kind of terms the length of the alleys.

If, as we have assumed, the automatic behavior of the rat in the maze is governed by distinctions lying within the kinæsthetic impulses themselves, we are in a position to understand the situation presented to the rat when it is introduced into the maze at some one of these positions. The animal must perforce run up and down the alleys until it experiences some one or several of these characteristic motor situations which would give rise to the necessary stimulations to release the old automatic movement. The rat may run the length of the alleys, around corners, or traverse several alleys before getting the cue. Moreover, on this basis, one can conceive why at times the cue should be gradually attained. At such times, a summation of stimuli would be required.

On the other hand, it may with justice be argued, as we ourselves above suggested, that if the cue is received through data from some distance sense, the animal must still run about at random until it receives some one or several such characteristic stimuli. This argument cannot be met wholly, but if our own behavior under similar circumstances can by analogy be made to apply to the case of the rat, we should be allowed to assume, when our elimination experiments are considered, that this period of random activity would be much shorter when distance sense data are employed than when kinæsthetic are used. It must be frankly admitted that the purpose of our work was to see whether the facts of orientation offered insuperable difficulties to our theory rather than to attempt to rule out all possibility of the rats' receiving aid from extraorganic sense data.

This assumption granted us, our argument may now be stated as follows. If the animals orient themselves in the maze in the majority of cases by running at least the full length of one alley, by rounding corners into a second alley, or by running through several alleys before picking up the cue, the facts will be explicable in terms of the kinæsthetic hypothesis, and consequently there will be no theoretical difficulty in supposing that the rat's automatic movements in the maze as a whole are controlled by kinæsthetic impulses alone. If, on the other hand, the rats orient themselves in the majority of cases with a minimum of random movement, the facts will not be so easily explicable in terms of our hypothesis, as in terms of some other which would admit that control is inaugurated by data from some distance sense and consequently, that automatic behavior in the maze may be guided and controlled effectually as occasion demands by such means.

In order to make a careful test of the facts of orientation, several conditions must be observed in the experiment: (1) The alleys of the maze into which the rats are introduced should be relatively long and should differ markedly in their length. (2) When placed in the maze, the animals naturally tend to spring from the hand on the run, and go for a short distance before attempting to adjust themselves to the situation. This tendency should be minimized as much as possible by holding them in position for a short time, or by allowing them to nibble a crumb of bread when released. (3) Since, with successive attempts, the rats will gradually learn to make immediate orientation, only a few trials for each position should be given. The series of tests, the results of which are given in the paper previously referred to, are faulty in the first and third respects. We have repeated the experiment in order to eliminate these possible sources of error.

In order to meet the conditions required under (1), a new maze was constructed the plan and dimensions of which are represented in the cut. The alleys are six inches wide and six inches deep. Finished lumber was used, the cracks in the floor were filled with putty, and the whole maze was given three coats of white paint. The maze was constructed so that it could be sawed across at the dotted lines and divided into three sections for the purpose of the second experiment. The maze was not so divided until the first experiment was completed. The cut represents the maze as used, with the exception that the opening into the cul-de-sac *B* was closed. The experiment was conducted out of doors in an enclosed yard. The rats were introduced into the maze at the positions x_1, x_2, and x_3. Two of these alleys are seven and one-half feet long, while the third is two feet shorter. This allows the animals to run a distance of two and one-half to three and one-half feet in either direction from the starting place before a turn is possible or necessary. The experiment was started with twelve rats, but four became sickly and unreliable in conduct and were discarded. The group consisted of three normal males, two blind males and three normal females. After the rats had been thoroughly trained, the experiment was started each day by giving them a preliminary run through the maze and then introducing each rat separately at x_1 with wrong orientation, at x_2 with correct orientation, and at x_3 with wrong orientation. By "wrong" orientation, we mean that the rats were headed back towards the starting box, *O*. This procedure was followed the second day with the exception that the orientation for each position was reversed. Thus three trials were given each rat per day, and the same orientation for any one position was repeated every other day. Not more than a total of twelve trials was given to any one rat. These varied conditions were designed to eliminate the possibility of *learning* to react immediately to a given position. An accurate account of the behavior of each rat was taken, including the changes in direction of movement, the distance traversed, the turnings inside the alleys, partial returns and the position where the rat seemed to pick up its cue. The conduct was noted by two, and sometimes by three observers. In all, 84 tests were made and the results were tabulated in statistical form.

No noticeable tendency for the rats to start in the direction in which they have been oriented was observed. They are just as likely to turn around immediately and start off in the opposite direction. Neither do they tend to start either toward the food-box, *W*, or back toward the original entrance, *O*. In other words, the direction of starting is apparently a matter of chance entirely. This fact of itself argues the lack of any immediate orientation. The situation in which they have been placed thus does not influence nor determine their conduct at the beginning of the test.

The movements in the latter part of the period of exploration are determined to some extent: The rats tend to migrate back toward the starting box, *O*. In 75 per cent of the trials, the cue was picked up somewhere between the position where they were released and the box *O*. The rats often explore on both sides of the position at which they are released, but 85 per cent of the distance traversed in the period of exploration is on the side toward *O*. This general fact may be difficult of explanation, but that some determining influence is at work is too evident to admit of doubt. The following explanation may be suggested as a possibility. In learning the maze originally, the rats explore for a distance from *O* and retrace their steps. This performance is repeated on successive trials with more extensive excursions. When the rats become lost or confused during any trial, although the maze is partially learned, they always run back toward *O*. It seems that the maze is learned in sections, as it were, and in case the rats become lost at any time, they are able to retrace their steps to more familiar surroundings. When the rat is now introduced at the position x_1 and begins to explore, the situation becomes familiar to some extent, and the rat acts as it has been accustomed to in order to get started correctly, *i. e.*, drifts back toward *O*. Such a conception, however, leaves much to be explained.

The general statement that the situation is not entirely novel during the period of exploration and that the behavior of the rats is influenced as a result, is also supported by the fact that few errors are made, *i. e.*, errors in the sense of running into cul-de-sacs during the period of exploration. Of the 84 trials, errors occurred in but eight. Four of these errors were made by one rat. Such a percentage of errors is possible in running the maze normally. In four cases, the error occurred after the orientation had apparently been secured. But two chances for error were offered in those parts of the maze traversed during the period of exploration. In 55 of the trials, the rats passed by one of these openings leading into a cul-de-sac before securing orientation; and they often passed by the same opening several times in the same trial. Yet out of these numerous possibilities, only four cases of error of this kind occurred. The exploring movements are thus confined almost exclusively to the true pathway.

On the average, the rats turned around 2.5 corners in each trial before being able to pick up the cue; in other words, they explored fully or in part three alleys per trial before becoming oriented. Their explorations averaged a distance of 12.6 feet per trial. Inside the alleys, they changed the direction of exploration 1.3 times per trial. In only ten trials out of the 84 was the exploration confined to the alley in which they were placed and in these cases the distance traversed averaged 2.8 feet per trial, while the direction of movement was changed at least once. In 57 cases out of the 84, they went outside of the alley into which they were introduced before becoming oriented. Immediate orientation apparently occurred in seventeen trials. It is extremely doubtful whether several of these are legitimate cases of immediate orientation. A rat may by chance run forward toward the food-box, *W*, and become oriented gradually. In four of these cases, the rat went forward to the food-box, but ran hesitantly, made stops, or entered some of the cul-de-sacs. It was our policy to record under the heading of immediate orientation every case that could possibly be interpreted in that manner. As may be seen, these four trials are exceedingly questionable. In four other cases the rats turned around several times in the alley before starting off. Nine trials were clear-cut, legitimate cases of immediate orientation. However, eight of the total number of immediate orientations were made by two rats, and the influence of the learning factor is evident in spite of the small number of trials allowed to each animal. *No immediate orientations were made during the first day. Only three cases occurred during the first half of the trials, while the remaining fourteen cases were made during the last half of the tests.*

There was a tendency for the rats to pick up the cue at distinctive points in the maze. In the 67 trials in which there was a period of exploration, the cue was picked up 13 times at *O*, 11 times at or near the corner *M*, 15 times at the turn *N*, 11 times at the corner *P*, 7 times at the corner *R*, 3 times at *S* and once at *T*. In only six trials was the orientation clearly effected near the middle of one of the alleys, to which number must be added the number of trials in which immediate orientation occurred. This fact, that the cue is picked up at distinctive positions, cannot be explained on the hypothesis that each rat would finally learn to orient itself at some one of these positions and hence that all of the 15 orientations at *N*, for example, belong to that one rat, as might very well be the case, if such a point offered a distinctive visual or olfactory cue. As a matter of fact, the greatest number of orientations per rat at any position was four out of a total of twelve trials. The 67 trials give an average of 8.37 per rat, and on the average, these 8.37 orientations occurred at 4.75 positions—less than two orientations per position. For any one rat, the greatest average number of orientations per position was 2.2. This general fact that orientation is secured at such distinctive positions as the *turns* supports our general contention.

The statistical results show no differences between the blind and the normal rats in any respect. The females have better records than the males. Their period of exploration is shorter, fewer turns are made inside the alleys, fewer corners are turned, and the percentage of immediate orientations is much higher. Whether this difference is a matter of chance, or whether the results represent individual or sex differences, it is impossible to say.

These various results of the experiment speak for themselves. They can be easily interpreted in terms of our theory. We do not mean to assert that they furnish conclusive and indubitable proof of our contention, but we do maintain that they can be more readily explained on the basis of our conception[6] than in terms of a theory which assumes that orientation is secured mainly through some distance sense.

EXPERIMENT II.

THE EFFECT OF SHORTENING AND LENGTHENING CERTAIN ALLEYS IN THE MAZE.

1. The Effect of Shortening the Maze.—For the second experiment, the maze was divided into three sections by sawing it across at the dotted lines. By removing or replacing the middle section, the maze could be shortened or brought back to its original length. This change merely alters the length of four alleys without altering the relation of the turns leading to or from them. The maze was cut very carefully so that the two end sections would fit quite snugly together after the middle section had been removed. For reasons presently made known the cul-de-sac, *B*, remained open during Experiment II.

The trained rats formerly used were employed in this experiment with the exception of the second blind one. This animal became somewhat feeble and refused to work consistently from day to day. After the maze had been sawed through but before the middle section was removed, the animals were allowed to run the maze for seven days. Four trials per day were given each rat. All disturbances of their old habits due to the new smell factors introduced by sectioning of the maze, to the opening of cul-de-sac *B*, and to the tests described above were thus eliminated. After their reactions became thoroughly automatic, the maze was shortened and the behavior of the rats in the new situation was noted. Each rat was given four trials per day for five days.

As above outlined, our theory assumes that the rats make the correct turns in the maze in response to some internal (kinæsthetic) impulse. If the assumption is not true, the rounding of the corners must be in response to some extraorganic stimulation received there. That is, the wall at the end of the runways and the opening into the next alley must contribute data through some distance sense. The experiment is designed to test the relative

[6] With the exception of the cases of immediate orientation. Since two out of eight animals made eight of the nine unquestioned immediate orientations we are willing to admit the possibility of the use of distance sense data in their cases.

efficiency of these two possible modes of stimulation in determin ing the rats' behavior at the turns. If the animals run at full speed against the ends of the shortened alleys at *I*, *II*, *IV* and *V*, evidently the assumption that they receive extraorganic stimulation there of functional value to them is most improbable. If the rats succeed in making the turns as correctly as usual, we must conclude that such conduct is determined wholly by extraorganic stimulations and is not influenced effectively by kinæsthetic ones. The experiment is decisive in estimating the relative efficiency of the two possible modes of stimulation, because it brings them into functional opposition.

The results obtained from this experiment justify our assumption that the turns are made in response to differences lying in the kinæsthetic impulses themselves. Marked disturbances of conduct were noticed in every rat. On the average sixteen trials per rat were necessary wholly to eliminate these disturbances, *i. e.*, to secure accurate, automatic adjustment to the shortened maze. Rats can often learn a maze of this complexity *de novo* in this number of trials. This fact is evidence of the profound disturbances effected by the change.

The time for running the maze was increased despite the shortened length. The increase of time was hardly proportionate to the degree of disturbance as reflected in the nature of their behavior. Table I gives the average time in fractions of a minute. The normal time for running the maze in its shortened form was secured by averaging many individual records of trips made after the reactions of the animals had become thoroughly automatic. The records of the seven animals made after the maze was shortened were averaged for each trial. The time increases for the first trials, and then gradually decreases toward the norm.

The disturbances consisted of (1) running squarely into the ends of the alleys at *I*, *II*, *III*, *IV* and *V*; (2) errors, such as partial returns or entering into some of the cul-de-sacs; (3) slow, hesitant and careful movements; (4) stopping here and there and "nosing" around the sides of the alleys, and (5) compensatory adjustments. By the last phrase, we refer to the fact that, after running into the end of an alley for several trials, the rats often attempted to make that turn too soon and would come in contact with the inner corner of the turn. This tendency was most evident at *IV*. The alley *IV* in the shortened maze occupies the position of cul-de-sac *G* in the lengthened maze. After "bumping" into the wall at *IV* several times, the rats tended to turn too soon and consequently failed to round the turn. As a consequence they formed the habit of running into cul-de-sac *F*. This error was very characteristic and was difficult to eradicate.

TABLE I.

Average time for successive trials in running the shortened maze. (Based upon records of 7 rats).

Normal	.21 min.	(5)	.33 min.	(10)	.25 min.
(1)	.39 "	(6)	.33 "	(11)	.25 "
(2)	.45 "	(7)	.33 "	(12)	.30 "
(3)	.45 "	(8)	.27 "	(13)	.22 "
(4)	.37 "	(9)	.25 "	(14)	.22 "

The following record of Female III, which may be considered typical of the series, furnishes the best description of their behavior.

Sept. 6. (1) Ran into *I* with all her strength. Was badly staggered and did not recover normal conduct until she had gone 9 ft. Ran against *IV* hard and then touched *V* lightly with nose.

(2) Ran into *I* and "nosed" *IV*.

(3) Hesitated at *I* and *IV* but did not touch walls with nose.

(4) Perfect.

Sept. 7. (5) Ran into *I* with sufficient force to land her whole body against the wall. Did not recover normal behavior until after passing *IV*. Stopped at *IV*.

(6) Ran very slowly and hesitantly. Did not gather any momentum. Hesitation at the four crucial corners.

(7) Hugged inner wall at *I*. Stopped at *IV*.

(8) Perfect.

Sept. 8. (9) Slowed up and hesitated at *I* and hugged inner wall at *IV*.

(10) Stopped and "nosed" at *I*, *IV* and *V*.

(11) Perfect.

(12) Perfect.

Sept. 9. (13) Perfect. Ran rapidly.

(14) Perfect.

(15) Entered cul-de-sac *F*.

(16) Perfect.

Sept. 10. All four trials were correct.

One result was obtained which is rather peculiar and is difficult of explanation. The six normal rats found little difficulty with the turn at *II*. Three of these animals effected this turn accurately in every trial. One rat touched the wall lightly on the first trial but made the turn accurately thereafter. The fifth rat struck the wall lightly on the ninth trial, but made the turn perfectly thereafter. The sixth rat hesitated at the turn on the fifth and sixth trials. Out of a total of 120 trials, the rats touched this wall lightly twice, and hesitated momentarily three times. In the remaining 115 cases, the turn was made accurately and unhesitatingly. On the other hand, the blind rat found as much difficulty with this corner as with any of the others. He ran into the wall quite hard the first trial, touched it lightly on the second trial and hesitated there the third trial. On the second day, he ran into the wall twice and made the turn correctly thereafter. It may be supposed that this difference between the conduct of the blind rat and that of the normal rats indicates that the latter effected this turn with the aid of visual data. This assumption is hardly legitimate, inasmuch as the normal animals failed to use vision effectively at the other three corners. Neither can one assume that the turn at *IV* presented visual distinctions not possessed by the other corners, because, if such visual differences exist, they are too minute for the human eye to detect, and, in case the rat possesses a visual acuity superior to that of human beings, it ought to be able to detect a solid wall sufficiently well to refrain from running headlong into it time after time. Again, one may suppose that the normal rats were accustomed to see the opening *B* before reaching the turn at *II*, and made the correct adjustment in response to this visual cue. On this basis, the normal animals should have had no trouble at the turn *V* because the opening *H* bears the same relation to the turn *V* as does *B* to the turn *II*. However, this assumption may be supported by the fact that the cul-de-sac *H* has been open during the previous experiment, while *B* has been open only some eight days. One may argue that the normal rats had neglected the opening *H* as a visual cue in the course of the long series of trials which was given them in the learning maze from the first, while the recent opening of *B* had attracted their "visual attention" and they had learned to utilize it as a visual cue. Such a conception is possible, but the argument is based upon a rather precarious foundation. If the rats can see the opening *B* so as to react to it, it seems that they ought to be able to see the opening into any alley at the turn and utilize it as a visual cue, inasmuch as there is no reason why they should neglect this cue throughout the course of the long series of tests. When the fact was noticed that the normal animals turned the corner *II* correctly, it was suggested that the shortened alley leading up to *II*, which is five feet long, possessed the same kinæsthetic characteristics as some alley in the lengthened maze. As a matter of fact, the alley leading from the box *O*, four and one-half feet long, is very similar to alley *II*. Hence it could be argued that, since the alleys possess the same motor peculiarities, the turns would be made in a similar manner. The conception is ingenious, and it would support our thesis, but on this basis, the blind rat should have had no trouble at *II*. Consequently, we are forced to admit that the phenomenon remains inexplicable so far as the present experiment is concerned.

With the exception noted above, no difference between the behavior of the blind and the normal rats could be detected.

2. *The Effect of Lengthening the Maze.*—After the above series of tests had been completed, the rats were forced to continue running the shortened maze for a period of three weeks, at the end of which time their reactions to it had become thoroughly automatic. The maze was then lengthened by replacing the middle section, and the behavior of the animals under these conditions was observed. In the previous experiment, this middle section had been thoroughly explored by the animals and it should now have presented a minimum of possible sensory disturbance.[7]

The conditions are again such that they bring into functional opposition the influences of kinæsthetic cues and any possible distance sense cues which might be involved in rounding the corners of the alleys. If the rats turn in response to kinæsthetic cues, they should now attempt to turn in the extended alleys at the positions corresponding to the length of the alleys in the shortened form. In the first alley, this position is at *Q′*. In the remaining alleys, the cul-de-sacs *B*, *G* and *H* now occupy these crucial positions. For example, the distance *S-B* in the extended maze equals the distance *S-B′* in the shortened maze. According to our theory, the rats should now run into the *wall at Q′* and into the cul-de-sacs *B*, *G* and *H*.

The results again support our contention. Marked disturbances in conduct occurred for twelve trials (three days). After this time, the disturbances occurred occasionally, though they may be regarded as practically eliminated at the end of this period.

The time for running the maze was noticeably increased in the first trials, but it was gradually decreased thereafter (Table II).

[7] The blind rat whose behavior had become erratic was not used in the shortened form of the maze. We utilized this animal, however, by allowing him each day to run several times through the lengthened form of the maze. In this way, we kept the middle section constantly in use during the experiments in the shortened maze. By this means, the original smell values of this middle section were retained unaltered, for the males at least, since this blind rat was a male, and was kept in the same living cage with all the other males used in the experiment.

These times, as before, are expressed in fractions of a minute. The normal time was secured by averaging a number of trial records taken immediately before Experiment I was made.

TABLE II.

Average time for running the lengthened form of maze after becoming habituated to shortened form (Based upon 7 animals.)

Normal .28 min.	(3) .52 min.	(6) .34 min.
(1) .59 "	(4) .31 "	(7) .35 "
(2) .65 "	(5) .49 "	(8) .34 "

As the best description of their behavior, we give as typical the record of Male I for eight successive trials.

Oct. 2. (1) Came to a full stop at *Q′* and "nosed" along the wall. Ran into and traversed the full length of alleys *B*, *G* and *H*.

(2) Slowed up at *Q′*. Entered *B* its full length. On coming out of *B*, ran back into *A*, started from *A* in the right direction, slowed up at *Q′* and partly entered *B*, *G* and *H*.

(3) Turned into the wall at *Q′* and became badly confused. Ran back and forth between *Q′* and *A* three times. On coming to *Q′* the third time, reared upon the wall and "nosed" about. A slight error was made at *B*. Ran the full length of *G* and made a slight error at *H*.

(4) Ran rapidly to *Q′* and then went slowly until turning the corner. Ran past *B* but hesitated at *G* and *H*.

Oct. 2. (5) "Nosed" along the wall at *Q′* until turning the corner. Slowed up at *B*, ran with full speed against the end of *G* and partially entered *H*.

(6) Ran past *Q′* correctly, and went into *B* its full length. On coming out of *B*, went back to *A*, started from *A* in the right direction, and "nosed" around the wall at *Q′*, went back again to *A*, turned and came to *Q′* and "nosed" about; continued but hesitated at *B*, *G* and *H* but did not enter them.

(7) Slight hesitancy at *B* and *H*.

(8) Merely slowed up at *Q′*, *B* and *H*.

All the animals ran into the wall at *Q′* and into all of the crucial cul-de-sacs. These errors had been eliminated to a great extent by the end of the first four trials (first day's experience), but were again prominent during the first trials of the second and third days. *On entering the crucial cul-de-sacs, the rats frequently ran full speed into the end of the alley.* This is evidence that the cul-de-sacs were mistaken for the true pathway. After a few trials, the cul-de-sacs were entered only part way, and finally the disturbance manifested at these positions consisted of hesitations or of a swerve in the direction of the openings without any decrease in speed. At first, the rats actually attempted to turn through the wall at *Q′* at the definite position at which they would have had to turn in the shortened maze. Striking the wall at an angle, the rat would slide along it for eight to ten inches and would then go on until it stumbled upon the opening at the end of the alley. This turn occurred relatively accurately (*i. e.*, with respect to old habit) during the first five trials on the average. After this number had been given, the animal often struck the wall at a point slightly further on between *Q′* and the corner *Q*. It seemed that the attempted turn was a resultant of two impulses, one to turn at *Q′* and the other to go on to *Q* at the end of the extended alley. Failure to find the opening at *Q′* often caused the rat to stop and go back in the maze for a new start, or to go ahead slowly until it stumbled upon the opening. In later trials, the animals ran rapidly past *Q′* without stopping or hesitating, but a *deflection of an inch or two toward Q′* could be noted; the same behavior was noted as the animals passed the crucial cul-de-sacs. In spite of these various disturbances, *i. e.*, hesitations, entering the cul-de-sacs, running into the wall and partial returns over the true pathway, it is a noteworthy fact that very rarely was the confusion so great that the animals ran into any cul-de-sac other than the three crucial ones.

No differences between the behavior of the normal animals and that of the blind rat could be detected.

The results of these two experiments, combined with those reported in the previous paper, form rather conclusive proof of the contention as to the fundamental importance of the kinæsthetic factor in the rat's adjustment to the maze.

CONCLUSION.

In concluding this paper, it may be well to reformulate our contention even at the expense of repetition, by contrasting the habits of the rats in the maze with the habits of human beings in a similar environment.

Human beings can form habits of the type we have been discussing (kinæsthetic-motor) which may become absolutely automatic. When this latter stage has been reached the "movement to come" is released at the proper time by the afferent (kinæsthetic) impulses aroused by the movement which has just been made. So far, these statements apply alike to the behavior of rat and man.

When an automatic series of movements in man is disturbed, the "movement to come" can no longer be released by the afferent impulses arising from the movement just effected. Visual, auditory or tactual impulses (cues) are then utilized, *i. e.*, the adjustment becomes, *e. g.*, momentarily visual-motor. A few movements made in response to these distance sense cues may suffice to restore the kinæsthetic-motor character of all the ensuing adjustments.

Likewise, when an automatic series of acts in the rat is disturbed, the "movement to come" can no longer be released by impulses arising from the movement just preceding. But at this point the analogy between the behavior of rat and man breaks down. The former apparently has no well developed distance sense cues, consequently he must utilize some method other than the one above described to reëstablish the automatic character of his acts. Our hypothesis provides the rat with such a method. According to it, the rat has the possibility of receiving kinæsthetic cues which function for "control" exactly as do visual cues in man. These kinæsthetic cues are ordinarily not needed by the rat for controlling his movements any more than visual cues are needed by man for controlling his. But the moment a break occurs in the series of the acts of the rat a cue is needed which will lead to the reëstablishment of the automatic character of the movement. The rat receives this cue by traversing at random any "unit" of the maze. The group of afferent impulses (kinæsthetic) which are aroused by traversing this unit releases the proper adjustment (*i. e.*, the old movement which has been synergized on many past occasions with this particular group of impulses) and the automatic character of the movements is again restored.

On this supposition, man's kinæsthetic-motor habits would differ from the rat's mainly in this, that whereas the former utilizes distance sense cues for reëstablishing automatic adjustments, the latter utilizes kinæsthetic cues.

STUDIES ON NERVE CELLS

I. THE MOLLUSCAN NERVE CELL, TOGETHER WITH SUMMARIES OF RECENT LITERATURE ON THE CYTOLOGY OF INVERTEBRATE NERVE CELLS.

BY

W. M. SMALLWOOD AND CHARLES G. ROGERS.

WITH PLATE I AND THIRTEEN FIGURES IN THE TEXT.

I. INTRODUCTION.

The purpose of this paper is twofold: First, to summarize and correlate the more important contributions on the structure of invertebrate nerve cells exclusive of the neuro-fibrillæ (a special problem which cannot be adequately treated in the space allotted to this review); and, second, to present our own studies on the structure of the gasteropod nerve cell with special reference to the problem of the so-called NISSL bodies, whose nature is still in controversy. It has been maintained that these bodies are artefacts. Inasmuch as we have been able to cause them to appear by feeding experiments and have been able to photograph them in the living and unstained nerve cells, we feel reasonably sure of their actual existence and shall make suggestions as to the manner of their development, as well as their probable function. In a

[1] Contributions from the Zoölogical Laboratory, Syracuse University, C. W. HARGITT, Director.

later paper we hope to show structural and physiological similarities between the nerve cells of invertebrates and vertebrates.

The various terms employed to describe the stainable and non-stainable substances of the cytoplasm of vertebrate nerve cells have been in large measure carried over to the description of the invertebrate nerve cells. Since the NISSL bodies were discovered and known as the visible or stainable part of the cytoplasm, the following words have been used for similar structures; the chromatic substance, the chromophile part of the cytoplasm, the chromatophile elements, the chromophilic particles, the basophile constituents, the tigroid substance, the sigroid substance, the collagenous substance, granules or granular substance. The names of most of the authors who have created this confusing and unnecessary terminology may be found in ROBERTSON's review. In a similar manner the non-stainable substance is designated as the achromatic, fundamental, invisible, not formed, unstainable, acidophile substance, trophoplasm, or kinetoplasm. The fibrillar substance is included in these terms although it is a distinct structure and whether it is considered as a part of the stainable or non-stainable substance depends largely on the writer.

The pigment found in nerve cells of the central nervous system is deposited in masses distinct from the NISSL bodies and is pale yellow or dark brown. These seem to be unlike, the brown appearing early in life and ceasing to increase after a few years. It is not blackened by osmic acid. The yellow appears in man during the sixth year, increases with age and is blackened by osmic acid. Some writers maintain that the yellow is not fat, but that it undergoes fatty degeneration. In certain mental diseases there is an accumulation of this pigment and a breaking down of the structure of the cytoplasm. Whether the two processes are related or not is unknown. A golden yellow pigment is found in the nervous system of certain gasteropods and a yellow pigment in other classes of invertebrates, the origin and use of which are somewhat problematic.

A further modification of the cytoplasm of nerve cells is found in the presence of vacuoles, lymph spaces and the actual though infrequent penetration of nerve cells by capillaries. The vacuoles occur in the cytoplasm, nucleus, and nucleolus and are probably in each case formed in a similar manner even when the exciting cause is different. The vacuoles which occur in the nucleolus are similar to those that occur in this structure in ova in most animals during their growth. The vacuoles that occur in the nucleus are not as common and it is doubtful whether they are normally present. So far as we are aware they have not been seen in the living nerve cells, but are common in cadaveric specimens. Nerve tissue poorly fixed may also exhibit them, which renders it all the more probable that they are artefacts.

The vacuoles in the cytoplasm are present in the nerve cells of many animals both vertebrate and invertebrate. *They can be seen in the living nerve cells of Gasteropods* and have been reported in some vertebrate nerve cells. In well fixed and stained sections, vacuoles are very commonly found which agree in form and appearance with the conditions in the living cells. Considerable work has been done to determine the question whether or not these cytoplasmic vacuoles have a definite wall. It is necessary in this connection to distinguish the vacuoles from the lymph spaces and capillaries. The vacuoles are usually small and irregularly distributed throughout the cytoplasm. They contain a homogenous fluid or differential bodies, and their presence is, we believe, intimately associated with the metabolism of the cell and probably with its constructive phases. These vacuoles vary in number in the same animal and in the same species. This would indicate that they are transitory structures which appear when certain chemical changes occur, and then disappear. A very critical study of the cytoplasm in contact with the vacuoles fails to show any evidence of a separate wall. The vacuole in the living nerve cell forcibly reminds one of the food vacuoles in protozoa which appear to have a wall; but this appearance is really due to the contact of fluids of different refractive index. In stained specimens the vacuoles look as if they were limited by a more deeply staining border, but this may be explained as due to the accumulation of cytoplasmic granules about the enclosed liquid. We believe that it is no more proper to speak of a wall for these vacuoles than it is to say that the numerous vacuoles in a protozoan have walls.

The lymph spaces are of a different character and are usually located in the periphery of the cytoplasm. They are intimately associated with the circulatory system and may contain blood. In some of the larger invertebrate nerve cells the periphery is richly supplied with lymph canals which may occasionally contain corpuscles. These canals or spaces can in many instances be traced directly into the surrounding neuroglia tissue and appear to be of a more permanent character than the vacuoles. We are inclined to believe that these lymph canals are supplied with definite walls.

A sufficient number of cases has been described to show that occasionally nerve cells are actually penetrated by capillaries. We have observed one instance in Helix. These capillaries terminate in finger-like branches or pass through the cell or even through two or three adjacent cells. They have a definite wall and contain blood corpuscles.

The question as to how the nerve cell is nourished, and how it maintains itself during long periods of excitation, long fasts or hibernation is one which has attracted the attention of scientists and will continue to do so. The appearance and disappearance of the granular particles in the cells at once gives evidence that they are temporary structures. It is natural to think of nerve cells as performing *one* function, and we frequently lose sight of the fact that the cell has a protoplasmic structure which must be nourished just as truly as that of any other cell. The activities of a nerve cell are not all of a nervous character; metabolic processes must go on here just as truly as in the muscle cell or the gland cell. But these processes may be overshadowed or concealed by the more specialized activities of the cell.

We shall attempt to show that these metabolic processes actually take place within the nerve cell, that certain food substances are stored up within the nerve cell, that these substances may remain in the cells for long periods, and that they may be called upon at any time of want or stress to supply material out of which new protoplasm may be built or to act as a source of energy.

Twenty years after the admirable work of NANSEN, we can do no better than to quote from him the following sentence. "If we look through the modern literature having special reference to the invertebrate nervous system, and compare the many different views of the structure of the ganglion cells, we meet with a confusion on the subject which is far from encouraging."

II. MORPHOLOGY OF THE GASTEROPOD NERVOUS SYSTEM.

Much of the work on nerve cells where a direct stimulation has been employed has been on a certain ganglion through a specific nerve passing to that ganglion. The nervous system of gasteropods does not permit of any such treatment, as the following description and diagram shows. The nervous system of Limax may be taken as typical of the common snails. It makes its first appearance on the sixth or seventh day after the eggs are laid (HENCHMAN '90) and is derived entirely from the ectoderm. The several ganglia which constitute the nervous system of Limax arise separately to become secondarily joined by commissures.

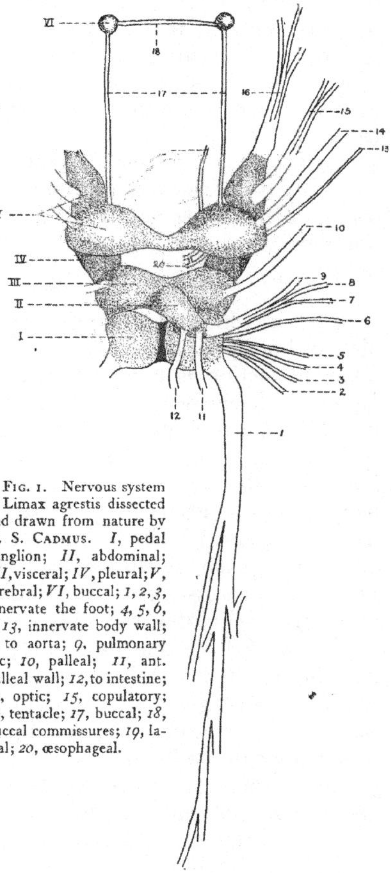

FIG. I. Nervous system of Limax agrestis dissected and drawn from nature by H. S. CADMUS. *I*, pedal ganglion; *II*, abdominal; *III*, visceral; *IV*, pleural; *V*, cerebral; *VI*, buccal; *1, 2, 3*, innervate the foot; *4, 5, 6, 7, 13*, innervate body wall; *8*, to aorta; *9*, pulmonary sac; *10*, palleal; *11*, ant. palleal wall; *12*, to intestine; *14*, optic; *15*, copulatory; *16*, tentacle; *17*, buccal; *18*, buccal commissures; *19*, labial; *20*, œsophageal.

In the adult stage, the central nervous system consists of five pairs of ganglia and a single ganglion asymmetrically placed. The relative position of the ganglia can be appreciated from the view shown in Fig. I. "In passing from behind forward, the ganglia are encountered in the following order: (1) The pair of pedal ganglia, which lie under the radular sac, and are joined to each other by an anterior and a posterior commissure; (2) one abdominal ganglion a little to the right of the median plane (which is intimately fused with the right visceral, and is also in close connection with the left visceral ganglion, p. 199); (3) a pair of visceral ganglia occupying the posterior angle formed by the outgrowth of the radular sac from the œsophagus. They are separated by the abdominal ganglia from which connectives pass to them; (4) a pair of pleural ganglia, not joined by a commissure and not giving off nerves. They are united by means of connectives to the pedal, visceral, and cerebral ganglia of the same side; (5) a pair of cerebral ganglia with their supra-œsophageal commissure and connectives to the pleural, pedal, and buccal ganglia; (6) a pair of buccal ganglia, with a commissure under the œsophagus posterior to its connection with the sac of the radula." (Quoted from HENCHMAN '90, p. 193.)

A comparison of this drawing with those of pond snails by LACAZE-DUTHIERS shows a number of differences in respect to the origin of the nerves and the announcement of two nerves that are not shown in his figures.

III. LYMPH CANALS.

Structures known as lymph canals we differentiate from vacuoles, although both have a similar appearance in the fixed cell. This distinction is made after a study of the living nerve cell. In a subsequent section on vacuoles it is suggested that in certain instances the lymph canal, trophospongium, etc., are not real lymph spaces, but isolated and independent vacuoles. That lymph canals do really exist in nerve cells seems to be well established, as the accompanying review indicates. Our study of fixed material in Helix and Aplysia shows that the outer border of the cytoplasm is frequently penetrated by spaces, as well as numerous processes from the neuroglia. Many of the drawings of RHODE and HOLMGREN indicate a similar state of the cytoplasm so that we believe that these lymph canals have a rather general distribution in invertebrate nerve cells. HOLMGREN in his several papers has given an elaborate account of lymph-spaces. Apparently the same class of structures had been previously described under the caption "intercellular neuroglia" by RHODE. RHODE observed these structures in various animal classes, making a special study of Aplysia, Helix, and Doris. His results are interpreted in terms of his theory of work on the part of the neuroglia cells. The neuroglia cells are not considered as intruders but as cells which by their activity build up the nerve cell.

In order to give some conception of the extent and importance of the work on lymph canals, the following rather full review is made

Our review of the work of HOLMGREN can give us at best but an inadequate conception of its amount and quality. His numerous papers, while somewhat controversial, contain a large range of observations on *fixed* nerve cells, both vertebrate and invertebrate. His main contention seems to be centered around the character of the cytoplasm. Whence come the numerous spaces in it, a ' what of their character? It seems to us necessary to include here a review of some of his studies upon the nerve cells of vertebrates, since he makes this his starting point. A good summary of HOLMGREN's ideas concerning the structure of the nerve cell may be found in vol. 11 of MERKEL and BONNET's Ergebnisse.

HOLMGREN ('01) makes the first mention of the "Saftkanälchen" in the spinal nerve cells in his paper on Lophius piscatorius, where he makes the following statement, "localized endocellular nets of 'Saftkanälchen' are seen especially well in the rabbit." A thick network of fine tubules is to be seen in the cytoplasm surrounding the nucleus, and usually near the poles of the cell. The sectioned lumina of the tubules are always circular in outline and are always sharply marked off. Here and there one can find how these networks of tubes are connected with the pericellular tubes. In these places the walls are clearly marked. Within the cells the author could see no definite walls to the canals. Most of the cells of the spinal ganglia possess such networks, but they do not always seem to agree with each other with respect to the breadth of the lumen or the wall of the canal.

In the cells the author distinguishes two cytoplasmic zones, an inner canalicular and an outer extra-canalicular zone. These canals are supposed to have walls—at any rate something which appeared to be a wall stained red with erythrosin. In addition to the observations just cited upon the rabbit, the author studied the d g, cat and various birds. In these animals he found remarkably strong dilated canals winding in a corkscrew manner through the ganglion cells. From the peri- or extra-cellular tubes more or less numerous canals force their way into the ganglion cells. Inside of the cell they often divide in the characteristic finger-form manner, and they turn in manifold ways, not infrequently in spirals. By this means there exist glomerulus-like collections of tubes in the cell. In the case of the birds there were seen canals so strongly dilated that the protoplasm appears only as islands or thread-like heaps between the tubes. These dilations or tubes are not localized in any particular part of the cell but may be found in any part. He says that these tubes must correspond to the bands which were described by NELIS, with the exception that NELIS did not make any mention of bands going out of the cell. Such connections do not exist in all cases, but are nevertheless general. HOLMGREN could find these cells in the sympathetic and central nervous system of birds. He considered the canals which may be continued beyond the limits of the nerve cell as lymphatic passages.

As opposed to STUDNICKA, HOLMGREN says that the lymph canals come from the anastomosis of vacuoles or alveoli, and again he states that the canals in the case of Petromyzon are bounded by intensely staining walls which continued rectly outside of the nerve cell into the walls of the extracellular paths.

If one stimulates the spinal ganglion cells by means of weak induction currents, almost all parts of the whole canal are strongly widened. This agrees with the statement of NELIS that the bands occur in altered cells. "The nerve cells are permeated with a very rich canal system hitherto unsuspected, and only the more dilated parts of these networks are the passages which I was able to see befo ."

The great dilations of the canals are certainly only accidental, and so one can understand without anything further the great variability of the canals.

After working upon a variety of animal forms both vertebrate and invertebrate, and especially upon Lophius, the author concludes that his former position in harmony with that of FRITSCH ('86) is a mistaken one and that the vessels are not blood vessels within the nerve cells but are to be considered as lymphatic in their nature, and that they press their way into the nerve cells and there branch about. Among the invertebrates he finds Astacus and Palæmon, next to Lophius, excellent material for clearing up the true nature of the lymph canals.

In very young animals he finds the canal net to be remarkably simpler than in the case of older animals. Often this net is to be found at one pole of the very eccentric nucleus. The sympathetic nerve cells of the mammals show the canal nets only within the cell body. The same nerve cells of the bird, like the central nerve cells of all the vertebrates studied, possess continuations of the net also within the dendrites. An electrically stimulated nerve cell of the bird will show, according to HOLMGREN, the presence of the lymph canals in the neurites.

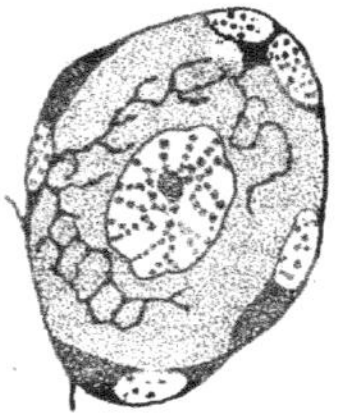

FIG. 2. The intracapsular cells surround the nerve cell. The trophospongium branches as a net of coarse threads through the endoplasm and at two points reaches the surface connecting itself with the colored bodies of the intracapsular cells. After HOLMGREN ('04, Fig. 1).

The question as to the morphological and genetic character of the lymph canals has been much discussed and various opinions held. NELIS considered them as achromatic hyaline bands, but he seems to be somewhat uncertain in his meaning. To him they are riddles as to morphology and function. HOLMGREN and his followers believe them to be canal-like, fluid carrying structures. According to HOLMGREN and his followers the bands of NELIS are only modified parts of the lymph-canals. HOLMGREN opposes the view that they are formed out of the nerve cells but holds that they press into the substance of the nerve cells from without in the form of hollow processes (Kapselfortsätse). He further claims to have seen unmistakable nuclei-bearing capsule processes in the spinal nerve cells of Lophius and other teleosts, also in the gastric ganglion cells of the Crustacea, within which there were sap spaces. According to his view these canals do not represent drainage tubes but are rather the morphological expression of certain phases of the penetration of nerve cells and the intracapsular cells belonging to them. The trophospongium has pseudopodia-like mobility whose intensity is supposed to depend upon intracellular chemical processes (Fig. 2).

The lymph-canals are of a lymphatic nature and are certainly associated with the nourishment of the cell. NELIS claims that as the nerve cells change there is a decrease of the tigroid substance which is accompanied with an increase in the amount of the transparent bands. HOLMGREN believes that the localization of the tigroid substance should coincide with the appearance of the canals, that the canalicular zone of the cell should be the tigroid layer free of ectoplasm. The tigroid substance stands in a causal relation to the lymph clefts and is associated with their activities. Where the clefts are especially dilated, a rich accumulation of tigroid substance takes place. In more protracted periods of activity the clefts become smaller and the tigroid substance vanishes; but in such places where the tigroid substance remains, the clefts remain dilated. Electric stimulation points to the same conclusion. The nerve cells, as a result of such a stimulus, receive new supplies of tigroid substance and at the same time become somewhat larger; accompanying this, there is a dilation of the lymph clefts. This leads one to believe that the electric current calls forth an alteration of the circulatory relations. HOLMGREN cites a number of investigators whose work bears directly on the interpretation of these structures as follows:

ADAMKIEWICZ ('86) from his researches with injections could have made the same report, that the nerve cells are furnished with their own blood vessels and that the nuclei of these cells should present venous spaces, but these discoveries have nothing to do with the sap canals which do not carry blood. From the work of other investigators it is evident that blood vessels very rarely enter nerve cells.

FRITSCH ('86) found that blood vessels were constantly to be found in the giant ganglion cells of Lophius piscatorius. HOLMGREN uses the results of FRITSCH to confirm his own belief that lymph spaces exist in the cell, but makes the additional statement that the blood capillaries are supposed to be drawn into the cell through endocellular branching processes. In 1900 HOLMGREN came to the conclusion that these spaces in the cells were not to be considered as blood vessels but rather as lymph spaces in so much as they do not carry corpuscles. STUDNICKA in the same year expressed the same belief, though more indirectly.

NELIS ('99) describes in nerve cells homogeneous non-staining bands of a skein-like appearance found within the cell. These appear in various places in the cell body. They exhibit various forms, half moon, spiral, corkscrew, and hang together at the ends, but do not form a true reticulum. They are to be found in the cells of the spinal and sympathetic systems as well as in the brain. They are particularly prominent in animals which have been poisoned. HOLMGREN claims that these structures are the same as are called "Saftkanälchen."

STUDNICKA ('99) held that the canals are formed from the running together of vacuoles which had formed in the cell in a row.

BETHE ('00) opposes this view on account of the fact that he had observed single canals which passed completely through several nerve cells and their capsules at the same time.

FRAGNITO ('00) regarded the canals as the remains of the interstices between the neuroblasts, through whose melting together the single nerve cells are supposed to come into existence.

PUGNAT ('97) believes that the canals force their way into the nerve cells from without, as lymph capillaries.

PEWSNER-NEUFELD ('03) studied the finer anatomy of the nerve cells in the nervous system of the white rat and guinea pig. He does not find that there are distinct zones in the plasma of the cell. Small canals are scattered throughout the cytoplasm, no region being free from them. They do exist in the nucleus (Fig. 3). The canals may or may not occur in the protoplasmic processes of the cell. The canals run about the NISSL flakes, sometimes passing through them, at other times merely surrounding the flakes, or they may be free in the cytoplasm. Some of the small canals approach the nuclear membrane, but in no place were they seen to penetrate it. The size and extent of the canals is dependent on the physiological state of the cell. The canals do not have a distinct wall but a linear boundary due to the arrangement of the cytoplasmic granules. The intracellular lymph canals of the central ganglion cells open into channel-like spaces.

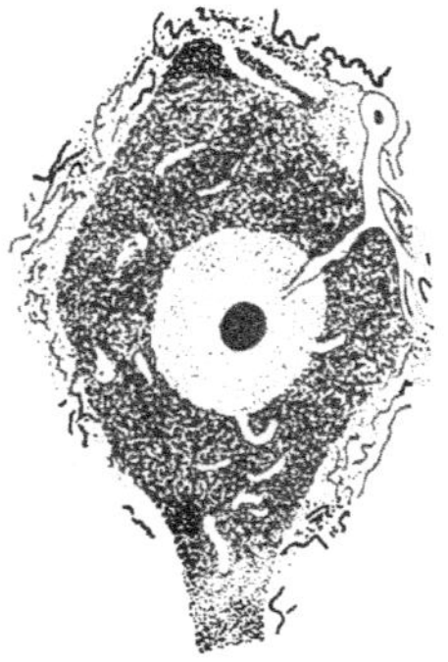

FIG. 3.

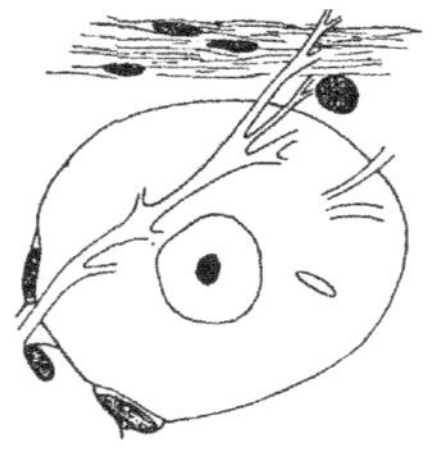

FIG. 4.

FIG. 3. Ganglion cell of white rat. Illustrates penetration of cytoplasm and nucleus of nerve cell by sap canals. The isolated clear spaces are the cut ends of sap canals. After PEWSNER-NEUFELD ('03, Fig. 3).

FIG. 4. A large ganglion cell with tubes formed from the capsule extending entirely through it. After BETHE ('00, Fig. 2).

STUDNICKA ('99) presents a discussion of the origin and use of the canals in ganglion cells. The little canals can very often be followed in the body of the cell some distance, indeed, often through the half of the entire cross section of the cell. They are seen in such a study to branch freely. These little canals which are identical with those described by HOLMGREN, arise very likely through the union of a row of vacuoles. Many of the canals have smooth outside walls. Some separate vacuoles are found which are explained as being the cross sections of the branches of such vacuoles as have not yet fused into canals. He is unable to define the contents of the canals and alveoli, but suggests that they are during life, no doubt filled with a fluid which may be identical with that in the pericellular space with which the little canals are united. Some of the greater alveoli contain a homogeneous substance which colors more intensely with eosin and is to be considered a special deposit.

BETHE ('00). We have here to do only with dependent canals (blood vessels) which can be proven only by injections. No nuclei are to be found in the walls of these canals. The canals result from the fusion of separate vacuoles. The canals have nothing in common with the neuro-fibrillæ (Fig. 4).

IV. VACUOLES.

The presence of vacuoles or vacuolar-like structures in the cytoplasm of nerve cells is a common structural character. They have been recorded as follows:

HODGE'S ('92, '94) work is of great importance to all interested in the question of fatigue and the accompanying structural changes in the nerve cells. The spinal ganglion cells of the frog, cat and dog, under electrical stimulation and the spinal ganglion and brain cells of English sparrow, pigeon and swallow show the following changes. The nucleus undergoes a marked decrease in size and changes from a smooth and rounded structure to one having a ragged outline. Its reticulate appearance is changed and the whole structure takes a denser stain. The cell protoplasm gives evidence of slight shrinkage and the formation of vacuoles. These vacuoles appear quite constantly in the ganglion cells of birds. The vacuoles have a sharp outline and a definite shape in the rested animal but are indistinct in the bird that has been at work during the day. Vacuoles also appear in the honey bee under the following conditions. Honey bees were collected in a raspberry patch as soon as they appeared in the morning. The first six bees were quickly decapitated, the brains removed, and three were dropped into one-half per cent osmic acid, and three into saturated mercuric chloride solution. At about seven o'clock at night six more bees were captured and treated in the same manner. After the morning and evening bees had been paired at random, each pair was stained and studied and an attempt was made to measure the nuclei and work out the amount of shrinkage. The minimal shrinkage was 9 per cent, and the maximal 75 per cent. The author does not attach much value to these figures, although they express the fact that a wide difference exists between the two. The average in diameter of the morning bees is more uniform than for the evening bees. These results indicate first, that the nerve cells of a number of bees' brains are in a more uniform condition in the morning than in the evening. Secondly, they differ in appearance, or condition, from one another, somewhat in the morning and a great deal in the evening.

MONTGOMERY ('97) finds in the nemerteans, Cerebratulus and Lineus, chromophilic corpuscles under the following conditions: The cytoplasm of the medium sized cells is of a coarsely vacuolar structure; sometimes the hyaloplasm fills the whole proximal portion of the cell as far as the nucleus. But a thin, peripheral layer of spongioplasm is always present, and a similar layer envelops the nucleus. These cells are much larger in Cerebratulus and the cytoplasm is much denser, *i. e.*, there is a proportionately greater amount of spongioplasm, and a coarsely vacuolar structure is seldom found. The large cells of the brain are of an elongated pyriform shape, largest and rounded proximally, seldom nearly spherical. It may be noted that while the cell bodies vary considerably in size, their nuclei remain of nearly uniform dimensions. The cytoplasm is, as a rule, coarsely vacuolar (vesicular), especially so toward the distal pole. A thin peripheral layer of finely granular cytoplasm is always present. The vacuoles do not seem to have any definite grouping, but such groupings as exist are explained as corresponding to the different physiological states.

Certain bodies occur in these ganglion cells in Lineus which are absent in all of the cells in Cerebratulus. These bodies are frequently larger than the nucleolus and of a spherical or oval shape, and are not refractive. After the use of a double stain they stain usually with eosin, sometimes with hæmatoxylin, but always more intensely than the surrounding cytoplasm, though seldom as deeply as the nucleolus. Structurally, they are homogeneous, with a peripheral membrane, which may be scarcely discernible or in other cases, of considerable thickness; this membrane always stains more intensely than the enclosed portion, and forms a boundary against the surrounding cytoplasm (Fig. 5). These bodies do not occur in all cells, but only in about one-sixth of the total number; when they are present, it may be but a single one, more frequently four or five, apparently never more than

Fig. 5.

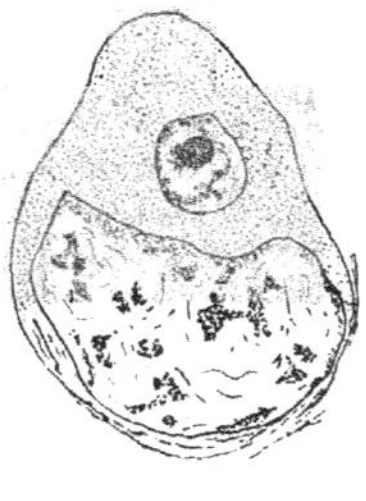

Fig. 6.

Fig. 5. Lineus gesserensis. Ganglion cell of the third class, showing the presence of vacuoles, some of which contain differentiated granules. After Montgomery ('97, Fig. 9).

Fig. 6. Nereis, brain cell of sixth class. Nucleus lies in narrow end surrounded by granular cytoplasm, while in the other end there is a large vacuolar space. After Hamaker ('93, Fig. 17).

fifteen. There is also no regularity in their distribution, such as a concentric or radial arrangement, and in the same cell they are usually of various sizes and of different staining power. To these cytoplasmic bodies may be applied the term chromophilic corpuscles, to distinguish them from the chromophilic granules in the ganglion cells of other animals.

Rand ('01) reports vacuoles and gives an analysis of the cytoplasm as follows: Very little can be said as to the finer structure of the cell protoplasm in the Lumbricidæ. The most careful examination fails to reveal its precise nature. It varies in degrees of homogeneity somewhat according to the size of the cell. In the smaller cells, it usually appears compact and fairly homogeneous. In larger cells, it is much less homogeneous, and there is a tendency toward the formation of large vacuolar spaces. The substance of the fixed cytoplasm, as it appears to the eye, may be said to be of four kinds. There is (1) a perfectly homogenous "ground," represented by the lightest areas in the figures; (2) material which gives the impression of being very finely granular; in the smaller cells this is quite evenly distributed, while in the larger cells it tends to concentrate in regions, giving the cytoplasm a blotchy appearance; (3) rather conspicuous granules or masses staining fairly deeply and often surrounded by an area within which the material of the second class is less dense; (4) a fine fiber irregularly distributed through the cell body, but often appearing to be associated with the more conspicuous granules and sometimes occurring about granules as centers of radiations.

HAMAKER ('93) shows in one type of the nerve cells in Nereis the following: In the posterior half of the brain there are several pairs of very large cells which have a very striking characteristic. The nucleus lies in the narrow end of the cell, and is surrounded by the granular cytoplasm. At the other end of the cell, there is a large vacuolar space containing a number of deeply staining bodies of irregular form, embedded in an indistinct coagulum (Fig. 6). Other cells have very finely granular substance occupying a similar position, the granules being much smaller and staining less deeply than those of the body of the cell. In these cases the nucleus shows no signs of degeneration.

LENGENDRE ('05, '06) in a series of short papers during the years 1905 and 1906 has given us reports of an investigation on nerve cells of Gasteropods. He has studied the cell from the physiological point of view, with the idea of determining whether the structures described by HOLMGREN and others are in any way related to the nutritive functions of the cell. He follows the work of HOLMGREN, BOCHENEK, McCLURE, RHODE, and others. A study is made of the effects of various fixing reagents and he finds that RABL's solution is a very poor reagent for the study of nerve cells. Consequently many of the results which have been obtained through the use of this fluid are to be considered as artefacts and not as actual structures which exist in the living cell. He questions the work of RHODE and does not believe that the fibrils of the nerve protoplasm are continuations of the processes of the neuroglia cells on account of the difference in size and staining qualities. He finds in the cells of Helix pomatia vacuoles of various sizes, arranged in various ways in the cell. Sometimes they communicate with one another and sometimes open to the outside of the cell. These vacuoles are without definite walls and contain a homogeneous fluid without granules. The chromophile granules are always found in the protoplasm when present at all and never appear in the vacuoles. LEGENDRE does not admit the theories of HOLMGREN concerning the nutritive functions of the nerve cells. He advocates in his first paper that the vacuoles represent accumulations of excretory products and that they are in no way connected with the constructive metabolism of the cell (Fig. 7).

In these papers he calls attention to the following points: He describes the appearance of living nerve cells that have been immersed in water for a considerable time. The result is a rapid increase in size due to osmotic exchange. In the protoplasm of the cells thus treated the meshes of the spongioplasmic net become greatly enlarged and more clearly visible. The nucleus becomes large and numerous vacuoles appear in the periphery of the cytoplasm. He also advances the idea that the HOLMGREN canals in the trophospongium are to be interpreted as pathological rather than nutritive and that they act more like the phagocytes in that they destroy cell substance rather than build it up.

PFLÜCKE ('95) notes the presence of a few vacuoles in the cell plasma which he does not regard as true vacuoles but as accumulations of unstainable substance.

EWING ('98) takes an extreme position in regard to the presence of vacuoles, claiming in the majority of cases that they are cadaveric or artificial products.

The formation of vacuoles has long been recognized as one of the necessary imperfections in most methods of fixing of nerve cells. The writer cannot agree with the statement often seen that the vacuolation may be regarded as pathological only when it is found in advanced degree. Among the present cases, extreme vacuolation when found, was always plainly referable to post-mortem processes. The study of cadaveric changes in ganglion cells indicates that vacuoles are one of the most constant of post-mortem products; and that they frequently form in considerable numbers and of large size within a few hours, often preceding other post-mortem changes. Especially when the brain and meninges are œdematus, or when the patient has suffered from general sepsis, vacuolation of cells may be expected unless the tissues are fixed very shortly (one half hour) after death. The above observations, as well as the circumstances under which vacuoles are usually found in stained specimens, indicate that in the great majority of instances vacuola-

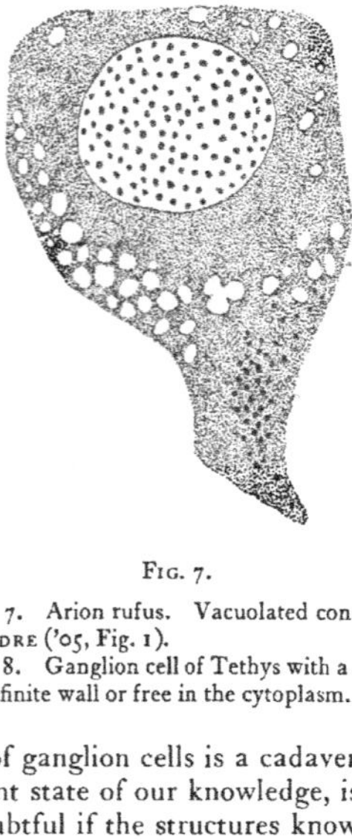

FIG. 7.

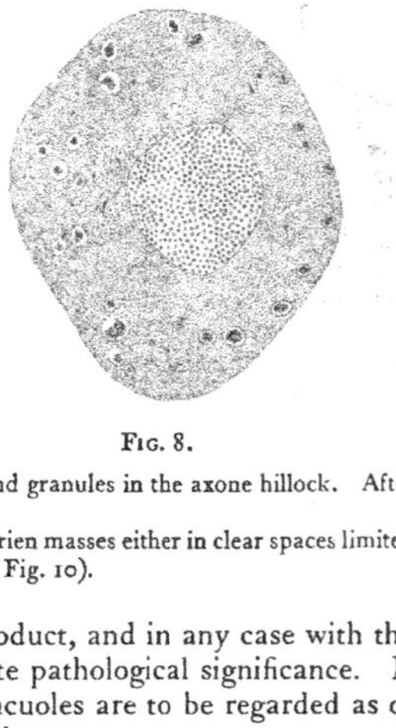

FIG. 8.

FIG. 7. Arion rufus. Vacuolated condition of cytoplasm and granules in the axone hillock. After LEGENDRE ('05, Fig. 1).

FIG. 8. Ganglion cell of Tethys with a number of mitochondrien masses either in clear spaces limited by a definite wall or free in the cytoplasm. After RHODE ('04a, Fig. 10).

tion of ganglion cells is a cadaveric or artificial product, and in any case with the present state of our knowledge, is devoid of definite pathological significance. It is doubtful if the structures known as nucleolar vacuoles are to be regarded as of a similar character with the vacuoles of the cytoplasm.

RHODE presents numerous facts in his several papers in regard to the structure of the ganglion cell. The sphere referred to in the following is a differentiation of the cytoplasm of a distinct character and should not be confused with the sphere associated with the centrosome. According to RHODE ('04a) the sphere in the ganglion cell of Tethys consists of a central part surrounded by a clear layer having the granules arranged compactly and in a radial manner. The clear layer is made up of a homogenous or fine granular substance which colors intensely (Fig. 8). The outermost bodies in the peripheral layer of granules may fuse completely so that there is the appearance of a thick membrane which seems to separate the sphere from the cytoplasm. The clear region may be encroached upon and occupied by radially arranged granules which vary in size. All stages in the origin of the sphere may easily be seen in the same ganglion cell. In the frog these same spheres have a nuclear origin, *i. e.*, they are derived from the smallest bodies in the nucleus. In the same manner as in the frog, arise the spheres in Tethys with this difference, that the origin does not take place within, but without the nucleus. The various stages in the development of the spheres are seen in the cytoplasm, which may be compared to similar stages in the development of the spheres in the ganglion cells of the frog.

When the spheres attain a certain size, their destruction occurs as follows: The central body becomes indistinct and the radial zone breaks up into large or small pieces, finally becoming so small that they cannot be distinguished from the cytoplasmic granules so far as their shape is concerned, but they retain their avidity for stain, which gives them prominence everywhere. Some of the large spheres do not go through these regular changes and are described as vacuoles (Bläschen) with a thin wall and a clear center. In the transformation of the sphere into a vacuole this stage corresponds to the term "Mitochondrien." When the peripheral layer of the sphere is broken up into a number of loose threads the term "Chondromiten" is applied to them. The largest spheres are as a rule the oldest and arise out of the smallest, structureless globules (Küglchen) of the cytoplasm. These may be seen to grow and to differentiate themselves into a light inner zone and a dark outer band. The larger the sphere, the more plainly the granules, which finally assume a radial arrangement in the outer zone, appear. The last stage in the formation of the sphere shows the central body assuming its complete shape and size.

SMALLWOOD ('06) reported the presence of numerous vacuoles in Haminea, Venus, Planorbis, Limax, Helix, Littorina, Melantho, Montagua and Aplysia which were designated as lymph spaces. A more extended study suggests that this term should be reserved for the larger peripheral spaces and that the term vacuoles more correctly describes them. There is no definiteness about their position or size in the cell (Fig. 7). Animals examined during all seasons of the year show them to be present in living nerve cells.

From this review, we learn that the nerve cells of Nemerteans, Annelida, Crustacea, Insecta and Mollusca among invertebrates exhibit a highly modified cytoplasm. A sufficient number of specimens have been examined in each of these great groups to indicate the very general appearance of differential structures in the cytoplasm other than fibrillar. In the introduction eleven different terms are cited as having been given to this stainable substance in the cytoplasm which of itself suggests that the problem is one of great difficulty; certainly a doubt must have existed in the minds of the various workers who have coined these terms as to their significance and relationship.

It is rather hard to make a classification of these structures as described by the various authors because in most instances the cytological study was not followed or preceded by an examination of living nerve cells. Our results have been so clear and satisfactory that we are tempted to try to correlate some of the previous facts with them. Probably the commonest structure present in the cytoplasm of the invertebrate nerve cells is the *vacuole.* These vacuoles are present in all of the great groups already cited, although usually described under the terms "lymph space," "Netzapparate," "Saftkanälchen," "Trophospongien," etc. The vacuole can be determined in the following manner in the living cell: Isolate a nerve cell and study it in a 1–500 solution of methylene blue or neutral red in normal salt solution under the oil immersion lens. At first, but little can be determined; but as the stain progresses the vacuoles become more distinct and their contents often take on a differential stain. The experienced worker can make out these vacuoles without any stain. The time that it takes to stain these vacuoles will vary; but usually from 5 to 20 minutes will be the limit, as after that time the nerve cell is apt to become overstained and undergo some changes in its general appearance and the character of its parts. This gives about 15 minutes when a critical study may be made. During this time the vacuoles are readily made out as isolated spherical bodies containing a fluid. It is impossible to trace any connection between vacuole and vacuole. The size is also further evidence of their individuality, for they range from the very minutest bodies recognizable with the oil immersion lens to structures a third the size of the nucleus. Studying these vacuoles in Planorbis and Limax for two years, in which we examined almost weekly the living nerve cells from hundreds of specimens, we are convinced that these vacuoles are transitory structures, that they vary in number from time to time, and that they are not limited by a distinct wall. The vacuoles move about in the cytoplasm when the nerve cell is put under pressure, which would be impossible if they were part of lymph spaces that had grown in from the surrounding neuroglia tissue.

The Chronodromiten and Mitrochondrien of RHODE, the Trophospongien of HOLMGREN as interpreted by BERGEN, present in Helix but not figured by MCCLURE, the chromophilic corpuscles of MONTGOMERY, the vacuolar spaces of HAMAKER, the granules within clear spaces of RAND, the numerous vacuoles described in Arion by LEGENDRE, all, we believe, are to be classified as nerve cell vacuoles. The significance of these vacuoles is discussed further on.

V. THE NISSL BODIES.

RHODE ('04a) has called attention to certain similarities of structure in the ganglion cells of vertebrates and invertebrates. Both have the following facts in common: (1) a homogeneous hyaloplasm, (2) a spongioplasmic groundwork which consists of coarse and fine fibrils, (3) a *stainable substance* which in the case of the invertebrates and a part of the vertebrates is lodged in the coarse fibrillar spongioplasm. In the remainder of the vertebrates it clumps and forms the NISSL bodies, which are, indeed, independent of the spongioplasm, which appears between them in almost colorless fibrils.

The structures known as NISSL bodies or granules furnish a most interesting field of research. The great degree of variability in the appearance of nerve cells from different animals has led to the belief that structures existing in one nerve cell may have no counterpart in another. Among the invertebrates the failure of some authors to identify structures closely similar to those found in vertebrates has led to the supposition that such structures were lacking. It seems evident that such bodies as NISSL granules must be present in the cell for some specific purpose. The nerve cells of invertebrates have fundamentally similar functions to perform as the cells of vertebrates. If this be true, may we not expect to find some structure, perhaps even morphologically and chemically different, which takes the place of that structure known as the NISSL granule? We are of the opinion that such bodies do exist.

The *stainable structures* of the cell, referred to above, have received various names, as they have been observed and described by different authors under dissimilar conditions. The terms chromatic substance, chromophile substance, tigroid substance, sigroid substance, basophile constituent, etc., have all been employed to designate the structures recognized by us as NISSL bodies or NISSL granules. Various authors have recognized the fact that these bodies may vary in size, in number and in capacity for taking up various staining agents.

NANSEN ('87) described the structure of the nerve cells of Patella vulgata, Nereis, Lumbricus, Homarus vulgaris, Nephropa norwegicus and six different Ascidians which he classed with the above. He found in the cells of the Nereidæ structures which correspond very closely in description to the granules commonly known as NISSL bodies. Some of the granules were very large and prominent and were situated in the mesial part of the protoplasm. In preparations fixed with osmic acid and stained with hæmotoxylin they were very dark, almost black in color, and consisted of a fatty (myeloid ?) substance (Fig. 9).

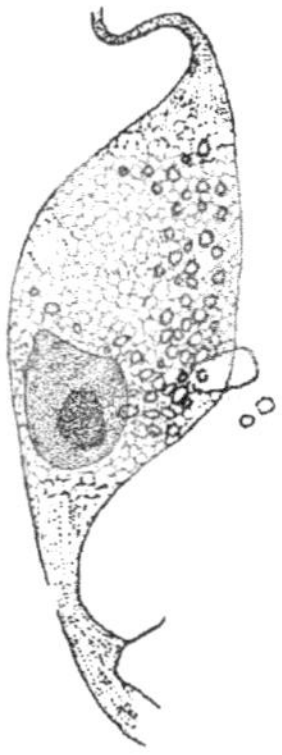

FIG. 9.

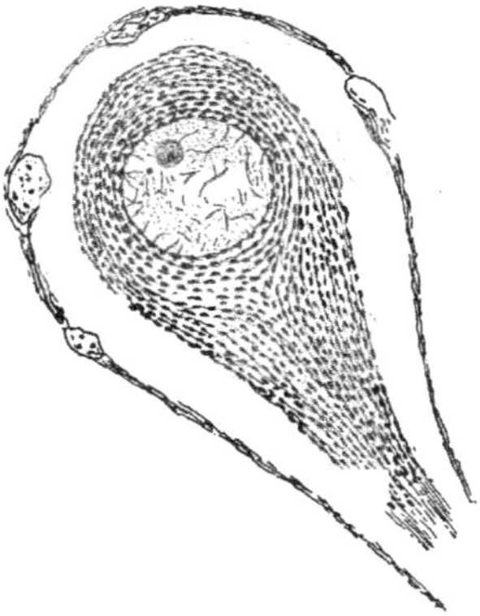

FIG. 10.

FIG. 9. The yellow granules are scattered through the cytoplasm and are drawn with heavy outlines. After NANSEN ('87, Fig. 54).

FIG. 10. Crayfish. Shows the chromophile bodies spindle shaped and apparently associated intimately with the fibers. After PFLÜCKE ('95, Fig. 10).

Nissl bodies in invertebrates.—The question as to the existence of NISSL granules in the nerve cells of invertebrates has more than once been raised. PFLÜCKE ('95) undertook the investigation of the finer anatomy of the nerve cells of the crab, snails and worms. In the crab he succeeded in demonstrating granules which appear like the commonly accepted NISSL bodies. In the snails and worms he failed to identify such structures (Fig. 10).

McCLURE ('97) found (chromophilous) granules in the nerve cells of Helix and Arion, and expressed the opinion that this chromophilous substance is homologous with that found in the nerve cells of vertebrates.

FLOYD ('03) was unable to differentiate by means of methylene blue any NISSL bodies in the ganglion cells of the common cockroach. In well fixed material, however, he found varying quantities of deeply staining granules and masses.

Distribution.—These deeply staining granules were found by RHODE ('04a) in both vertebrates and invertebrates to occupy a zone of the cytoplasm surrounding the nucleus but not extending out to the cell wall. A rather broad zone (the spongioplasm) at the periphery of the cell is free from these bodies, so that the ganglion cell resembles the Amœba in that it has a light ectoplasm and a dark entoplasm. Only the finely granular hyaloplasm enters into the axis cylinder.

McCLURE ('97) found the granules to be arranged chiefly in rows, but at certain points in the cell body they appeared to be collected into spindle-shaped groups, having their long axes parallel to the periphery of the cell (see Fig. 11). A statement of McCLURE's is of particular interest: "The cell bodies stain a deep blue, while the axis cylinder processes are only partially affected by the stain, and thus appear light in color. The cause which produces this difference is fundamentally the same in both cases: namely that the intense staining capacity of the cell body, and the lack of the same for the axis cylinder process in Limax are due respectively to the presence and absence of the chromophilous granules. The Flemming-iron-hæmotoxylin preparations are especially interesting for the reason that they show with great clearness, not only the same chromophilous granules but also certain spindle shaped structures in the cell body, which in all probability are collections of some small chromophilous granules. The above results concerning the presence of chromophilous granules in the nerve cells of Gasteropods point toward the acceptance of the view that this chromophilous substance is homologous with that found in the nerve cells of vertebrates (NISSL bodies)."

PFLÜCKE ('95) found that in the crab the chromophile granules of the nerve cells are arranged in rows, and in the nerve processes they were few in number. The granules were especially numerous about the nucleus, being regularly distributed. Under high magnification they were found to be spindle-shaped and to be arranged in parallel concentric rows.

FLOYD ('03) finds the granules disposed in areolar fashion in the cell, deposited upon the cyto-reticulum.

Physical constitution.—Among vertebrates the NISSL bodies have been found by FLEMMING, VON LENHOSSÉK, MARINESCO, VAN GEHUCHTEN, HELD, CAJAL, PFLÜCKE, EWING, CARRIER and others to have a granular structure—to be in reality aggregations of minute particles of deeply staining substance. FLOYD and McCLURE have presented evidence of the same structure for the NISSL bodies of the invertebrates.

Resistance to degenerative change.—The work of EWING ('98) upon cadaveric changes taking place in the ganglion cells of brains and cords of rabbits which were allowed to decompose in the air from 48 to 72 hours may give evidence as to the function of the NISSL granules. During the first twenty-four hours there was noticed a granular disentegration of the chromatic substance. This disintegration was evidently due to the separation from each other of the granules which made up the NISSL bodies. As the degenerative changes proceeded, the granular disintegration became more and more marked. During this time the individual granules retained all of their natural capacity for stains. Later when putrefaction changes were set up in the cells the NISSL granules exhibited a remarkable resistance to the action of the bacteria and still retained distinct outlines even when the cells were becoming filled with vacuoles or when the cell consisted merely of a nucleus with a narrow fringe of granules (see Figs. 12–13).

Do NISSL granules exist in the living cell? The existence of the NISSL granules in the living cell has been seriously questioned by several prominent observers and various answers have been published. DOGIEL, HELD, RUZICKA, FLEMMING hold to the view that they are an aggregation of material produced in the cell at the time of fixation, by the reagents employed. OLMER ('01) contends that the material of which the NISSL bodies are composed is scattered through the cells, and that these particles are clumped and precipitated by the fixing agent.

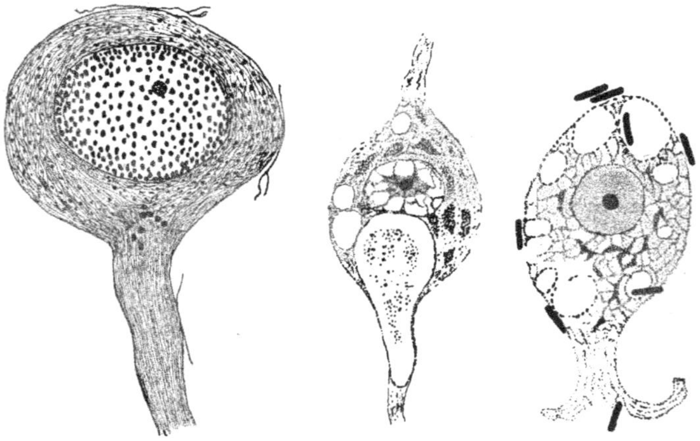

FIG. 11. FIG. 12. FIG. 13.

FIG. 11. Helix. Cell from infra-œsophageal ganglion. FLEMMING's sol., prog. iron-hæm. Concentric arrangement of fibrils and granular rows. Spindles. Pigment granules at base of process. After McCLURE ('97, Fig. 12).

FIG. 12. Medullary stichochrome of infant, 3 hours after death. LANG's fluid. Methylene blue. Very rapid and extreme vacuolation. Coarsely granular appearance of chromatic bodies. After EWING ('98, Fig. 1, plate 2).

FIG. 13. PURKINJE cell of rabbit, after 48 hours exposure to air. LANG's fluid. Methylene blue. Extreme vacuolation. Growth of putrefactive bacteria. The chromatic reticulum and bodies are reduced to a series of coarse dark granules. Complete nuclear chromatophilia. Shrinkage and destruction of dendrites. After EWING ('98, Fig. 3, plate 2).

The admirable work of CARRIER ('04) gives strong evidence for the belief that the NISSL bodies are not due to postmortem changes but actually exist in the living cells. In support of the same view may be mentioned the results by ARNOLD, VON LENHOSSÉK, CAJAL, TURNER, etc.

Our own work, done upon the living nerve cells, has convinced us beyond any possibility of doubt of the actual existence of these structures in the living cell. The detailed discussion of this feature appears later in the paper.

Studies upon the molluscan nerve cells.—In a previous paper in this *Journal* SMALLWOOD ('06) described certain morphological characters of molluscan nerve cells. Without indulging in vain repetition, it is perhaps well to call attention here to certain of these facts. There were found in the nerve cells of Haminea solitaria, Venus mercenaria, Planorbis and Limax cytoplasmic vacuoles containing a colorless, transparent liquid, also solid bodies of various sizes, irregularly rounded forms of varying numbers. The solid bodies were of different appearance in the different genera named, and somewhat different in distribution, those occurring in Limax being always found within the limits of the vacuoles, while in the other forms the bodies or granules were only rarely to be seen in the vacuoles. Attention was called to the fact that these bodies could be seen in the living nerve cell, and hence could not be considered as artefacts. The fact that the number of these bodies present in a given cell varies from time to time convinced us that a morphological study could not satisfactorily account for their presence and variable appearance.

In our discussion of our work upon the bodies mentioned above we do not wish to be understood as maintaining that these structures are in every sense homologous with the NISSL granules of vertebrates. Morphologically and chemically they may not correspond to the NISSL granules of vertebrates, and may even differ much among themselves in these respects. We are convinced, however, that the question of function is more fundamental and believe that these structures will be found to fill the same place in the economy of the invertebrate nerve cell as does the NISSL body of the vertebrate nerve cell.

Since the bodies found in the cells of Limax more nearly correspond to those found in vertebrates we will first describe our experiments upon this form and later discuss our work upon the other forms under the caption "Pigment."

Experimental.—The experimental work in connection with our study of the molluscan nerve cells has been carried out with a view to determine, if possible, the nature of some of the structures which have been found to exist in the cells. In order to insure a certain degree of accuracy in the work it was found desirable to bring the animals into the laboratory and keep them under definite environmental conditions, which could be more easily controlled. The temperature and surroundings were generally more uniform than they would have been outside of the building. The animals were kept in moist boxes or glass dishes. Some were fed upon grass or chestnuts; others were starved. At intervals animals were taken from both the fed and starved groups and their nerve cells studied, either in sections or in the live condition.

Limax.—The remarkable appearance of the vacuoles and granulations in the nerve cells led us to make a series of tests with a number of fixing agents in order to assure ourselves that we were dealing with actual structures and not artefacts due to faulty fixation or preservation.

The following agents were employed: CARNOY's fluid, PETRUNKEVITCH's solution, picro-nitric acid, FLEMMING's strong solution, osmic acid and absolute alcohol. The vacuoles and the bodies contained within them appeared with a constancy that was remarkable.

Effects of starvation and feeding.—Specimens of Limax taken in the early spring as soon as they emerge from their hibernation exhibit in the cytoplasm of the nerve cells collections of vacuoles of various sizes, scattered about in various parts of the cell. Sometimes the whole cytoplasm appears to be peppered with them; sometimes they are packed together with their thin walls touching each other in such a way as closely to resemble in appearance a mass of soap bubbles. In some of the vacuoles small granules of various shape may be found. The granules are, however, not to be found in all of the vacuoles at this time. Fig. 1 of Plate I is a photograph showing the strongly vacuolated condition which may be seen in the cells and also indicates the presence of some of the granules mentioned in certain of the vacuoles.

The other cell structures do not show any especially important features. The nucleus is large and well defined. In some cases the cytoplasm appears to be somewhat shrunken, but this is far from being a constant character.

Later in the spring animals taken into the laboratory and studied, or animals which have been kept in the laboratory and fed upon grass show that the number of solid bodies within the vacuoles has increased. This increased number is found to hold throughout the summer and fall of the year.

Two possibilities for the increase in the number of bodies in the vacuoles present themselves, either the bodies are to be considered as storage products which may be called upon in time of stress to supply energy for the nerve cells or they are to be considered as degeneration products which have accumulated in the cell during the increased activity of the animal in the active season. If they are of the former class, any great increase in the activity of the animals should have the effect of breaking them down and should cause them to disappear. If they are of the second class, prolonged activity should bring about an increase in their number and size.

Fatigue.—HODGE and others have noticed that the nerve cells of animals respond to excessive stimulation in definite ways. HODGE found that the cytoplasm of the cells took on a different appearance and that the nucleus became shrunken. Our work upon Limax has failed to confirm these particular observations and has convinced us that we have here conditions which may have escaped notice.

An active, living, specimen of Limax was taken and by means of an induction current applied to the posterior part of the body forced to crawl until it could no longer draw itself away from the point of stimulation—a period varying somewhat with different specimens, but usually from one-half to three-quarters of an hour. The nerve collar was then dissected out, fixed, sectioned and stained in the usual manner. The nerve cells of an animal treated in this way differ in a very marked degree from those of the normal rested animal. In the periphery of the nerve cells are to be found the vacuoles to which attention has already been called. These vacuoles are usually numerous, but differ from those found in the normal, well fed, rested specimens in that they contain no dark solid bodies. The limits of the vacuoles are sharply marked. The vacuoles appear in various parts of the cytoplasm and may occupy nearly all the space between the nucleus and the cell wall, when at their greatest development. It is evident that these vacuoles are filled with a liquid substance, for when the cells are placed in a medium of higher concentration than the body liquids an osmotic action takes place which draws out water from the vacuoles and finally ends in their collapse.

One may say that the disappearance of the dark bodies from the vacuoles is not the result of the fatigue of the animal, but rather the effect of the current upon the nerve cells. To avoid the possibility of this criticism another experiment was devised.

A number of well fed snails which had been living upon damp earth were taken. A part of these were killed at once and the rest were fatigued by poking them with a sharp needle until they would no longer withdraw from the point of stimulation. This process was somewhat slower than the fatigue by means of the electric needle and took from two and a half to three hours. A comparative study of the two sets of animals revealed conditions entirely in harmony with the previous experiment. The nerve cells of the rested animals show the presence of the vacuoles with the bodies in them, the nerve cells of the fatigued animals show the presence of the vacuoles but the solid bodies have disappeared. Evidently there has been some change in the cells of the animal due to the excessive amount of work which the animal was called upon to do.

Up to this time in our work no attempt had been made to ascertain whether it were possible to see these structures in the living nerve cell. It was our good fortune to find that with care in manipulation and careful observation it was possible clearly to distinguish these several structures in the living nerve cell. A trace of methylene blue added to the salt solution in which the cells were examined brought out the bodies with great distinctness within a few seconds and they could be easily studied.

The next step in our work was to study the disappearance of the bodies in the nerve cell under the microscope—to watch the process. The results were even better than we had hoped. Small electrodes of platinum foil were attached to a slide and the nerve cells mounted between these electrodes. The electrodes were connected with the secondary coil of an inductorium. Under the $\frac{1}{12}$ inch oil immersion lens the bodies in the vacuoles showed with great clearness and sharpness of outline. When the current was first applied there was no change in the appearance of the bodies but within a few minutes a change appeared. The outline of the body lost its sharpness. The body seemed to grow larger in size. The line of demarcation between the solid body and the liquid became less and less distinct and finally disappeared. The substance of the body appeared to be going into solution in the liquid of the vacuole. At the same time, there was a slight change in color, the body taking on the color of the liquid. This process was allowed to continue until the body had been completely replaced by the more transparent liquid mass. At this time the current was stopped. A continued study of the cell showed that in the same vacuoles where the disappearance of the bodies had been noted there was later a reconstruction of the solid body. Within an hour or two the solid masses had again become established in the cells. These bodies were, however, not the same bodies as had existed in the cells previous to the stimulation, as they exhibited entirely different forms.

That we have here substances in the cell which are intimately connected with the normal activities of the cell seems to be demonstrated. As to the chemical nature of these bodies we have only little knowledge. The fact that they are more or less darkened by osmic acid would indicate that they are of a fatty nature. Further we can not say at the present time.

VI. PIGMENT.

NANSEN ('87) was, so far as we are aware, the first to give an accurate account of the yellow pigment granules existing in certain of the invertebrate nerve cells, although such granules had been observed before. He found in the nerve cells of Patella plenty of large yellow granules lying in the cytoplasm. These granules had a variable size, and no regular shape, being sometimes spherical, sometimes square or polyhedral. They looked as if they had been produced by the coagulation of a homogeneous yellow substance. The granules were sometimes found scattered through the whole mass of protoplasm, but more frequently were concentrated in special parts of the cells, especially in the neighborhood of the nucleus. Plenty of similar smaller and larger granules were also to be found outside the ganglion cell. They frequently occurred in such numbers that one, for a time, could feel disposed to believe that they belonged to a substance extending through the whole nervous system. NANSEN was convinced that they were either exuded from cells, or that they sprang from destroyed cells. He had observed such a substance exuded from the protoplasm of the cell. Fig. 9 represents such a case. The substance here occurred inside as well as outside the cell. The granules were concentrated toward the part of the cell surface where they were probably to be exuded. Outside the cell they were united into larger pieces of irregular shape. The granules were situated not only near the surface of the cell but also occurred in the mesial parts of the protoplasm. NANSEN recognizes that the granules gave the yellow color to the nervous system of Patella, as well as other molluscs. He thought the yellow color to be due to a substance allied to or similar to hæmoglobin, and also believed that the granules contained fat. He found difficulty in recognizing these granules in the sections of nerve cells. As to function, he believed them engaged in the nutrition of the cell.

MCCLURE ('97) in connection with his studies of the chromophile granules mentions the existence of pigment granules in the cells of gasteropods, but gives them no further attention.

Legendre ('06) found in the nerve cells of Helix aspersa, Helix pomatia and Arion rufus pigment granules of various sizes, sometimes isolated, sometimes grouped together in irregular masses. The granules were most frequently located in the cone of origin of the axone, though they were sometimes arranged in concentric rows in the peripheral layer of the cytoplasm. Frequently they extended out along the axis cylinder. Osmic acid alone or in combination attacked the granules and stained them black at times; at other times they were unaffected. Hæmatoxylin gave them a brown color. These reactions resembled those of the lipochrome pigments observed in the nerve cells of a large number of vertebrates and some invertebrates. The number of granules varied in different individuals, and the author had failed to establish any connection between their appearance and the physiological state of the animals. He says, "The rôle of the granules is not known. One may consider them as a food, a reserve material, a functional precipitate, a product of disassimilation, a degenerative product. The multitude of hypotheses tells us nothing concerning their composition, their variation or their functions."

Planorbis.—In the preliminary study of the nerve cells of Planorbis the same general methods were employed as in the case of Limax. A number of fixing fluids were used and their comparative effects carefully studied. The various cell structures appeared almost equally well in the cells fixed by all the different agents. From a study of a large number of sections it appeared that absolute alcohol was at least as good as any other. For clearness and sharpness of detail it could hardly he surpassed. One feature should be mentioned. A long continued stay in alcohol is not good for this material, as it tends to swell the pigmented bodies in the nerve cells and to remove from them a portion of their color, changing it from a bright golden brown to a lemon yellow. These bodies are, however, clearly distinguishable in our sections, even when the stay in the alcohol was somewhat prolonged.

The vacuoles which formed so constant a structure in the nerve cells of Limax are rarely found in the cells of Planorbis. When present they are usually located in the end of the cell farthest from the axone and very seldom contain pigmented granules, though specimens have been found in which even in the living cell it was possible to see these bodies within the limits of the vacuole. The contents of the vacuole is a liquid of low viscosity, for the little brown granules could be seen dancing with the characteristic Brownian movements. In other parts of the cell where the bodies do not appear in the vacuoles they lie perfectly at rest.

Effects of starvation and feeding.—In Planorbis the number and size of the pigmented granules depends upon the general nutritive conditions under which the animal is placed. The changes in appearance, however, of the nerve cells are so slow that it has been necessary for us to extend our observations over a period of two years in order to satisfy ourselves of their correctness. Specimens have been taken from their natural habitat at various times of the year, have been kept in the laboratory under fairly constant conditions, have been fed or starved as we wished, and have finally been killed and their nerve ganglia examined.

In the summer and autumn specimens, these golden brown bodies are rounded granules of somewhat irregular shape, and varying in diameter from 1 to 5 μ. In specimens kept in the warm laboratory for a considerable time (up to three months) without feeding a distinct change is noticed in the appearance of the cells. The pigment bodies become distinctly smaller, in some cases becoming so small as not to be easily distinguished from each other even with the $\frac{1}{12}$ inch oil immersion lens. Under these conditions it is of course impossible to measure them. No bodies as large as 5 μ were found at all and only an occasional one so large as 2 μ. The substance is in the process of being broken down. It becomes very finely divided and seems to become actually less in amount in the cells. These changes take place so slowly that it is a difficult matter to follow them and only by a long series of observations can one be at all sure of any change in the condition of the pigmented matter.

In specimens taken early in the spring, which have passed the winter in hibernation, the bodies are not as a rule numerous, though some still remain in the cells. Judging from the appearance of the nerve cells of animals which had been kept in the warm laboratory during most of the winter and those which were taken early in the spring, it would seem that the processes of metabolism had been much greater in the specimens kept in the warm room and that the total amount of matter stored up was in both cases somewhat in excess of the amount which would ordinarily be needed for the use of the cells.

Fatigue.—On account of the fact that this snail, like many others, withdraws into its shell when disturbed, it was found impossible to subject it to the same conditions for producing fatigue as in the case of Limax. It was possible to remove the ganglia, to place them upon a slide between electrodes and to stimulate the nerve cells directly by means of induction currents. As a result of such stimulation it was found that, unlike the granules found in Limax, these bodies are extremely resistant and would not change in appearance during the time which the cells would live under these unusual conditions. This fact, as well as their different appearance in the cell, indicates that they are of a different nature from those in Limax. They are, however, a storage product and have to do with the nourishment of the cells during times when proper food is unavailable.

Nature of the bodies.—Many experiments have been made to determine the chemical nature of the golden brown bodies, and while we cannot say definitely just what the substance is we are in a position to state to which general class of substances it belongs. It is even possible that the bodies are not of constant composition. Most of the tests used require a long time for their action, and in some cases even failed to act at all. Osmic acid blackens the bodies after a long time. In many of the specimens the blackening was merely superficial, indicating that the substance is a highly resistant or that it is not a fat but some substance which may break up into a fat and some other substance. The tests with Sudan III and with cyanin indicate the same thing. With Sudan III the bodies assume an orange color for a short time. The color soon disappears, however, and leaves the body a sort of yellow lemon color. With cyanin the action is slow. The bodies stain a deep blue, which is sometimes temporary and sometimes more lasting. In ether the bodies swell up and clump together, becoming gradually dissolved and diffused throughout the cell. The resistance of the granules is shown by the fact that it requires frequently an hour or more to dissolve a granule 1 μ in diameter.

On account of the difficulty in making these tests it was thought for a time that they might be proteid in character, but all attempts to digest them with pepsin have so far failed. The results of the tests seem to indicate that they are some sort of a fat.

Further tests with concentrated sulphuric acid indicate that the pigment is one of the lipochrome group, the bodies assuming a bright blue color as soon as the acid touches them.

Venus.—Our experiments upon the nerve cells of the edible clam, Venus, have been few in number and serve only to add emphasis to what has already been stated. We find in the nerve cells certain yellow spots, whether solid or semifluid in character we are at present uncertain. The color is not the same as that of the bodies in Planorbis and they are of larger size. When tested with Sudan III and cyanin they give the colors which are characteristic for fats.

The cells of *Limnea* contain granules so closely similar to those of Planorbis that we have yet to find any way of distinguishing them. The pigmented granules are of the same color, size, and position in the cell. They also react in the same way to the various tests. We have not had opportunity to observe any seasonal changes in their appearance.

In the cells of *Melantho* we find a pigment of a light yellow color. The granules are generally smaller than those found in Planorbis and Limnea. This is evidently a different sort of substance, for it does not give a blue test with sulphuric acid. We have not yet made sufficient study to make a definite statement as to its chemical nature.

VII. THE CENTROSOME IN NERVE CELLS.

A few years ago the centrosome was all the fashion among biological works. The question of its origin, use and fate furnished the basis for many papers. With the accumulation of a considerable number of facts, it became evident that no general homology was to be established for the centrosome; nor did its detailed structure permit of reducing all centrosomes to a common form. About the only feature generally agreed upon was that the centrosome was at the center of radiation. In order to be sure that the dark staining granule or granules or vesicle when found in various parts of the cytoplasm has any claim to be regarded as a centrosome, it must have astral radiations. The question of the sphere substance which immediately surrounds the centrosome is more indefinite and less clearly defined than that of the centrosome. It may assume a variety of appearances and probably plays an unimportant part.

While centrosomes were being recognized in a great variety of cells, VON LENHOSSÉK ('95) was the first definitely to announce the presence of centrosomes in nerve cells. His observations were on the moderate sized spinal ganglion cells of the frog. He found the nucleus occupying in some cells an eccentric position and flattened or slightly concave on the side nearest the cell center. In this larger region of the cytoplasm there was a concentric figure in the center in which he located minute granules.

LEWIS ('96) describes in the giant ganglion cells of an annelid centrosomes on one side of the nucleus—the one toward the center and the one which tends to be flattened or concave. The sphere varies somewhat in size, but its diameter is approximately one-third that of the cell. In some cases it is quite sharply marked off from the surrounding protoplasm of the cell; in other cases the transition to the surrounding protoplasm is so gradual that it is impossible to define its limits with precision. In the center of the sphere there is a highly refractive body, or occasionally two or three such bodies. From this central corpuscle there are in many preparations radiations which transverse the whole sphere. The rays are due to the close arrangement in radiating lines of granulations of the ordinary size. Some of the rays are very distinct, others much less clear. They are few in number, usually separated by rather uniform intervals, but often interrupted over an arc of many degrees. The central corpuscle (or corpuscles) is very distinct. It is sometimes spherical, sometimes elongated so as to look like a short rod. It shows a remarkable affinity for stains, being always colored much more deeply than any other part of the sphere.

McClure ('97) finds in certain cells in the ganglia of Helix structures which he has been pleased to designate as centrosomes. In certain unipolar cells of Helix which have a transverse diameter ranging between 17 and 22 μ, the nucleus was found in longitudinal sections to have an eccentric position. In addition to this, in such cells the side of the nucleus directed toward the axis cylinder pole of the cell was often flattened, or more frequently invaginated, so that the nucleus presented a kidney-shaped appearance. The flattened or invaginated side of the nucleus was never found to be directed exactly opposite to the base of the axis cylinder process, but always to a point one side of it. In the body of the cell, directly opposite the invagination, a disk-shaped structure was found. The contents of the disk was finely granular but so far as could be determined there was no evidence of radiation. At about the center of the disk two or three small granular bodies were present which stained much deeper than the surrounding granules and which are taken to be centrosomes (Mikrocentrum).

Hamaker ('98) described in the nerve cells of Nereis structures to which the term centrosome was given. He found from two to as many as ten in a single cell, each one consisting of a deeply stained granule. *No radiations were seen.*

Kolster ('00) represents in Cottus scorpius deeply stained granules with no radiations, which are designated as centrosomes.

Rand ('01) states that there is commonly present in the nerve cells of Lumbricidæ a centered system consisting of centrosome and radiations. The single centrosome (or rarely two, or even three, small granules lying close together) is found in the axis of the cell, on the side of the nucleus opposite the nerve process, and therefore on the side of the greatest cytoplasmic mass. It is generally not far from the nucleus and approximately at the center of the cell as a whole. Radiations consisting of fibrils bearing minute granules extend from the centrosome toward the periphery of the cell. Calling these "primary radiations," there may also be distinguished secondary radiations, which arise from certain of the large granules in the course of the primary radiations. In rarer cases tertiary radiations may be found arising from granules in the secondary radiations. The centered system is, therefore, a complex one, consisting of a chief center or centrosome, and numerous inferior centers situated throughout the cytoplasm, all with their corresponding sets of radiations, the whole system forming a network whose complexity increases toward the periphery of the cell. In most cases no structure which could be called a centrosome is present. The centrosome, when present, as well as each of the inferior centers, is generally surrounded by a small clear space.

The structure which LENHOSSÉK designated as a centrosome received its correct interpretation only when the toad was studied during hibernation. LEVI ('98) in describing the changes in the nerve cells during hibernation gives a minute account of the so-called concentric figure or vortex as it occurs in the toad. During hibernation the deeply staining granular substance does not appear and the other parts appear more clearly. The centrosome is nothing more than a transverse section of the axis of the vortex which is composed of fibrils. These results of LEVI throw serious doubts on the correctness of other observations which were published soon after LENHOSSÉK's. Furthermore, we do not believe in the light of all that has been recently discovered in the cytoplasm of nerve cells that the structures described by MCCLURE, HAMAKER, and KOSTER are centrosomes at all, but probably belong to one of the classes of granules. The fewness of the radiations in the results of LEWIS and RAND is of itself enough to suggest a reasonable doubt as to their actual presence, while the secondary and tertiary systems of radiations as figured and described by RAND are not in harmony with the ordinary aster structure. That the centrosome is not usually found in adult nerve cells is abundantly shown by numerous investigations; that it does appear in some nerve cells cannot be doubted, as HATAI ('01) has shown in the young rat. The centrosome is more easily seen in the young nerve cell than in the adult, which he believes indicates a slight tendency to the degeneration of this structure. Most of the results referred to above are so questionable that we are inclined to believe that there is very little positive evidence in favor of the centrosome in adult nerve cells.

SUMMARY.

1. The nervous system of gasteropods does not permit of direct stimulation of a specific ganglion because of the compactness of the nerve collar and the numerous nerves arising from the different ganglia.

2. Lymph canals are not identical with the cytoplasmic vacuoles. They really exist, and have a rather general distribution among the nerve cells of invertebrates.

3. Vacuoles are present in the cytoplasm of nerve cells of Nemerteans, Annelida, Crustacea, Insecta, and Mollusca. The vacuoles can easily be seen in the living cells as independent structures filled with a fluid or differential bodies. They are transitory structures, vary in number and are not limited by distinct walls.

4. NISSL bodies exist in invertebrate as well as vertebrate nerve cells. They are found to occupy a zone of cytoplasm next to the nucleus but not extending out to the cell wall in most instances. They are chiefly arranged in rows or in spindle-shaped groups.

The Nissl bodies are aggregates of extremely minute particles and exhibit marked resistance to degenerative changes. They actually exist in the living nerve cell. Those occurring in Limax are always found within the limits of the cytoplasmic vacuoles. They can be caused to appear in the cell by rest and feeding and can be made to disappear through hibernation, fatigue and electrical stimulation. They are probably of a fatty nature.

5. Pigment granules are found very generally in molluscan nerve cells. They do not readily respond to starvation experiments, can be increased in size and number through feeding, are practically unchanged by fatigue or electrical stimulation, but do show occasional variations in size and number. These bodies respond to the tests indicated for lipochrome substances or fats.

6. The centrosome has been described in many of the invertebrate nerve cells, but there is considerable doubt as to its persistent presence in adult nerve cells.

BIBLIOGRAPHY.

Adamkievicz, A.
Der Blutkreislauf der Ganglienzelle. *Hirschwald, Berlin.* 1886.

Altmann, R.
Die Struktur des Zellkerns. *Arch. f. Anat. u. Entwickel.* 1889.
Elementar Organismen und ihre Beziehung zu den Zellen. *Leipzig.* 1890.
Ueber Granula und Intergranular Substanzen. *Arch. f. Anat. u. Physiol.*, p. 360. 1895.
Ueber das Wesentliche in der Zelle. *Arch. f. Ana . u. Physiol.*, p. 423. 1896.

Ambronn, H. und Held, H.
Beiträge zur Kenntnis des Nervenmarkes. Ueber Entwickelung und Bedeutung des Markes. *Arch. f. Anat. u. Entwickl.*, p. 202. 1896.

Anfimow, J. A.
On changes in the central nervous system of animals due to varnishing the skin. *Vrach., St. Petersburg*, vol. 7, p. 689. Rus. 1886.

Anglade, D.
Sur les alterations des cellules nerveuses de la cellule pyramidale en particular dans la paralysie generale. *Ann. Med. Psychol.*, Ser. VIII, vol. 8, pp. 40-46. 1897.

Apáthy, S.
Ueber Neurofibrillen und ihre nervös Leitendenatur. *Proc. 4th Intern. Zoöl. Congress. Cambridge*, pp. 126-141. 1899.

Apolant, R.
Ueber die sympatischen Ganglienzellen der Nager. *Arch. f. Mikr. Anat.*, Bd. 47, p. 461. 1896.

Arnold, Julius.
Bemerkungen eines Beteiligten über Spiralfasern und Fadennetze an den Ganglienzellen des Sympaticus. *Anat. Anz.*, vol. 5, p. 204. 1890.
W. Flemming und die "Mitomlehre." *Anat. Anz.*, vol. 16. p. 607. 1899.
Weitere Beobachtungen über "vitale" Granularfärbung. *Anat. Anz.*, vol. 16, p. 568. 1899.
Ueber Struktur und Architektur der Zellen. I, II, III. *Arch. f. Mikr. Anat.*, Bd. 52. 1898.
Ueber "vitale" Granularfarbung in den Knopelzellen, Muskelfasern und Ganglienzellen. *Arch. f. Mikr. Anat.*, Bd. 55, pp. 479-488. 1900.

Athias, M.
La vacuolization des cellules des ganglions spinaux chez les animaux a l'état normale. *Anat. Anz.*, Bd. 27, p. 9. 1905.

AUERBACH, L.
Ueber die protoplasmatische Grundsubstanz der Nervenzellen insbesondere der Spinalganglienzellen. *Monatschr. f. Psychol. u. Neurol.*, Bd. 34, p. 31. 1898.
AXENFELD, TH.
Ueber die feinere Histologie der Thränendrüse, besonders über das Vorknommen von Fett in den Epithelien. *Ber.* 28. *Vers. opht. Ges. Heidelberg*, p. 160.
BARBACCI, OTTONE.
Die Nervenzellen in ihren anatomischen, physiologischen, und pathologischen Beziehungen nach den neusten Untersuchungen. *Centralbl. allg. Path.*, Bd. 10, pp. 865–936. 1900.
BERGEN, VON FREDRIK.
Zur Kenntnis gewisser Strukturbilder (Netzapparate, Saftkanälchen, Trophospongium) im Protoplasma verschiedener Zellenarten. *Arch. f. Mikr. Anat.*, Bd. 64, pp. 49–574. 1904.
BERKLEY, H. J.
Studies on the lesions produced by the action of certain poisons on the cortical nerve cell. *Johns Hopkins Hospital Report.* Vol. 6, pp. 1–108.
BETHE, A.
Ueber die primitive Fibrillen in den Ganglienzellen vom Menschen und Wirbelthieren. *Schwalbe's Morphol. Arbeiten*, Bd. 8, H. 1. 1898.
Einige Bemerkungen über die "intercellulären Kanälchen" der Spinalganglienzellen und die Frage der Ganglienzellenfunktion. *Anat. Anz.*, Bd. 17, p. 304. 1900.
Ueber die Neurofibrillen in den Ganglienzellen von Wirbelthieren und ihre Beziehung zu den Golginetzen. *Arch. f. Mikr. Anat.*, Bd. 55, p. 513. 1900.
BIRSCH-HIRSCHFIELD.
Beiträge zur Kenntnis der Netzhautganglienzellen unter physiologischen und pathologischen Verhältniss. *Arch. f. Ophthal.*, Bd. 50, pp. 100–240. 1900.
BOCHENEK, A.
O Budowie Komorki nerwowej slimaka Helix Pomatia. *W. Krakowie.* 1901.
L'Anatomie fine de la cellule nerveuse de Helix pomatia. *Compt. rend. de l'assoc. des Anatomists.* 3 *Session, Lyon.* 1901.
Contribution a l'étude du systeme nerveux des Gasteropodes (Helix pomatia). *Le Nevraxe*, vol. 3, fasc. 1, pp. 83–105. 1901.
BOVERI, TH.
Beiträge zur Kenntnis der Nervenfasern. *Abhangl. d. k. baier. Akad. d. Wissensch. München.* 1885.
BROWICZ, T.
Intracelluläre Gallenkanälchen und ihr Verhältniss zu den Kupferschen Sekretionsvacuolen und gewissen Formen pathologischen Vacuolization der Leberzelle. *Anz. d. Akad. Wiss. Krakau*, März. 1897.
BRUCKNER, J.
Note sur la structure fine de la cellule sympathique chez l'homme. *C. R. Soc. Biol.* (10) V. 162–163.
BÜHLER, A.
Untersuchungen über den Bau der Nervenzellen. *Verh. Ges. Würzburg*, 31, 285–392. 1898.
CAJAL, RAMÓN Y.
La fine structure des centres nerveux. *Proc. Roy. Soc.*, vol. 55, 1894.
Estruttura del protoplasms nervioso. *Rev. Trimestral Micrografica*, vol. 1, pp. 1–30.
Allgemine Betrachtungen über die Morphologie der Nervenzelle. *Arch. Anat.*, pp. 187–201. 1896.
Die Struktur des nervösen Protoplasmas. *Monatschr. f. Psychol. u. Neurol.*, Bd. 1, pp. 156 u. 210.
CARRIER, HENRI.
La cellule nerveuse normale et pathologique. *Paris, Baillière et Fils.* 1904.
COX, W. H.
Beiträge zur pathologischen Histologie und Physiologie der Ganglienzellen. I. Die Spinalganglienzellen. *Inter. Monatschr. f. Anat. u. Physiologie*, Bd. 15. 1898.
Der Feinere Bau der Spinalganglienzellen des Kaninchens. *Anat. Hefte.* I. Abth. 10B, pp. 73–103. 1900.

DEHLER, A.

Der feinere Bau der Sympathischen Ganglienzellen. *Arch. mikr. Anat.*, Bd. 46, p. 724. 1895.

DELAGE, YVES.

La structure du protoplasma et les théories sur l'heredite et les grands problėms de la biologie generale. *Reinwald, Paris.* 1895.

DOGIEL.

Die Struktur der Nervenzellen der Retina. *Arch. mikr. Anat.* Bd. 46, pp. 394–413. 1895.

Der Bau der Spinalganglienzellen bei den Saügethieren. *Anat. Anz.*, vol. 12, p. 140. 1896.

DONAGGIO, A.

Nuovo osservazioni sulla struttura della cellula nervosa. *Riv. Sperim. di Freniat.*, vol. 24, pp. 772–778. 1899.

Contributo alla conoscenza dell' intima struttura della cellula nervosa nei vertebrata. *Ibid.*, pp. 60–464. 1898.

EHRLICH, R. UND LAZARUS, A.

Ueber die Darstellung und Bedeuten der Zellgranula. Aus *Nothnagel, Spec. Pathol. u. Therapie*, 8 Bd.

EMBDEN, G.

Primitivfibrillenverlauf in der Netzhaut. *Arch. mikr. Anat.*, Bd. 57, pp. 570–583. 1900.

EVE.

Sympathetic nerve cells and their basophil constituent in prolonged activity and repose. *Jour. Physiol.*, vol. 20, pp. 334–353. 1896.

EWING, JAMES.

Studies on ganglion cells. *Arch. Neurol. Psychopathol.*, vol. 1, pp. 263–463. 1898.

FISH, P. A.

The action of strong currents of electricity upon nerve cells. *Jour. Nerv. and Mental Disease. New York*, vol. 23, pp. 6–11. 1896.

FLEMMING, W.

Vom Bau der Spinalganglienzellen. *Festgabe für J. Henle.* 1882.

Nervenzellen. *Ergebn. d. Anat. u. Entwick, Merkel und Bonnet*, vol. 5, pp. 273–284. 1895.

Ueber den Bau der Spinalganglienzellen bei Saügethieren, und Bemerkungen über den centralen Zellen. *Arch. mikr. Anat.*, Bd. 46, p. 379. 1895.

Morphologie der Zelle. *Merkel und Bonnet, Ergebn.*, vol. 6, pp. 184–283. 1896.

Ueber den Bau der Bindegewebezellen und Bemerkungen über die Struktur der Zellsubstanz *Allgemein. Zeit. f. Biol.*, Bd. 34. 1897.

Ueber die Struktur centraler Nervenzellen bei Wirbelthieren. *Anat. Hefte*, Bd. 6, pp. 561–570. 1896.

Die Struktur der Spinalganglienzellen der Säugtiere. *Arch. f. Psychiatrie*, Bd. 29, Hefte 3. 1897.

FLOYD, R.

A contribution to the nervous cytology of Periplanata orientalis, the common cockroach. *Mark Anniversary Volume*, p. 339. 1903.

FRAGNITO, O.

Lo sviluppo della cellula nervosa, e i canalicoli del Holmgren. *Ann. di Nevrologia*, Anno 18. 1900.

FREUD, S.

Ueber den Bau der Nervenfasern und Nervenzellen beim Flusskrebs. *Wiener Sitzb.* 1832.

FRITSCH, G.

Ueber einige Bemerkungswerte Elemente des Centralnervensystems von Lophius piscatorius. *Arch. mikr. Anat.*, Bd. 27. 1886.

FÜRST, C. M.

Ringe, Ringreihen, Fäden und Knäuel in den Kopf- und Spinalganglienzellen bei Lachse. *Anat. Hefte*, Bd. 19, h. 2. 1902.

GEHUCHTEN, A. v.

L'Anatomie fine de la cellule nerveuse. *Rapport prés au 12. Congrès intern. de Med. Moscou.* 1897.

L'Anatomie fine de la cellule nerveuse. *La Cellule*, vol. 13, pp. 313–390.

Modes de conservation du tissu nerveux et technique de la méthod de Nissl. *La Belgique Médicale.* 1898.

GEHUCHTEN, A. v. AND NELIS.

Quelques points concernant la structure des cellules des ganglions spinaux. *Bull. de l'Acad. de Médecin de Belgique.* 1898.

GOLDSCHEIDER UND FLATEAU.
Normal und Pathologische Anatomie der Nervenzellen auf Grund der Neusten Forschungen. *Berlin.* 1898.
GOLGI, C.
Untersuchungen über den feineren Bau des centralen und periphereischen Nervensystems. *Jena.* 1894.
Intorno alla struttura delle cellule nervose. *Boll. Soc. Med. Chir. Pavia,* Nr. 1. 1898.
Appunti intorna alla struttura delle cellule nervose. *Rend. R. Ist. Lomb. Sc. Lett.* S. 2, vol. 31. 1898.
Sulla struttura delle cellule nervose dei gangli spinale. *Boll. Soc. Med. Chir. Pavia.* 1898.
Di nuovo sulla struttura delle cellule nervose dei gangli spinali. *Acrh. Ital. de Biol.* 1899.
Sur la structure des cellules nerveuses de la moëlle épinière. *Cinquant. de la Soc. de Biol. de Paris.* 1899.
Intorno alla struttura delle cellule nervose della corteccia cerebrale. *Verh. d. Anat. Ges.* 1900.
Ueber die feinere Struktur der Zellen mancher Drüsen bei Säugethieren. *Verh. d. Anat. Ges.* 1900.
Le reticulum intracellulaire et la structure fibrillaire periphérique de la cellule nerveuse. *C. R. 13 Congr. Intern. Med. Paris,* p. 83. 1900.
GRAF, A.
On the use and properties of a new fixing fluid (chrome-oxalic) with preliminary notes upon the fibrillar structure of the ganglion cells and introductory remarks upon the methods of fixation in general. *N. Y. State Hospital's Bulletin,* vol. 2, pp. 368–384. 1897.
GUERRINI, G.
De l'action de la fatigue sur la structure des cellules nerveuse de l'ecorce. *Arch. Ital. Biol.,* vol. 32, pp. 62–64. 1899.
Modifications de structure de la cellule nerveuse corticale dans la fatigue. *Rev. Path. Nerv. Ment.* vol. 4, pp. 1–18. 1900.
HAMAKER, J. I.
The nervous system of Nereis virens Sars. *Bull. Mus. Comp. Zool.* vol. 32, No. 6. 1898.
HATAI, S.
The finer structure of the spinal ganglion cells in the white rat. *Jour. Comp. Neurol.,* vol. 11, p. 1. 1901.
On the presence of the centrosome in certain nerve cells of the white rat. *Jour. Comp. Neurol.* vol. 11, p. 25. 1901.
On the mitosis in the nerve cells of the cerebellar cortex of fœtal cats. *Jour. Comp. Neurol.,* vol. 11, p. 277. 1901.
Effect of partial starvation on brain of the white rat. *Am. Jour. Physiol.,* vol. 12, p. 116. 1904.
HAVET, J.
Note preliminaire sur le systéme nerveux des Limax. *Anat. Anz.,* Bd. 16, p. 241. 1899.
L'état moniliforme des neurones chez les Invertébrés avec quelques remarques sur les vertébrés. *La Cellule,* vol. 16, pp. 37–44; 45–46. 1899.
HEIMANN, E.
Ueber die feinere Struktur der Spinalganglienzellen. *Fortschr. d. Med.,* Bd. 16. 1898.
Beiträge zur Kenntnis der feineren Struktur der Spinalganglienzellen. *Virch. Arch.,* Bd. 152. 298. 1898.
HELD, H.
Beiträge zur Struktur der Nervenzellen und ihrer Fortsätze. I. Abhandl. *Arch. Anat.* [*u. Phys.*]. 1895–1897. supl. 1897.
Beiträge zur Struktur der Nervenzellen und ihrer Fortsätze. *Arch. f. Anat.* [*u. Physiol.*], p. 204. 1897.
Beiträge zur Struktur der Nervenzellen und ihre Fortsätze. *I. Arch. f. Anat.* [*u. Physiol.*]. 1895.
Ueber experimentelle Reifung der Nervenmarks. *Ibid.,* p. 222. 1895.
HENCHMAN, ANNIE P.
The origin and development of the central nervous system in Limax maximus. *Bull. Mus. Comp. Zool.,* vol. 20, No. 7. 1890.
HENSCHEN, S.
Ueber Trophospongienkanälchen sympathischer Ganglienzellen beim Menschen. *Anat. Anz.,* Bd. 24, p. 385. 1904.

HODGE, C. F.
Some effects of stimulating ganglion cells. *Am. Jour. Psych.*, vol. 1, p. 479. 1888.
Some effects of electrically stimulating ganglion cells. *Am. Jour. Psych.*, vol. 2, p. 376. 1889.
The process of recovery from the fatigue occasioned by the electrical stimulation of ganglion cells. *Am. Jour. Psych.*, vol. 3, p. 530. 1891.
A microscopical study of changes due to functional activity in nerve cells. *Jour. Morph.*, vol. 7, pp. 95–168. 1892.
A microscopical study of the nerve cell during electrical stimulation. *Jour. Morph.*, vol. 9, pp. 449–463. 1894.
Die Nervenzelle bei der Geburt und beim Tode an Alterschwäche. *Anat. Anz.*, Bd. 9, p. 706. 1894.

HOLMGREN, E.
Zur Kenntnis der Spinalganglienzellen des Kaninchens und Frosches. *Anat. Anz.*, Bd. 16, pp. 161–171. 1899.
Zur Kenntnis der Spinalganglienzellen von Lophius piscatorius. *Merkel und Bonnet. Anat. Hefte*, Bd. 9, H. 1. 1899.
Weitere Mitteilungen über den Bau der Nervenzellen. *Anat. Anz.*, Bd. 16, pp. 388–397. 1899.
Noch Weitere Mitteilungen über den Bau der Nervenzellen verschiedener Tiere. *Anat. Anz.*, Bd. 17, pp. 113–129. 1900.
Studien in der feineren Anatomie der Nervenzellen. *Anat. Hefte*, Bd. 15, H. 1. 1900.
Weitere Mitteilungen über den "Saftkanälchen" der Nervenzellen. *Anat. Anz.*, Bd. 18, p. 290–296. 1900.
Von den Ovocyten der Katze. *Anat. Anz.*, Bd. 18, p. 63–69. 1900.
Beiträge zur Morphologie der Zelle. I. Nervenzellen. *Anat. Hefte*, Bd. 18, H. 2. 1901.
Weiteres über das "Trophospongium" der Nervenzellen und der Drüsenzellen des Salamander-Pankreas. *Arch f. mikr. Anat.*, Bd. 60. 1902.
Ueber die "Trophospongien" der Darmepithelzellen nebst einer Bemerkung in betreff einer von Prof. Browicz neulich publizierten Abhandlung über die Leberzellen. *Anat. Anz.*, Bd. 21, p. 477–484. 1902.
Ueber die Saftkanälchen der Leberzellen und der Epithelzellen der Nebentiere. *Anat. Anz.*, Bd. 22, pp. 83–86. 1902.
Ueber die "Trophospongium" der Nebenhodenzellen und der Lebergangzellen von Helix pomatia. *Anat. Anz.*, Bd. 22, pp. 83–88. 1902.
Weiteres über die "Trophospongien" der Leberzellen und der Darmepithelzellen. *Anat. Anz.* Bd. 22, pp. 313–323. 1902.
Weitere Mitteilungen über die Trophospongienkanälchen der Nebenniere vom Igel. *Anat. Anz.*, Bd. 22, pp. 476–481. 1902.
Ueber die Trophospongien der Nervenzellen. *Anat. Anz.*, Bd. 24, pp. 225–244. 1904.

HUNTER, G. W.
Notes on the finer structure of the nervous system of Cynthia partita. *Zool. Bull.*, vol. 2, pp. 99–115. 1898.

JOLY.
Ueber die Ganglienzellen des Rückenmarkes. *Zeit. f. wiss. Zool.*, Bd. 18, p. 443. 1868.

JOSEPH, H.
Bemerkungen über der Bau Der Nervenzellen. *Sitzungsber. d. deutschen naturw. med. Vereins. f. Böhmen. "Lotos."* Nr. 6. 1898.

JULIUSBURGER.
Bemerkungen zur Pathologie der Ganglienzelle. *Neur. Centralbl.*, Bd. 15, p. 386. 1896.

KOLSTER.
Studien über das Centralnervensystem. I. *Acta. Soc. Scient. Fennie.* T. 29. 1900.

KRIEGER, K. R.
Ueber das Centralnervensystem des Flusskrebses. *Zeit. f. wiss. Zool.*, Bd. 33, p. 527. 1880.

KRONTHAL, P.
Von der Nervenzelle und der Zelle im Allgemeinen. *Jena, Fischer*, 1902.

LACAZE-DUTHIERS.
Du système nerveux des Mollusques Gastéropodes pulmonés aquatiques. *Arch. Zool. Exp. et Gén.*, T. I., pp. 437–500. 1872.

LAHOUSSE, E.
La Cellule nerveuse et la Névroglie. *Anat. Anz.*, Bd. 1, p. 114. 1886.

LANG, A.

Die Polycladen des Golfes von Neapel, etc. Fauna und Flora d. Golfes v. Neapel. (Zool. Stat. Neapel) 11, Monog. Nervensystem, pp. 169–190. 1884.

Untersuchungen zur vergleichenden Anatomie und Histologie des Nervensystems der Platyhelminthen. II. Trematoden. *Mitteil. aus der Zool. Stat. Neapel*, Bd. 2, p. 28. 1881.

LEGENDRE, R.

Note sur la nature des Canalicules de Holmgren des Cellules nerveuses d'Helix. *Bull. de la Société Philomathique de Paris.* 1905.

De la Nature pathologique des Canalicules de Holmgren des Cellules nerveuses. *Compt. Rend. d. Seances de la Soc. de Biol.*, vol. 59, p. 687. 1905.

Nature Pathologique des Canalicules de Holmgren des Cellules nerveuses. *Compt. Rend. Acad. Scien.* Dec. 26. 1905.

Sur divers aspects de neurofibrilles intracellulaires obtenus par la méthode de Bielschowsky. *Anat. Anz.* Bd. 29. 1906.

Sur un nouveau detail de la structure des cellules d'Helix pomatia. *Compt. Rend. Soc. Biol.*, vol. 60, p. 488. 1906.

Sur les modifications des cellules nerveuses d'Helix pomatia pendant l'asphyxie par immersion. *Comp. Rend. Soc. Biol.*, vol. 60, p. 388. 1906.

LENHOSSÉK, M. v.

Das feinere Bau des Nervensystems im Lichte neuster Forschungen. *Berlin.* 1895.

Ueber Nervenzellenstrukturen. *Verh. Anat. Ges., in Anat. Anz.*, Bd. 12, pp. 15–20. 1896.

Ueber den Bau der Spinalganglienzellen. *Neur. Central.* 1898.

Ueber den Bau der Spinalganglienzellen des Menschen. *L'Année Biol.*, vol. 3, p. 648. 1897.

LEVI, G.

Considerazione sulla struttura del nucleo delle cellule nervose. *Riv. Pathol. Nerv. e Ment.*, vol. 3, p. 289–295. 1898.

Sulle modificazione morfologiche delle cellule nervose di animali a sangue freddo durante l'hibernazione. *Riv. Pathol. Nerv. e Ment.*, vol. 3, p. 337–359. 1898.

LEWIS, M.

Centrosome and sphere in certain of the nerve cells of an invertebrate. *Anat. Anz.*, Bd. 12, p. 291–299. 1896.

LUGARO, E.

Sulla patologia delle cellule dei gangli sensitivi. *Riv. di Patologia nervosa e mentale*, vol. 5. 1900.

Sulle modificazioni delle cellule nervose nei diversi stati funzionali. *L'Année Biol.*, vol. 1, p. 624. 1895.

MANN, G.

Changes in nerve cells during functional activity. *Jour. Anat. and Physiol.*, vol. 29, p. 100–107. 1894.

Die fibrilläre Struktur der Nervenzellen. *Anat. Anz.*, Bd. 14, Ergänzungsheft. Verh. Anat. Ges. 1898.

MACCALLUM, A. B.

Some points in the micro-chemistry of nerve cells. *British Med. Jour.*, vol. 2, p. 778. 1898.

On the detection and localization of phosphorus in animal and vegetable tissues. *Pro. Roy. Soc. Lon.*, vol. 63, p. 467. 1898.

MCCLURE, C. W. F.

On the presence of centrosomes and attraction spheres in ganglion cells of Helix pomatia, with remarks on the structure of the cell body. *Princeton Coll. Bull.*, vol. 8, p. 38. 1896.

The finer structure of the nerve cells of invertebrates. I. Gasteropoda. *Zool. Jahrb. Abt. Morph.*, Bd. 11, p. 13–60. 1897.

MERUS, F.

Ueber den von LaVallette St. George entdeckten Nebenkern (Mitochondrienkörper) der Samenzellen. *Arch. f. mikr. Anat.*, Bd. 61.

MARINESCO, M. G.

Recherches sur l'histol. de la cellule nerveuse etc. *Compt. Rendu.* 12 Avril, 1897.

Pathologie de la cellule nerveuse. *Rapport prés au Congrès intern. de Med. Moscou.* 1897.

MARTINOTTI, C.
Sur quelques particularités de structure des cellules nerveuses. *Arch. Ital. Biol.*, vol. 32, p. 293. 1899.
MARTINOTTI ET TIRELLI.
La Microphotographie appliquée à l'étude des cellules nerveuses des ganglions spinaux. *Anat. Anz.*, Bd, 17, p. 369. 1900.
MATHEWS, A.
A contribution to the chemistry of cytological staining. *Am. Jour. Physiol.*, vol. 1, No. 4. 1898.
MENCL, E.
Ueber das Verhältniss der Lymphocyten zu den Nervenzellen. *Sitzungsber. d. Kgl. böhm. Ges. d. Wissenschaften. Prag.* 1903.
MERCK, L.
Vom Fett in Allgemeinen, von Hautfett in besonderen. *Biol. Cent.*, Bd. 18, No. 12, p. 425. 1898.
MICHAELS, H.
Beschreibung des Nervensystems von Orcytes nasicorum im Larven, Puppen, und Käferzustande. *Zeit. f. wiss. Zool.*, Bd. 34, p. 641. 1880.
MINOT, C. S.
Senescence and rejuvenescence. *Jour. Physiol.*, vol. 12, p. 97. 1891.
On certain phenomena of growing old. *Proc. Am. Assoc. Adv. Sci.*, vol. 39. 1890.
MONTGOMERY, T. H.
Studies on the elements of the central nervous system of the Heteronemertini. *Jour. Morph.*, vol. 13, No. 3. 1897.
MONTI, RINA.
Sur les gradulationes du protoplasma de quelques ciliés. *Arch. Ital. de Biol.* T. 24, F. 2. 1895.
MOSSO, A.
Les lois de la fatigue étudiés dans les muscles de l'homme. *Travaux de Lab. de Physiol. de la Universite de Turin.* 1899.
MÜLLER, E.
Untersuchungen über den Bau der Spinalganglien. *Nord Med. Arkv.* N. F., vol. 1. 1891.
MÜHLMANN, M.
Die Veränderungen der Nervenzellen in verscheidenen alter beim Meerschweinchen. *Anat. Anz.*, Bd. 19. p. 377. 1901.
MÜNDEN, M.
Dritter Beitrag zur Granulafrage. *Arch. f. Anat. u. Physiol.*, p. 370. 1897.
NANSEN, F.
Preliminary communication on some investigations upon the histological structure of the central nervous system in the Ascidia and in Myxine glutinosa. *Ann. Nat. Hist.*, Ser. V. 18, p. 209. 1886.
The structure and combination of the histological elements of the central nervous system. *Bergens Museums Aarsberetning, Bergen.* 1887.
Die Nervenelemente, ihre Struktur und Verbindung im Centralnervensystem. *Anat. Anz.*, Bd. 3, p. 157. 1888.
NELIS, CH.
Un nouveau détail de structure du protoplasma des cellules nerveuses (état spirémateux du protoplasme). *Bull. de l'Acad. R. de Belgique.* T. 37, pp. 102–125. 1899.
NISSL, FR.
Ueber die Untersuchungmethoden der Grosshirende. *Tagebl. d. Versam. deutsch Naturforsch. Strasburg.* 1885.
Die Hypothese der Spezifischen Nervenzellenfunktion. *Zeit. f. Psychiatr.*, Bd. 54.
Ueber die Nomenklatur in den Nervenzellenanatomie und ihre nächsten Ziele. *Neur. Cent.*, Bd. 14, pp. 66–75, 109–110. 1895.
Nervenzellen und graue Substanz. *Munch. Med. Wochenschr.*, Bd. 45, pp. 988–992, 1023–1029, 1060–1063. 1898.
Die Beziehungen zu den thätigen und ermüdeten Zellzustande. *Neur. Centralbl.*, Bd. 15, p. 39. 1895.
Kritische Fragen der Nervenzellenanatomie. *Neur. Centralbl.*, Bd. 15, p. 157. 1896.

OLMER, D.

Recherches sur les granulationes de la cellule nerveuse. *Thesis, Lyon*, 93 pp. 1901.

Note sur le pigment des cellules nerveuse. *C. R. Soc. Biol.*, vol. 53, p. 506. 1901.

OWSJANNIKAU.

Recherches sur la structure intime du systéme nerveux des Crustacés. *Ann. Sc. Nat.*, vol. 15.

PACKARD, A. S.

On the structure of the brain of the sessile eyed crustacea. *Nat. Acad. Sci.*, vol. 3, *Washington*, Ap. 1884.

PALADINO, G.

Sur la constitution morphologique du protoplasme des cellules nerveuses dans la moelle épenière *Arch. Ital. Biol.*, vol. 29, p. 60.

PFLÜCKE.

Zur Kenntnis des feineren Baues der Nervenzellen bei Wirbellosen. *Zeit. f. wiss. Zool.*, Bd. 60, pp. 500–542. 1895.

PEWSNER-NEUFELD.

Ueber die "Saftkanälchen" in den Ganglienzellen des Rückenmarkes und ihre Beziehung zum pericellulären saftlüchen system. *Anat. Anz.*, Bd. 23, p. 424. 1903.

POIRIER, J.

Contribution à l'histoire des Trematodes. *Arch. Zool. Expér. et Gen.*, Ser. II, T. III., pp. 465–625. 1885.

PUGNAT, CH.

Recherches sur la structure des ganglions spinaux de quelques reptiles. *Anat. Anz.*, Bd. 14, p. 89. 1897.

Sur les modificationes histologiques des cellules nerveuses dans l'état de fatigue. *C. R. Acad. Sci.*, vol. 125, p. 736.

Des modificationes histologiques de la cellule nerveuses dans ses divers états fonctionells. *Bibliog. Anat.* 1898.

Sur les modificationes histologiques des cellules nerveuses dans la fatigue. *Jour. Phys. Path. Gen.*, vol. 3, p. 183.

RAND, H. W.

The regenerating nervous system of the Lumbricidæ and the centrosome of its nerve cells. *Bull. Mus. Comp. Zool. Harvard*, vol. 37, p. 86–164. 1901.

RETZIUS.

Das Nervensystem des Lumbricus. *Biol. Unters.*, N. F. 1892.

RHODE, E.

Histological investigations upon the nervous system of the Chætopoda. *Ann. Nat. Hist.*, Ser. V., vol. 18, p. 311. 1886.

Histologische Untersuchungen über das Nervensystem der Chætopoda. *Zool. Beiträge Schneider*, vol. 2. 1890.

Ganglienzelle und Neuroglia. *Arch. mikr. Anat.*, Bd. 42. 1893; Bd. 45, Bd. 47. 1895.

Die Ganglienzelle. *Zeit. f. wiss. Zool.*, Bd. 64. 1898.

Untersuchungen über den Bau der Zelle. I. Kern und Kernkörper. *Zeit. f. wiss. Zool.*, Bd. 73, pp. 497–682. 1903.

The same. II. Ueber einartige aus der Zelle wandernde "Sphären" und "Centrosome" ihre Entstehung und ihre Zufall. Bd. 75. 1903.

The same. III. Die Entstehung von Mitochondrien und Chondromiten aus eigenartigen intra- und extrazellularen Sphären. Bd. 76. 1904 a.

The same. IV. Zum histologischen Wert der Zelle. Bd. 78. 1904 b.

ROBERTSON, F.

Critical digest, normal and path. histology of the nerve cell. *Brain*, vol. 22. 1899. (A very excellent account of the histology of the nerve cell of vertebrates for the years between 1893 and 1898. There accompanies the article a bibliography of 523 different titles. The review aims to present much that is of direct interest to physicians.)

RUZICKA, V.

Untersuchungen über die feinere Struktur der Nervenzellen und ihrer Fortsätze. *Arch. f. mikr. Anat.*, Bd. 53, p. 485. 1898.

SADOVSKI, S.

On the changes of the nerve centers caused by peripheral irritation. (Russian.) *St. Petersburg.* 1899.

SCHULTZ, H.
Die fibrillare Struktur der Nervenelemente bei Wirbellosen. *Arch. mikr. Anat.*, Bd. 16.
SCOTT, F. H.
The structure, microchemistry, and development of nerve cells with special reference to their nuclein compounds. *Trans. Can. Inst.*, vol. 6, p. 405. 1898–1899.
University Toronto Studies, Phys. Ser., 1. 1900.
SIMON, CH.
Recherches sur la cellule des ganglions sympathiques des hirudinées. *Jour. Internat. d'Anat. et de la Physiol.* T. 13. 1896.
SIMROTH.
Ueber die Sinnesorgane unserer einheimischen Weichtiere. *Zeit. f. wiss. Zool.*, Bd. 26, p. 227–349. 1876.
SJÖBRING.
Ueber das Formol aus Fixierungsflüssigkeit. *Anat. Anz.*, Bd. 17. 1900.
SMALLWOOD, W. M.
Preliminary Report of the cytology of molluscan nerve cells. *Jour. Com. Neu. and Psych.*, vol. 16, p. 183. 1906.
SMALLWOOD, W. M. AND ROGERS, C. G.
Some observations on gasteropod nerve cells. *Science*, N. S., vol. 22, No. 588. 1906.
SMIRNOW.
Einige Beobachtungen über den Bau der Spinalganglienzellen bei einen viermonatlichen menschlichen Embryo. *Arch. f. mikr. Anat.*, Bd. 59. 1901.
SOLBRIG, A.
Ueber die feinere Struktur der Nervenelemente bei den Gasteropoden. *Leipsig.* 1872.
SOLGER, B.
Struktur von Nervenzellen. *Med. Verein. Griefswald. Sitz.* 1. 1897.
STUDNICKA, F. K.
Ueber das Vorkommen vom Kanälchen und Alveolen im Körper der Ganglienzellen und in dem Achsencylinder einiger Nervenfasern der Wirbeltiere. *Anat. Anz.*, Bd. 16, pp. 397–461. 1899.
Beiträge zur Kenntnis der Ganglienzellen. I. Ein neuer Befund von Centrosome: die intracellulären Kanälchen. *Sitz. d. k. Ges. d. wiss. Prag.*, Mai, 1900.
II. Einige Bemerkungen über die feinere Struktur der Ganglienzellen aus dem Lobus electricus von Torpedo marmorata. *S. B. Böhm. Ges.*, Bd. 15.
SZCZAWINSKA, W.
Sur la structure réticulaire des cellules nerveuses centrales. *C. R. Ac. Sci.*, vol. 122, pp. 379–380.
Recherches sur le systéme nerveux des selachiens. *Arch. de Biol.*, vol. 15, p. 463. 1897.
TIMOFEEW, D.
Beobachtungen über den Bau der Nervenzellen des Spinalganglien und des sympathicus beim Vogel. *Intern. Monat. f. Anat. u. Physiol.*, Bd. 15, p. 259. 1898.
TURNER.
A method of examining fresh nerve cells. *Brain*, vol. 20 p. 450. 1897.
Substance chromophile normale et path. Réaction acide ou alcaline du cortex. *Brain*, vol. 22, p. 100. 1899.
VAS, F.
Studien über den Bau des Chromatin in der Sympathischen Ganglienzellen. *Arch. f. mikr. Anat.*, Bd. 40, p. 375.
VERATTI, E.
Ueber die feinere Struktur der Ganglienzellen des Sympathicus. *Anat. Anz.*, Bd. 15, pp. 190–195. 1899.
VIGNAL, W.
Note sur l'Anatomie des centres nerveux du mole, Orthagoriscus mola. *Arch. de Zool. Exp. et Gen.*, T. 9, p. 369. 1881.
Arch. de Zool. Exp. et Gen., Ser. II, T. 1, p. 267–412. 1883.
WAGNER, G.
Ueber den Zusammenhang des Kerns und Kernörpers der Ganglienzellen mit dem Nervenfaden. *Zeit. wiss. Zool.*, Bd. 8. 1857.

WARRINGTON.
On the structural alterations observed in nerve cells. *Jour. Physiol.*, vol. 23, pp. 112–129. 1898.
WHITWELL, J. R.
Nuclear vacuolation in nerve cells of the cortex cerebri. *Brain*, vol. 12, p. 520. 1899–1900.

PLATE I.

FIG. 1. Photograph of a section of ganglion of Limax, fixed in absolute alcohol and stained by iron-hæmatoxyln. The cells show many vacuoles of various sizes in the cytoplasm, some of which contain solid bodies.

FIG. 2. Same as above, under higher magnification, $\frac{1}{12}$ inch oil immersion lens used in making photograph.

FIG. 3. Photograph of a living nerve cell of Planorbis under $\frac{1}{12}$ inch oil immersion lens. Note the very large nucleus and mass of pigment granules at the axone hillock of cell.

FIGS. 4 AND 5. Photographs of sections of ganglion of Planorbis, fixed in osmic acid, unstained. The dark bodies are the same as those shown in Fig. 3.

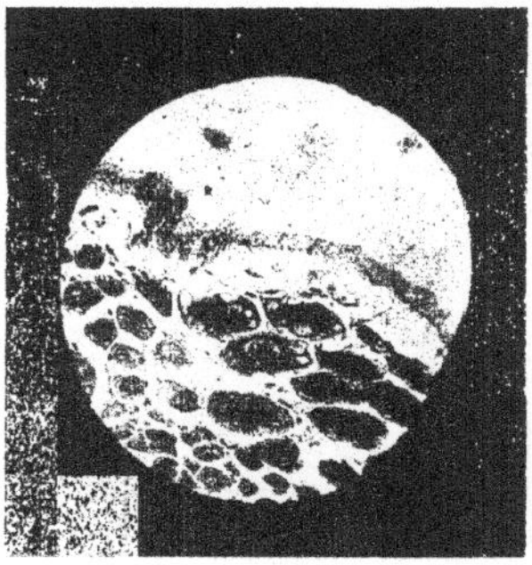

Fig. 1.

Fig. 2.

Fig. 3.

Fig. 4.

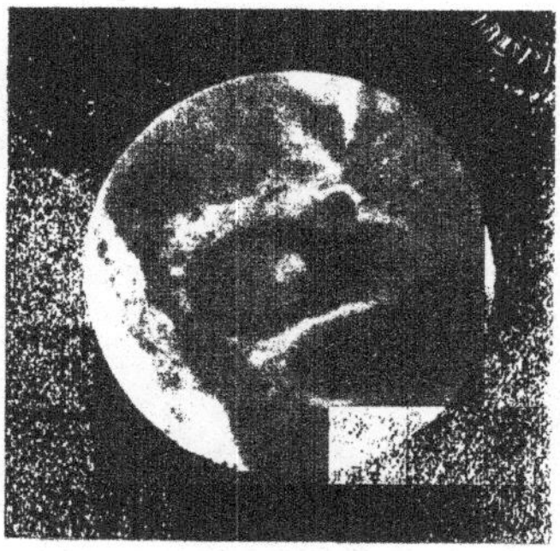

Fig. 5.

DOCUMENT I OF THE REPORT OF THE PRESIDENT OF THE BRAIN COMMISSION (Br. C.).

In the spring of 1906, the Imperial Academy of Sciences at Vienna issued to the Associated Academies a report on the condition of the interacademic institutes for brain study. This report appeared in two documents (A) and (B), the former of which (A) contained the earlier correspondence relating to the institutes for brain study, the names of those at that time members of the Central Commission for brain study, and the provisional order of business and program proposed for the meeting of the Brain Commission (Br. C.) to be held in Vienna at the end of May, 1906.

Document (B) gave a brief historical review of the progress thus far made in the establishment of Interacademic Institutes for brain study.

I now add a report on the further course of affairs and especially on the first session of the Brain Commission, together with a report of the meeting of the Committee of the Associated Academies.

The meeting, called conjointly by the undersigned and the President of the Vienna Academy, took place on the twenty-seventh and twenty-eighth days of May at Vienna. All the members of the Brain Commission were invited. There were present besides the undersigned, the Messrs. DONALDSON (Wistar Institute, Philadelphia), EHLERS (Göttingen), FLECHSIG (Leipzig), LANGLEY (Cambridge, England), v. MONAKOW (Zürich), HERMANN MUNK (Berlin), OBERSTEINER (Vienna), and G. RETZIUS (Stockholm), GOLGI (Pavia), Dr. GREENMAN, Director of the Wistar Institute, was present at the deliberations. Detailed minutes of the proceedings were taken by Dr. MARBURG, of Vienna. These are in the possession of the President of the Central Commission for brain study. (The President at that time, and until 1909 inclusive, being Dr. WALDEYER, Berlin, N. W. 6, 56 Luisenstrasse, Anatomisches Institut). These minutes are, however, open to inspection by the several academies, as well as by the institutes for brain study and the members of the Commission.

At these sessions there was drafted a constitution, together with an order of business, and later this was revised for publication by the undersigned, with the approval of the Executive Committee. This appears as Document II of the present report, and in its revised form must be presented for final acceptance and ratification at the next meeting of the Brain Commission (Br. C.), which will occur in the spring of 1909, previous to the next session of the Committee of the Associated Academies Moreover, the paragraphs in this document which regulate the relations of the Brain Commission to the Association of Academies, must also be ratified at the regular meeting of the Association of Academies at Vienna in the spring of 1907. In the meantime, the undersigned President, with the coöperation of the Executive Committee, will conduct the business in accordance with the provisional constitution, as given in Document II.

The accompanying Document III contains information concerning the relation of the Brain Commission to the Association of Academies.

By vote of the members, the Central Commission was enlarged. It consists at present of the following members:

Egypt	ELLIOT SMITH	Cairo.
Belgium	VAN GEHUCHTEN	Louvain.
Denmark	F. C. C. HANSEN	Copenhagen.
	1. EDINGER	Frankfurt a/M.
	2. EHLERS	Göttingen.
Germany	3. FLECHSIG	Leipzig.
	4. H. MUNK	Berlin.
	5. WALDEYER	Berlin.
	1. LANGLEY	Cambridge.
England	2. SHERRINGTON	Liverpool.
	3. V. HORSLEY	London.
France	1. DEJERINE	Paris.
	2. RAYMOND	Paris.
Japan	SHUZO KURE	Tokio.
	1. GOLGI	Pavia.
Italy	2. LUCIANI	Rome.
	3. ROMITI	Pisa.
Holland	WINKLER	Amsterdam.
Norway	GULDBERG	Christiania.
	1. S. EXNER	Vienna.
Austria-Hungary	2. OBERSTEINER	Vienna.
	3. v. LENHOSSÉK	Budapest.
Russia	1. BECHTEREW	St. Petersburg.
	2. A. DOGIEL	St. Petersburg.
Sweden	1. HENSCHEN	Stockholm.
	2. G. RETZIUS	Stockholm.
Switzerland	v. MONAKOW	Zürich.
Spain	S. RAMÓN Y CAJAL	Madrid.
	1. F. P. MALL	Baltimore, Md.
U. S. of North America	2. C. S. MINOT	Boston, Mass.
	3. H. H. DONALDSON	Philadelphia, Pa.

Thirty-one members in all.

At the suggestion of the members assembled in Vienna, the undersigned will put himself in communication with Australia, with the purpose of selecting a member of the Commission to be elected from that country. (A letter has been sent to Dr. WILSON in Sydney).

The large number of members, and their selection from different countries and nations will serve only to advance the cause. Besides this a large membership is necessary in order that at the triennial meeting of the Brain Commission, the desired number of members may be present. It may be remarked that only 8 of the 18 members were able to be present at the last meeting. The same conditions may be looked for in the future. Suggestions as to further elections, particularly from countries not yet represented, will be gladly received. These may be forwarded to the present President.

The proposals, especially those concerning the Executive Committee, made by the P esident in the previously mentioned Document (A), were accepted.

The Executive Committee is composed for the present of the following gentle-men:

1. W. WALDEYER, Berlin, President of the Br. C.
2. H. OBERSTEINER, Vienna, Vice-President.
3. E. EHLERS, ⎫
4. P. FLECHSIG, ⎬ Members.
5. H. MUNK, ⎭

The reason why the members of the Executive Committee have been chosen thus far solely from Germany and Austria, is merely to facilitate communication with the present President whose home is in Berlin, but this arrangement must not be regarded as establishing a precedent.

Owing to vacancies caused by death, certain changes became necessary in the special commissions established in London in 1904 for the several departments of brain research. These changes were made at the meeting in Vienna in May, 1906, and are as follows:

RETZIUS was made President of the Commission on Embryology, and DONALDSON was named on this commission in the place of SCHAPER, deceased On the Commission on the Pathology of the Brain, MINGAZZINI was named in place of WEIGERT, deceased.

At present the seven special commissions are constituted as follows:

I. Commission on descriptive Anatomy. WALDEYER (President), CUNNINGHAM, MALL, MANOUVRIER, ZUCKERKANDL.

II. Commission on comparative Anatomy. EHLERS (President), EDINGER, GIARD, GULDBERG, ELLIOT SMITH.

III. Commission on histological Anatomy. GOLGI (President), RAMÓN Y CAJAL, DOGIEL, VAN GEHUCHTEN, LUGARO.

IV. Commission on Embryology. RETZIUS (President), BECHTEREW, DONALDSON, v. LENHOSSÉK, C. S. MINOT.

V. Commission on Physiology. H. MUNK (President), V. HORSLEY, LUCIANI, MOSSO, SHERRINGTON.

VI. Commission on pathological Anatomy and Physiology. OBERSTEINER (President), DEJERINE, v. MONAKOW, LANGLEY, MINGAZZINI.

VII. Commission on clinical Neurology. FLECHSIG (President), HENSCHEN, FERRIER, LANNALONGUE, RAYMOND.

The following were recognized as interacademic Institutes for brain study.

1. The Neurological Institute of the Madrid University conducted by RAMÓN Y CAJAL.

2. The Neurological Institute of the Leipzig University conducted by P. FLECHSIG.

3. The Neurological Institute of the Vienna University, conducted by H. OBERSTEINER.

4. The Neurological Institute of the Zürich University, conducted by v. MONAKOW.

5. The neurological department of the The Wistar Institute, Phila., U. S. A., conducted by H. H. DONALDSON; M. J. GREENMAN, Director of the Institute.

6. The Neurological Institute in Frankfort a/ M., conducted by EDINGER.

In addition to the Institutes already recognized, The Wistar Institute in its neurological department was accepted as a Central Institute for brain study, and consequently will be regarded by the Brain Commission as the Central Institute for brain study in the United States of North America.

The Messrs. v. MONAKOW and OBERSTEINER proposed that the Central Commission should request the proper authorities in the countries mentioned below, to recognize the Neurological Institute at Zürich, as the Central Institute for Switzerland, and the Neurological Institute at Vienna, as the Central Institute for Austria. This request will be made at an early date.

In addition, reports were made on the Neurobiological Institute of the Berlin University, under the direction of O. VOGT, and on the condition of affairs in Norway, Sweden, Holland, England, Italy and Hungary.

In Sweden, Professor LENMALM is ready to undertake work of this sort.

In Norway, Professor GULDBERG is prepared to utilize his Institute for the same purpose.

In Holland, Professor WINKLER has taken steps to organize there an Institute for brain study. Assent has also come from Italy and Hungary. The Imperial Academy at St. Petersburg has also reported to the undersigned that the proposition for the establishment of an Institute for brain study will be favorably considered there. Finally, Messrs. BECHTEREW and A. DOGIEL in St. Petersburg, and Messrs. DARKSCHEWITSCH in Kasan, and ROTH in Moscow, have announced their willingness to place their laboratories or clinics at the service of this cause.

The question whether or not a Central Imperial Institute should be organized in Germany, was considered at the session of the Associated Academies in Göttingen in October, 1906. For various reasons such a Central Institute for the united German Empire was rejected by the authorities, who much preferred to leave the organization of the Institutes for brain study to the individual states. This, however, does not prevent the larger states from establishing Central Institutes as well as local Institutes. As previously stated, the Institutes under the direction of FLECHSIG and of EDINGER, have been already recognized as Interacademic Institutes for brain study.

The Institute in Berlin, directed by O. VOGT, has not yet become connected with the Brain Commission.

As regards the recognition as Interacademic Institutes for brain study, see section xvii of the constitution. In regard to the Central Institutes and their recognition and arrangement, see section xxi.

It is not to be expected that immediately upon their inception the proposed organizations shall at once exhibit a complete activity, but by degrees, a closer union of the separate Institutes will develop, and through experience, that form of organization will be found which will make possible effective coöperation.

In accordance with Professor LANGLEY's proposal, made at the meeting of the Central Commission, one of the first steps taken will be towards the further revision of the nomenclature, with the purpose of obtaining international uniformity.

Moreover, we beg the Academies still to lend their powerful support to this undertaking which has developed through their initiative, for without such support we shall find it hardly possible to induce the several governments, in view of the many demands made upon them, to grant with the desired promptness, the means necessary for the establishment of specially planned and suitably arranged Institutes for brain study.

Finally, we beg the above mentioned Institutes, in accordance with the constitution, to furnish the Central Commission with the necessary reports as to their condition and activities, and at the same time, to assist one another through an interchange of material and publications.

By this means, in the course of time, the Institutes may hope to attain the desired completeness in the matter of collections and reference libraries.

(Signed) WALDEYER,
President of the Brain Commission.

NEUROLOGY AT THE PHYSIOLOGICAL CONGRESS, HEIDELBERG, 1907, AND AT THE CONGRESS FOR PSYCHIATRY, NEUROLOGY, PSYCHOLOGY AND THE NURSING OF THE INSANE, AMSTERDAM, SEPTEMBER, 1907.

At both of the congresses named above considerable attention was paid to topics that are of special neurologica interest. Owing to the difference in the membership, the character of the papers differed in the two congresses, the papers at the Physiological Congress being largely of a purely scientific character, while those at the Amsterdam Congress treated more especially matters in connection with human diseases. In the later congress more attention was devoted to the anatomy of the nervous system than in Heidelberg, although at both the functional study was very prominent. At Heidelberg about half of the sessions of one section were occupied with papers concerned with the central nervous system and the special senses, and at Amsterdam half of the time of the section of neurology and psychiatry and some of the time of the section of psychology and psychophysics were so devoted. Many of the most prominent physiological neurologists were present at the Physiological Congress and few attended the Congress at Amsterdam, although the representation of the clinical neurologists at Amsterdam was large and most important. In addition to those whose papers are abstracted below may be mentioned: BETHE, EDINGER, EXNER, GOTCH, HERING, LUCIANI, MUNK, NAGEL, NISSL, RICHET, SCHÄFER, v. TSCHERMAK, v. UEXKULL, and VERWORN at the Physiological Congress; and v. BECHTEREW, CAJAL, v. GEHUCHTEN, v. JELGERSMA, LANGELAAN, MOTT, OBERSTEINER, OPPENHEIM, WINKLER, and WESTPHAL at the Congress of Neurology and Psychiatry.

Among so many papers it is almost impossible to select those that are of most importance to the readers of the *Journal*, but the following abstracts give a fair idea of the diversity of subjects and of the character of the work presented at the two meetings.[1]

Professor GASKELL (Cambridge), *A*, gave a general account of his views on the evolution of the vertebrate nervous system, which are already known to some of the readers of the *Journal*. He considers the vertebrate central nervous system to be developed phylogenetically from the cœlenterate type of oral nervous ring. Ontogenetically there are two types of tissues in the body, the master tissues, connected with nervous system, and the free cells of the body which arose as modifications of the germ cells. The central nervous system has been developed from the combination of the nervous and the alimentary systems; the infundibulum is the relic of the development from the early œsophagus, the crura represent œsophageal commissures, the spinal cord is the ventral chain of ganglia, and so on. All the principal parts of the vertebrate type of nervous system were compared with parts

[1] In the report of each paper will be found after the name of the man presenting the paper a letter indicating the congress at which the communication was read: *A*, for Amsterdam; *H*, for Heidelberg.

of the nervous system in the arthropod types. The reason why ontogeny has not so far revealed this form of evolution is said to be because all the energies of the embryologists have been bent to the study of the matter from the standpoint of the germ layer theory. A new embryology is, therefore, necessary, according to GASKELL. Numerous charts and diagrams illustrated the paper, but in so limited time it was not possible to go into the matter in sufficient detail that the reader could point out the relation of the hypothesis to conditions of disease, but the main points were well shown.

"The salts of nerve, their importance to its function" was discussed by Prof. J. S. MACDONALD (Sheffield), *H.* The paper gave an account of experiments to determine the changes in the chemical composition of nerves, especially when injured. MACDONALD found that when a nerve is injured there results a precipitation of some of the colloid substance of the intramyelin material, the precipitation being accompanied by the appearance of potassium salts capable of reacting with cobalt nitrate, and of chlorides capable of reacting with silver nitrate. This change in composition is taken to indicate the explanation of the current of "injury" which is found in injured tissues, and the inference was drawn that the nerve current is about the same sort of change in the composition of the nerve, though in the normal uninjured nerve the salts do not leave the nerve, but change their position. The potassium salts are normally deposited about the nodes of RANVIER, which act as cathodes by which the electrical current leaves the fiber. There is also a deposition of the chloride salts at these points.

Further experiments to indicate the chemical character of the nerve-muscle activity were reported by LANGLEY (Cambridge), *H.* He gave the results of work to indicate that the effect produced by a motor nerve depends upon the nature of some receptive substance or substances formed by the cell in the region of the nerve ending. In such a muscle as the sartorius of the frog it can be seen that after the application of a dilute solution of nicotine the muscle contracts, but that the greatest thickening is in the regions where nerve endings are most numerous. When the nicotine is applied to points the response is found only from the parts where the nerve endings are. Other experiments with nicotine, curari, sodium chloride, and adrenalin show that there are in the muscle probably two substances, radicles, one causing the slow the other the brief quick contraction. These substances are believed by LANGLEY to be radicles of the contracting molecule in the neighborhood of the nerve ending. The functions that have been attributed to the nerve endings are in reality, according to the author, functions of the muscle plasm, and the "motor nerve endings are not organs with specific properties." It is difficult to understand the last statement in a literal manner, for the motor nerve endings must have some relation to the production of the contraction of the muscle, if it be only, and the writer believes this to be Professor LANGLEY's opinion also, that of starting the chemical change in the muscle protoplasm which we call "contrac ion." Curari, according to the work already finished, may no longer be considered to act on the nerve endings, but is active on the "receptive radicles" of the muscle substance, and is partly antagonistic to the action of nicotine.

Dr. R. HÖBER (Zürich), *H*, read a paper "Der Erregungsvorgang als Kolloidprozess" in which he brought forth strongly the chemical view of nerve and muscle activity. The alterations in the excitability which are produced in mu cles and nerves by the application of various salts were formerly attributed by HÖBER to the effect of the salts on the colloid protoplasm. Later experiments of the effects of salts on albumen and lecithin have shown that there is a close relation between the effects on colloid and on the excitability. The excitability alterations produced by salts resemble corresponnding (i. e., reversible) electromotor phenomena, and a connection can be shown between the alterations of excitability and the condition of the colloid protoplasm. The conclusion that the reader further drew from the results of his experiments is that normal production of excitation by the usual electrical means is accompanied by alterations in the condition (composition) of the colloid. This it will be noted is the same conclusion that was reached by MACDONALD in the paper mentioned above. This view of the nature of excitation is corroborated by the fact that the current of rest produced by the action of salts is, like the action current, retarded by various narcotics. According to the view of excitation and the action of various narcotics the nature of narcosis must be or must depend upon some sort of retardation of the colloid process normally accompanying excitation.

Dr. N. A. BARBIERI (Paris), *H*, denied that there is any regeneration in nerve fibers, after they have been sectioned. His paper contained statements not in accord with the experience of the return of function in man and other animals, and it was difficult to get the author's point of view. The abstract of his paper, "Cycle d'evolution des nerfs sectionnes" gives the following conclusions: There exists no autoregeneration of nerves. In strictly physiological evolution the peripheral end of a sectioned nerve remains inexcitable and always degenerates; the central end does not regenerate, but remains excitable and its structure remains normal. If suppuration exists the central end of the nerve also undergoes retrograde degeneration. If there be no regeneration of a divided nerve it is difficult, or perhaps impossible, to explain how the animal recovers the motor and sensory functions after an interval of time. The experimental evidence adduced by BARBIERI in support of his conclusions is not of the best, I believe, for he waited only three months for the regeneration of the vagus nerve. Had he extended his experiments over a longer period of time he would doubtless have been compelled to conclude that regeneration is the rule in respect to the peripheral nerves.

Certain nerves and parts of nerves have been long known to have an inhibitory function. Examples of this are the vagus and the so-called vasodilator nerves. The inhibitory function for the muscular nerves was shown by Professor NICOLAIDES (Athens), *H*, by demonstrations on the frog. The nerve fibers supplying the gastrocnemius muscle of the frog are from two bundles of the lumbar plexus. When the upper one of the bundles was stimulated with a tetanizing current the gastrocnemius contracted, but if immediately after the application of the current to the upper bundle the lower bundle was stimulated with feeble currents the contraction gave way to a relaxation. When strong currents were used for the stimulation of the lower bundle at the time the upper was being stimulated, the relaxation did not take place, but the original rise was accentuated. These findings can be explained only on the supposition that there are in the muscular nerves of the vertebrates inhibitory as well as excitor fibers. This conclusion, as has been hinted at above, is in accord with results from other parts of the nervous system.

Against the views of BETHE, Professor F. B. HOFMANN (Innsbruck), *H*, considered some evidence regarding the nerve endings in his paper, "Zur Frage der peripheren Nervennetze." The histological studies of the nerves going to the heart and the smooth muscles of the vertebrates as well as to the muscles in molluscs show that in these muscle systems the nerves end not in free fibrils but in end nets. Each nerve can form a closed net for itself or there may be a continuous net formed by an anastomosis between the various nerve filaments. These nets are limited to the final branches of the nerves and are, according to HOFMANN, entirely independent of the presence of ganglion cells. The appearance of the nerve nets with nuclei is an artifact. Physiologically the innervation of the smooth muscles in vertebrates and molluscs, in so far as there are no ganglion cells present, is a localized one, and there is no general radiation of the excitation aroused in the central nervous system. Certain conclusions that would follow from this view of the matter were referred to by the speaker, and the paper was discussed by BETHE and LANGLEY.

In spinal animals, Professor SHERRINGTON (Liverpool), *H*, demonstrated the effect of "removal of stimulus from the stepping reflex of the spinal dog" and the "influence of strychnine on the reflex inhibition of skeletal muscles." A cat was shown in which all the nerves of the four feet were severed, but the animal was able to walk well and accurately. In this animal burning the feet did not produce a reflex withdrawal and there could have been no nerve conduction to the spinal cord. This suggested that an important source of stimuli for the reflexes of walking or stepping is in the proximal part of the limbs. To confirm this supposition SHERRINGTON divided the spinal cord in a dog at the tenth thoracic vertebra (the animal shown at the congress had the operation performed almost three years ago), and when the limbs of the dog were held from the ground they executed the stepping reflex. When one thigh was gently lifted the reflex immediately ceased in both legs. On allowing the thigh to hang again the reflex began immediately with the same activity as before. The reflex stepping was inhibited by pinching the tail, but on releasing the tail it began with ncreased activity, quicker and with greater amplitude. The antagonistic action of strychnine on the reflex inhibition of skeletal muscles was shown by SHERRINGTON in the following manner: In a decerebrized or spinal cat the vasto-crureus muscle was prepared for examination. All the other muscles of the leg were paralyzed by severing their nerves or their attachments. After this was done it was found that stimulation of the internal saphenous nerve below the knee always caused reflex relaxation of the vasto-crureus, which in normal action produces an extension of the leg. After the inhibition was obtained strychnine was injected and then stimulation of the internal saphenous nerve was followed by reflex contraction of the vasto-crureus. In some way the strychnine acted on the spinal cells to change the central inhibition into excitation.

Dr. M. PHILLIPSON (Brussels), *H*, demonstrated the movements of a spinal dog and considered the subject, "Sur les réflexes croisés chez le chien." The dog had been shown to the congress in 1904 after a complete section of the spinal cord in the dorsal region, and at that time it showed the following: Numerous direct and crossed reflexes, principally of direct extension, direct flexion, and crossed extension; when the animal was suspended vertically the feet of the animal were moved rhythmically; when hung horizontally the feet moved faster and the movements were seen to be those of walking, trotting, and galloping; when placed on the ground the feet moved to bring about the propulsion of the animal, and the feet movements were correct in point of view of coördination but strongly ataxic. In the same animal the dorsal columns of the cord in the lumbar region were extirpated to determine the part played by each of the types of reflexes, the direct and the crossed. After the second operation when the dog was suspended vertically rhythmic movements were not produced, nor were they when the animal was suspended horizontally. The left leg was moved when it was stimulated, the right not. The right leg did not contribute to locomotion. It can be said, therefore, that the direct and the crossed reflexes may be preserved independently; that the direct reflex is necessary for the foot to be kept in a normal position, but that for the rhythmic movements, especially those of locomotion, the crossed reflexes are indispensable.

A general study of the ontogenetic course of some human reflexes was reported by BYCHOWSKI (Warsaw), *A*, and from this some phylogenetic conclusions were drawn. The reflexes studied in detail were the knee kick, the tendo Achillis, and the abdominal in new-born children and during the first few months of life He found that the knee kick was constantly present from birth, and that this reflex is more lively than in adults, which is to be explained by the lack of cerebral control. In the first month the Achillis reflex is seldom obtained. From the middle of the first year until the second year it comes more often until it is a constant occurrence. Similarly with the abdominal reflex, althought it is not so constant as the Achillis reflex. These facts are taken to indicate that the Achillis and the abdominal reflexes are later phylogenetically than the knee kick; that the knee kick is purely spinal in origin; that the Achillis reflex is controlled by the midbrain, and that the abdominal reflex is under the control of the cerebrum.

Dr. NOVOA SANTOS (Santiago, Spain), *A*, reported results and conclusions of a study to determine reflex and conscious time. The time taken up by the purely mental part of a reaction has been calculated by the author from a formulathat he has manufactured for the purpose, and he concludes that the mental time varies for the different senses, as follows: touch .01 second; vision, .027 second; hearing, .013 second, and so on. From the abstract and the paper it is impossible to properly criticise the work, but it is most interesting that we should find a thoroughgoing interaction hypothesis at the basis of the work.

A paper of some anatomical interest is that of Dr. S. J. DE LANGE (Amsterdam), *A*, "Sur l'anatomie du faisceau longitudinal postérieur." The author gave the results of his studies on this bundle, made on rabbits, cats, and guinea pigs. Lesions were made in different parts of the medulla oblongata, in the posterior longitudinal bundle, in the nuclei of DEITERS and DARKEWITSCH. Most of the material was examined twenty days after the operation by the MARCHI method, and a few specimens after three or four days by the NISSL method for nerve cells. In addition to personal material the author had access to material showing the effects of lesions of the cochlear nerve, the vestibular, and the trigeminal, and embryological series of the cat and rabbit. The results of the examination of this material are that the principal fibers of the posterior longitudinal bundle are descending fibers, having their origin in the nucleus of DARKEWITSCH. There are some ascending fibers at the most distal portion of the bundle, with cell bodies in the medulla oblongata, which go to the nuclei of cranial nerves. Some fibers of the vestibular nerve go by way of the bundle to motor nuclei, but there are more crossed fibers than homolateral ones. There are also some fibers from the cochlear nerve, but none of the trigeminal fibers go by way of the posterior longitudinal bundle.

Professor WINKLER (Amsterdam), *A*, reported on "Labyrinthtonus." Immediately after the extirpation of the labyrinth on one side or after section of the eighth nerve in rabbits there is found: the eye on the same side is turned down and inward as though the internal and inferior rectus functioned with the other muscles weakened. The contralateral eye is fixed outward and upward as if the abducens muscle were paralyzed. There is a fixation of the head toward the operated side and at times the neck is so much turned that the cheek or the head touches the floor. There i a decided atony of the extremities. After a time all the phenomena decrease in severity even after complete destruction of the labyrinth. Incomplete extirpation of the labyrinth as well as the extirpation of the cochlea produce the main symptoms noted above, but less completely. Bilateral extirpation of the labyrinths or the eighth nerves produces a strong atony in nearly all muscles; protrusion of the eyes which are level, but with nystagmus; the head is erect but wobbles and is often thrown back in paroxysms; the ears hang down; the back is sunk in; the legs can no longer bear the weight of the body; the animal crawls rather than walks, with the legs apart and the extremities extended. There is, therefore, a normal tonus control by the labyrinth. The removal of the influence produces inexactitude in movement, not paralysis.

An interesting report of work on the anatomical relations of the cerebellum was that of Dr. L. J. J. Muskens (Amsterdam), *A*, on cerebellar connections. Animals that had part of the cerebellum injured or destroyed were examined by the Marchi method and the results were given in the paper. In the rabbit the flocculus cerebelli (lobulus petrosus cerebelli) contains cortical matter, but also a part of the dentate nucleus; after this whole lobe had been removed no degeneration was found in the restiform body or in the spinal cord, but there was a coarse degeneration of the middle third of the superior crus cerebelli. This peduncle, therefore, is not connected with the spinal cord, but is made up of strands of fibers similar to the fibers in the internal capsule. The ventrothalamic bundle of Probst was also found degenerated in all cases. In the squirrel the flocculus contains only cortical matter and fibers, but no part of the dentate nucleus. In this animal after destruction of the flocculus the degeneration stops in the dentate nucleus. In the cat the superior crus cerebelli was found to be the seat of degenerations, but there were none in the inferior crus or in the cord. In cats, after section of the superior peduncle in front of its decussation caudal to the red nucleus, no degeneration was found in the reticular nucleus and the predorsal region, but in one animal after lesion of the tegmentum (the instrument passing through the middle peduncle) there was some degeneration of transverse fibers, which ran through the substantia reticularis, sweeping dorsally across the raphé and ascending to the red nucleus on the other side. Dr. Muskens concluded that the majority of the fibers of the ventral cerebello-thalamic bundle may be considered as a part of the decussation of the superior crus; the only difference is that they cross the raphé far more distally in the pons, and in the rabbit at least a number of the fibers appear to run in the crus cerebelli ad pontem.

On the physiology of the cerebellum, van Rynberk (Rome), *H*, reported some experiments. This was a continuation of the work upon which he had formerly been engaged, but instead of dogs the author used sheep. The cerebellum of the sheep, it will be remembered, differs from that of the dog in that the posterior median lobule of the dog is inconsiderable, and the ansiform lobule is large, while in the sheep there is a large posterior lobule and a small ansiform lobule. Localized results followed different lesions of these parts of the cerebellum, especially those concerned with movements of progression. When the ansiform lobule was extirpated on one side there was no observable effect. When this sort of lesion was combined with the destruction of the posterior median lobule there was ambulatory dysmetria in the homolateral forefoot (*Hähnetritt* of LUCIANI). After simple extirpation of the posterior median lobule there was always an inability to move, which was transient but for a time complete. After extirpation of the paramedian lobule there was a turning of the animal about the long axis.

Dr. W. A. JOLLY (Edinburgh), *H*, read an account of "the effects of lesions of the ascending parietal convolution in monkeys." He told of experiments that he had performed in which lesions of the ascending parietal convolution were made by the cautery, which were followed by distinct degenerations in the posterior limb of the internal capsule. When the lesion embraced all of the convolution the animals exhibited a preference for the use of the limb on the homolateral side, indicating in general that there was not so good control of the limb innervated or supplied by the nerves going to the ascending parietal (nerves for muscle sensation, probably). There was, however, no definite ataxia noted, but this is not surprising in view of the experiments of SHERRINGTON that are recorded above and other experiments by the same investigator. It is interesting that the ability to salute at the word of command remained unimpaired after the whole ascending parietal region on the opposite side had been destroyed, but the reader gave no indication of how long before the operation the habit had been formed.

The paper of LEWENDOWSKI (Berlin), *A*, was one of the most interesting at either congress, if he has excluded all other explanations for the condition he reported. The title of his paper is "Abspaltung des Farbensinnes durch Herderkrankung des Gehirns." In this he gave an account of a patient who had hemiplegia and hemianopia, but in the field of vision still remaining colored stimuli evidently did not mean color, for he could not name a color that was shown to him, or state the color of an object that was given, or select a color when the name was spoken, or match colors. According to other tests there was no definite color blindness, but there was no connection between the colors and the names of colors.

On account of the recent disputes regarding aphasia, which have been due to the investigations and writings of MARIE, the discussion of the subject by Professor VON MONAKOW (Zürich), A. PICK (Prague), LIEPMANN (Berlin) and HARTMANN (Graz) was most welcome (*A*). PICK in discussing "Asymbolie und Apraxie," dealt in a very general way with the problems, but referred especially to the meanings of the terms used to designate the different forms of disturbance in the appreciation of sensations dealing with social intercourse, that is, in the naming, understanding and in general the appreciation of the things used by all for the conveyance of ideas. He noted the three ways, all different, in which the term asymboly has been used, and urged that the term should be employed in its original sense and should be used to imply what FINKELBERG had first used it to mean, disturbances of the means of expression. If it be used in this sense it would include many of the forms of so-called aphasia, but not all. Agnosia, in WERNICKE's sense, and apraxia would not be included but considered separate subjects. VON MONAKOW entitled his communication "Aphasie und Apraxie." Aphasia, apraxia and asymboly, he said, are names for groups of conditions accompanyiny disturbances in a motor or sensory sphere. The conditions are large and can only be roughly outlined but they fall into two main groups: (1) in which especially the use and understanding of the signs of language have been lost, and (2) in which orientation in space and time, recognition (by each sense for itself), is included, i. e., *sensory asymboly* and *agnosia;* or in which there is lost the ability to realize coördinated movements directed to an end, i. e., *motor asymboly* and *apraxia.* Both these groups VON MONAKOW would join under the general term *asemia.* He showed diagrams of fifty-two published cases of different forms of aphasia, with lesions in but not limited to BROCA's convolution, of which eight were permanent without improvement, two were permanent with little improvement, thirteen temporary with complete recovery, ten acute, and five with pure subcortical aphasia, while fourteen cases were negative as regards speech defects. His summary of the results is that aphasia, apraxia and asymboly are usually produced by lesions, more or less indefinitely localized, in the left hemisphere, but that sometimes, though seldom, the lesions are sharply defined. Some of these left residual conditions, while others showed the phenomena only temporarily and the disorder disappeared after a greater or less length of time. The latter type of cases fall into two groups: (1) in which the localized symptoms disappear nearly simultaneously with the general phenomena, and (2) in which the symptoms disappear after some weeks or months, perhaps years, although the form of the lesion remains unaltered. The disappearance may be gradual or all the symptoms may disappear at the same time. Some remain as permanent defects. This view of the conditions in aphasia is of special interest as compared with the views of MARIE who, to give the situation in brief, believes that all aphasic disorders are of the nature of mental defects, more or less permanent, and who does not believe that the aphasias are caused by well defined lesions in the cerebrum. VON MONAKOW believes that the sharply defined clinical forms of aphasia and apraxia are due less to the injury as such, i. e., disturbance of any number or quality of neurones, than to what he calls *diachesis*, and that the better the differentiation of the symptoms the more does the principle of diachesis come in. Diachesis, it should be noted, is the term used by VON MONAKOW to indicate the lack of stimulation of certain centers by impulses from other centers, which normally act by their impulses as stimuli to the others. In other words, it is the condition of inability of a secondary center to function because of the destruction or paralysis of the primary center connected with the given secondary center. This is placing the blame one step further back than has usually been done. Dr. HARTMANN took up the subject of what problems are to be solved for a proper understanding of the various speech defects. In regard to aphasia it may be taken as settled that the pathological çonditions of asymboly and apraxia appear when both sides of the cerebrum are diseased or when one side is affected with complications of the fiber system of the corpus callosum. It is at present impossible to refer the different forms of aphasia and apraxia to definite lesions in the brain, but careful study of the residual symptoms and of those that are temporary, with minute consideration of the related and general symptoms will help toward a better understanding of the relation of the different parts of the cerebrum to the speech functions. At present we know little regarding the normal physiology of the nerve processes as compared with our anatomical knowledge and we must have more information on the functional side of the associational processes before we shall have an understanding of the complex associations which may be called aphasia, or asemia, asymboly, apraxia, etc.

From both the scientific and the social sides the two congresses were very valuable. At both congresses special bronze medals were given to each member, at Heidelberg one from the Grand Duke of Baden with the portrait of HELMHOLTZ, and at Amsterdam one with the portrait of the Queen. The social part of both congresses was well conducted, and the short résumés of papers that are given above can do no more than indicate the diversity and interest of the full programs in both series of meetings. In Heidelberg many of the papers were chemical in character and not of special interest to comparative or human neurologists. In Amsterdam there were some few psychological papers and discussions, and one section was devoted to the consideration of questions dealing with the care of the insane. It is expected that the proceedings of the congress of psychiatry, neurology, psychology and the care of the insane will be published, but there will be no official report of the proceedings of the congress of physiologists.

SHEPHERD IVORY FRANZ.

BOOKS AND PAMPHLETS RECEIVED

Sterzi, G. Il sistema nervoso centrale dei vertebrati, vol. i. Ciclostomi. *A. Draghi, Publisher, Padua.* 1907.

Kappers, C. U. Ariens. Untersuchungen über das Gehirn der Ganoiden Amia calva und Lepidosteus osseus. Reprinted from *Abhandl. Senkenbergischen Naturforschenden Gesellschaft*, vol. 30, no. 3. 1907.

Kohnstamm, O. and Wolfstein, J. Versuch einer physiologischen Anatomie der Vagusursprünge und des Kopfsympathicus. Reprinted from *Journal für Psychologie und Neurologie*, vol. 8. 1907.

Cole, F. J. Notes on Myxine. Reprinted from *Anatomischer Anzeiger*, vol. 27, nos. 12 and 13. 1905.

Cole, F. J. and Dakin, W. J. Further observations on the cranial nerves of Chimæra. Reprinted from *Anatomischer Anzeiger*, vol. 28, no. 23. 1906.

Cole, F. J. A monograph on the general morphology of the myxinoid fishes, based on a study Myxine. Part I. The anatomy of the skeleton. *Transactions of the Royal Society of Edinburgh*, vol. 51, part 3, no. 30. 1905. Part II. The anatomy of the muscles. *Ibid.*, vol. 55, part 3, no. 26. 1907.

Santee, Harris E. Anatomy of the brain and spinal cord, with special reference to mechanism and function. 4th ed. Revised. *Philadelphia, P. Blakiston's Son & Co.* pp. xxxvi + 453. 1907.

Edinger, L. and Wallenberg, A. Anatomie des Centralnervensystems. 1905 and 1906. *Leipzig, Verlag von S. Hirzel.* 1907.

Waldeyer, W. Ueber Gehirne menschlicher Zwillings- und Drillings-fruchte verscheidenen Geschlechtes. Reprinted from *Sitzungsberichte der Königlich Preussischen Akademie der Wissenschaften.* vi. 1907.

Fragnito, O. Le fibrille e la sostanza fibrillogena nelle cellule ganglionari dei vertebrati. Reprinted from *Annali di Nevrologia*, anno 25, fasc. 3. 1907.

Ikegami, K. and Yagita, K. Ueber den Ursprung des Lungenvagus. Reprinted from *Okayama-Igakkwai-Zasshi* (*Contributions from the Medical Society of Okayama*), no. 206. 1907.

Bender, Otto. Die Schleimhautnerven des Facialis, Glossopharyngeus und Vagus. Studien zur Morphologie des Mittelohres und der benachbarten Kopfregion der Wirbelthiere. Reprinted from *Semon's Forschungsreisen in Australien u. s. w.* 110 pp., 9 pl. Jena, 1906.

Winkler, C. The central course of the nervus octavus and its influence on motility. Reprinted from *Verh. Kon. Akad. van Wettenschappen, Amsterdam* (2 sec.), Deel 14, no. 1, 202 pp., 24 pl. 1907.

Holmes, S. J. Regeneration as functional adjustment. Reprinted from *Journal of Experimental Zoölogy*, vol. 4, no. 3. 1907.

Holmes, S. J. The behavior of Loxophyllum and its relation to regeneration. Reprinted from *Journal of Experimental Zoölogy*, vol. 4, no. 3. 1907.

Holmes, S. J. Observations on the young of Ranatra quadridentata Stal. Reprinted from *Biological Bulletin*, vol. 12, no. 3. 1907.

Williams, S. R. Habits and structure of Scutigerella immaculata (Newport). Reprinted from *Proc. Boston Society of Natural History*, vol. 33, no. 9, pp. 461–485. 1907.

Dodge, Raymond. An experimental study of visual fixation. Studies from the Psychological Laboratory of Wesleyan University, vol. 1, no. 1. *Psychological Review, Monograph Supplements*, vol. 8, no. 4. 1907.

Sumner, Francis B. Further studies of the physical and chemical relations between fishes and their surrounding medium. Reprinted from *Am. Journal of Physiology*, vol. 19, no. 1. 1907.

Kellogg, Vernon L. Darwinism today. *New York, Henry Holt & Co.* 1907.

Bovard, J. F. The structure and movements of Condylostoma patens. *Publications Univ. of California*, vol. 3, no. 14, pp. 343–368. 1907.

Krafft-Ebing and Obersteiner. Die progressive allgemeine Paralyse. *Wien und Leipzig, A. Hölder.* 1907.

Frankl-Hochwart, L. v. Die Tetanie der Erwachsen. 2d ed. *Wien und Leipzig, A. Hölder*, 141 pp. 1907.

The Journal of
Comparative Neurology and Psychology

Volume XVIII | APRIL, 1908 | Number 2

THE ARCHITECTURAL RELATIONS OF THE AFFERENT ELEMENTS ENTERING INTO THE FORMATION OF THE SPINAL NERVES.

BY

S. WALTER RANSON, PH.D., M.D.

(From the Anatomical Laboratory of the University of Chicago.)

With One Figure.

INTRODUCTION.

Some rather surprising observations are recorded in a paper recently published on "Retrograde degeneration in the spinal nerves" (Ranson '06).

It was found that after the division of a nerve, containing 1500 medullated afferent fibers, there occurred a complete degeneration of 4500 spinal ganglion cells and that this was accompanied by little or no degeneration of the dorsal roots. It was at once apparent that these results would be very difficult to explain on the basis of the usual conception of the spinal ganglion. Accordingly, the literature dealing with the architecture of the spinal nerves and of their dorsal root ganglia has been carefully reviewed in the hope of finding some observations that would be of assistance in interpreting these facts.

Another reason for presenting the normal relation of the sensory elements of the spinal nerves is the fact that in order to obtain a norm for the second cervical nerve of the white rat (the nerve studied in this series of experiments) it was necessary to make a study of the numerical relations in that nerve and these observations have some value from the anatomical point of view

This work was begun under the direction of Dr. H. H. DONALDSON, to whom the writer is indebted for many suggestions.

THE SPINAL GANGLION.

I. The distinction between the large and the small cells and the functional significance of the two forms.—It has long been known that there exist in the spinal ganglion two well marked types of cells, which differ from each other both in size and staining reaction. As early as 1886 v. LENHOSSÉK made a careful study of the small cells and expressed an opinion concerning their functional sig-

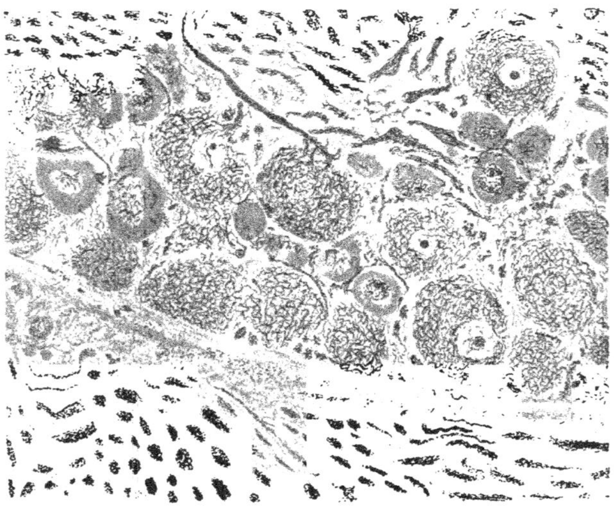

FIG. 1. The drawing represents a section 5μ thick from a spinal ganglion of a white rat, prepared by a modification of DONAGGIO's Method VII, Zeiss, ocular 4, Objective $\frac{1}{12}$.

nificance. According to his description, which relates in this instance to the spinal ganglion of the frog, these cells are very small, sometimes not more than 5μ in diameter; they are often angular and possess a relatively small amount of cytoplasm surrounding a large nucleus. In 1895 he adds to the previous description that the small cells are characterized not only by their size but also by the fact that they stain more intensely with the diffuse protoplasmic dyes. Among those who have confirmed these observations of v. LENHOSSÉK may be mentioned FLEMMING ('95) and COX ('98). These small cells correspond to LUGARO's Type III and HATAI's Type II.

In connection with another investigation the writer has obtained preparations of the spinal ganglion by a slight modification of DONAGGIO's Method VII[1] (DONAGGIO '04) which demonstrate in a very striking manner a difference, probably chemical but possibly structural, between the large and the small cells. Since no other method presents so marked a contrast between the two cell types it is worth while to note the peculiarities of these preparations (see Fig. 1). The large cells present an absolutely colorless cytoplasm, throughout which there is a network of deep blue threads. These are largely absent from the nucleus. The small cells, on the contrary, present a cytoplasm of a deep violet which is almost entirely free from the blue threads, while the nucleus contains them in abundance. These same threads are seen in the axis cylinders. After a careful study of the literature it has not been possible to identify these threads with any known structure; but since the granular reticulum of CAJAL, the NISSL-bodies, the GOLGI-intracellular-net, the canals of HOLMGREN, the neurofibrils of BETHE and the still different fibrils of DONAGGIO and CAJAL together with the remaining protoplasm and the nucleus must occupy nearly all the space in one small cell, it does not seem probable that the threads just described are new structures. There can be no question however concerning the clear distinction which these preparations show between the large and the small cells, since the difference is a constant one and the picture is always the same. The distinction so strongly emphasized in these preparations is probably a chemical one and has its counterpart in the functional differences about to be mentioned. It should be noticed that there is a certain number of transitional cells which partake of the qualities of both large and small cells and are usually of medium size. Several are represented in the

[1] DONAGGIO's *Method VII (modified).*—Pieces 2 to 3 mm. in thickness are fixed for 24 hours in a saturated solution of mercuric chloride in 10 per cent formalin to which has been added 1 per cent of glacial acetic acid; iodine-water 24 hours; distilled water 2 hours; pyridine 48 hours (change once); distilled water 24 hours; ammonium molybdate 24 hours; distilled water 1 hour; pyridine 48 hours (change once); an aqueous solution of thionin $\frac{1}{15000}$, prepared at least two weeks previously (change once and stain for 48 hours); dehydrate and embed in paraffine; cut sections 5 to 7μ thick.

drawing; they correspond to HATAI's Type III. These intermediate cells represent the stages through which the small cells pass while developing into the larger ones, a process which, as we shall see, is constantly going on in the growing animal.

RAWITZ ('80), in studying the spinal ganglia of various animals, had his attention drawn to these small deeply staining cells and came to the conclusion that they were young developing ganglion cells, the immediate result of a supposed—but confessedly undemonstrated—cell division. As proof he advances his observations that they are seldom found in the grown animal, but, on the contrary, are relatively frequent in the young. VON LENHOSSÉK ('86) does not agree with these statements of RAWITZ, for, while he admits that these cells are found more abundantly in young than in adult animals, he has also found them in large numbers in the full grown frogs. "I believe," says v. LENHOSSÉK, "that one may account for the presence of these little cells through the following consistent explanation: while, in the course of embryological development, the majority of ganglion cells become very much enlarged, a part of them as well as their associated nerve fibers stop at lower stages of development; such undeveloped nerve cells represent the little cells under discussion. According to this conception the cells in question would not be young and capable of further development, but represent ganglion cells remaining permanently at primitive stages of evolution." In 1895 v. LENHOSSÉK returned to the subject of the significance of the small cells. "It is not superfluous to insist that the smaller cells, even indeed the smallest cells, are not to be regarded as functionless rudimentary structures, but as elements which just as truly as the large cells are functional parts of the nervous mechanism: we find them associated just like the large cells with a process which divides in the typical way" into a central and a peripheral fiber. "Still less is it justifiable to look upon them as young elements still undergoing development. We are dealing here therefore not with cells which will further divide or otherwise develop . . . . . but with cells which are formed small once for all."

Evidence, to be presented in a succeeding paragraph, supports v. LENHOSSÉK in his contention that the small cells are not young in the sense of RAWITZ; but that all, large and small alike, being derived from a cell division at an early embryonic period, may be designated as old. It would seem, however, that the point concerning their incapacity for further development is not so well taken. We will return to these points in another paragraph, and will now consider BÜHLER's conception of the *raison d'être* of the small cells.

BÜHLER ('98) noticed that under physiological conditions in the toad, the frog and the rabbit there occurred a degeneration of a few isolated large ganglion cells, which were however not described. The degeneration is, to all appearances, not very rapid; in a spinal ganglion of a frog about 20 or 25 at a time, in rabbits relatively much fewer. He assumes that these disappearing ganglion cells are recruited from the ranks of the small cells, which develop into large cells as they are needed. "Since after the earliest stages a proliferation of ganglion cells no longer occurs, in order to remain capable of functioning throughout the period of life, the spinal ganglion must receive for its portion in the anlage sufficient reserve material in the form of undeveloped cells." HATAI ('02) has argued against this assumption on the ground that the number of spinal ganglion cells is approximately constant throughout the life of the individual. However the recent observations of KÖSTER on the spinal ganglia of cats, dogs and rabbits give some support to BÜHLER's statement (KÖSTER '03, p. 1098). "One recognizes, in every section of a normal ganglion, cells with all possible appearances of degeneration. One can see cells with eccentric swollen or fragmented nuclei, coarse and fine chromatolysis, and all the changes which one may look upon as the reactional manifestations of the cells to the physiological degeneration found by SIGMUND MAYER in the peripheral nerves. We can, therefore, speak of a physiological degeneration of nerve cells." From these observations it would seem not impossible that a certain very slight amount of degeneration is going on constantly in the normal ganglion; and the question, whether or no the small cells are, as BÜHLER assumes, capable of replacing the cells lost in this way, is a question worthy of some consideration.

HATAI ('00) has given some attention to the significance of the small darkly staining elements, which with their scanty cytoplasm and large nuclei present many of the characters of embryonic cells, and concludes that they are "in a growing state or in a more or less permanently immature condition." In order to test this assumption he ('02) counted the number of large and small cells in the spinal ganglia of the VI C., IV T., and II L. nerves of four white rats, ranging in weight from ten to one hundred sixty-seven grams, and found that, while the total number of cells in each ganglion remained approximately constant, there was a constant increase in the number of large cells and a corresponding decrease in the number of the small cells. This can only mean that the small cells are developing into large ones, and that therefore a considerable number of the former retain their capacity for development at least during the growing period.

It is of interest to note in this connection the observations made by HODGE ('89) that after electrical stimulation of nerves it is chiefly the large cells in the associated spinal ganglia that show the effect of fatigue. Considering all the cells large which have one diameter 50μ or over and those small which have not, a count gives the following results:

TABLE I.

Effect of Stimulating Ganglion Cells (HODGE).

In 100 Large Cells, Nuclei			In 100 Small Cells, Nuclei	
Shrunken.	Normal.		Shrunken.	Normal.
5	95	Resting	0	100
94	6	Stimulated	8	92

HODGE did not attempt an explanation of these interesting results; but in the light of the preceding discussion there seems to be little room for doubt that these small unworked elements are the immature cells of HATAI.

In summing up this discussion concerning the functional significance of the small cells of the spinal ganglion, it may be said that the absence of mitosis in the spinal ganglia during extra-uterine life excludes the possibility of their being young cells in the sense of RAWITZ. No more acceptable is the view of v. LENHOSSÉK that they are elements, the development of which has been permanently arrested; we must rather agree with HATAI that they retain for a long time their capacity for development, that, in fact, some of them are always in the process of transformation throughout the growing period of the animal. During the time that they are still undeveloped they do not show fatigue when the nerve is stimulated electrically. It is not yet satisfactorily determined whether they may serve as reserve cells capable of replacing the mature neurones destroyed by trauma or disease.

2. *Classification of the spinal ganglion cells according to the number and character of their processes.*—Since in this paper we are not directly concerned with the form of the spinal ganglion cells, we need only mention the most important points under this heading. That the cells of the spinal ganglion were all associated with a single T-shaped process was the accepted view until 1896, when DOGIEL published his important work on the form of the elements in the spinal ganglion. To DOGIEL belongs the credit of having first clearly differentiated the following cells in the spinal ganglia of mammals.

A. Unipolar cells. Type I. The well known unipolar cells, both large and small, with the typical T-shaped processes of RANVIER.

Type II. A new form first seen by DOGIEL, the single process of which breaks up into numerous fine branches that end in pericellular baskets within the ganglion itself.

B. Bipolar cells—very few, only one or two in each ganglion.

C. Multipolar cells with two nerve processes, one centrally, the other peripherally directed, and many dendritic processes arising from the angles of the irregularly shaped cell body. These dendrites penetrate the capsule and end among the cells of the ganglion.

The observations of DOGIEL were made upon preparations stained by his modification of the methylene blue technique. More recently ('05) CAJAL has published a preliminary account of his studies on the spinal ganglion with his new silver method. One of his cell-types is distinctly new and may be described here since it serves to emphasize the wealth of connections within the ganglion. This is an unipolar cell, possessed of very fine dendrites which take origin, sometimes from the surface of the cell itself, sometimes from the origin of the axis cylinder. These dendrites gradually enlarge and terminate in spheres, encircled by an entire system of concentric capsules. These dendrites sometimes bifurcate and give rise to a pair or more of terminal globes. He distinguishes two varieties among these cells: in one the terminal spheres are found beneath the capsule of the cell of origin and are in relation with the pericellular "nests" of CAJAL and DOGIEL, in the other the terminal globes are lodged in the intercellular spaces sometimes far distant from their point of origin.

We return now to a point more directly in keeping with the general purpose of this paper, namely, to the form of the small cells. HATAI ('01), apparently quoting from DOGIEL, says that "the number of these cells from which no axon can be traced is large." HARDESTY ('05) agrees that "a larger portion of these extra cells belong probably to the anaxonic type of neurone, latent cells which have not yet developed processes." Both HATAI and HARDESTY had in mind only the fact that the small cells were not connected with medullated fibers in the dorsal root or peripheral nerve—a fact which stands uncontested—but the conclusion that these cells are necessarily anaxonic is unnecessary and misleading. I have not been able to verify the citation from DOGIEL, and there seems every reason to believe that instead of being numerous such apolar cells do not occur at all in the spinal ganglion. In his extremely careful study of these structures, which lead him to insist on the presence of bipolar and multipolar cells, although never more than two or three such were found in one ganglion, DOGIEL does not mention the presence of these "anaxonic neurones." On the other hand, he describes in detail the single process of the small cell as being a typical T-shaped process with two branches, one directed toward the spinal cord, the other toward the periphery. These processes are usually destitute of myelin, but a few are medullated for a part of their course. He was able to trace these non-medullated processes of the small cells into the dorsal roots and into the peripheral nerves as far as the junction of the afferent and efferent fibers.

The absence of apolar cells is again the implication of v. LENHOSSÉK in the quotation already given. "We find them (the small cells) just as truly as the large cells associated with a process which divides in a typical way." But v. LENHOSSÉK does not leave the question in this obscure way, but says, in another place (Bau des Nervensystems, p. 268), "If we study the spinal ganglion of one of the more highly developed vertebrates or even the frog with suitable isolation, teasing or staining methods, we find in it, in addition to the interstitial connective tissue, blood vessels and nerve fibers, also numerous nerve cells of varying size of which the typical form is unipolar. *There are no apolar cells.*"

We have also to note the negative findings of HODGE ('89), who, having obtained physiological results that lead him to expect large numbers of apolar cells in the spinal ganglia of frogs, undertook to demonstrate their presence in teased preparations but came to the conclusion that "Apolar cells do not occur in the spinal ganglia of frogs in any considerable numbers, none having been found."

3. *Interrelations among the spinal ganglion cells.*—The spinal ganglion is not to be regarded as an aggregation of more or less spherical cells each independent of the others and connected only with its central and peripheral processes; but is in reality a complicated mass containing the ramifications of dendrites and axis cylinders, forming exceedingly intricate intercellular meshworks and pericellular baskets, the cells in this way being brought into close functional relations with each other. Moreover there are sympathetic fibers which enter the ganglion via the ramus communicans to join in the formation of these baskets.

ARONSON ('86) was the first to describe the pericellular baskets in the spinal ganglion and his observations were confirmed by CAJAL ('90). The latter investigator regarded them as ramifications of fibers from the sympathetic system. It was DOGIEL however who first cleared up our notions on this point by describing a variety of cell ("Type II") which has for its sole function the establishment of intraganglionic connections.

SPIRLAS ('95) called attention to the existence of collaterals arising from the processes of the embryonic spinal ganglion cells. The observations were confirmed upon adult material by DOGIEL in 1896: "from the processes of many large and small ganglion cells, before their division into two fibers, one, two or three collaterals of varying thickness are given off which at a greater or less distance from their cells break up into fine threads." LEVI ('05) has followed the embryological formation of these collaterals.

In 1896 HUBER described a variety of spinal ganglion cell from the axon of which recurrent collaterals are given off. These run back and end in disks upon the cell from which the axon arose.

Still another means of intercommunication between the spinal ganglion cells is found in the dendritic processes of the multipolar cells and the more numerous unipolar cells of CAJAL, possessing fine dendritic branches with spherical endings which may either be in connection with the immediate pericellular basket or may run for considerable distances in the intercellular spaces to make connections in other parts of the ganglion.

Expressed in other words, the relations are as follows. Sympathetic fibers enter the ganglion and break up about the cells, especially those of DOGIEL's Type II. The single processes of these cells of DOGIEL break up within the ganglion into a multitude of little twigs which form baskets about still other cells, while the stem process of many of the latter, i. e., the ordinary spinal ganglion cells, gives off delicate collaterals, which also take part in the formation of the fiber complex of the ganglion. All this wealth of axonic ramifications, together with the dendritic branches of some of the cells, forms a basis of intercommunication which argues for a close functional relationship among the individual spinal ganglion cells.

THE DORSAL ROOTS.

1. The relation of the fibers of the dorsal roots to the cells in the substantia grisea of the spinal cord. Examination of NISSL's *view that they are axons of such cells.*—A very peculiar observation noted in a paper recently published on "Retrograde Degeneration" (RANSON '06), namely, that after half of the spinal ganglion cells have disappeared as the result of section of the associated nerve, there are to be found in the dorsal root very nearly if not quite the normal number of fibers—and all this according to careful counts made on a considerable number of animals—has led to a careful consideration of NISSL's ('03) recently published conception of the dorsal roots, as a possible though improbable explanation of these results. Had the idea that the dorsal root fibers are independent of the spinal ganglion cells been advocated by any lesser authority we might indeed pass it by as unthinkable; but we cannot so lightly treat a statement by FRANZ NISSL. According to him ('03, p. 334), "The posterior root fibers are united with the cells of the spinal cord and especially with the cells of the substantia gelatinosa and only pass through the ganglion; and . . . . the cells of the spinal ganglion also send fibers toward the periphery." The facts on which he bases this remarkable assertion are that after section of the posterior roots the cells of the spinal ganglia do not show change, while certain cells in the spinal cord, especially the cells of the substantia gelatinosa, undergo chromatolysis and even complete degeneration. It may be said however that, while true axonal reaction does not occur after section of the root, yet very considerable changes are induced in the ganglion cells

(KLEIST and KÖSTER); and there are plenty of cases of degeneration in neurone chains similar to the disappearance of the cells of the substantia gelatinosa—e. g., the degeneration of the motor cells with their peripheral motor fibers after section of the dorsal roots (BRAEUNIG '03). He also cites, as bearing on this point, the anomalous trigeminus found by v. GUDDEN in a calf, which showed no sensory root fibers. The ganglion itself was normal, as were also the fibers of the peripheral nerve arising from it, although these peripheral fibers were greatly decreased in number. According to NISSL's view the fibers whose cells were located within the brain had failed to develop, while those whose cells were located in the Gasserian ganglion had developed normally.

It would seem that these are rather insufficient grounds for revising our conception of the dorsal root, based as it is on such a large number of careful investigations; and it should be remembered, in this connection that those who have worked with the histology of the spinal ganglion, whether in teased preparations or GOLGI-material, have considered the dorsal root fibers identical with the central branches of RANVIER's T-fibers. Of the existence of some fibers in the dorsal root whose cells lie in the cord, there can be no doubt, at least so far as certain of the lower forms are concerned (CAJAL '90; v. LENHOSSÉK '90; and VAN GEHUCHTEN '93).

According to the experiments of JOSEPH ('87) on the second cervical nerve of the cat, these fibers are also found in mammals. After section of the dorsal root, he has observed that some fibers in the central portion remain normal and in the nerve a small number degenerate. At first opposed by SINGER and MÜNZER ('90), these results of JOSEPH have been confirmed bythese same authors in a later paper ('95) and anticipated by the earlier work of KAHLER ('84). KOPCZYNSKI ('06) has been unable to verify these observations.

It is to be borne in mind that, while these direct observations show the presence of fibers of passage in the spinal ganglion, they also indicate that only a few fibers are of this category and that these are efferent; and by no means do they support the view of NISSL that all the fibers of the dorsal root arise from cells situated in the spinal cord. And, as we shall see, the condition described in the paper on retrograde degeneration, namely, the presence of a normal dorsal root associated with a ganglion which has lost one-half of its cells, is susceptible of another explanation than that suggested by NISSL's theory. All in all, then, while the evidence requires that we should be open-minded on this question, it is not sufficient to overthrow the belief that the dorsal roots are predominately composed of the central branches of RANVIER's T-processes. The evidence from the silver preparations, that one branch of the stem process of the spinal ganglion cell runs through the dorsal root, is very convincing. This evidence has been well summarized by VAN GEHUCHTEN ('92).

2. *Numerical relations between the spinal ganglion cells and the medullated fibers of the dorsal roots.*—FREUD ('78), working on Petromyzon, found a considerable excess of fibers in the dorsal roots over the cells in the spinal ganglion, due to the fact that the cell bodies of many of the afferent neurones are located in the spinal cord. HODGE ('89) counted the dorsal root fibers and the cells in the associated ganglion in the frog, and found about three cells for each fiber. BÜHLER ('98) has shown that the number of cells in the spinal ganglia increases as the test-animal is higher in the zoölogical series; least for fish, it is greatest in mammals. He also found that in the frog there were about five ganglion cells for each dorsal root fiber. GAULE and LEWIN ('96) found in the rabbit a ratio between cells and fibers of 6 to 1. HARDESTY ('05) found in the frog a ratio varying from 2.7 to 3.6 cells per fiber. HATAI ('02), working on the white rat, obtained the following results for the adult specimen of 167 grms. body weight.

TABLE II.

Ratio of Spinal Ganglion Cells to Dorsal Root Fibers (HATAI).

NERVE.	NUMBER OF CELLS.	NUMBER OF FIBERS.	RATIO
VI C.	12,200	4,227	1 : 2.8*
IV T.	4,406	1,522	1 : 4.8*
II L.	9,442	1,644	1 : 5.7

* The figures 2.7 and 4.3 given in the original are obviously misprints.

The writer, in studying the normal relations in the second cervical nerve of the white rat, has obtained results confirmatory of those of the authors already mentioned. In the three cases in which the dorsal root fibers and spinal ganglion cells were enumerated in the same individual nerve, a rather constant ratio of approximately 1 fiber to 3.2 cells was obtained. The first two specimens were 72 days old and weighed about 110 grams, the third was six months old and weighed 188 grams.

TABLE III.

Ratio of Spinal Ganglion Cells to Dorsal Root Fibers (RANSON).

SPECIMEN.	NUMBER OF CELLS.	NUMBER OF FIBERS.	CELLS PER FIBER.
72 days	7,721	2,472	3.1
72 days	8,116	2,394	3.3
6 months	8,624	2,689	3.2

The number of cells in a given spinal ganglion exceeds the number of medullated fibers in the corresponding root; this excess holds alike for frogs and mammals, although the actual percentage of the excess varies greatly. HATAI and HARDESTY ascribe to the anaxonic cells the responsibility for this condition; but, since we know that these cells do not exist in any appreciable number, we are thrown back upon the non-medullated fibers of the small cells as the chief source of the discrepancy. While it is possible that the majority of these non-medullated fibers do not push out into the dorsal root, it seems probable that the number of cells in the spinal ganglion does not exceed the number of axis cylinders in the dorsal root by so large a number as it does the number of myelin sheaths. A count of the dorsal root fibers obtained by a differential axis cylinder stain is the logical method of answering this question.

3. The increase in the number of medullated fibers in the growing animal.—HARDESTY ('05) has shown that, when his frogs were arranged in a series of increasing body weight, there was a general, though not very regular, increase in the number of fibers in the ventral and dorsal roots as well as in the peripheral nerves.

The white rat however in the hands of HATAI ('03) has given uniform results, showing a regular increase in the number of medullated fibers both in the ventral and dorsal roots.

The II C. nerve of the white rat shows more variability; in a general way however the number of fibers is increasing. The increase is, however, by no means as rapid, nor is there such a large number of fibers added as in the nerves studied by HATAI.

These observations are recorded in the accompanying table.

TABLE IV.

Rate of Medullation in the Second Cervical Nerve of the White Rat (RANSON).

AGE.	MEDULLATED FIBERS IN THE DORSAL ROOT.		MEDULLATED FIBERS IN THE VENTRAL ROOT.	
12 days	1608			
12 days	1521			
12 days (average)	—	1564.5		
72 days	2472		689	
72 days	2394		660	
72 days	1059		590	
72 days	2217		591	
72 days (average)	—	2261	—	632.5
6 months	2891		773	
6 months	2689		703	
6 months (average)	—	2790	—	736.5

THE NERVE.

1. The proportion of sensory and motor fibers.—All investigators have found a larger number of fibers in the dorsal than in the ventral root. According to INGBERT ('04) the ratio of all the motor and sensory fibers arising from the left side of the human spinal cord is 1 : 3.2, and from the second cervical segment alone 1 : 6. HATAI ('03) working with the C. VI, T. IV, and L. II nerves of the white rat finds an average ratio of 1 : 2.3. The normal relations in the C. II nerve of the white rat are expressed in the following table, representing the writer's enumeration of the ventral and dorsal root fibers for that nerve.

TABLE V.

Number of Ventral and Dorsal Root Fibers in the II C. Nerve of the White Rat (RANSON).

AGE.	VENTRAL ROOT.	DORSAL ROOT.	RATIO.
72 days (110 grms.)	689	2,472	1 : 3.6
72 days (110 grms.)	660	2,394	1 : 3.6
72 days (110 grms.)	590	1,959	1 : 3.3
72 days (110 grms.)	591	2,217	1 : 3.7

This table shows that the ventral root fibers are about 28 per cent as numerous as the dorsal root fibers.

2. The distal excess.—As has been said, considerably more sensory than motor fibers enter into the formation of the peripheral nerve. But, when we compare the sum of the fibers in the ventral and dorsal roots with the total number present on the distal side of the ganglion, we find a distinct excess on the distal side. The earlier investigators who undertook to compare the number of fibers on the two sides of the ganglion, either found them equal or else the peripheral count so little in excess that they regarded it merely as a matter of technical error and attached no significance to it. Later, BIRGE ('82) found an excess of fibers on the peripheral side of the II C. ganglion of the frog amounting to 16 per cent; and BÜHLER ('98), also working on the frog, found in one nerve an excess of 25 per cent. HARDESTY ('99, '00, '05) has spoken of these extra fibers as the "distal excess" and found that it varied in the frog from 5 per cent to 61 per cent. GAULE and LEWIN ('96) found a distal excess in three of the sacral nerves of a rabbit of 19 per cent, 11 per cent, and 15 per cent, respectively. My observations on the II C. nerve of the white rat are confirmatory of these previous results. Here we have to do with a distal excess of 8 or 10 per cent. This is of interest since DALE ('00) found in coccygeal nerves of cats an average distal excess of only 0.63 per cent.

TABLE VI.

Showing the Distal Excess in the II C. Nerve of the Adult White Rat (RANSON).

WEIGHT.	VENTRAL ROOT.	DORSAL ROOT.	SUM OF ROOTS.	DISTAL EXCESS.	PRECENTAGE OF D. E	SUM OF RAMI.	VENTRAL RAMUS.	DORSAL RAMUS.
302 grms.....	646	2386	3032	257	8	3289	887	2402
161 grms.....	672	2090	2762	276	10	3098	708	2390

HARDESTY has made a careful study of the possible explanations of this distal excess. It is much too complicated a question for us to enter upon here. It can only be said in passing that there is evidence for the presence of medullated fibers of sympathetic origin which pass through the nerve to end in the ganglion and, hence, would not be found in either of the roots. There is also evidence that both sensory and motor fibers may bifurcate at the level of the ganglion. But after a careful consideration of all the possibilities, HARDESTY does not think any one cause sufficient to explain the facts and believes that several factors must operate together in the production of the distal excess. The idea of NISSL, discussed in a previous section, that the dorsal root fibers pass through the ganglion without making any connections and are there joined by others arising in the ganglion, would, if it should be found correct, offer an adequate explanation of the distal excess, especially of those cases where the excess is large and amounts to more than 60 per cent, as in one case (HARDESTY '00) where 337 fibers on the proximal side were associated with 544 on the distal side of the ganglion. In those cases however, where the excess is not more than 5 per cent (2422 proximal to 2543 distal fibers, HARDESTY '05), the hypothesis of NISSL would require that very few fibers originate in the ganglion.

3. The presence of non-medullated fibers in the nerve.—It has already been said that DOGIEL was able to trace the non-medullated fibers as far as the junction of the afferent and efferent roots. So important is this work of DOGIEL's that a full quotation may be given.

> Ausser der Grösse besteht der einzige Unterschied zwischen den in Rede stehenden Zellen und den grossen Ganglienzellen darin, dass von einer jeden solchen Zelle immer nur ein einziger äusserst dünner und während seines ganzen Verlaufs myelinlos bleibender Fortsatz abgeht. Von der Zelle gewöhnlich in der Form eines kleinen Konus beginnend, bekommt der Hauptfortsatz das Aussehen eines äusserst dünnen, nicht selten varicösen Fadens, welcher noch unter der Zellkapsel, oder sofort nach dem Austritt aus ihr, 2–3 bogenförmige Biegungen macht, worauf er mehr oder weniger gradlinig oft eine sehr lange Strecke zurücklegt und sich endlich V- oder T-förmig in zwei dünne varicöse Fäserchen teilt. Soviel ich weiss, hat Retzius zuerst die Aufmerksamkeit auf das Vorkommen kleiner Ganglienzellen in den Spinalganglien der Säugetiere (Kaninchen) gelenkt, indem er sich über dieselben folgendermaasen ausdrückt: "Im Gegenteil geht, besonders bei kleineren Ganglienzellen, oft von einer schwach abgeschnürten Stelle der Zelle ein blasser Ausläufer aus, welcher zuweilen sich auf weite Strecken verfolgen lässt und dabei die marklose Beschaffenheit behält; länglich-ovale Kerne treten in gewissen Entfernungen an ihm auf, und er wird allem Anscheine nach zu einer gewöhnlichen myelinfreien Nervenfaser; wie sich diese im späteren Verlaufe verhält, konnten wir nicht ergründen. Einmal sahen wir indessen diesen blassen Ausläufer sich dichtomisch teilen." Es is mir gelungen diese Lücke in den Beobachtungen von Retzius auszufüllen und nachzuweisen, dass die Hauptfortsätze der kleinen Zellen und die aus ihrer Teilung hervorgehenden Fasern, soweit sie in den Ganglien und sogar in den hinteren Wurzeln und an deren Zusammentrittsstelle mit den vorderen Wurzeln zu verfolgen sind, überall den Charakter markloser Fasern bewahren, oder aber nur an einer gewissen Strecke von einer äusserst dünnen, frühner oder später wieder verschwindenden Markhülle umgeben werden.

It is important to note in this connection that WEIGNER has recently shown that there is a considerable number of non-medullated fibers in the nervus intermedius; and this should lead to an examination of other cranial and spinal nerves to determine if non-medullated nerve fibers had not been overlooked in them.

It will be shown in my next paper that the small cells of the spinal ganglion, which DOGIEL says possess non-medullated fibers show typical axonal reaction after section of the peripheral nerve and that it is these small cells that degenerate and disappear, facts which can be explained only on the assumption that these non-medullated fibers extend into the peripheral nerve. The existence of these fibers and the degeneration of the small cells offers a satisfactory explanation of the results presented in the former paper on Retrograde Degeneration in the Spinal Nerves.

BIBLIOGRAPHY.

ARONSON.

Beiträge zur Kenntnis der centralen und peripheren Nervenendigungen. Inaugural-Diss. *Berlin.* 1886. (Cited after DOGIEL.)

BETHE, A.

Allgemeine Anatomie und Physiologie des Nervensystems. *Leipzig.* 1903.

BIRGE, E. A.

Die Zahl der Nervenfasern und der motorischen Ganglienzellen im Rückenmark des Frosches. *Arch. f.* [*Anat. u.*] *Physiol.*, p. 435. 1882.

BRAEUNIG, KARL.

Ueber Chromatolyse in den Vorderhornzellen des Rückenmarkes. *Arch. f.* [*Anat. u.*] *Phys.*, p. 251. 1903.

Ueber Degenerations-vorgänge im motorischen Teleneuron nach Durchschneidung der hinteren Rückenmarkswurzeln. *Arch. f.* [*Anat. u.*] *Phys.*, p. 481. 1903.

BÜHLER, A.

Untersuchungen über den Bau der Nervenzellen. *Verhandlungen der Physik.-med. Gesellschaft zu Würzburg*, N. F. Bd. 31. 1898.

BUMM, A.

Die experimentelle Durchtrennung die vordern und hintern Wurzel des zweiten Halsnerven bei der Katzte und ihre Atrophiewirkung auf das zweite spinale Halsganglion. *Sitz. Ber. Ges. Morph. Physiol. Münchau*, p. 18. 1903.

CAJAL, S. RAMÓN Y.

Sobre la existencia de terminationes nerviosas pericelulares en los ganglios nerviosos raquidianos. Pequenas comunicaciones anatomicas. *Barcelona.* 1890. (Cited after DOGIEL.)

Types cellulaires dans les ganglions rachidiens de l'homme et des mammifères. *C. R. Soc. Biol.*, Paris, T. 58, p. 452–453. 1905.

CASSIRER, R.

Ueber Veränderungen der Spinalganglienzellen und ihrer centralen Fortsätze nach Durchschneidung der zugehörigen peripheren Nerven. *Deutsche Zeitschr. f. Nervenheilk.* Bd. 14, S. 150. 1898.

COX, W. H.

Der feinere Bau der Spinal Ganglienzellen des Kaninchens. *Anat. Hefte*, Abth. 1, Bd. 10. 1898.

Beiträge zur pathologischen Histologie und Physiologie des Ganglienzellen. *Internat. Monatsschrift für Anat. und Phys.*, Bd. 15, p. 240. 1898.

DALE, H. H.

On some numerical comparisons of the centripetal and centrifugal medullated nerve fibers arising in the spinal ganglia of the mammal. *Journ. of Physiol.*, vol. 25, p. 196. 1900.

DOGIEL, A. S.

Der Bau der Spinalganglien bei den Säugetieren. *Anat. Anzeiger*, Bd. 12, S. 140. 1896.

Zur Frage über den Bau der Spinalganglien beim Menschen und bei den Säugetieren. *Internat. Monatsschr. für Anat. u. Physiol.*, Bd. 15, p. 343. 1898.

Zur Frage über den feineren Bau der Spinalganglien und deren Zellen bei Säugetieren. *Internat. Monatsschr. für Anat. u. Physiol.*, Bd. 14, p. 73. 1897.

DONAGGIO, A.

The action of pyridin upon the nervous tissues. *Annali di Nevrologia.* vol. 22. 1904. Ref. *Rev. Neurol. and Psy.*, vol. 2, p. 635.

FLEMMING, WALTHER.

Vom Bau der Spinalganglienzellen. *Beiträge zur Anatomie und Embryologie als Festgabe für J. Henle von seinen Schülern*, S. 12. 1882.

Ueber den Bau der Sipnalganglienzellen bei Säugetiere, und Bemerkungen über den der centralen Zellen. *Arch. für Mikrosk. Anat.*, Bd. 46, S. 379. 1895.

FLEMMING, F.

Die Structur der Spinalganglienzellen bei Säugetieren. *Arch. für Psych. u. Nervenkrankheiten*, Bd. 29, S. 969. 1897.

FREUD, S.

Ueber Spinalganglien und Rückenmark des Petromyzon. *Sitzungs-berichte der k. Akad. d. Wissench.*, Bd. 78, Abth. 3. 1878.

GAD.
Anatomie und Physiologie der Spinalganglien. *Archiv f. [Anat. und] Phys.*, p. 570. 1887.
GAD AND JOSEPH.
Ueber die Beziehungen der Nervenfasern zu den Nervenzellen in den Spinalganglien. *Archiv f. [Anat. und] Phys.*, p. 199. 1889.
GAULE AND LEWIN.
Ueber die zahlen den Nervenfasern u. Ganglienzellen des Kaninchens. *Centrabl. f. Physiol.*, H. 15 u. 16. 1896.
HARDESTY, IRVING.
Further observations on the conditions determining the number and arrangement of the fibers forming the spinal nerves of the frog (Rana virescens). *Journ. Comp. Neurol.*, vol. 10, p. 323. 1900.
On the number and relations of the ganglion cells and medullated nerve fibers in the spinal nerves of frogs of different ages. *Jour. Comp. Neurol. and Psy.*, vol. 15, p. 17. 1905.
The number and arrangement of the fibers forming the spinal nerves of the frog (Rana virescens). *Journ. Comp. Neurol.*, vol. 9, p. 64. 1899.
HATAI, SHINKISHI.
Number and size of the spinal ganglion cells and dorsal root fibers in the white rat at different ages. *Journ. Comp. Neurol.* vol. 12, p. 107. 1902.
On the increase of the medullated nerve fibers in the ventral roots of the spinal nerves of the growing white rat. *Journ. of Comp. Neurol.* vol. 13, p. 177. 1903.
The finer structure of the spinal ganglion cells in the white rat. *Journ. Comp. Neurol.*, vol. 11, p. 1. 1900.
HODGE, C. F.
Some effects of electrically stimulating ganglion cells. *Am. Journ. Psychol.*, vol. 2, p. 375. 1889.
HOLL.
Ueber den Bau der Spinalganglien. *Sitzungsb. d. k. Akad. im Wien.* Bd. 70, Abth. 2. 1875. (Cited after HATAI.)
HORTON-SMITH, R., JR.
On efferent fibers in the posterior roots of the frog. *Journ. Physiol. (Foster).* 1897.
HUBER, G. C.
The spinal ganglia of Amphibia. *Anat. Anz.*, Bd. 12, S. 417. 1896.
INGBERT, C. E.
An enumeration of the medullated nerve fibers in the ventral roots of the spinal nerves of man. *Journ. of Comp. Neurol. and Psych*, vol. 14, p. 209. 1904.
JOSEPH, MAX.
Zur Physiologie der Spinalganglien. *Arch. f. [Anat. u.]* Phys., S. 296. 1887.
KAISER, O.
Die Functionen der Ganglienzellen des Halsmarkes. Haag. 1891.
KLEIST, KARL.
Experimental-anatom. Untersuchungen über die Beziehungen der hinteren Rüchenmarkswurzelen zu den Spinalganglien. *Arch. f. path. Anat. u. Phys.*, Bd. 175, p. 381. 1904.
KOPCZYNSKI, S.
Experimental Untersuchungen aus dem Gebiete des Anatomie und Physiologie der hinteren Spinalwurzeln. *Neurol. Centralbl.*, p. 297. 1906.
KÖSTER, G.
Ueber die verschiedene biologische Werthigkeit der hinteren Wurzel und des sensiblen peripheren Nerven. *Neurol. Centralbl.*, vol. 22, p. 1093. 1903.
Zur Physiologie der Spinalganglien u. der tropischen Nerven sowie zur Pathogenese der Tabes dorsalis. *Leipzig.* 1904.
LENHOSSÉK, M.
Beobachtung an den Spinalganglien und dem Rückenmark von Pristiurusembryonen. *Anat. Anz.*, Bd. 7, p. 528. 1892.
Centrosom und Sphäre in den Spinal Ganglienzellen des Frosches. *Arch. f. Mikr. Anat. und Entwicklungsgeschichte*, Bd. 46. 1895.
Der feinere Bau des Nervensystems. *Berlin.* 1895.
Ueber den Bau der Spinalganglienzellen des Menschen. *Arch. f. Psych. u. Nervenkrankheiten.* Bd. 29, S. 345. 1897.

v. LENHOSSÉK.

Ueber Nervenfasern in den hinteren Würzeln welche aus den Vorderhorn entspringen. *Anat. Anz.*, Bd. 5, p. 613. 1890.

Untersuchungen über die Spinalganglien des Frosches. *Archiv f. mikroskop. Anat.*, Bd. 26, p. 370. 1886.

LUGARO, E.

Sulla patologia della cellule dei gangli sensitivi. *Rivista Patol. nerv. e ment.*, vol. 5, nos. 4, 6, 9; vol. 6, no. 10; vol. 7, no. 3; vol. 8, no. 11.

LEVI, G,

Beitrag zur Kenntnis der Structur des Spinalgangliens. *Anat. Auzeiger*, Vol. 27, p. 158. 1905.

LEWIN AND GAULE.

Ueber die Zahlen der Nervenfasern und Ganglienzellen in den Spinalganglien des Kaninchens. *Centrabl. f. Physiologie*, Bd. 10, p. 437. 1897.

MARINESCO.

Essai de localisations dans les ganglions spinaux. *Revue Neurol.*, vol. 16. 1904.

NISSL, F.

Die Neuronenlehre und ihre Anhanger, p. 282, 333. *Jena.* 1903.

RANSON, S. W.

Retrograde degeneration in the spinal nerves. *Journ. of Comp. Neur. and Psy.*, vol. 16. 1906.

RAWITZ, B.

Ueber den Bau der Spinalganglien. *Archiv f. mikr. Anatomie*, Bd. 18, S. 283. 1880.

RETZIUS, G.

Weiteres zur Frage von den freiers Nervenendigungen und anderen Structurverhältnissen in den Spinalganglien. *Biol. Untersuchungen von Prof. Dr. G. Retzius. Jena.* 1900.

SHERRINGTON, C. S.

Regeneration of the afferent root fibers. *Schafer's Text-book of Physiology*, vol. 2, p. 804. 1900.

SINGER AND MÜNZER.

Beiträge zur Anatomie des Centralnervensystems. *Denkschr. der Kais Akad, d. Wiss. zu Wien*, Bd. 57. 1895. (Cited after VAN GEHUCHTEN.)

STEINACH, E.

Ueber die motorische Innervation des Darmtractus durch die hinteren Spinalnervenwurzeln. *Lotos*, N. F. Bd. 14. 1893.

VAN GEHUCHTEN, A.

Les éléments moteurs des racines postérieures. *Anat. Anz.*, Bd. 8, p. 215. 1893.

Nouvelles recherches sur les ganglions cérébro-spinaux. *La Cellule*, T. 8, p. 255. 1892.

WALLER, A.

Sur la reproduction des nerfs et sur la structure et les fonctions des ganglion spinaux. *Müller's Archiv*, p. 392. 1852.

WARRINGTON, W. B.

Further observations on the structural alterations in the cells of the spinal cord following various nerve lesions. *Journ. of Physiol.*, vol. 25, p. 462. 1900.

WARRINGTON, W. B., and GRIFFITH, F.

On the cells of the spinal ganglia and on the relationship of their histological structure to the axonal distribution. *Brain*, vol. 28, p. 297. 1904.

WEIGNER, K.

Ueber den Verlauf des Nervus intermedius. *Anat. Hefte*. Bd. 29, p. 97. 1905.

THE NERVOUS SYSTEM OF THE AMERICAN LEOPARD FROG, RANA PIPIENS, COMPARED WITH THAT OF THE EUROPEAN FROGS, RANA ESCULENTA AND RANA TEMPORARIA (FUSCA).

BY

HENRY H. DONALDSON.

(*Professor of Neurology at the Wistar Institute.*)

(*From the Wistar Institute of Anatomy and Biology, Philadelphia.*)

WITH SIX FIGURES.

With the advances which are being made in the correlation of function and structure, the need is felt on many sides for a more detailed, accurate and quantitative determination of the anatomical, physiological and chemical differences between closely related species, as well as between the same species from different localities.

The general notion of a physiological and chemical criterion for species has been discussed by DE VARIGNY ('99) and, although this is not the place to review the literature touching this topic, it is nevertheless appropriate to name CAMERANO's paper ('00) on the variation of the toad, which, among his many important contributions in this general field, is the one most closely related to the following investigation. Moreover, KELLICOTT's recent study of correlation and variation in internal and external characters in the common toad ('07) emphasizes relations which have a direct bearing on the interpretation of my own results.

In 1898 I made a study of the weight of the brain and of the spinal cord in the bull-frog R. catesbiana (DONALDSON '98). Two years later, in collaboration with Dr. D. M. SCHOEMAKER, a similar series of observations on the leopard frog, R. pipiens,[1] was published (DONALDSON and SCHOEMAKER '00).

In 1902, utilizing the data in both of these investigations, I was able to show that the weight of the central nervous system in both of these species could be calculated by a formula based on the body weight and on the total length of the frog (DONALDSON '02).

For comparison with these results the observations of FUBINI ('81) on the European frogs were alone available.

An examination of FUBINI's tables, which are discussed in part in my paper of 1898, referred to above, showed that his findings were so irregular and so different from my own, that it was fair to conclude that he had not worked with sufficient care.

In order to test this conclusion, I obtained in the spring of 1898, through the courtesy of the Zoölogical Institute at Zurich, Switzerland, a series both of R. esculenta and R. temporaria (fusca), all the specimens having been fixed in formalin by a uniform method. A comparison of these specimens with the fresh R. pipiens on the one hand, and on the other with R. pipiens fixed by the same method, indicated that the central nervous system in R. pipiens was heavier than in the European species, and at the same time did not support any of the peculiar findings of FUBINI, such as the relatively great weight of the spinal cord. Nevertheless, limitations in the range in size of the Zurich series and the possibility that the European and American species were differently affected by the fixation treatment, led me to delay publication on this point until fresh material could be examined. The opportunity to do this came in the summer of 1904. In July of that year, through the courtesy of Professor SHERRINGTON, I was able to examine a series of R. temporaria (fusca) in the physiological laboratory of University College at Liverpool, England; and in August, through the courtesy of Professor GAULE, a corresponding series of R. esculenta was examined in the Physiological Institute of the University at Zurich, Switzerland.

In order to eliminate as far as possible, the influence of *season* on this comparison, Dr. HATAI examined for me, also in August, a series of R. pipiens in the Neurological laboratory of the Uni-

[1] In previous papers on the leopard frog, published from my laboratory, this species has been designated as Rana virescens brachycephala, Cope. It now appears that this name is not correct, and that the species in question should be designated Rana pipiens (Schreber) as given above (DONALDSON, *Science*, vol. 26, p. 655, 1907.)

versity of Chicago. It is the results of these three series of observations which are now to be compared.

As the foregoing shows, this investigation was undertaken primarily to test the correctness of FUBINI's observations. It has resulted however in bringing to light several differences between the nervous systems of the species compared, and these differences seem worth recording. At the same time, FUBINI's observations have been found untrustworthy. This, however, is a matter of small importance, and the brief discussion of FUBINI's work will be deferred to an appendix.

Before presenting the data on the nervous system, it will be desirable to record some of the characters in which these three species of frogs closely resemble one another. The resemblances important for our present purpose are enumerated below:

(1) In external appearance and shape; color markings excepted.

(2) In the range in body weight (the heaviest specimens are always females).

(3) In the ratio obtained by dividing the body weight by the total length, that is, the average amount of body weight for each running millimeter of total length.

It will be necessary to interrupt the enumeration for a moment in order to elaborate this point (3).

In the full tables are given the body weight and the total length of each individual examined. In the condensed Table 10, these same data are arranged to give the averages for *groups of three.* Thus for each species in the latter table there are four entries, and in each entry both the average body weight and the average total length are given. If the former be divided by the latter

$$\frac{\text{Body weight}}{\text{Total length}}$$

we obtain a number which represents the average amount of weight for each running millimeter. Since the increase in body weight is more rapid than the increase in total length, this value of course changes with the absolute weight of the frog, increasing as the frog becomes absolutely heavier. The values thus obtained are given in Table 1.

These values are better understood when thrown into a curve, as in Chart 1.

It appears from the chart that the curves are nearly parallel,

TABLE 1.

	Body Weight in Grams.	Body Weight per Millimeter in Grams.
R. pipiens	14.9	.102
	23.2	.135
	30.8	.168
	43.2	.218
R. esculenta	15.9	.114
	22.0	.134
	35.0	.199
	40.2	.208
R. temporaria	15.9	.107
	23.1	.137
	28.0	.162
	31.3	.177

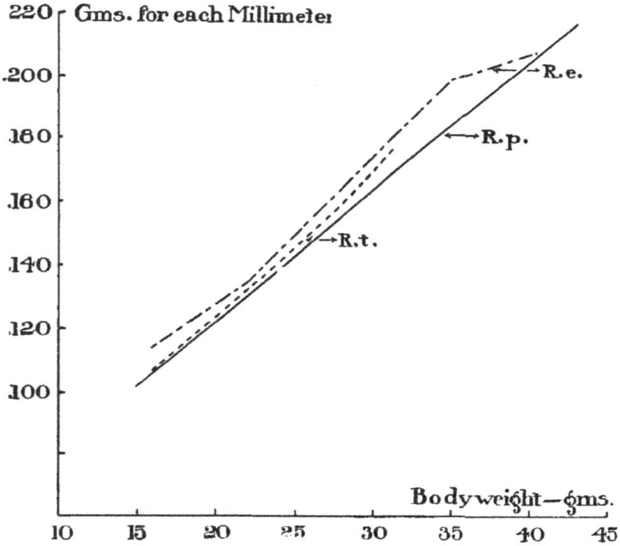

CHART 1.

Showing the average amount of body weight for each running millimeter of total length.
R. p. = Rana pipiens.
R. e. = Rana esculenta.
R. t. = Rana temporaria.

and that on the average, the lower figures differ only 4.1 per cent (R. pipiens) and 2.3 per cent (R. temporaria) respectively, from the highest figures, given by R. esculenta. The relation of body weight to total length is therefore nearly the same in all three species.

(4) In the fraction of the total length represented by the combined lengths of the leg bones.

Table 2 gives these figures in their final form.

TABLE 2.

Percentage of the total length represented by the combined lengths of the leg bones.

	No. of Specimens.	
R. pipiens	12	68.7%
R. esculenta	5	70.7%
R. temporaria	6	69.4%

The percentages in the foregoing tables were obtained as follows: That for R. pipiens from an average of twelve records on individuals ranging from 14.85 to 42.54 grams in body weight (DONALDSON and SCHOEMAKER, '00, Table VII); that for R. esculenta from five specimens of the Zurich series of 1898, having a body weight of 12.3–20.4 grams; and that for R. temporaria (fusca) from six specimens of the same series ranging in body weight from 17.9–34.7 grams.

Table 2 serves to show that in this character the three species are nearly alike.

(5) In the proportional lengths of the several leg bones.

TABLE 3.

	No. of Specimens.	Femur.	Tibia.	Foot (Tarsus and Pes).
R. pipiens	12	25.5%	29.3%	45.2%
R. esculenta	5	26.3%	28.2%	45.5%
R. temporaria	6	26.1%	28.7%	45.2%

The figures in Table 3 are based on the same data as were used for Table 2. For comparison in the case of R. esculenta however, we have in addition, the measurements from BOULENGER ('97). These are taken both from his tables and from measurements made on the bones as represented in his plates.

The data from BOULENGER give the following:

TABLE 4.

R. ESCULENTA.	No. OF SPECIMENS.	FEMUR.	TIBIA.	FOOT (TARSUS AND PES).
Variety rabibunda	1 M.+1 F.	26.7%	29.2%	44.1%
Variety rabibunda	1 F.*	27.7%	28.9%	43.4%
Variety typica	1 M.+1 F.	26.8%	27.3%	45.9%
Variety lessonæ	1 M.+1 F.	25.9%	26.5%	47.6%

* Measured from Boulenger's Fig. 101, p. 280.

In my Zurich series the individual measurements correspond to those for the varieties *rabibunda* and *typica* as determined by BOULENGER. The average for these from the above table (4) is:

AVERAGE VALUES FOR VARIETIES RABIBUNDA AND TYPICA.

Femur	27.0
Tibia	28.3
Foot	44.7

And these values are close to those given for R. esculenta in Table 3.

For comparison in the case of R. temporaria, an average of two determinations, one male, one female, by BOULENGER is available. These give

Femur	25.6%
Tibia	28.4%
Foot	46.0%

which is in fair agreement with the values given in Table 3.

(6) In the relative length of the entire central nervous system (that is, the length of the brain plus the length of spinal cord), in relation to the total length of the frog.

This relation is of course not a constant one, because the total length of the frog increases more rapidly than the length of the entire nervous system. To make the comparisons, therefore, the percentages must be recorded in relation to the total length found for each individual or group. The data used in this determination were the following:

Ten (10) specimens of R. pipiens ranging in total length from 124 mm. to 185 mm. inclusive. The data being taken from the research of DONALDSON and SCHOEMAKER ('00). These cases are averaged in groups of five.

Twelve (12) specimens of R. pipiens, these being the same as are given in Table 7 and averaged in groups of four (observations by Dr. HATAI). Both of the foregoing series of measurements were made on the fresh specimens.

Ten specimens of R. pipiens, after fixation in formalin, averaged in groups of five, and ranging in total length from 128–174 mm. This was the series used to control the measurements on the European frogs received from Zurich in 1898.

From the Zurich series of 1898, there were taken one group of five esculenta, ranging in total length, after fixation, from 128–170 mm., and also a series of fifteen R. temporaria, averaged in groups of five, and ranging in total length, after fixation, from 151 to 179 mm.

It seems fairest to tabulate the fresh, separately from the fixed material, so the first lot, entirely R. pipiens, is given in Table 5.

TABLE 5.

R. pipiens. Percentage values of length of the entire central nervous system, the total length of the frog being taken as the standard. Measurements on fresh specimens.

AVERAGES OF	TOTAL LENGTH MM.	PER CENT VALUE OF THE LENGTH OF ENTIRE CENTRAL NERVOUS SYSTEM.
4	150	17.2
5	153	17.8
5	175	17.2
4	177	16.2
4	196	16.3

TABLE 6.

Showing the same relation as Table 5. All measurements on material fixed in formalin.

SPECIES.	AVERAGE OF	TOTAL LENGTH MM.	PER CENT VALUE OF THE LENGTH OF THE ENTIRE CENTRAL NERVOUS SYSTEM.
R. pipiens	5	133	18.2
R. esculenta	5	145	17.9
R. temporaria	5	158	16.9
R. temporaria	5	166	16.2
R. pipiens	5	167	16.4
R. temporaria	5	173	16.0

Table 5 shows that the value in question ranges in the fresh specimens from 17.8 per cent to 16.2 per cent, and also tends to diminish as the total length of the frog increases. The same relations are shown in Table 6, in which all three species are represented, and these form as satisfactory a series as is given in Table 5. We therefore conclude that in this character—the relative length of the entire central nervous system—the three species resemble one another closely.

It should be pointed out here that it follows from this that the smaller weight of nervous system which we find in the European forms (see below) must be associated with a diminution of one or both the transverse diameters, since the foregoing shows that it is not associated with variations in total length.

(9) In the arrangement of the main branches of the crural and sciatic nerves (DUNN '00 and '02). In the papers to which reference is here made, this point is fully discussed.

In view of the fact that the several species are similar in the foregoing characters, we might expect a high degree of similarity in the weight and structural relations of the central nervous system. Such however is not the case, and we turn therefore to a statement of the differences which have been observed.

The technique of weighing, measuring and dissecting, was uniform for the three species. This has already been described (DONALDSON, '98, DONALDSON and SCHOEMAKER, '00).

It may, however, be well to repeat here that the body weight was taken in a closed box; the weight of the contained ova being deducted from the body weight of the unopened specimen, in the case of the females. Also in both sexes correction was made for the stomach contents.

In taking the total length, the frog was suspended by the lower jaw, and the distance between the tip of the nose and the longest toe, the legs being fully extended, was measured with vernier calipers. The central nervous system was removed immediately after death, and the brain separated from the spinal cord by a section at the level of the tip of the calamus scriptorius. Both brain and cord were separated from their nerves by severing the latter at the points of their attachments to the central structures. To obtain the percentage of water, the material was dried for several days until it maintained a constant weight. For this a water bath ranging from 85° to 95° C. was used.

Although this is probably not the best method, it was uniformly applied in the case of all three series, so that the results are at least comparable, though the absolute values for the percentage of water may be open to question, until it has been shown that this material dried in vacuo, gives similar results.

Material examined.—The specimens of R. pipiens were 12 in number (10 males and 2 females) ranging in body weight from 11.6–47 grams. They were taken in the neighborhood of Chicago in the month of August, and examined between the twenty-third and thirty-first of August. For the data which are presented in Table 7 I am indebted to Dr. S. HATAI.

TABLE 7.

Data on R. pipiens.

BODY WEIGHT IN GRMS.	SEX.	TOTAL LENGTH IN MM.	WEIGHT IN GRAMS OF			RATIO OF BRAIN WEIGHT TO CORD WEIGHT.	PERCENTAGE OF WATER.	
			C. N. S	BRAIN.	SP. C.		BRAIN.	SP. C.
11.6	M.	130	.0918	.0666	.0252	2.64	84.4	79.4
16.0	M.	150	.1148	.0796	.0352	2.26	85.2	80.7
17.0	F.	159	.1054	.0714	.0340	2.10	84.0	80.6
20.8	M.	170	.1232	.0844	.0388	2.17	85.2	81.6
22.5	M.	162	.1165	.0807	.0358	2.25	84.5	80.4
26.4	M.	180	.1372	.0946	.0426	2.22	84.4	78.4
27.6	F.	179	.1416	.1014	.0402	2.52	84.8	80.1
30.6	M.	180	.1454	.0998	.0456	2.18	84.6	79.8
34.2	M.	190	.1518	.1056	.0462	2.28	85.6	81.6
41.8	M.	197	.1652	.1146	.0506	2.26	86.9	82.2
43.9	M.	200	.1708	.1210	.0498	2.42	85.8	80.7
47.0	M.	198	.1664	.1140	.0524	2.17	84.4	80.5

The specimens of R. esculenta were eleven in number (3 males and (8 females) ranging in body weight from 12.4–45.03 grams. They were taken near Zurich on July 31, and were examined between August 1 and 5. The data are given in Table 8.

The specimens of Rana temporaria (fusca) were twelve in number (8 males and 4 females) ranging in body weights from 14.05–32.81 grams. They were taken near Liverpool shortly before July 11, and were examined July 11 and 12. The data are given in Table 9.

In all the foregoing series there is considerable individual variation in the characters observed, and so for the purposes of comparison, the complete tables have been condensed by taking the averages for each three successive individuals, thus giving four entries in each of the condensed tables.

The only exception to this statement is in the case of R. esculenta, with but 11 records, and there the third entry in the con-

TABLE 8.
Data on R. esculenta.

Body Weight in grms.	Sex	Total Length in mm.	Weight in Grams of			Ratio of Brain Weight to Cord Weight.	Percentage of Water	
			C. N. S.	Brain.	Sp. C.		Brain.	Sp. C.
12.40	F.	131	.0818	.0577	.0241	2.39	84.2	78.4
16.75	F.	144	.0926	.0634	.0292	2.17	83.4	79.1
18.43	F.	144	.0928	.0650	.0278	2.34	83.2	78.2
20.00	F.	161	.1103	.0756	.0347	2.17	82.5	79.2
22.00	F.	164	.1107	.0769	.0338	2.27	84.0	79.0
24.10	M.	167	.1217	.0841	.0376	2.23	83.4	78.4
33.85	M.	175	.1327	.0895	.0432	2.07	83.2	78.2
36.30	M.	177	.1478	.1004	.0474	2.11	83.4	78.6
37.56	F.	188	.1490	.0993	.0497	1.99	82.9	78.8
37.96	F.	194	.1427	.0953	.0474	2.01	82.8	77.8
45.03	F.	196	.1578	.1078	.0500	2.15	83.9	78.4

TABLE 9.
Data on R. temporaria.

Body Weight in grms.	Sex	Total Length in mm.	Weight in Grams of			Ratio of Brain Weight to Cord Weight.	Percentage of Water.	
			C. N. S.	Brain.	Sp. C.		Brain.	Sp. C.
14.05	F.	144	.0881	.0596	.0285	2,09	82.3	78.2
16.10	F.	151	.0991	.0690	.0301	2.22	82.7	79.0
17.65	M.	154	.0916	.0618	.0298	2.07	83.0	78.5
21.75	M.	171	.1045	.0671	.0374	1.79	82.8	78.2
23.45	M.	162	.0947	.0628	.0319	1.96	82.1	77.0
24.17	F.	173	.1333	.0864	.0469	1.84	81.9	76.5
27.05	M.	173	.1298	.0874	.0424	2.06	82.4	77.1
28.15	M.	168	.1018	.0687	.0331	2.07	82.5	76.7
28.95	M.	174	.1324	.0813	.0511	1.59	81.3	76.8
28.95	M.	178	.1485	.0928	.0557	1.66	81.3	76.8
32.15	M.	173	.1321	.0890	.0431	2.06	80.9	78.6
32.81	F.	178	.1161	.0766	.0396	1.93	82.7	78.0

densed tables is based on only two individuals, numbers 7 and 8 in the series. The condensed tables are given in connection with each character discussed. That for the weight of the entire central nervous system follows:

TABLE 10.

Weight of the central nervous system in grams, averages from groups of three.

	BODY WEIGHT	WEIGHT OF CENTRAL NERVOUS SYSTEM.	TOTAL LENGTHS IN MM.
R. pipiens	14.9	.1040	146
	23.2	.1256	171
	30.8	.1463	183
	43.2	.1674	198
R. esculenta	15.9	.0890	139
	22.6	.1142	164
	35.0	.1402	176
	40.2	.1498	193
R. temporaria	15.9	.0929	149
	23.1	.1108	167
	28.0	.1213	173
	31.3	.1323	176

The data in the foregoing tables are also presented in Chart 2 and the relations are more easily discussed by reference to the chart, which shows R. pipiens to have the heaviest central nervous system, R. esculenta the next heaviest, and R. temporaria the lightest.

By measuring the differences between the several entries for the two European species, and the corresponding points on the curve for R. pipiens, which is taken as the standard, it appears that on the average, the central nervous system of R. esculenta weighs about 89 per cent, and that of R. temporaria 88 per cent, of that found in R. pipiens.

It is known that dry air and starvation (DONALDSON '98, DONALDSON and SCHOEMAKER '00) tend to reduce the weight of the living frog, and probably of the central nervous system, also that the weight of the latter is increased in frogs which are moribund, and have consequently taken up an excessive amount of water.

There is no reason to think however that the foregoing observations are seriously modified by any of these influences.

Moreover unpublished observations on R. pipiens, in my possession, indicate a variation in the weight of the central nervous system with season. Nevertheless from the middle of June to the middle of September, such variations as occur, are hardly significant, and the observations here used were made within fifty-one days (July 11 to August 31) and fall within the general

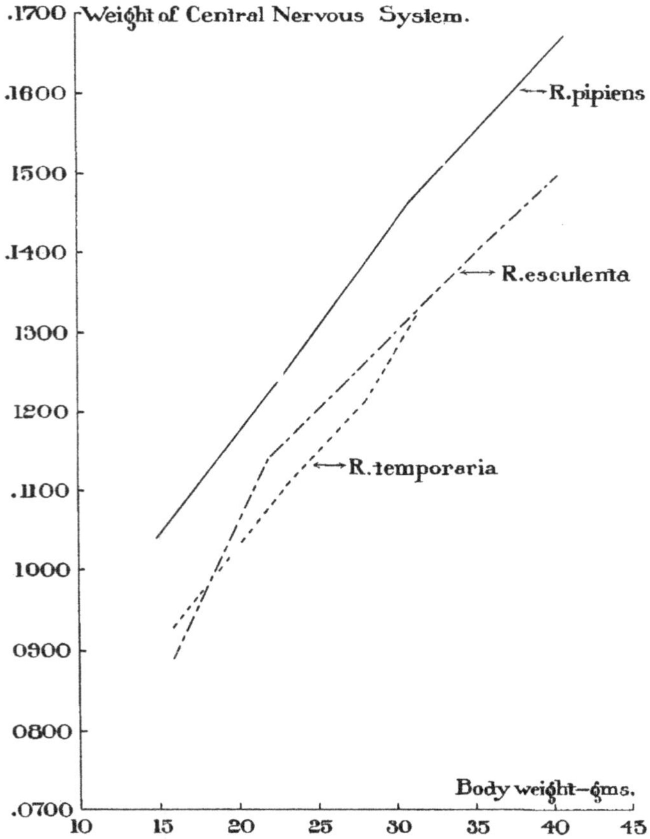

CHART 2.

Showing the weight of the entire central nervous system.

seasonal limits for constancy, as given above. This summer period is moreover the one during which the central nervous system shows its greatest weight. At the same time it is worthy of note that the results obtained on R. pipiens in August, 1904, correspond with the *lowest* weights found during August, 1901, in the case of the unpublished series.

In one sense this is perhaps fortunate, because it shows that the values here reported for R. pipiens are minimal, and if those for the European forms are also minimal, then the differences are approximately normal. If, on the other hand, the values for the European species are higher than the minimal, then the differences here given are less than they should be. In any case, and this is the main point, it follows that the differences here given are not exaggerated. I conclude therefore that the European species have a central nervous system which weighs from 11 per cent to 12 per cent less than that of R. pipiens.

As the Chart 2 shows, the curves for the weight of the central nervous system run nearly parallel, and as in a previous study (DONALDSON '02) R. pipiens has been found to conform to the formula for the determination of this weight, which is based on the body weight and total length of the frog, it follows that the European species would also conform to this same formula.

The formula contains a constant, C., which is different for each species, and which is modified by the general condition of any series. In the series of R. pipiens of 1902, the value of the constant C. was 28. In the present series of R. pipiens which, as has been noted, yields a low weight for the central nervous system, the value of C. is 26, and we should anticipate that it would be less for the two European species.

On making the calculations, I find the following values for C.

R. pipiens	C= 26
R. esculenta	C= 24
R. temporaria	C= 23

Our expectation then that the formula for the European species would have smaller values of C. is shown to be warranted.

On separating the weight of the brain from that of the spinal cord, and recording them separately, we have the relations given in Table 11.

Presenting these results in a form of a chart (Chart 3) it is seen that the brain weights for the several species follow the same order as that of the weight of the entire central nervous system, the superiority of R. pipiens being even more marked. The weights for the spinal cords however run much closer together.

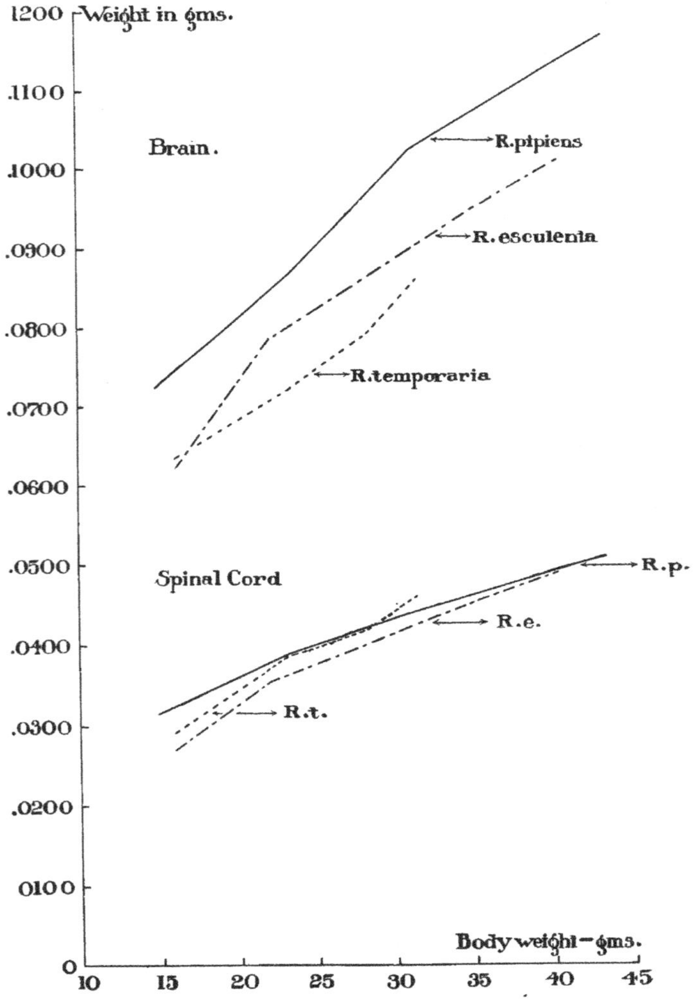

CHART 3.

Showing the weight of the brain and of the spinal cord.

R. pipiens has still the heaviest cord, but R. temporaria, with the lightest entire central nervous system, has a spinal cord nearly as heavy as that of R. pipiens, while the cord in Rana esculenta is distinctly lighter than in the other two species, having on the average 94 per cent of the weight of the cord in R. pipiens.

TABLE 11.
Weight of the brain and spinal cord in grams. Averages from groups of three.

	Body Weight.	Brain.	Spinal Cord.
R. pipiens	14.9	.0725	.0315
	23.2	.0866	.0390
	30.8	.1023	.0440
	43.2	.1165	.0509
R. esculenta	15.9	.0620	.0270
	22.0	.0788	.0354
	35.0	.0949	.0453
	40.2	.1008	.0490
R. temporaria	15.9	.0635	.0294
	23.1	.0721	.0387
	28.0	.0791	.0422
	31.3	.0862	.0461

To show the relative weight of the brain as compared with that of the cord in these three species, we may use the ratio obtained by dividing the brain weight by the cord weight. These ratios are given in the following table:

TABLE 12.
Ratios of the weight of the brain to the weight of the spinal cord. Averages from groups of three.

	Body Weight.	Ratio.
R. pipiens	14.9	2.33
	23.2	2.22
	30.8	2.32
	43.2	2.28
R. esculenta	15.9	2.29
	22.0	2.22
	35.0	2.09
	40.2	2.05
R. temporaria	15.9	2.15
	23.1	1.86
	28.0	1.87
	31.3	1.87

Putting these data in the form of curves in Chart 4, we see that the relative brain weights follow the order of the absolute weights of the entire nervous system and of the brain, the highest ratios being given by R. pipiens.

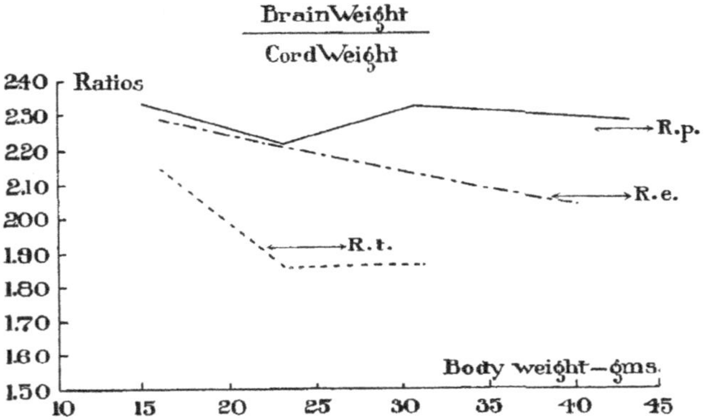

CHART 4.

Showing the ratios of brain weight to spinal cord weight.
R. p. = Rana pipiens.
R. e. = Rana esculenta.
R. t. = Rana temporaria.

TABLE 13.

Showing the percentage of water in the brain and in the spinal cord. Averages from groups of three.

	Body Weight.	Percentage of Water in	
		Brain.	Spinal Cord
R. pipiens	14.9	84.5	80.2
	23.2	84.7	80.1
	30.8	85.0	80.5
	43.2	85.7	81.2
R. esculenta	15.9	83.6	78.6
	22.0	83.3	78.9
	35.0	83.3	78.4
	40.2	83.2	78.3
R. temporaria	15.9	82.7	78.6
	23.1	82.3	77.2
	28.0	82.1	76.9
	31.3	81.6	77.8

In addition to the several weights, a determination of the percentage of water was made in the case of both the brain and spinal cord. The method has been described already on p. 128.

The condensed results are given in Table 13.

On putting the data in Table 13 in the form of curves (Chart 5) it becomes evident at a glance that the percentages found in the three species are different, and also that they follow the order of the weight of the entire central nervous system and of the brain.

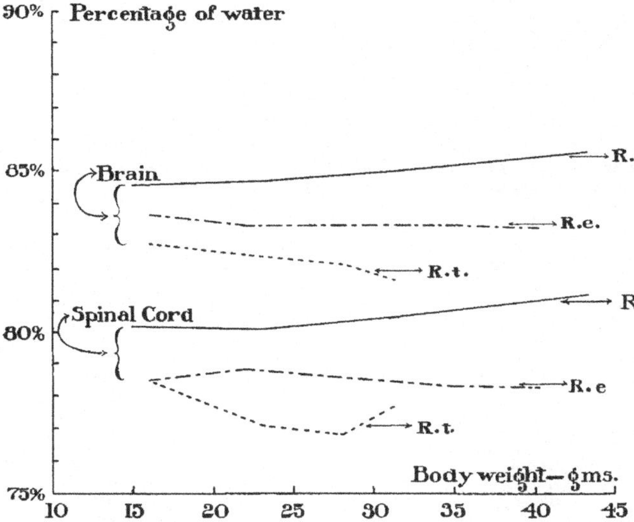

CHART 5.

Showing the percentage of water in the brain and in the spinal cord.

R. p. = Rana pipiens.

R. e. = Rana esculenta.

R. t. = Rana temporaria.

These differences in a character in which we might expect a high degree of similarity, call for some comment touching the trustworthiness of the results.

First, as to R. pipiens; the percentages in this species are the highest. The best evidence for their general correctness is furnished by the following table, extracted from the unpublished observations (1901) previously mentioned. At each date given in Table 14 the percentage of water was determined in a group of eight frogs (4 males and 4 females) and seventeen such records are here reported. The body weight of the frogs in each group was the same, that is, 25 grams, approximately.

The range in the entire series here given is for the brain from 84.2 per cent to 86.2 per cent, and for the spinal cord from 78.8 per cent to 81.6 per cent, while the average of the four entries for the month of August, is for the brain 85.1 per cent and for the spinal cord 80.5 per cent. These are very close to the values for the second and third groups of R. pipiens (with a body weight near 25 grams) as given in Table 13.

TABLE 14.

Percentages of water in the brain and in the spinal cord of R. pipiens. Average body weight is 25 grams. Each entry is based on a determination for eight specimens (4 males and 4 females). From an unpublished study on the influence of season made in 1901.

Date.		Percentage of Brain.	Water in Cord.
June	5	86.2	81.2
	12	85.8	79.5
	19	85.4	81.3
	26	85.8	81.2
July	1	85.7	81.2
	8	84.3	79.8
	15	84.4	79.8
	22	85.1	80.9
	29	85.2	81.2
Aug.	5	84.6	79.6
	12	85.2	80.6
	19	85.1	81.6
	26	85.3	80.0
Sept.	2	84.8	80.3
	9	84.2	78.8
	16	85.6	81.2
	23	84.4	80.1

It seems probable from this comparison that we have obtained a generally correct value for the percentage of water in R. pipiens. In the case of R. temporaria, the specimens were dried at Liverpool, but not weighed until I reached Zurich. There they were further dried for 24 hours in the oven that was also used for drying the R. esculenta material, and then were weighed. They were found to give (see Table 13) the smallest percentages of water. This naturally raised the question as to whether they had been completely dried. The evidence that the drying was complete is only indirect. It is as follows:

The brains and spinal cords of R. esculenta dried and weighed at Zurich, were left in the original weighing bottles from the summer of 1904 to the spring of 1907, when, after careful redrying, they were weighed at the Wistar Institute in Philadelphia.

The last series of weighings made at Philadelphia, differed from those made in Zurich in 1904, by an average of plus 0.1 per cent. The fact that there was a trifling *gain*, is probably to be credited to the different balances used. But whatever the explanation of this gain may be, it seems to show that the drying in Zurich was complete, and thus to warrant the use of the values for R. esculenta and R. temporaria as entered in Table 13.

Assuming that in any given locality, the humidity of the atmosphere might be a factor influencing the amount of water in the body of a frog, I made an examination of the humidity records from July 1 to September 1, 1904, taken by the weather bureaus at Liverpool, Zurich and Chicago. For the data with which to do this, I am indebted to the officials of the U. S. Weather Bureau, whose courtesy I desire to acknowledge with thanks.

The matter is far too complex to permit us to make here more than the most general statements, but I feel justified in stating that the humidity conditions at Liverpool in July, 1904, and at Zurich and Chicago in August, 1904, were not unusual. Further, that broadly-speaking, the humidity is greatest at Liverpool, intermediate at Zurich, and least at Chicago. It is to be noted that the percentage of water in the several species follows the inverse order, being most in the Chicago specimens, where the humidity is lowest, and least at Liverpool, where it is greatest; a suggestive result which invites further inquiry.

Two more comments are however desirable before leaving this general topic.

From previous studies, we should expect that the percentage of water in the brain and in the spinal cord would diminish with increasing age, for the measure of which we here take the body weight.

This decrease is clear and regular for the brain of R. esculenta and R. temporaria, is indicated though less regular, in the case of the spinal cords of these two species, but in R. pipiens is regularly reversed in the case of the brain, and irregularly reversed in the case of the spinal cord. This makes it highly probable that some disturbing influence has modified the percentage of water in the brain and cord of R. pipiens, so as to mask the effect of age (size), but it is to be added that the disturbance thus produced is relatively small, and not sufficient to affect the distinctive differences between R. pipiens and the species here compared with it.

If the average values for the percentage of water in the brain and spinal cord of the three species are calculated from Table 13 we obtain the following:

TABLE 15.

Average values for the percentage of water in the brain and spinal cord of all three species, together with the difference between that for the brain and for the spinal cord in each species, and the relative amount of water in the spinal cord, that in the brain being taken as a standard.

	Percentage of Water in		Difference.	Percentage Value of Cord Determination.
	Brain.	Sp. Cord		
R. pipiens	84.97	80.50	4.47	94.7%
R. esculenta	83.35	78.55	4.80	94.2%
R. temporaria	82.17	77.62	4.55	94.4%

It appears from this table that the absolute differences in the percentage values for the brain and cord are similar in the three species, and that the determinations for the brain being taken as the standards the relative values of the determinations for the spinal cord are about alike, ranging from 94.2 per cent R. esculenta, to 94.7 per cent R. pipiens. The similarity in these relations speaks for the correctness of the general results.

In this connection it is natural to enquire how the weight relations of the central nervous system or its parts, might be affected if the percentages of water in R. esculenta and R. temporaria were raised to that found in R. pipiens. Calculations have been made, and the results show that the superiority of the entire central nervous system and of the brain in Rana pipiens would be diminished only slightly. On the other hand, the weights of all the spinal cords would be brought together, and R. temporaria given the heaviest cord.

Moreover, in general, the weight values in the two European species would be brought closer to one another.

These alterations would however not essentially modify any of the differences on which we have had occasion to lay emphasis.

For the foregoing comparison of the central nervous system and its parts, together with the determination of the percentage of water, data on all three species have been available. But before commenting on the results just given, I wish to present some observations based on the comparison of two species only.

These additional observations are on the peripheral nervous system and relate first, to the number of medullated fibers in the spinal nerve roots; comparing R. esculenta with R. pipiens (there being no corresponding observations on R. temporaria). Second, to the length of the internodal segments; comparing R. temporaria with R. pipiens (there being no corresponding observations on R. esculenta).

The number of medullated nerve fibers in the spinal nerve roots of R. pipiens compared with the number in R. esculenta.—In a female R. pipiens weighing 48.2 grams, HARDESTY ('99) reported 14,582[2] medullated nerve fibers in both roots of the ten spinal nerves of one side. This was a much larger number than had been found by BIRGE ('82) in a specimen of R. esculenta of greater body weight. To reduce BIRGE's figures for the specimen of R. esculenta, weighing 63 grams, to those for a specimen weighing only 48.2 grams, we have proceeded as follows:

The smallest frog in BIRGE's series, with a body weight of 1.5 grams, in which he enumerated 2992 motor fibers in the ventral spinal roots of one side, was selected as one limit, and to this frog the same proportion of sensory fibers as was found in the 63 gram specimen, was allotted, a concession which probably makes the number of sensory fibers somewhat too large.

The number of fibers corresponding to each gram of body weight between 1.5 grams and 63 grams was then determined. By this method, it was found that when the number of fibers in the spinal nerves of the 63 gram frog was reduced to the number for a 48.2 gram frog, it amounted to 92.8 per cent of that found in the 63 gram frog, or 8925 fibers. Thus the difference between the two species is (14,582 – 8925) 5657 fibers, or put in another way, R. esculenta has only about 61 per cent as many medullated nerve fibers in the spinal nerve roots as has R. pipiens. On reducing the original observations of BIRGE for the number of fibers in the

[2] By a clerical error the number was printed on p. 78 (HARDESTY '99) as 14,783. It should be 14,582, and consequently I shall use the corrected number subsequently, even when referring to HARDESTY's paper.

several nerve roots to 92.8 per cent of their value, we obtained the following table.

TABLE 16.

Number of medullated nerve fibers in the dorsal and ventral roots of the spinal nerves of one side.

Number of Nerve.	R. pipiens (Hardesty) 48.2 Grams.		R. esculenta (Birge) Original Weight 63 Grams Reduced to 48.2 Grams.	
	Dorsal.	Ventral.	Dorsal.	Ventral.
II	132	1045	115	727
III	2496	1478	1530	905
IV	329	379	245	446
V	371	163	179	98
VI	299	127	208	106
VII	350	251	171	148
VIII	1108	377	521	132
IX	2108	1295	1022	807
X	1171	721	921	409
XI	61	321	38	197
Totals	8425	6157	4950	3975
Sums	14582		8925	
Ratios	1.36—1		1.24—1	

When the data in Table 16 are thrown into the form of curves, we have the relations exhibited in Chart 6.

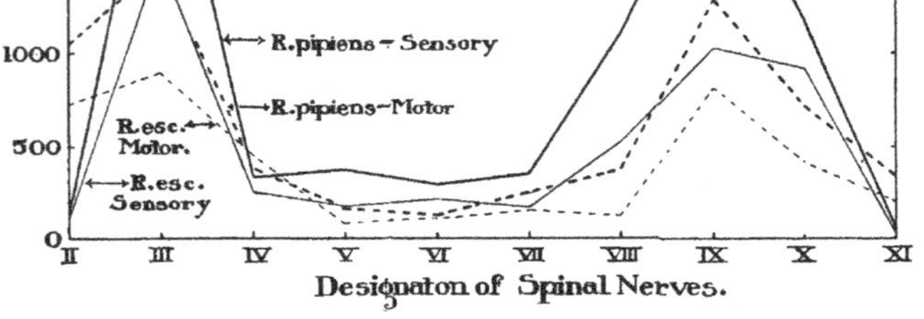

CHART 6.

Showing the numbers of medullated fibers in the spinal nerves of Rana pipiens and Rana esculenta.

From a study of Chart 6 we see that the form of the curves for the two species is very similar, even in details, although there are two evident differences.

In the first place, R. pipiens has regularly more fibers in each instance except in the ventral root of the IV spinal nerve, in which it has only 379 fibers against 446 in R. esculenta. In the second place, there is in R. pipiens a marked excess of fibers in the dorsal root of the IX nerve. In R. esculenta, the corresponding excess is distributed between the IX and X nerves. As a second specimen of R. esculenta, weighing 23 grams, examined by BIRGE ('82) shows a similar distribution, the possibility is suggested that this arrangement may be characteristic for R. esculenta.

The foregoing Table 16 also brings out the fact that the number of sensory, as compared with the number of motor fibers, is relatively greater in R. pipiens. Thus

	MOTOR.	SENSORY.
R. pipiens	1	1.36
R. esculenta	1	1.24

However a further analysis of this relation shows that in the lumbar nerves VIII, IX and X the proportions of motor to sensory are nearly alike in the two species.

	MOTOR.	SENSORY
R. pipiens	1	1.833
R. esculenta	1	1.827

Moreover for the III nerve these proportions have been assumed as similar (see HARDESTY '99, p. 78), so that the difference which is found when the total number of fibers is compared, must depend on differences in this relation, which exists in the roots of the II, IV, V, VI, VII - - - XI nerves.

The ratio in this group of roots is

	MOTOR.	SENSORY
R. pipiens	1	0.674
R. esculenta	1	0.555

Thus showing R. pipiens as superior in this last group, although in both species the ratio is less than unity.

As a consequence of these relations, it appears that while R. pipiens has everywhere a better sensory innervation, because there are absolutely more afferent fibers present for the same area of skin and weight of muscle, the relative sensory supply is superior only in the head and trunk, but not in the skin and muscles of the limbs.

Such anatomical differences as these just described, suggest corresponding physiological differences between the two species. In pursuance of this suggestion, I sent out a letter of inquiry to my physiological colleagues in May of 1907. I take this opportunity of thanking my numerous correspondents for their courteous replies, but at the same time must report with regret that there do not appear to be any data bearing on the possible physiological differences, concerning which inquiry was made. The number of medullated nerve fibers in the spinal nerve roots of R. temporaria, has still to be determined.

A comparison of the length of the internodal segments in the fibers of the sciatic nerve of R. pipiens and R. esculenta.—As the heading indicates, the comparison will here be limited to fibers from one nerve. BOYCOTT ('04) has determined the length of the internodes in fibers taken from the sciatic nerve just at the point where it divides into the nervus tibialis and the nervus peroneus. The length of the internodes at this locality depends on two factors, first the size (length) of the frog, and second the diameter of the fiber; the internodes becoming longer, the larger the frog and the greater the diameter of the fiber examined.

By a study of specimens of R. temporaria of different lengths, BOYCOTT was able to show that the average length of the internodes of fibers of all diameters taken from the sciatic, increased in the same proportion as the length of the sciatic, the curves representing the two series of measurements running parallel. Accepting this result, it is possible to calculate the average length of the internodes at this point for frogs of different sizes.

Below is given a table containing the data on the six largest specimens of R. temporaria examined by BOYCOTT ('04, p. 375). These are arranged in the order of increasing body length, the measurement being made from the tip of the nose to the end of the urostyle.

The body weight here given was taken as usual. The length of the sciatic in millimeters is defined by BOYCOTT (*loc. cit.*, p. 371), as follows: "The upper end has been taken throughout as the point of emergence from the vertebræ of the upper of the two larger branches of the plexus.

There is no good fixed point for the lower end, the one which has been adopted as the cut end, obtained by cutting across the leg through the knee joint at right angles to the axis of the leg when it is in full extension."

TABLE 17.

R. temporaria.

No. of Specimen.	Body Weight in Grams.	Length of Sciatic in mm.
XXI	15.10	46.0
XXII	20.15	45.0
XXIII	16.45	45.5
XXIV	18.15	50.0
XXV	20.80	49.0
XXVI	24.80	53.5
Average	19.20	48.2

To compare with these, we have observations on four specimens of R. pipiens, made by Mr. TAKAHASHI[3] in the Neurological Laboratory of the University of Chicago.

The specimens examined by TAKAHASHI, and in which the internodes were studied in the same locality as that selected by BOYCOTT, were four in number, and the measurements made on them are given in the following table:

TABLE 18.

No. of Specimen.	Body Weight in Grams.	Sciatic Length in mm.* (Calculated.)
III	26	49.
V	34	53.3
VI	37	60.2
VIII	63	65.8
Average	40	57.3

* This measurement was not made by Mr. TAKAHASHI, but has been calculated from other data in his tables.

In accordance with BOYCOTT's results, we should expect in this series of R. pipiens, with an average sciatic length of 57.3 mm. to find longer internodes than in the series of R. temporaria, with an average sciatic length of only 48.2 mm., but on the contrary, the internodes in R. pipiens are much shorter. To make the comparison fair however it is necessary to reduce the measurements on R. pipiens to the measurements of the R. temporaria series,

[3] Mr. TAKAHASHI kindly allows me to use the data from his forthcoming paper on the internodes in R. pipiens.

which is taken as the standard. To do this, we divide the observed values for the R. pipiens series by 1.188, since 57.3 mm., the average length of the sciatic in the series of R. pipiens is 118.8 per cent of 48.2 mm., the average length of the sciatic in the series of R. temporaria.

The observations thus reduced to the same standard are given in the following table.

TABLE 19.

Giving the lengths of the internodal segments in μ on the medullated fibers of the sciatic nerve, for frogs with a sciatic length of 48.2 mm., arranged according to the diameter of the fibers.

Number of Observation	R. temporaria (Boycott) Length of Internodes	Diameter of Fibers.	R. pipiens (Takahashi) Length of Internodes.	Number of Observations	Relative Value of Lengths in R. pipiens.
In all about 1050	767	5-5.9μ	500	159	65%
	*1186	6-6.9μ	586	107	49%
	1102	7-7.9μ	705	92	64%
	1159	8-8.9μ	826	16	71%
	1288	9-9.9μ	917	.4	71%
	1399	10-10.9μ	929	47	66%
	1416	11-11.9μ	942	14	66%
	1536	12-12.9μ	1027	5	66%

* As will be seen, Boycott's value for the length of the internodes in fibers 6-6.9μ in diameter, is plainly aberrant, and therefore the percentage value for the internodes of fibers having this diameter in R. pipiens, is excluded from the general average.

The foregoing table shows that when grouped according to diameters, the internodal lengths in R. pipiens range between 64 per cent and 71 per cent of that in R. temporaria, the average being 67 per cent.

It follows from this that R. pipiens has three sheathing cells on a fiber, where R. temporaria has only two, and therefore more cells in the length of the sciatic.

Consequently R. pipiens has the finer and more complete construction, although it is not possible to say what physiological advantage goes with this difference in structure. There are no observations on R. esculenta to compare with those just given.

Conclusions. From the observations presented, we conclude that the three species studied are similar in general form and proportions, but that R. pipiens has:

1. A heavier central nervous system.

2. A heavier brain and spinal cord.
3. A heavier brain in proportion to the weight of the spinal cord.
4. A greater percentage of water in both the brain and spinal cord.
5. A larger number of both sensory and motor medullated fibers in the spinal nerves (when compared with R. esculenta).
6. A slightly greater proportion of sensory fibers in the spinal nerves (when compared with R. esculenta).
7. Shorter internodes, and therefore a greater number of sheathing cells (when compared with R. temporaria).

With the possible exception of the percentage of water, the interpretation of which is not yet clear, all these characters may be counted to the credit of R. pipiens as indicating a higher development of its nervous system, and if we may make these characters a basis for physiological predictions, we should expect the American leopard frog, R. pipiens, when compared with the European, R. esculenta and R. temporaria, to give (1) more perfect general reactions associated with (2) less perfect reflex ones, and also to be both (3) stronger and (4) more sensitive.

APPENDIX.

The observations of Fubini, '81.

In 1881 FUBINI published, under the title "Gewicht des Centralen Nervensystems im Vergleich zu dem Körpergewicht der Thiere bei Rana esculenta und Rana temporaria," a study of the weight of the brain and spinal cord in the two European species commonly used for experiment. His data are comprised in eight tables, each sex being represented by four tables, and the records on twelve specimens entered in each table. His main object in this study was to show that in the female frog, the weight of the central nervous system was less than in the male. As I have elsewhere explained (DONALDSON and SCHOEMAKER, '00), he does not show this, having fallen into error by reason of his failure to appreciate that the relative weight of the central nervous system diminishes with the increasing body weight of the frog.

Despite this failure in the interpretation of his records, it was desirable to examine further his original tables in order to determine what he had recorded concerning the weight of the brain and spinal cord.

The weight of the brain and of the entire central nervous system is given in all the tables. The weight of the spinal cord can be obtained therefore by subtracting the former from the latter. Having the weight of the brain and spinal cord, we can find the ratio between them.

There are moreover two tables, one for each species, in which we have the body weights of males to compare with the weight of the central nervous system. In the other six tables, the body weights for the males (two tables) are given "after evisceration" and for the females (four tables) without correction for ova. In these cases the body weights can only be estimated.

These data have been carefully worked over, with a view to determining how they compare with my own.

In the first instance, FUBINI'S observations on the brain weights in unopened males of R. temporaria, are closely similar to mine. He obtains, however, weights for the spinal cord nearly double mine; thus his brain cord ratio is abnormally low. This is shown in the following table.

TABLE 20.

Showing the ratios of brain weight to the cord weight as determined by FUBINI, and by me.

Rana temporaria

DONALDSON REPETITION OF TABLE 12. BODY WEIGHT.	RATIO.	FUBINI BODY WEIGHT.	RATIO.
15.9	2.15	23.1 (male observed)	1.26
23.1	1.86	25.0 (male estimated)	1.70
28.0	1.87	31.0 (female estimated)	1.14
31.3	1.87	35.0 (female estimated)	1.77

Rana esculenta.

DONALDSON REPETITION OF TABLE 12. BODY WEIGHT.	RATIO.	FUBINI BODY WEIGHT.	RATIO.
15.9	2.29	20.0 (male estimated)	1.76
22.0	2.22	28.2 (male observed)	1.25
35.0	2.09	30.0 (female estimated)	1.80
40.2	2.05	35.0 (female estimated)	1.62

In the same way his observations on the weight of the brain in R. esculenta run only 10 to 15 per cent below mine, but the weights for the cords are much higher than mine, and the ratios as seen in the above table, are quite impossible and hopelessly irregular.

In view of these relations of brain to cord, I conclude that FUBINI's results are in general not trustworthy, and therefore do not require further discussion.

BIBLIOGRAPHY.

BIRGE, E. A.

1882. Die Zahl der Nervenfasern und der motorischen Ganglienzellen im Rückenmark des Frosches. *Archiv f. Anat. u. Physiol.*, Part 5 and 6, p. 435–570.

BOULENGER, G. A.

1897. The tailless Batrachians of Europe. Published by the Ray Society. *London.*

BOYCOTT, A. E.

1904. On the number of nodes of Ranvier in different stages of the growth of nerve fibers in the frog. *J. of Physiol.*, vol. 30, p. 370–380.

CAMERANO, L.

1900. Richerche intorno alla Variazione del "Bufo Vulgaris" Laur. *Mem. d. r. Accad. d. sc. di Torino*, S. 2, vol. 50, p. 81–153.

DONALDSON, H. H.

1898. Observations on the weight and length of the central nervous system and of the legs, in bull-frogs of different sizes. *J. of Comp. Neurol.*, vol. 8, no. 4, p. 314–335.

1902. Weight of the central nervous system of the frog. *Decennial Publications, University of Chicago*, vol. 10, p. 3–15.

DONALDSON, H. H., and SCHOEMAKER, D. M.

1900. Observations on the weight and length of the central nervous system and of the legs in frogs of different sizes (Rana virescens brachycephala, COPE). *J. of Comp. Neurol.*, vol. 10, no. 1, p. 109–132.

1902. Observations on the post-mortem absorption of water by the spinal cord of the frog (Rana virescens). *J. of Comp. Neurol.*, vol. 12, no. 2, p. 183–198.

DUNN, E. H.

1900. The number and size of the nerve fibers innervating the skin and muscles of the thigh in the frog (Rana virescens brachycephala, COPE). *J. of Comp. Neurol.*, vol. 10, no. 2, p. 218–242.

1902. On the number and on the relation between diameter and distribution of the nerve fibers innervating the leg of the frog (Rana virescens brachycephala, COPE) *J. of Comp. Neurol.*, vol. 12, no. 4, p. 297–328.

FUBINI, S.

1881. Gewicht des centralen Nervensystems im Vergleich zu dem Körpergewicht der Thiere, bei Rana esculenta und Rana temporaria. MOLESCHOTT's *Untersuchungen zur Naturlehre des Menschen und der Thiere*, Bd. 12.

HARDESTY, I.

1899. The number and arrangement of the fibers forming the spinal nerves of the frog (Rana virescens.) *J. of Comp. Neurol.*, vol. 9, p. 64–112.

KELLICOTT, W. E.

1907. Correlation and variation in internal and external characters in the common toad (Bufo. lentiginosus americanus, Le. C.) *J. Exper. Zoöl.*, vol. 4, p. 575-614.

DE VARIGNY, H.

'99. Sur la notion physiologico-chimique de l'espèce. *Volume jubileum de la Soc. de Biol de Paris*, p. 598.

PRELIMINARY NOTE ON THE SIZE AND CONDITION OF THE CENTRAL NERVOUS SYSTEM IN ALBINO RATS EXPERIMENTALLY STUNTED.

BY

SHINKISHI HATAI, Ph.D.

(*Associate, The Wistar Institute of Anatomy and Biology.*)

For these observations five litters of rats were so divided into two groups that the average body weight was nearly the same in both and one group was given the full laboratory ration, while to the other was fed the minimal amount of bread, corns and cereals. The normally fed group constitutes the "first controls," and the underfed rats, the "stunted group." For further comparison, young rats with the approximately same body weight as the "stunted group," bût much younger, were taken for the "second controls."

Beginning at the age of thirty days, the underfeeding considerably retarded the growth of the stunted group so that when they were, on the average, 170 days old they weighed 91.5 grams, whereas the "first controls"—of the same average age—weighed 146.5 grams. The younger rats from 80 to 100 days old which formed the "second controls," weighed on the average 86.3 grams. It must be remembered that during the time the behavior experiments were carried on (for nearly thirty days), the experimented rats were fed with normal diet and as a consequence these rats gained somewhat rapidly in body weight. Therefore the possibility of obtaining permanently stunted rats by means of underfeeding is still undetermined. All the rats were killed and weighed immediately after the behavior experiments were ended.

The main results obtained from the present experiments are exhibited in the following table.

External characters.—The most conspicuous external differences between normal and stunted rats as shown by the stunted rats are in the length of the body and of the tail, both of which were considerably reduced with respect to the body weight.[1] This peculiar difference, as is seen from the table, holds true in every case. Further, the ratio between the length of the body and

TABLE I.

Sex	Age.	Body Weight.	Body Length.	Tail Length.	Ratio, Body Length to Tail Length.	Brain Weight Observed.	Brain Weight Calculated.	Spinal Cord Weight.	Percentage of Water, Brain.	Percentage of Water, Spinal Cord	Remarks.
	days.	*grams.*	*mm.*	*mm.*							
F......	215	186.2	189	158		1.8996		.5597	78.365	70.573	1st control
F......	215	106.2	155	112		1.6325		.4030	78.623	72.641	stunted
F......	80–100	105.7	155	131		1.6313		.4049	78.900	72.925	2d control
M.....	182	186.7	190	154		1.6578		.4862	78.447	70.547	1st control
M.....	182	113.7	155	109		1.5678		.3756	78.772	73.003	stunted
M	80–100	99.7	157	121		1.7029		.4059	78.971	73.121	2d control
M	164	119.1	167	130		1.6671		.4524	78.315	71.353	1st control
M	164	77.6	137	100		1.5884		.3573	78.374	71.760	stunted
M	80–100	87.4	154	119		1.7156		3579	78.870	72.981	2d control
F......	164	126.9	170	139		1.6859		.4941	78.118	71.623	1st control
F.......	164	89.3	140	113		1.6053		.3730	78.477	72.193	stunted
F......	80–100	61.7	128	113		1.4880		.2975	78.568	*71.092	2d control
F......	127	113.2	160	135		1.7698		.4459	78.675	72.303	1st control
F......	127	70.9	132	105		1.7394		.3404	78.836	73.413	stunted
F......	80–100	77.2	139	115		1.5110		.3359	79.139	73.861	2d control
					Averages.						
F + M	170	146.4	175	143	1 : 0.823	1.7360	1.772	.4877	78.384	71.280	1st control
F + M	170	91.5	144	108	1 : 0.750	1.6267	1.646	.3699	78.616	72.602	stunted
F + M	80–100	86.3	147	120	1 : 0.816	1.6098	1.629	.3604	78.889	72.796	2d control

* This is the only exceptional case, where percentage of water in the second control is less than that of the stunted.

that of the tail is considerably less in the stunted rats than in the control rats. The ratio just mentioned is found to be on the average 1 : 0.82 in both "first" and "second controls" while

[1] The measurement taken from the tip of the nose to the anus is designated as "body length" while that from the anus to the tip of the tail is designated "tail length."

in the stunted rats the ratio is 1 : 0.75. Underfeeding therefore produces short tailed individuals. The nature of this result has still to be investigated.

Central nervous system.—The weight of the brain and spinal cord and the percentage of water in both were separately determined according to the usual procedure.

The weight of the encephalon was found to be normal to the body weight in both the controls and stunted rats, the brain weight of the first controls is heaviest and that of the "stunted" and "second controls" follow in the order named. The relation between body and brain weights was tested by the formula,

$$\text{Brain weight} = 0.554 + 0.569 \log (\text{body weight} - 8.7).$$

This formula has been developed through the study of our laboratory records and gives us the theoretical weight of the brain for any body weight. The former was found to be normal even in the stunted group. As seen from the table, the difference between calculated and observed brain weights on the average was about 1.5 per cent, indicating a normal relation of the brain weight to the given body weight. Thus we conclude that the normal relation between the body and brain weights was not disturbed by stunting.

We have as yet no satisfactory method for determining the normal weight of the spinal cord in respect to either body weight or any other characters. Nevertheless the proportional weight of the spinal cord in the experimented and in the second control rats with respect to the brain and body weights suggests that it also has grown normally (see Table I). Therefore we conclude that the weight of the spinal cord and brain are similarly related to the body weight. Consequently so far as the weight of the central nervous system is concerned, the normal relation to the body weight is still maintained by the stunted rats.

It is interesting to note in this connection that the definite relation between the body and brain weights is not disturbed even when the growth of the body has been considerably accelerated by means of the lecithin[2] or when the rats have been once starved and then returned to normal diet so that the final body weights

[2] The effect of lecithin on the growth of the white rat. *American Journal of Physiology*, vol. 10, no. 1. 1905.

become normal to the given age.[3] Whether or not this definite relation between the brain and body weights can still be maintained even when we modify the conditions in other ways will be the subject of further investigations.

Percentage of water in the central nervous system.—The percentage of water in the central nervous system was always higher in the stunted rats than in the first controls, despite the fact that ages of the two groups were the same. On the other hand, this value in the stunted rats—though slightly less—was very close to that of the second controls which were much younger. It has been established in the laboratory that among rats of the same age, those with heavier brains have a smaller percentage of water than those with lighter brains. Therefore the higher percentage of water in the stunted rats as compared with the first controls indicates "a usual" rather than "an unusual" condition, since we should expect to find a somewhat higher percentage of water in the rats with less heavy brain at a given age. We conclude therefore that the percentage of water in the central nervous system in both the controls and stunted rats is normal, in the latter of course having due regard for age and body weight, as well as weights of the brain and spinal cord.

Since the percentage of water and that of the extract are inversely related we may infer that somewhat greater percentage of water found in the central nervous system in the stunted rat indicates with highest probability relatively smaller development of the medullated nerve fibers in that organ when compared with that of the first controls. This statement is correct at least for the peripheral system, as a recent investigation[4] by Mrs. J. W. Hayes shows that the number of medullated fibers in the second spinal nerve in heavier albino rats is greater than that in the less heavy rats of the same age. A further discussion of this general point is however reserved for a future publication.

Conclusions.—Our final conclusions are, then, that aside from the shorter length in the body and tail, which is not only absolute but relative also, the stunted rats differ from the normal rats only in the absolute magnitude of the measured characters, while, on the other hand, when differences in the central nervous system

[3] Effect of partial starvation followed by a return to normal diet on the growth of the body and central nervous system of albino rats. *American Journal of Physiology*, vol. 17, no. 5. 1907.

[4] As yet unpublished.

are compared with the growth of the entire body the growth of the stunted rats may be considered just as normal as that of the controls.

The stunted rats were made the subject of tests, by Mr. JOHN W. HAYES, fellow of psychology in the University of Chicago, to determine whether their behavior was modified by their arrested growth, and the results will be published by him later.

ON THE PHYLOGENETIC DIFFERENTIATION OF THE ORGANS OF SMELL AND TASTE.

BY

C. JUDSON HERRICK.

(From the Anatomical Laboratory of the University of Chicago.)

There are in vertebrates two systems of sense organs adapted to respond directly to peripheral chemical excitation, the organs of smell and taste. In this respect they are in contrast with the other sense organs of the body; but when we come to compare the two chemical senses with one another we find it difficult to discover any objective difference between their stimuli or any explanation for the development of two chemical senses in the primitive aquatic vertebrates. And yet the very lowest vertebrates exhibit important morphological differences between the peripheral organs of smell and taste, a complete separateness in the nervous pathways to the brain and still more important differences in the central reflex connections within the brain. In view of the similarity in the nature of the stimuli to which the peripheral organs respond, these fundamental central differences have thus far baffled explanation.

Let us first consider briefly the criteria by which in the case of human beings the modalities of sense may be distinguished. (1) Doubtless the most important criterion for us is direct introspective experience, the *psychological* criterion. (2) The adequate stimuli of the various senses exhibit characteristic physical or chemical differences, the *physical* criterion. (3) The data of anatomy and physiology may differentiate structurally the receptive organs and conduction paths of the several types of sensation, the *anatomical* criterion. (4) The type of response varies in a characteristic way for the different senses, the *physiological* criterion.

It is impossible in the present state of our knowledge to frame adequate definitions of all of the senses in terms of any one of these criteria alone. Thus, we are not able introspectively to discriminate between olfactory and gustatory sensations, but rather elaborate physiological experimentation is necessary to enable us to effect the analysis of these two sets of stimuli. Again, the anatomical and physiological bases of several of the senses are still very imperfectly known and in still other cases we are almost wholly ignorant of the distinctive chemico-physical qualities of the stimuli which call forth diverse sense modalities. The latter point is notably true for the senses of smell and taste. The common statement that we smell substances only in the gaseous state and taste liquids (solutions) is only approximately true, if at all, in the mammals, and certainly cannot hold for the lowly aquatic vertebrates where the differentiation of these two sense organs in practically their definitive form first occurred.

Attention has been drawn to the fact that, while tastes can be classified under the four subjective qualities, sweet, sour, bitter and salty, the innumerable odors are apparently quite incapable of any such classification. To this it may be added, on the one hand, that ZWAARDEMAKER claims to be able to classify the known odors into some nine groups which he compares with the four classes of taste, and, on the other hand, that some recent studies on the chemical physiology of taste[1] go to show that it is a reaction between the receiving organ and the ions of the sapid substances and that the ions belonging to a given group, such as those giving "salty" tastes, do not all produce the same sensation quality. In other words, the four groups of taste qualities, like the nine groups of smell qualities, are more or less ill defined both from the standpoints of their psychological and their physico-chemical criteria. It is to be expected that future research will shed additional light on the physical and psychological criteria of smell and taste, but it will not eliminate their strong similarity.

These considerations suggest that smell and taste have originated phylogenetically from a common undifferentiated chemical sense, a conclusion which is supported by the morphological relations of their cerebral centers. The details of this anatomical evidence are far too complex to be summarized here and the reader

[1] L. KAHLENBERG: The action of solutions on the sense of taste. *Bul. Univ. Wisconsin, Science Series*, vol. 2, pp. 1–31. 1898.

T. W. RICHARDS: The relation of the taste of acids to their degree of dissociation. *Am. Chemical Journal*. 1898.

is referred to the exposition and discussion of the cerebral centers for smell and taste given by JOHNSTON and HERRICK.[2]

But despite these fundamental similarities, it still remains true that the organs of smell and taste are topographically widely separated and structurally very different both peripherally and centrally. Their central neural pathways and connections are in fact as different as are those for hearing and vision, two senses whose psychological and physical criteria are most clearly defined. The anatomical relations of the gustatory system are known in lower vertebrates and those of the olfactory system are well understood and are tolerably uniform throughout the vertebrate series. It is possible to determine by experiment to which one of the peripheral sense organs an animal responds when given a chemical stimulus. The anatomical criteria of smell and taste are therefore clearly defined.

As far as vertebrates are concerned, we may define taste in accordance with the anatomical criterion as the reaction or sensation arising from the appropriate chemical stimulation of the organs known as taste buds (wherever found in the body), and smell as the reaction or sensation arising from the appropriate chemical stimulation of the termini of the olfactory nerve. (See the Addendum, p. 165.)

These definitions cannot be extended to the invertebrates unless homologous organs can be discovered among them. It may well be that there are no such organs in the invertebrates, a single chemical sense alone serving their needs; or two or more chemical senses may be present among the invertebrates which are wholly unlike either of the vertebrate senses.

In this discussion it will be observed that I take a somewhat different standpoint from that of NAGEL,[3] who defined taste and smell in terms of the state of physical aggregation of the stimulus. Smell, he says, is the faculty of perceiving vaporous (*dampfförmige*) substances and taste is the faculty of perceiving liquid substances. It follows from this, he argues, that it is not proper to attribute to aquatic animals a sense of smell in addition to a sense of taste, but both functions fuse into a single one.

[2] J. B. JOHNSTON: The nervous system of vertebrates. *Philadelphia,* 1906, chap. 10.
C. JUDSON HERRICK: The central gustatory paths in the brains of bony fishes. *Journ. Comp. Neurol. and Psych.*, vol. 15, 1905, pp. 450–454.

[3] W. A. NAGEL: Vergleichend physiologische und anatomische Untersuchungen über den Geruchs- und Geschmackssinn und ihre Organe, mit einleitenden Betrachtungen aus der allgemeinen vergleichenden Sinnesphysiologie. *Bibliotheca Zoologica, Stuttgart,* Heft 18. 1894.

His argument for the absence of smell in all aquatic animals is based upon the definition of smell as the perception of gaseous or vaporous stimuli. He adduces evidence that when air is dissolved in water it is incapable of absorbing the vapors given off by volatile substances unless these vapors are soluble in the water itself, stating that they cannot be dissolved in the air contained in the water. They affect the organs, therefore, as true solutions, not as gases dissolved in water. He says (p. 60), "All substances which pass over into water from an object lying in the water, say a decomposing organic body, diffuse themselves in the water in accordance with the laws of the diffusion of liquids, not those of gases and vapors, even though the object in question when brought into the air may have vaporous emanations."

It is unnecessary to summarize here his elaborate argument for the absence of smell in fishes based upon anatomical differences in the receptive olfactory organs between fishes and air breathing vertebrates; for when examined closely in the light of our present knowledge these differences are seen to be trifling when compared with the broad resemblances of both peripheral and central organs of smell throughout the whole vertebrate phylum.

NAGEL's conclusion is expressed on p. 62: "We can with the greatest probability assume that the end-buds of the glossopharyngeus in the mouth of fishes and amphibians serve the chemical sense, viz: taste, and thus function in eating. We can with some probability assume that the sense organs of fishes and aquatic amphibia supplied by the N. olfactorius likewise serve the chemical sense; but this is certainly no olfactory organ in the sense of that term in the land animals. What the occasion of its chemical excitation may be is quite unknown. The method by which it is excited is with highest probability similar to the excitation of the taste buds in the mouth, i. e., the excitation follows through substances dissolved in water."

This conclusion, to my mind, simply illustrates the fact that it is impossible in the present state of our knowledge to interpret these two senses in terms of the physical stimuli. It is not meant to imply that there is no difference between the physical stimuli of smell and taste; for I think it probable that further research will bring such differences to light. But these differences are apparently very small in aquatic animals, whereas the structural differences between the nervous apparatus involved are very great indeed. even in the lowest fishes.

Our argument thus far leads to an apparent *impasse*. The physical and psychological criteria of smell and taste seem inadequate to account for the definite and fundamentally different anatomical peculiarities of the organs in question. But we have not yet considered the fourth line of evidence mentioned at the beginning, that which we called the physiological criterion; viz: the characteristic responses normally following the stimulation of these organs of sense.

A suggestion made by Professor SHERRINGTON in his recent Lectures on the Integrative Function of the Nervous System seems to me to put the matter in a perfectly clear light. As is well known, SHERRINGTON classifies the sense organs (receptors) into (1) exteroceptors, adapted for response to stimuli arising from without the body; (2) proprioceptors, sense organs lying within the body adapted to report to the central nervous system the physiological state of the organs of somatic response themselves (typified by muscle spindles, neuro-tendon organs, etc.); (3) interoceptors, organs set to to guard the receptive surfaces of the body—enteron, lungs, etc. Exteroceptors which are excited by stimuli arising at a distance from the body are termed by SHERRINGTON distance receptors.

The physiological analysis here outlined is full of helpful suggestion in the morphology of the nervous system. Putting SHERRINGTON'S analysis into correlation with that of the new school of functional morphologists, we recognize his first two types of receptors as falling within the somatic sensory group, for the chief organs of response (effectors) in both cases are the somatic or skeletal muscles. SHERRINGTON'S third type is the visceral sensory system, calling forth reflexes in the visceral musculature (including the specialized striated visceral muscles of the branchial arches and their derivatives in the higher vertebrates).

The taste buds lying within the mouth of vertebrates are typical interoceptors, and they with their nerves and cerebral centers are classified as specialized visceral sensory organs. They are in gnathostome vertebrates usually stimulated by food contained within the mouth and the effectors with which they are most directly connected are the visceral muscles of the jaws, gills, œsophagus, etc. In the protochordate vertebrate ancestry it is probable that there was but one chemical sense, and that feebly developed; for these animals probably did not masticate their food, and the undifferentiated primordial chemical sense may have been as important in determining the chemical character of the environing water as of the food eaten.

Be that as it may, with the appearance of teeth which pierce or crush the food, the organs of chemical sense within the mouth and pharynx assumed an important function as guardians of the entrance to the œsophagus, an interoceptive function which they perform in all gnathostome vertebrates—the organs of taste. Parallel with this differentiation within the mouth, the organs of chemical sense lying outside the mouth at the rostral end of the body would assume more and more importance as organs for detection of chemical differences in the surrounding water, differences resulting usually from the presence of sources of chemical alterations of the water lying outside the body of the fish. These external organs of chemical sensation in the leading segments of the body were finally aggregated as the organ of smell.

The differences in the character of the stimulus applied to these two organs may have been very slight at the beginning (and indeed may be so still); but in the case of any organism possessing the power of free locomotory movement the physiological significance of the stimulation of the two sense organs may be very different indeed. The object which acts as a stimulus to taste buds is already within the mouth. The appropriate reaction is typically a contraction of the visceral musculature of the mouth and pharynx adapted either to masticate and swallow or to eject the object, as the case may require. The somatic musculature is not necessarily brought into play. The olfactory organ, on the other hand, has become a distance receptor and the appropriate reaction is a movement, usually locomotor in type, of the somatic muscles, taking the animal toward or away from the source of the stimulus. Even though the stimuli in the two cases were identical, it is evident that the difference in the character of the response would bring into play a very different central reflex apparatus for the distance reaction from that for the mastication or swallowing reflex.

This difference between the characteristic reaction of the interoceptor and the distance receptor is in my opinion the sufficient explanation for the most important structural differences between the olfactory and gustatory systems of vertebrates. This same feature involves, it is true, a certain degree of difference between the physical stimuli and the psychical qualities of odors and savors, especially in the higher vertebrates; but these are in all animals quite subordinate to the type of reaction involved.

A critical examination of the central conduction paths for smell and taste supports this view of the case. The central olfactory apparatus is very constant throughout the vertebrate phylum. The organ of smell, as befits a distance receptor, is located in the leading segments and its central connections are with the extreme tip of the neural tube; indeed in all of the true vertebrates it has grown out rostrad beyond the primary neural tube, the entire rhinencephalon lying in the telencephalon, or ultra-terminal brain. The path extends from the olfactory bulb to the tuberculum olfactorium and other structures in the base of the forebrain, thence directly back to the olfactory centers in the thalamus or else first to the olfactory cerebral cortex (hippocampal formation, etc.) and then to the thalamus. The two principal olfactory centers in the thalamus lie in the epithalamus and hypothalamus respectively. Each of these thalamic centers receives in higher vertebrates olfactory tracts from both the basal and cortical olfactory centers of the forebrain; and each sends a strong tract to reach the motor centers. These tracts are the tr. habenulo-peduncularis (fasc. retroflexus or bundle of MEYNERT) and the fasciculus pedunculo-mammillaris (tr. mammillo-bulbaris). In lower vertebrates both of these tracts can be traced far downward into the medulla oblongata, where they come into relation directly with the motor nuclei of the cranial nerves and the evidence is that either directly or indirectly they pass still farther into the spinal cord for the somatic motor reflexes characteristic of olfactory reactions.

The central gustatory path is well known only in fishes. Here there are much more direct reflex connections with the visceral motor nuclei of the cranial nerves than the olfactory system shows, and in most fishes no important connections with somatic motor nuclei save by way of the hypothalamus and tractus mammillo-bulbaris. There are certain fishes, however, in which taste buds have been developed secondarily in the outer skin of the general body surface. Here they have been shown to function as exteroceptors[4] and in these cases the central connections of the cutaneous taste

[4] C. JUDSON HERRICK: The organ and sense of taste in fishes. *Bul. U. S. Fish Commisssion for 1902. Washington, 1904.* The central gustatory path in the brains of bony fishes. *Journ. Comp. Neurol. and Psych.*, vol. 15, no. 5. 1905.

buds are very different from those of the phylogenetically older taste buds within the mouth. In the catfish and carp the primary cerebral center for all of the cutaneous taste buds is the facial lobe, from which secondary gustatory tracts of the typical sort pass out to the visceral motor centers, and in addition a direct secondary path to the funicular nuclei where these gustatory impulses are coördinated with tactile impressions from the same areas of skin.[5] A single path leaves the funicular nuclei for the somatic motor centers, thus serving as a common reflex path for both tactile and gustatory impulses from the skin. In the cod[6] the cutaneous taste buds effect somatic motor connections in an entirely different way, passing directly from the equivalent of the facial lobe into the fasciculus longitudinalis medialis and thence to the somatic motor nuclei, indicating that the cenogenetic connection of the taste buds which act as exteroceptors with somatic motor centers has been acquired independently in the gadoids and the Ostariophysi.

The interesting point in this connection is that within the group of teleosts taste buds, which typically in fishes act as interoceptors, have secondarily acquired exteroceptive functions, and parallel with this change a new central reflex path has been established between the primary centers of cutaneous (exteroceptive) taste and the somatic motor centers. It is probable that at a much more ancient period in the phylogeny of vertebrates an analogous differentiation took place in the primordial unspecialized chemical sensory apparatus, one part becoming a typical interoceptor (gustatory apparatus) and establishing its most direct central reflex connections with the visceral muscles of mastication, deglutition, etc., and another part becoming a typical exteroceptor (olfactory apparatus) and early establishing direct central reflex connections with somatic muscles of locomotion, eye movements, etc., in addition to the visceral motor reflexes characteristic of a visceral system.

It should be expressly stated that the claim is not made that all anatomical differences between the organs of smell and taste are explained by this principle, but only that in this way the direction

[5] C. Judson Herrick: On the centers for taste and touch in the medulla oblongata of fishes. *Journ. Comp. Neurol. and Psychol.*, vol. 16, no. 6. 1906.

[6] C. Judson Herrick: A study of the vagal lobes and funicular nuclei of the brain of the codfish. *Journ. Comp. Neurol. and Psych.*, vol. 17, no. 1. 1907.

of the original phylogenetic differentiation was determined and that this is still the dominant feature of the two systems in question.

The conclusion is that the agencies which acted to produce the differentiation from each other of the senses of smell and taste are not to be sought primarily in the stimuli calling forth the reflexes, but rather in the character of the response evoked by the stimulus.

ADDENDUM. As these pages pass through the press an abstract of the very interesting experiments of PARKER appears in the Proceedings of the American Society of Zoölogists (*Science*, n. s., vol. 27, no. 690, March 20, 1908, p. 453). PARKER has previously determined that the skin of the body of the frog and of various other aquatic animals is sensitive to chemical stimuli. Quite in accord with those results, he now finds that the same is true for the common fresh water catfish, Ameiurus. This fish possesses taste buds innervated by the nervus facialis scattered in the skin over practically the whole body surface. If the nerves supplying these taste buds on the trunk are cut, the fish no longer reacts to a bait in the normal way (by turning to snap at the bait) when it is presented to the flank of the body. Nevertheless such operated fishes are sensitive to sour, saline and alkaline solutions when applied to the skin of the trunk.

These results, together with the control experiments described, demonstrate that the spinal nerves of this teleost, like those of the frog, are sensitive to certain external chemical stimuli. The important question at once arises, are these responses to chemical stimulation of the spinal nerves transmitted by the same nerve fibers which transmit the tactile stimuli, or by some other component of the spinal nerves? We know from abundant physiological and clinical experience that the cutaneous rami of the spinal nerves of man transmit impulses which are perceived introspectively as very diverse sensation qualities (touch, temperature, etc.). There is evidence that some at least of the different functions of the sensory spinal nerves are served by anatomically different neurone systems; but whether the ability to respond to direct peripheral chemical stimulation is limited to one or more of these systems or common to all of them, further experiment alone can determine.

Chemical irritability may prove to be more far-reaching and fundamental in nervous excitation than is commonly recognized. However this may be, two special reflex mechanisms have been very elaborately differentiated in vertebrates along quite diverse lines for precise and rapid response to special external chemical stimuli, the organs of smell and taste; and the explanation offered in the preceding pages for the phylogenetic differentiation of these two functional systems is not directly dependent upon any theory regarding the ultimate nature of the primordial undifferentiated sensory type from which they have sprung.

Professor PARKER concludes the note to which we have referred with the remark, "From these experiments it is to be concluded that the sense of taste in horn-pouts is complex and involves not only the seventh nerve, but also the spinal nerves." Assent to this proposition will be readily granted only if we define the sense of taste in accordance with the "physical criterion" (see p. 157) as NAGEL does. In the opinion of the writer neither this criterion nor the "anatomical criterion" (as I have used it on p. 159) alone is adequate *in the present state of our knowledge* to serve as the basis for generally acceptable definitions of all of the so-called senses. Pending the extension of our knowledge in these fields, fruitless controversy may be avoided by a clear recognition of the fact that harmonious conclusions can be expected only on the basis of an explicit understanding regarding the standpoint chosen in every discussion.

SOME CONDITIONS WHICH DETERMINE THE LENGTH OF THE INTERNODES FOUND ON THE NERVE FIBERS OF THE LEOPARD FROG, RANA PIPIENS.

BY

KATASHI TAKAHASHI, *Rigakushi.*

(*From the Neurological Laboratory of the University of Chicago.*)

WITH SEVEN FIGURES.

INTRODUCTION.

In the winter of 1903–04, the following study of the growth of the internodes on the nerve fibers of the leopard frog, was begun, in order to determine whether on a lengthening nerve fiber the number of internodes increased or remained constant. While this study was in progress, the interesting paper by BOYCOTT '04, "On the number of nodes of Ranvier in different stages of the growth of nerve fibers in the frog," was published. The species of frog used by BOYCOTT was the common Rana temporaria (fusca) of England.

After briefly referring to the scanty literature on the subject of the internodes (see KÖLLIKER '96), which shows that they have different lengths in different species of animals, are longer in old than in young animals, and longer in fibers of great than in fibers of small diameter, BOYCOTT presents evidence which demonstrates beyond reasonable doubt, that in the growing sciatic nerve, at the point where it divides into the nervus tibialis and nervus peroneus, the *average length* of the internodes increases very nearly as does the length of the nerve itself. It would seem from this to follow that the number of internodes should remain constant. The calculations show however a very slight but regular increase in the estimated number of the internodes as the frogs become larger. This result, noted but not explained by BOYCOTT, and touched on later in this paper is, I believe, susceptible of an explanation, which at the same time leaves BOYCOTT's main conclusion intact.

The second important point brought out by Boycott, although not especially commented on by him, is illustrated in the accompanying Table 1, which is copied from Boycott's paper, with a slight change made by putting the "sciatic length" in the column where the body lengths are given by him.

TABLE 1.

Average internodal lengths (μ) corresponding to each diameter. Rana temporaria (fusca). Copied from Boycott, *Journal of Physiology* (Foster), vol. 30, p. 373, 1904.

Diameter.......		2μ	3μ	4μ	5μ	6μ	7μ	8μ	9μ	10μ	11μ	12μ	13μ	14μ	15μ
Group.	Sciatic length.														
	mm.														
I	18.2	205	339	450	524	535	631	657	660	667	797				
II	25.1		428	525	592	659	701	772	742	789	846	865	890		
III	34.5					770	819	770	878	968	1007	1069	1043	1177	
IV	46.6		570	1000	766	1186	1102	1106	1243	1358	1361	1490	1511	1576	1766

It is here seen that on fibers of a given diameter from small (young) frogs, the internodes are shorter than on fibers of the same diameter, taken from large (old) frogs, the size being indicated by the sciatic length. If we apply the notion of growth to the interpretation of this table, and remember that a fiber of a given diameter in the small frog, becomes a fiber of greater diameter in the large frog, then it is found that the average of the measurements in Group I of fibers from 2μ to 11μ in diameter, which is 546μ, compared with the average of the measurements for fibers from 6μ to 15μ in diameter, in Group IV, which is 1370μ, gives an increase in the length of the internodes amounting to 2.51, and this corresponds very nearly to the increase in the length of the sciatic nerve, from 18.2 to 46.6 mm., which is 2.56. As will be observed, the average difference in diameter in the two series compared is 4μ.

This method of comparison is admittedly crude, but under the conditions, furnishes a satisfactory confirmation of Boycott's general conclusion that the number of internodes is not increased during growth, but that their average length increases as does that of the nerve in which they are found.

In view of the results obtained by BOYCOTT, it was thought best in the present study to examine especially some points which he has left untouched.

These will be presented under the following heads:

1. The average length of the internodes at different levels along the nerves to the leg.
2. The length of the internodes at different levels on fibers of like diameter.
3. The length of the internodes on fibers in the roots of the spinal nerves.
4. The number of medullated fibers at different levels in the legs of tadpoles of increasing size.
5. A comparison of the length of the internodes in the American frog, Rana pipiens, with their length in the English frog, Rana temporaria (fusca).

Before proceeding to the discussion of the special topics, I desire to state that this study was made under the direction of Prof. H. H. DONALDSON, to whom I am indebted also for the revision of my manuscript. Moreover, I wish to thank both Dr. E. H. DUNN and Dr. S. HATAI for their aid and suggestions given to me during the conduct of this investigation.

MATERIAL AND TECHNIQUE.

For this study, the common leopard frog, Rana pipiens (SCHREBER) was used, the specimens having been obtained from a local dealer and probably collected in the country about Chicago. The frogs were killed with chloroform; the body weight, corrected for ova in the case of the females, was taken in a closed box, and the total length, i. e., the length from the tip of the nose to the tip of the longest toe, as well as the body length, i. e., from the tip of the nose to the tip of the urostyle, were both recorded. In some cases also, the length of both the dorsal and ventral roots of the III and IX nerves (GAUPP's numbering) was determined. The data thus collected are given in Table 2.

In preparing the material, the following methods were employed: A short piece of the fresh nerve was cut out and laid on a wedge-shaped strip of cardboard, the piece of nerve being extended to its normal length. This was fixed, and at the same time macerated, by being placed for twenty-four hours in the following solution (A):

Osmic acid	1.00 per cent, 5 parts
Chromic acid	0.25 per cent, 3 parts
Hydrochloric acid	0.10 per cent, 2 parts

After washing for twenty-four hours in running water, the specimen was transferred for twenty-four hours to the following solution (B):

Glycerine	10 parts
50 per cent alcohol	20 parts
Hydrochloric acid	0.09 parts

After this treatment, the specimens are preserved in solution (C):

Glycerine	10 parts
50 per cent alcohol	20 parts

This last solution (C) should be renewed once or twice at intervals of twenty-four hours. Thick nerves were slit longitudinally with a razor, after they had been in solution (A) for two or three hours. This was done to assist the penetration of the fluid. The specimens were teased in solution (C).

TABLE 2.

Data on the specimens of Rana pipiens used in this investigation. Entries arranged in the order of increasing body length. V, ventral; D, dorsal.

No.	Sex.	Body Weight	Total Length.	Length of Body.	Length of III Spinal Roots.	Length of IX Spinal Roots.	Date of Killing.
		grms.	*mm.*	*mm.*	*mm.*	*mm.*	
1	M.	5.5	104	39	V .85 D .64	2.6 2.1	Aug. 29, '05
2	M.	23.5	169	71	V 2.6 D 2.5	7.0 6.5	Jan. '04
3	F.	26.0	166	72			Jan. 29, '06
4	F.	27.2	180	78	V 2.4 D 2.4	7.7 7.1	Mar. '04
5	M.	31.0	192	80			Dec. '03
6	M.	37.0	204	80			Nov. '03
7	M.	61.1	226	89		V 9.5 D 9.0	July 16, '04
8	M.	63.0	222	89.4	V 2.0 D 1.6	5.9 5.4	Aug. 3, '05

The technique just given, fails however to yield satisfactory results when applied to the nerve roots of the III or IX nerve as the fibers become brittle and distorted. Moreover, the roots of the III nerve do not yield to the technique which proved fairly satisfactory in the case of the roots of the IX nerve, so that the technical problem is complicated. In the case of the IX nerve, I used in the first instance solution (D):

Osmic acid	0.10 per cent, 4 parts
Chromic acid	0.02 per cent, 1 part

The specimen was left in this solution for twenty-four hours, then washed in running water for twenty-four hours, and finally preserved and teased in solution (E).

50 per cent glycerine.

This should be renewed several times.

Later, in place of solution (D), I used solution (F):

Osmic acid	0.100 per cent, 5 parts
Chromic acid	0.025 per cent, 1 part
Acetic acid	0.100 per cent, 1 part

This gave somewhat better results than solution (D) but none of these solutions acted upon the roots of the III nerve sufficiently well to justify an extended study of its fibers, hence only one III nerve was examined.

It is fundamental to the following argument, that the treatment of the nerves should not materially alter the length or the diameter of the fibers which are to be measured. It was necessary therefore to examine the effect of the solutions here employed, and this was done by measuring samples of the nerve as they were passing through the solutions.

Sixteen samples from different levels along the nerves to the leg were first measured, after having been for two or three hours in solution (A) and then finally measured after treatment in solution (C) when they were ready to be teased. The measurement showed an average loss in length of 3.6 per cent and an average loss in diameter of 12.8 per cent.

In the case of eight other specimens (four from the III nerve, and four from the IX nerve) examined in the same way, the loss in length was 1 per cent, and in diameter, 8.6 per cent.

The loss in length is trifling; that in diameter however seems large. It is probable nevertheless that it is to be mainly credited rather to a diminution in the connective tissue sheath, and to the compacting of the fibers, than to a diminution in their individual diameters, and it is therefore not thought that the normal diameters of the fibers are as much modified as the above measurements would indicate.

The samples of nerve were teased with fine needles under a dissecting microscope, and measured directly with a compound microscope, using lenses and eyepieces (with micrometer scales), suited to the determination of length on the one hand and diameter on the other.

The full series of individual measurements will not be printed here, but the original records remain in my posession, and a complete copy of them has been put on file at the Wistar Institute of Anatomy and Biology in Philadelphia, where it may be examined at any time. In the case of the condensed tables which follow, it should be stated here, once for all, that the averages used are always "weighted for the number of cases," while in those instances where it seemed important, there is printed in parentheses along with the average value, the number of measurements on which it is based. The value for the internodes is always given in thousandths of a millimeter (μ).

I. THE AVERAGE LENGTH OF THE INTERNODES AT DIFFERENT LEVELS ALONG THE NERVES TO THE LEG.

As has already been stated, BOYCOTT '04, limited his observations to the average length of the internodes taken from one locality, the distal end of the sciatic nerve. The attempt was therefore made to determine the average length of the internodes at various localities along the nerves supplying the leg.

Fig. 1 gives the arrangement of the nerves to the leg, based on a dissection made by Dr. DUNN '02. The levels from which samples of the nerve were taken are indicated by interruptions in the drawing, and designated by letters, S_1, S_2, S_3, T, T_1, T_2, T_3.

The first four are from the nerve in the thigh, the fifth and sixth from the nerve in the shank, and the seventh from the nerve in the foot.

In each instance a bit of the nerve was prepared according to the method already described, teased as completely as possible, and fifty or more measurements made on the internodes of the fibers, always preferring the larger to the smaller fibers in each instance.

Table 3 gives the results of this sampling; the average at each level being based on the fifty largest fibers which were found.

The method used is sufficiently accurate to justify the statement that on passing peripherally along the nerves to the leg, the fibers of larger diameter become less frequent, and the average

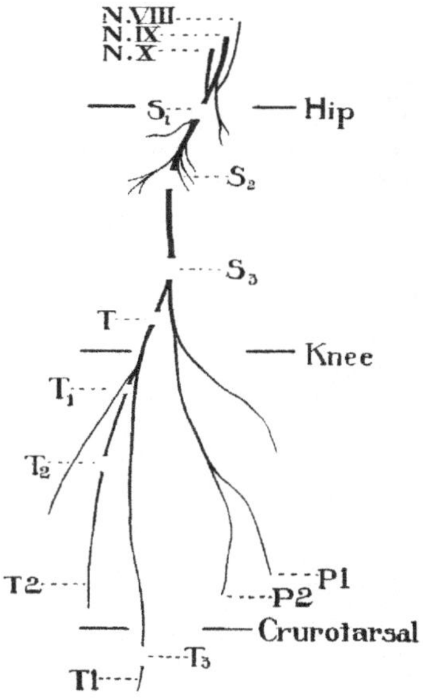

FIG. 1. Giving the main trunks in the nerve to the right leg of the Leopard Frog, as seen from the dorsal aspect. The levels of the several joints are indicated, and also the localities from which pieces of the nerve were taken. These latter are indicated by interruptions, and designated by the letters used in the text. Based on Fig. 1, DUNN '02. *P1*, n. peroneus lateralis. *T1*, n. tibialis r. superficialis. *P2*, n. peroneus medialis. *T2*, n. tibialis r. profundus.

length of the internodes diminishes correspondingly. With the exception of the level S_1, in which, owing possibly to the large number of fibers present, the sampling is less representative than at the lower levels, the internodes show a steadily diminishing length as indicated in the last column of Table 3.

The relations of the diameter and the length of the internodes at the several levels are shown in Fig. 2. The levels are indicated in the figure in their relative positions, and the locations of the hip, knee and crurotarsal joints are shown.

It would follow from this, of course, that if we attempted to determine, as BOYCOTT did, the number of internodes characteristic of the nerve between its origin and any distal point, we should find this number to increase as the sample of the nerve was taken nearer and nearer to the foot. This is exactly what we should

TABLE 3.

Showing the average diameters of the fibers and the lengths of the internodes at the several levels in the nerves to the leg of Frog 6. Body weight, 37 grams; total length, 204 mm. For the identification of the levels, refer to Fig. 1. The number of measurements is given in parentheses above the length of the internodes to which it applies.

Average Diameter. in μ Range	5.0	6.6 (6.2–) (7.5)	8.7	10.2 (10.0–) (11.2)	12.0	12.5	12.8	13.1 (13.0–) (13.5)	15.0	16.2	GENERAL AVERAGES. Diam. in μ	Internodes in μ.
Levels S_1				(13) 987	(15) 1240	(5) 1020		(12) 1225	(4) 1200	(1) 1740	12.1	1154
S_2			(6) 1021	(33) 1111	(9) 1438		(2) 1520				10.4	1175
S_3				(44) 1043	(6) 1283						10.4	1072
T			(2) 938	(40) 1044	(5) 1120			(1) 1051	(2) 1160		10.5	1052
T_1			(20) 816	(30) 793							9.6	802
T_2	(27) 684	(22) 858	(1) 1110								5.8	747
T_3	(34) 588	(16) 731									5.5	632

expect, as Dr. DUNN '02 has shown that the fibers of larger diameter run the shorter courses.

The foregoing result however does not inform us whether the fibers of a given diameter in the same animal have internodes of like length throughout their course. To determine this it is necessary to measure series of fibers of like diameter at different levels, and compare the results with one another. In Frog 6, which furnished the material for Table 3 there is not a sufficient number of measurements of small fibers at the upper levels to make possible such a comparison. It was therefore necessary to examine other specimens.

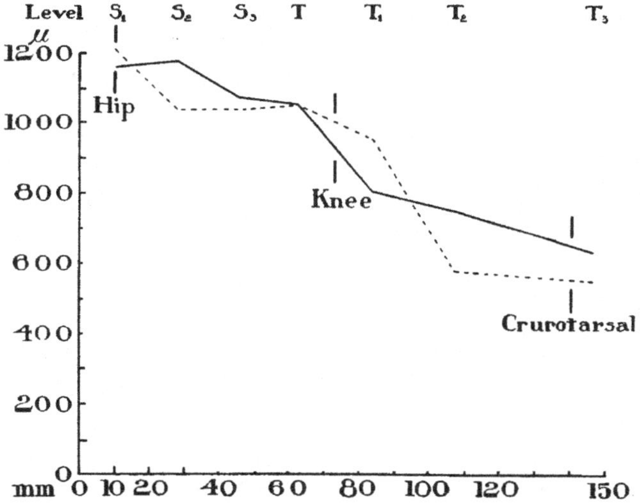

FIG. 2. Showing the average length of the internodes at the several levels in Frog. 6. The values for the diameters have been multiplied by 100, and as this frog is in the growth phase in which the length of the internodes is about 100 times the diameter of the fibers, the two curves run close together. The positions of the hip, knee and crurotarsal joints are also shown.

2. THE LENGTH OF THE INTERNODES AT DIFFERENT LEVELS, ON FIBERS OF LIKE DIAMETER.

For the determination of the length of the internodes at different levels, on fibers of like diameter, Frog 5, weighing 31 grams and having a total length of 192 mm., was used. At each of the seven levels, over 100 internodes on fibers from 5μ to 7.5μ in diameter were measured. For presentation, the fibers have been divided into three classes, having an average diameter of approximately 5.3μ, 6.3μ and 7.3μ, respectively. The results thus condensed are given in the accompanying Table 4.

TABLE 4.

Showing the average length of the internodes on fibers of like diameter at the several levels in the nerves to the leg of Frog 5. Body weight, 31 grams; total length, 192 mm. At each level the measurements are grouped in three classes according to diameter. The number in parentheses indicates the number of measurements.

Level.			Internodes.		Internodes.		Internodes.
S_1	Diameter.........	5.2μ (27)	611	6.4μ (47)	623	7.3μ (34)	805
S_2	Diameter.........	5.4μ (42)	645	6.3μ (38)	706	7.4μ (40)	889
S_3	Diameter.........	5.3μ (79)	769	6.3μ (36)	902	7.2μ (17)	1000
T	Diameter.........	5.2μ (38)	646	6.3μ (30)	787	7.3μ (50)	836
T_1	Diameter.........	5.2μ (58)	636	6.4μ (31)	870	7.4μ (26)	900
T_2	Diameter.........	5.1μ (74)	608	6.3μ (32)	787	7.3μ (18)	744
T_3	Diameter.........	5.2μ (72)	818	6.3μ (25)	917	7.1μ (9)	1011

When the data in this table are read horizontally, we observe that in all but one instance out of the twenty-one (T_2, 6.3μ and 7.3μ) the fibers with greater diameter have the longer internodes. When the measurements are read vertically however the length of the internodes on a fiber of a given diameter varies irregularly from level to level. Fig. 3 illustrates these relations.

In order to plot these internodal lengths fairly, the diameters of the classes at each level must be made exactly equal, hence they are all reduced for the purposes of this figure to precisely 5.3μ, 6.3μ and 7.3μ. The reduction is made by the method of simple proportion. By reason of this reduction, the internodal values are in some cases slightly different in the figures from those in the tables but at most these differences are slight however and hardly detectable on figures of the size here used.

As it will be necessary for comparison with Tables 6 and 7, to have the measurements from Frog 5 for the levels S_1, T and T_3 brought together, we now present the data in the accompanying Table 5.

It is to be noted however, in the case of all three diameter classes, that the fibers at T_3, the level of the foot, give higher values than

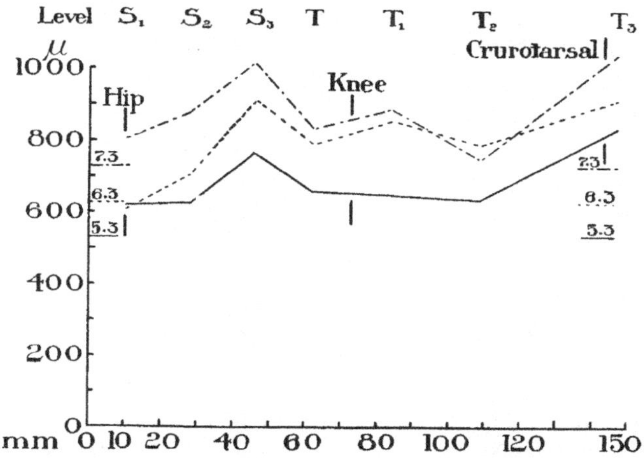

FIG. 3. Showing the average lengths of the internodes at the seven levels in Frog 5. The data for each of the classes has been reduced to the diameters 5.3μ, 6.3μ and 7.3μ. The diameters multiplied by 100 are indicated on the verticals.

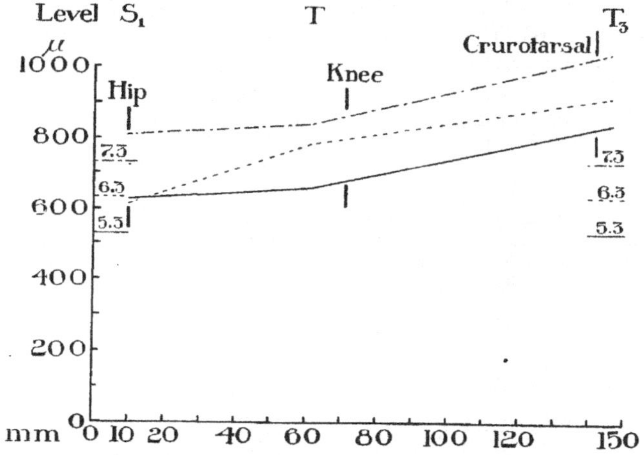

FIG. 4. Showing the average length of the internodes at the levels S, T_1 and T_3 on the fibers 5.3μ, 6.3μ and 7.3μ from Frog 5. The data are the same as those used in Fig. 3. The diameters multiplied by 100 are indicated on the verticals.

appear at S_1; that in the mid-position T and T_1 give intermediate values. This can also be seen by an examination of Fig. 3.

TABLE 5.

Showing the average length of the internodes on fibers of like diameter at the levels S_1, T and T_3 in the nerves to the leg of Frog 5. Otherwise, this table is similar to Table 4.

Level.		Internodes.		Internodes.		Internodes.	
S_1	Diameter.........	5.2μ (27)	611	6.4μ (47)	623	7.3μ (34)	805
T	Diameter.........	5.2μ (38)	646	6.3μ (30)	787	7.3μ (50)	836
T_3	Diameter.........	5.2μ (72)	818	6.3μ (25)	917	7.1μ (9)	1011

In view of this result, it was thought necessary to determine this same relation in other specimens. Before commenting on the foregoing results, therefore, the additional observations on this point will be presented.

A large specimen, Frog 8, body weight 63 grams, total length 222 mm., was examined. Three localities on the nerves to the leg, namely S_1, T and T_3 were selected, and more than 100 internodes measured at each level. The measurements were made on fibers from 5μ to 7.5μ in diameter. These have been arranged as before, in three classes, approximately 5.3μ, 6.3μ and 7.3μ in diameter, and the data thus condensed, are given in Table 6.

TABLE 6.

Showing the average length of the internodes at the several levels in the nerves to the leg of Frog 8. Body weight, 63 grams; total length, 222 mm. At each level, the measurements are grouped in three classes according to diameter. The numbers in parentheses indicate the number of measurements.

Level.		Internodes.		Internodes.		Internodes.	
S_1	Diameter.........	5.3μ (45)	711	6.3μ (31)	828	7.5μ (44)	909
T	Diameter.........	5.1μ (61)	589	6.4μ (33)	723	7.4μ (44)	923
T_3	Diameter.........	5.1μ (79)	806	6.3μ (56)	963	7.4μ (48)	1015

When read horizontally, the records in Table 6 show that the length of the internodes increases with the increasing diameter of the fibers. When read vertically however it appears that while at T_3 the internodes are always longer than at S_1, yet the internodes for fibers with the diameters 5.3μ and 6.3μ at the level T are shorter than those either above or below this level. Comment on this result will be made later.

Fig. 5 also represents these relations, the measurements at all three levels having been reduced to exactly the same diameter, namely, 5.3μ, 6.3μ and 7.3μ.

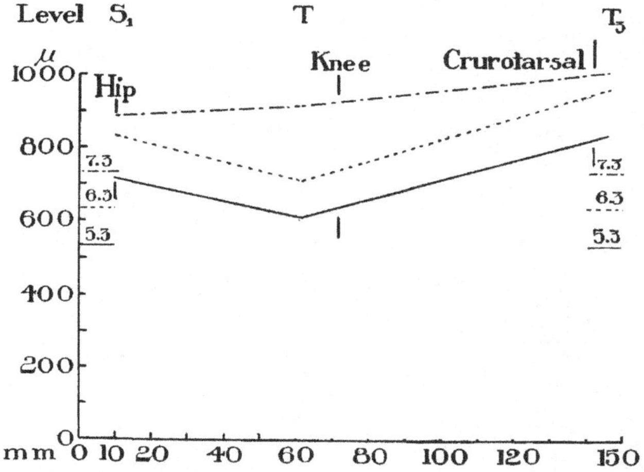

FIG. 5. Showing the lengths of the internodes at the levels S_1, T, and T_3, on fibers 5.3μ, 6.3μ and 7.3μ in diameter from Frog 8. The diameters multiplied by 100 are indicated on the limiting verticals.

In addition to Frog 8, still another specimen, Frog 3, body weight 26 grams, total length 166 mm., was examined in the same way. More than 100 internodes on fibers ranging in diameter from 3.75μ to 6.3μ were measured at each of the three levels S_1, T and T_3.

The measurements are treated as before, and are presented in Table 7.

The table reads regularly, both horizontally and vertically, and thus shows a steady increase in the length of the internodes, as the fibers increase in diameter, and also along a given fiber from S_1 towards the foot T_3.

TABLE 7.

Showing the average length of the internodes at the several levels in the nerves to the leg of Frog 3. Body weight, 26 grams; total length, 166 mm. At each level the measurements are grouped in three classes according to diameter. The numbers in parentheses indicate the number of measurements.

Level.		Internodes.		Internodes.		Internodes.
S_1	Diameter.........	3.9μ (12)	416	5.2μ (72)	520	6.3μ (25)	578
T	Diameter.........	4.0μ (13)	435	5.2μ (64)	601	6.3μ (23)	692
T_3	Diameter.........	4.0μ (24)	578	5.2μ (53)	701	6.3μ (63)	805

Fig. 6 exhibits these relations in the form of curves.

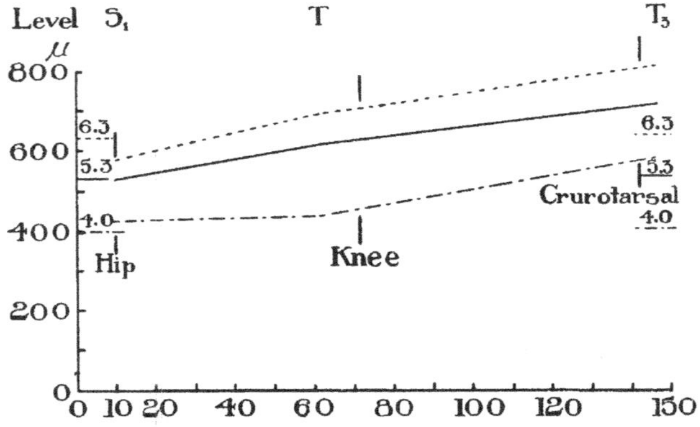

FIG. 6. Showing the average length of the internodes at the several levels, S_1, T and T_3, on fibers 4μ, 5.3μ and 6.3μ in diameter, from Frog 3. The other indications as in Fig. 5.

As the foregoing represents all the data collected in connection with this question, we return to a discussion of the fundamental point, namely, the length of the internodes on fibers of a given diameter through their entire extent from S_1 towards the foot T_3. An examination of Tables 5, 6 and 7 shows:

First, that on fibers of a given diameter, the internodes are not of the same lengths at the several levels;

Second, that in general, the internodes become longer as we pass toward the periphery;

Third, that they are markedly elongated at T_3, the level of the foot.

In attempting to explain these relations, we naturally call to mind the fact that in Rana pipiens, the average proportional lengths of the leg bones are

Femur	26.1
Tibia	29.6
Tarsus and pes	44.3

These figures are the averages from Table XI in DONALDSON and SCHOEMAKER '00.

As these relative values remain practically unchanged during the growth of the leg in length, it follows that the increments in length must be in the same proportion, and therefore a lengthening of 100 units in the femur, is accompanied by a lengthening of 113.4 units in the tibia, and 169.7 units in the tarsus and pes. If, for the moment, we assume that the portion of the nerve in each segment of the limb is so linked with that segment that it lengthens at the same rate, then we should expect a corresponding relation in the length of the internodes; provided, of course, they were of equal length when first laid down. It appears worth while to put this conclusion to the test, so far as the data in hand will permit.

Before this can be done however several adjustments and corrections must be made in the raw values. In the first place, as the intermediate level T is within the limits of the thigh, and hence associated with the femur, the measurements at T are excluded from the following comparisons, and we contrast only the length of the internodes at S_1 with that found at T_3, to determine whether these lengths stand in the same relation as the increments of growth in these segments of the limb, namely, as 100 : 169.7. In order to do this, it is necessary to compare the internodal lengths belonging to classes of fibers having exactly the same diameters. We choose as the standards for the diameter classes, 4μ, 5.3μ, 6.3μ and 7.3μ, since the observed values can be reduced to these standards by alterations which never amount to more than $.2\mu$.

In doing this, we assume that the reduction can be made by simple proportion. The results based on the reduced values are given in the following Table 8.

TABLE 8.

Showing the relative length of the internodes at T_3 compared with those at S_1 as a standard, in the case of the several diameter classes in all three frogs (Frog 3, Frog 5 and Frog 8).

	Diameter in μ.	Internodes at S_1	Internodes at T_3	Relative Value at T_3 for Each Diameter Class.	Relative Value at T_3. Average for Each Frog.
Frog 3..........	4.0	425	578	136.0	
	5.3	530	714	134.7	
	6.3	578	805	139.2	
					136.6
Frog 5..........	5.3	623	834	133.8	
	6.3	645	917	140.9	
	7.3	805	1039	129.1	
					134.6
Frog 8..........	5.3	711	837	117.7	
	6.3	828	963	116.3	
	7.3	885	1001	113.3	
					115.7

A study of Table 8 reveals several points of interest. First, the internodes at T_3 are always considerably longer than at S_1.

Second, the relative value at T_3 ranging from 136.6 to 115.7, is always much less than the relative growth of the foot, namely, 169.7.

Third, it appears that this proportional excess of the internodes in the foot tends to diminish in the larger frogs. Frog 8, the largest, showing a value of 115.7, whereas Frog 3, the smallest frog, shows 136.6, and Frog 5, intermediate in weight and length, gives an intermediate value.

We conclude from these relations, that while the length of the internode along the fiber is probably influenced by the elongation of the segment in which it is found the effect of the local elongation is more or less distributed over the entire length of the nerve fiber. If we wish therefore to discover what is really taking place as regards the lengthening of the internodes, we must study the changes over the entire extent of the fiber

For this purpose it is necessary to determine the average lengthening of the internodes during the period of growth on the nerve fibers taken from Frog 3, Frog 5, and Frog 8. To do this fairly, the diameter classes must be made exactly similar. In addition, due account must be taken of the fact that not only do the internodes increase in length, but the fibers to which they belong, increase at the same time in diameter, and therefore a diameter class of given size in the smaller frog must always be compared with a class of greater diameter in the larger frog. To make this, comparison it is necessary to obtain some notion of the amount of change in diameter which may be expected to occur in the cases we are examining.

Finally, for comparison, it is necessary to determine in the several frogs compared the proportional lengthening of the nerves to which these fibers belong.

In the absence of direct observations, we assume that the lengthening of the fibers which pass from the intervertebral foramina to the foot, is proportional to the lengthening of the leg itself. To determine what this is, we proceed as follows:

Since in the case of the frogs in question, the length of the legs is always a constant fraction of the total length of the frog, it follows that the increase in the length of the legs will be in proportion to the increase in the total length of the frog.

Treating the data in this way, we obtain the results shown in Table 9.

TABLE 9.

Showing the relative length of the legs in Frog 3, Frog 5, and Frog 8, based on a comparison of the total lengths of these same frogs. Group (A). Frog 3 taken as the standard. Group (B). Frog 5 taken as the standard.

Frog.		Total Length.	Ratio for the Legs.
		mm.	
Group (A)	3	166	100.0
	5	192	115.6
	8	222	133.7
Group (B)	5	192	100.0
	8	222	115.6

From Table 9 it appears that when Frog 3 is taken as the standard in Group (A) the length of the leg in Frog 5, is 15.6 per cent greater, and in Frog 8, 33.7 per cent greater, while in the second instance, Group (B) where Frog 5 is taken as the standard, the leg in Frog 8 is 15.6 per cent greater.

Our next step is to make an approximate determination of the increase in the diameter of the growing fibers in the frogs in which the nerves to the leg increase 15.6 per cent over the standard.

To determine the increase in diameter which probably occurs when the nerve increases 15.6 per cent in length, we proceeded as follows:

By comparing the sum of the internodal lengths of the 4μ, 5μ and 6μ fibers in Group I of BOYCOTT's table (reprinted as Table 1 on p. 168) with the corresponding sum of the 5μ, 6μ and 7μ fibers in Group II and these in turn with the sum of the 6μ, 7μ and 8μ fibers in Group III, it was found that for an increase of 1μ in diameter, there was an average increase in internodal length of 25.9 per cent. Since we assume in the case of our own frogs that the internodal length will increase in proportion to the increase in the length of the nerve, and since the latter amounts to 15.6 per cent, it follows that if an increase of 25.9 per cent in internodal length, calls for an increase of 1μ in the diameter of the fiber, then 15.6 per cent increase in internodal length, will call for approximately 0.6μ increase in the diameter of the fiber.

This result is based of course on BOYCOTT's measurements made on R. temporaria. It seems justifiable to apply it to R. pipiens however because, although DONALDSON '08 has shown that the internodes in R. pipiens are shorter than in R. temporaria, he has also shown that the proportional differences in length are nearly the same for the several diameter classes, and hence any given change in the diameter, is associated with the same relative change in length of internode in both species.

Accepting therefore this determination of the diameter increase, the next step is to compare the internodes on the fibers of a given diameter of one specimen of R. pipiens, with the internodes in another specimen, on fibers which are 0.6μ greater in diameter. To do this, we select from the foregoing Tables 5, 6 and 7, the internodal lengths on fibers for the diameter classes 5.3μ, 6.3μ and 7.3μ from all three levels. This permits us to make nine comparisons.

Thus in each of these comparisons, as for instance in the first one, in Table 10 the average internodal length in the diameter class 5.3μ at S_1 in Frog 3, is compared with the average internodal length on fibers 5.9μ in diameter at S_1 in Frog 5, and the same procedure is followed in each of the other eight comparisons.

The foregoing tables, 5, 6 and 7, however, show the internodal values only for the diameter classes 6.3μ and 7.3μ, while for our present purpose, it is necessary to use those for 5.9μ and 6.9μ. The desired values are obtained by the simple proportional reduction of the internodal length of the 6.3μ class to that for 5.9μ, and of the 7.3μ class to that for 6.9μ.

In view of all conditions, the values thus determined, are probably nearly correct, although the method is open to some theoretical objections. The comparisons which are thus made possible are given in Table 10.

TABLE 10.

Showing the growth of the internodes on the fibers 5.3μ and 6.3μ in diameter, in nerves which increase 15.6 per cent in length. Internodes from Frog 3, compared with those from Frog 5, and from Frog 5 compared with those from Frog 8.

LEVEL.	FROG.	DIAMETER IN μ.	INTERNODES.	PERCENTAGE INCREASE.	AVERAGE PERCENTAGE INCREASE FOR EACH LEVEL.
S_1	3 5	5.3 5.9	530 596	12.4	
	5 8	5.3 5.9	623 773	24.1	
	5 8	6.3 6.9	635 834	31.0	22.5
T	3 5	5.3 5.9	601 738	22.7	
	5 8	5.3 5.9	646 667	3.2	
	5 8	6.3 6.9	787 863	9.6	11.8
T_3	3 5	5.3 5.9	714 856	20.0	
	5 8	5.3 5.9	834 903	8.2	
	5 8	6.3 6.9	917 945	3.0	10.4

As Table 10 shows, the average growth of the internodes at S_1 is 22.5 per cent, which is greater than that at T, 11.8 per cent, or at T_3, 10.4 per cent. Also the percentage increase at the level S_1 is greater in the larger than in the smaller frogs. These results accord with those previously noted in the examination of the internodal lengths on fibers from the same frogs, in which the length of the internodes in the foot becomes proportionally less as the frog becomes larger (see Table 8).

To determine the average growth of the internodes on individual fibers, it is necessary to measure the fibers of a given diameter class taken from the same frog, at all three levels, and Table 11, based on the data in Table 10, gives the values found.

TABLE 11.

Showing the percentage increase in the average length of the internodes on fibers of a given diameter, when all three levels from the same frog are included. The averages used are those given in Table 10.

	Frog 3	Frog 5	Frog 5
Diameter	5.3μ	5.3μ	6.3μ
S_1	12.4	24.1	31.0
T	22.7	3.2	9.6
T_3	20.0	8.2	3.0
Averages for each frog	18.4	11.8	14.5
Grand average	14.9 per cent increase.		

It is seen from the foregoing, that the average increase in the length of the internodes in Frog 3, fibers 5.3μ in diameter, is 18.4 per cent, Frog 5, fibers 5.3μ in diameter, 11.8 per cent, and Frog 5, fibers 6.3μ in diameter, 14.5 per cent; the grand average for the three frogs being 14.9 per cent. The nerves to which these fibers belong have lengthened in each case 15.6 per cent, so that the accordance is fair in each instance, except in the case of the 5.3μ group in Frog 5. It should be recalled however that in the 6.3μ group in Frog 8, at the level T, a very low value was obtained (Table 6). Because this value is less than that at S_1, it may be considered aberrant, and it is the presence of this value which causes the low percentage, 3.2 per cent, in Frog 5, at the level T. If this observation is excluded, the value for the 5.3μ group in

Frog 5 becomes 16.1 per cent or nearly that for the lengthening of the nerve, and the grand average becomes 16.4 per cent, or a little greater than 15.6 per cent, which represents the lengthening of the nerve.

It seems allowable therefore to conclude that the internodes in the 5.3μ and 6.3μ diameter classes, grow, on the average, at approximately the same rate as does the nerve in which they are found. Nevertheless on passing distally along the nerve, the length of the internodes in a given diameter class, tends to increase in such a way as to suggest that it is influenced by the growth of the segment of the limb to which the internodes belong, although this influence becomes less marked as the frog becomes larger.

3 THE LENGTH OF THE INTERNODES ON FIBERS IN THE ROOTS OF THE SPINAL NERVES.

Touching this point we have observations on the roots of the IX nerve in five frogs of different sizes. In his plate VI, HARDESTY ('99) has given some excellent drawings of the nerve roots in this frog. The species used by HARDESTY was designated Rana virescens but is the same as that here designated, Rana pipiens (see DONALDSON '07).

The present data are brought together in Table 12. Each specimen is given the number which it bears in Table 2 but the series is arranged in the order of the increasing length of the nerve roots.

Table 12 shows that as the nerve roots increase in length, the internodes on the fibers in these roots also increase in length. The average length of the internodes is somewhat less in the dorsal roots than in the ventral, and this, in each instance, goes along with a smaller average diameter of the fibers measured. Fig. 7 shows these relations also.

Using the data in Table 12, we may form the supplementary Table 13 in which are compared the values for the length of the ventral or dorsal root, with the corresponding values for the internodes on fibers in this root. The series of ratios given in Table 13 indicate that the internodes on fibers of both roots lengthen in about the same proportions as the roots in which they appear. It is interesting to observe that this lengthening of the roots is quite independent of the increase either in the total length, or in the body length, of the frogs concerned. It appears from this that the internodes on the roots of the IX nerve grow as do the internodes in the nerve to the leg. Concerning the limits of the stretch of nerve which we have to examine, we may feel very sure that in the case of the dorsal root they have been correctly determined. This stretch lies between the spinal ganglion and the point of union of the root with the cord. In the case of the ventral root however the corresponding stretch appears to be between the cord as one limit and the junction of the ventral with the dorsal root as the other, although further observations are necessary to establish the latter limit beyond dispute.

TABLE 12.

Showing the length of the nerve roots of the IX nerve, and the average length of the internodes on the fibers in them. The averages were obtained from random sampling and are based on the measurement of 50 fibers in each case.

Serial Number from Table 2.	Sex.	Body Weight.	1 Total Length 2 Body Length.	Length of V (Ventral) D (Dorsal) Roots.	Diameter of Fibers. Ventral Root.	Length of Internodes. Ventral Root.	Diameter of Fibers. Dorsal Root.	Length of Internodes. Dorsal Root.
		grms.	*mm.*	*mm.*	μ		μ	
1	M.	5.5	10.4 3.9	V 2.6 D 2.1	8.6	389	6.4	303
8	M.	63.0	222.0 89.4	V 5.9 D 5.4	11.2	1054	10.2	961
2	M.	23.5	169.0 71.0	V 7.0 D 6.5	14.4	1131	11.7	955
4	F.	27.2	180.0 78.0	V 7.7 D 7.1	13.6	1325	12.1	1116
7	M.	61.1	226.0 89.0	V 9.5 D 9.0	14.7	1339	11.0	1153

In the relations between the diameter of the fibers and the length of the internodes in the dorsal and ventral roots, there are certainly no striking differences, since in the ventral roots the internodal length is on the average 84 times the diameter, while in the dorsal roots it is 87 times. If we compare the length of the internodes on fibers of a given diameter in the ventral and dorsal root and also in the sciatic nerve at the level S_1 in Frog 8—the only specimen in which the comparison can be made—it appears that

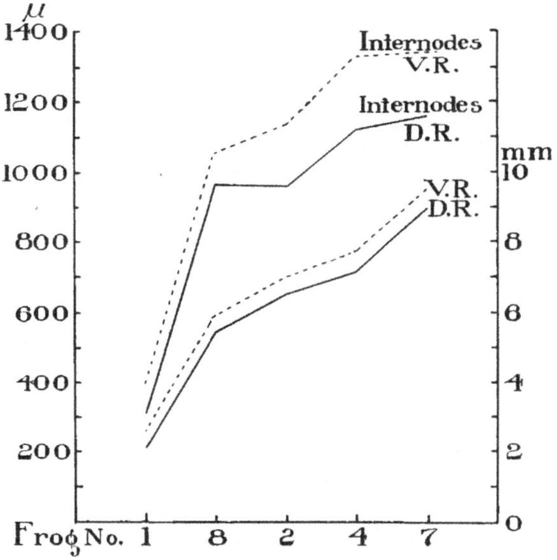

FIG. 7. To show the length of the dorsal (D.R.) and ventral (V.R.) roots respectively, in the IX nerve of the five frogs examined, and the corresponding lengths of the internodes on fibers from these roots, the measurements for the roots are given in millimeters on the vertical to the right, and for the internodes in μ, on the vertical to the left.

TABLE 13.

Based on data in Table 12 and comparing the relative increase in the length of the nerve roots with the relative increase in the length of the internodes as shown by a series of ratios.

SPECIMENS.	VENTRAL ROOTS.		DORSAL ROOTS.	
	RATIOS OF LENGTH OF ROOTS.	RATIOS OF LENGTH OF INTERNODES.	RATIOS OF LENGTH OF ROOTS.	RATIOS OF LENGTH OF INTERNODES.
Frog 1	1.00	1.00	1.00	1.00
Average of Frogs 8 and 2	2.48	2.80	2.83	3.16
Average of Frogs 4 and 7	3.31	3.42	3.83	3.74

the internodal lengths are greater in the sciatic than in the spinal roots. These values are given in Table 14.

TABLE 14

Showing in Frog 8 the lengths of the internodes on fibers 10μ, 11.25μ and 12.5μ in diameter from the ventral and dorsal roots of the IX nerve, and from the sciatic at the level S_1. The numbers in parentheses indicate the number of measurements made in each case, and apply to the internodal value above which they are placed.

Diameters in μ	Internodal Lengths.		
	IX Nerve Ventral Root.	IX Nerve Dorsal Root.	Sciatic Nerve Level S_1.
	(28)	(16)	(46)
10.00	1064	1037	1147
	(24)	(12)	(11)
11.25	1018	910	1238
	(24)	(10)	(51)
12.50	1110	1015	1316

Just what interpretation is to be given to the difference in the internodal lengths, from these several localities, must await the collection of a much larger number of observations.

We turn next to the comparison of the length of the internodes in the III, with those in the IX spinal nerve. Owing to the technical difficulties already mentioned, p. 171, we have only a limited number of observations on the ventral root of the III nerve of Frog 2, to be compared with those on the fibers in the ventral root of the IX nerve from the same frog. If we select the fibers 10μ to 15μ in diameter, inclusive, and tabulate the average values of the internodes for each diameter class, we get the results presented in Table 15.

TABLE 15.

Giving the length of the internodes on fibers 10μ to 15μ in diameter, in the ventral roots of both the III nerve and IX nerve of Frog 2. The numbers in parentheses indicate the number of observations.

Diameters in μ....	10	11.25	11.6	12.5	13.33	13.75	14.16	15
III nerve..........	(12)	(8)		(17)		(2)		(2)
	960	1052		1050		1210		1440
IX..............	(2)		(1)	(8)	(1)	(9)	(2)	(26)
	980		940	985	1000	1220	1130	1147

An examination of Table 15 shows that for fibers of the same diameter, the internodal lengths are nearly alike in the two nerves. If we make a general average, we find the relations given in Table 16.

TABLE 16.

Giving the average diameters and average length of internodes on the fibers from 10μ to 15μ in diameter, found in the ventral roots of the III and IX nerves of Frog 2.

	DIAMETER IN μ.	INTERNODAL LENGTH.
Nerve III	11.7	1051
Nerve IX	14.0	1117

In view of the small number of observations, we may look upon these values as similar, but the fact that they are similar, is the surprising result, since in this particular case (see Table 12) the length of the ventral root of the III nerve is 2.6 mm., while that of the IX nerve is 7.0 mm., giving a ratio of 1 : 2.7. If we assume that these roots had the same length when medullation began, it would appear that since the IX nerve had become 2.7 times as long as the III, the internodes should stand in a like relation.

The measurements show that such is not the case. Unfortunately, at the moment, it is not possible to explain this result. The discrepancy possibly arises through taking as the distal limit of the ventral roots the point of junction of the ventral with the dorsal root, and yet the assumption of this limit fitted perfectly with what has already been found in the case of the IX nerve. There are of course many suggestions which might be made, but it seems best to leave the question in abeyance, until a larger number of observations, especially on the dorsal root of the III nerve, has been made.

4 THE NUMBER OF MEDULLATED FIBERS AT DIFFERENT LEVELS IN THE LEGS OF TADPOLES OF INCREASING SIZE.

To fill out our information as to the way the medullary sheaths of the nerve fibers are acquired in the frog, the number of the medullated fibers at the level of the knee was counted, and compared with the number at the entrance to the foot, in the legs of tadpoles of increasing size. To prepare this material, the legs, or so much of them as was needed, were fixed in 1 per cent osmic acid, and then embedded and sectioned in the usual manner. The sections were made 12μ in thickness, and at the knee, the number of medullated fibers in the trunks of the nervus tibialis and nervus peroneous was counted. This number was contrasted with that found in the four trunks entering the foot, namely: the ramus superficialis and ramus profundus of the nervus tibialis, and the nervus peroneus lateralis and medialis.

At both levels the number of fibers in several successive sections was counted, and the average taken. The results of this examination are presented in Table 17.

TABLE 17

Showing the number of medullated fibers at the level of the knee and ankle in the leg of the tadpole. Tadpoles of Rana pipiens.

Number of Specimen.	Length of Shank	Length of Foot.	Number of Fibers in a Cross Section at the Level of		Ratios of Number at Knee to Number at Ankle.
			Knee.	Ankle.	
	mm.	*mm.*			
I	1.18	0.68	23	4	5.75-1
II	1.29	0.80	25	6	4.16-1
III	1.50	0.96	30	8	3.75-1
IV	1.88	1.36	85	25	3.40-1
V	3.68	2.71	286	118	2.42-1

It appears from this that both the absolute and relative number of medullated fibers entering the foot, increases as the leg of the tadpole becomes longer, rising from 1 : 5.75 to 1 : 2.42. In a large mature frog, Dr. Dunn ('02) has shown that the ratio is 1 : 1.66, an increase of $3\frac{1}{2}$ fold over the ratio in the smallest tadpole. At the moment however we have no data by which to determine when the ratio found in the largest tadpole's leg here examined passes over to that in the mature frog.

It is assumed that the medullated fibers which are counted at he level of the knee, represent fibers already medullated throughout their entire length, as well as fibers incompletely medullated, but having a sheath extending as far as the level of the section. The same assumption is made for the fibers at the level of the foot.

It is shown then that the medullation of fibers going to the shank, the more proximal segment of the limb, is more nearly complete than that of those passing to the foot, the more distal segment, and probably the greater part of this difference depends upon the fact that many of the fibers destined for the foot are not medullated at all.

This result, taken in conjunction with those of HARDESTY ('99) on the nerve roots of the frog, and HATAI ('01, '02), on the nerve roots of the rat, indicates very clearly that new medullated fibers are continually being added to the nerves during the period of growth.

5 A COMPARISON OF THE LENGTH OF THE INTERNODES IN THE AMERICAN LEOPARD FROG, RANA PIPIENS, WITH THEIR LENGTH IN THE ENGLISH FROG, RANA TEMPORARIA (FUSCA).

In his study, entitled "The nervous system of the American leopard frog, Rana pipiens, compared with that of the European frogs, Rana esculenta and Rana temporaria (fusca)," Dr. DONALDSON ('08) compared the measurements of the internodes made by me on Rana pipiens, with those made by BOYCOTT on Rana temporaria. Taking the same locality in both cases, and reducing the measurements on R. pipiens so that they apply to frogs of the same total length as those measured by BOYCOTT, a series of values was obtained for seven diameter groups. It appeared from a comparison of the results (see DONALDSON '08, p. 146, Table 19), that the internodal lengths in Rana pipiens, ranged between 64 and 71 per cent of those found in Rana temporaria, the average being 67 per cent.

The comparison appears to be a fair one, and if this is granted, it is evident that Rana pipiens has on its fibers three sheathing cells, where Rana temporaria has two. This result further draws attention to the fact that the character in question is subject to considerable variation, and that this appears not only in forms widely separated zoölogically, but also within the genus Rana, at least in the case of the two closely related species here compared.

CONCLUSIONS.

In the leopard frog, Rana pipiens, we have found:

1. The average length of the internodes on the fibers in the nerves to the leg diminishes towards the periphery. This diminution is accompanied by a corresponding diminution in the average diameter.

2. In the same frog, the length of the internodes at different levels on fibers of like diameter in the nerves to the leg, *increases* toward the periphery. This increase appears to be associated with the more rapid growth of the distal segments of the leg, but the influence of the segment on the portion of the nerve within it, is less marked as the frogs become larger.

3. When the average length of the internodes on fibers of a given diameter is compared with the average length on the fibers which represent them in a larger frog, it is found that the lengthening of the internodes corresponds with that of the nerve to which they belong, thus supporting BOYCOTT's ('04) general conclusion.

4. In the roots of the IX spinal nerve, the internodes lengthen in proportion to the lengthening of the nerve, but at the same time, the lengthening of these roots is only loosely correlated with the increase either in the total length or in the body length of the frog to which they belong.

5. When, in the same frog, the ventral root of the III nerve is compared with the ventral root of the IX nerve, it is found in both of them, that the fibers of the same diameter have internodes of the same length. In the case chosen, the ventral root of the IX nerve had become 2.7 times the length of the III nerve and we should therefore expect to find the internodes on the fibers of the IX nerve much longer than those on the corresponding fibers in the nerve III. The explanation of this result awaits further observations.

6. A determination of the number of medullated nerve fibers at the level of the knee and of the ankle in a series of tadpoles' legs of increasing length, shows that the relative number of medullated fibers at the ankle, increases as the leg becomes longer, thus proving that the fibers to the more distal divisions of the limb are medullated later.

7. It follows from the foregoing result that so long as the nerve receives new (young) fibers, there will always be internodes which

are relatively short, since they belong to fibers which have been subjected to the lengthening process for only a short time. The presence of these fibers reduces the average length of the internodes, and hence accounts in part at least, for BOYCOTT's observation that on the average the lengthening of the internodes in the sciatic nerve is slightly less than that of the nerve itself. It also accounts, in part at least, for the wide range in the length of the internodes found on fibers of the same diameter.

8. In the leopard frog, Rana pipiens, the length of the internodes at the distal end of the sciatic nerve, is on the average, only about two-thirds that of the corresponding internodes in Rana temporaria (fusca) as measured by BOYCOTT ('04)

SUMMARY.

The foregoing conclusions may be made more vivid perhaps if, in the light of our present knowledge, we attempt to picture the growth changes which affect the internodes on the nerve fibers of the leopard frog. From the observations of HIS ('86) HARRISON ('01, '04–'06), BARDEEN ('02–'03), and others, we know that the axone grows out from the cell body into the peripheral nerve, accompanied by its sheathing cells. There are no observations to show whether before the formation of the myelin the sheathing cells cover approximately the same length of fiber in all fibers, or at all periods of growth, but our observations as they stand, would favor such a view.

In the leg of the tadpole, the formation of myelin occurs first in the fibers which run the shorter course, and interpreting the findings of HARDESTY ('99) and HATAI ('03) showing a diminishing number of medullated fibers in the spinal roots as we pass away from the cells of origin, it appears that the development of the myelin progresses from the cell of origin toward the end of the axone.

When the axone has made its distal connection, and the myelin is formed, then the lengthening begins, and continues so long as the nerve to which the fiber belongs, continues to grow. In the nerves to the leg however this process is modified by the fact that the internodes have a tendency to lengthen at the same rate as the segment of the leg to which they belong; although this process is more marked in the younger than in the older frogs. Despite this however the average length of the internodes on fibers of a given diameter increases as does the nerve in which they occur.

The interpretation of the internodes, as we find them in a sample taken from any nerve, is complicated by the fact that for a long time during growth, new medullated fibers are appearing. As these new fibers start with very short internodes, and are late in appearing, they have been affected by the lengthening process for a shorter time than those fibers which were completely medullated at an earlier date. They must consequently exhibit internodal lengths shorter than would be expected, and since their absolute number increases as the frog becomes larger, and their presence lowers the average length of the internodes at any level, it will necessarily follow, as shown by BOYCOTT ('04), that the average length of the internodes increases a little less rapidly than that of the nerve to which they belong. This is our explanation of BOYCOTT's result.

While this change in the length of the internodes is taking place, there is also a change in the diameter of the fibers. In general, the increase in diameter is in advance of the increase in internodal length, so that, as BOYCOTT has shown, fibers of a given diameter have longer internodes in larger frogs.

The exact relation of these two processes has still to be worked out, but this relation, depending as it does on the medullation of the fibers at different dates, and on the fact that all fibers of small diameter are not destined to become fibers of large diameter (BOUGHTON '06), but may remain permanently small, seems to account for the great variation in the length of the internodes on fibers of the same diameter, quite aside from the fact that consecutive internodes on the same fiber may have very different lengths.

While the foregoing description is based on the study of the nerve to the frog's leg, we find that it applies also to the growth changes in the roots of the IX spinal nerve, when we take as the limits of the dorsal root, the spinal ganglion on one side, and the spinal cord on the other, and in the case of the ventral root, the spinal cord on one side, and the junction point of the ventral and dorsal roots on the other.

When however we compare the internodal lengths in the IX ventral root with those in the III ventral root of the same frog, taking the same limits, we get the surprising result that the internodal lengths are similar, although the lengthening of the IX nerve has been 2.7 times that of the III. This result still awaits an explanation.

BIBLIOGRAPHY.

BARDEEN, C. R.

'02–'03. The growth and histogenesis of the cerebrospinal nerves in mammals. *Amer. Journ. Anat.*, vol. 2, pp. 231–257.

BOUGHTON, T. H.

'06. The increase in the number and size of the medullated fibers in the oculomotor nerve of the white rat and of the cat at different ages. *Journ. Comp. Neurol. and Psychol.*, vol. 16, pp. 153–165.

BOYCOTT, A. E.

'04. On the number of nodes of Ranvier in different stages of the growth of nerve fibers in the frog. *Journ. Physiol.* (FOSTER), vol. 30, pp. 370–380.

DONALDSON, H. H., and SCHOEMAKER, D. M.

'00. Observations on the weight and length of the central nervous system and of the legs in frogs of different sizes (Rana virescens brachycephala—COPE). *Journ. Comp. Neurol.*, vol. 10, pp. 109–132.

DONALDSON, H. H.

'07. Rana pipiens. *Science*, n. s., vol. 26, no. 655, p. 78.

'08. The nervous system of the American leopard frog, Rana pipiens, compared with that of the European frogs, Rana esculenta and Rana temporaria (fusca). *Journ. Comp. Neurol. and Psychol.*, vol. 18, pp. 121-149.

DUNN, E. H.

'02. On the number and on the relation between diameter and distribution of the nerve fibers innervating the leg of the frog (Rana virescens brachycephala, COPE). *Journ. Neurol.*, vol. 12, pp. 297–328.

HARDESTY, I.

'99. The number and arrangement of the fibers forming the spinal nerves of the frog (Rana virescens). *Journ. Comp. Neurol.*, vol. 9, pp. 64–112.

HARRISON, R. G.

'01. Ueber die Histogenese des peripheren Nervensystems bei Salmo salar. *Arch. f. mikr. Anat.*, vol. 57.

'04. Neue Versuche und Beobachtung über die Entwicklung der peripheren Nerven der Wirbelthiere. *Sitz.-Ber. niederrhein. Ges. Nat. Heilk., Bonn*, pp. 55–62.

'06. Further experiments on the development of peripheral nerves. *Amer. Journ. Anat.*, vol. 5, pp. 121–131.

HATAI, S.

'03. On the increase in the number of medullated nerve fibers in the ventral roots of the spinal nerves of the growing white rat. *Journ. Comp. Neurol.*, vol. 13, pp. 177–183.

HIS, W.

'86. Zur Geschichte des menschlichen Rückenmarkes und der Nervenwurzeln. *Abhandl. d. math. phys. Cl. d. k. sächs Gesellsch. d. Wissensch., Leipzig*, vol. 13.

v. KÖLLIKER, A.

'96. Handbuch der Gewebelehre des Menschen. *Leipzig*. S. 2–5. (Gives a summary of the older observations on the internodes).

The Journal of
Comparative Neurology and Psychology

VOLUME XVIII JUNE, 1908 NUMBER 3

THE BEHAVIOR OF THE LARVAL AND ADOLESCENT STAGES OF THE AMERICAN LOBSTER (HOMARUS AMERICANUS).[1]

BY

PHILIP B. HADLEY.

(From the Biological Laboratory of Brown University and the Experiment Station of the Rhode Island Commission of Inland Fisheries.)

WITH TWENTY-TWO FIGURES.

OUTLINE OF CONTENTS.

[1] The present paper is the last of the series of four in which the author has attempted to analyze the behavior of the larval and early adolescent stages of the lobster. The papers already published are the following, references to which may be found in the bibliography at the end of the present work: (1) The relation of optical stimuli to rheotaxis in the American lobster; (2) Galvanotaxis in larvæ of the American lobster; (3) The reaction of blinded lobsters to light.

I. INTRODUCTION AND HISTORICAL SUMMARY.

Every year is bringing new and valuable additions to our knowledge of the behavior of the Crustacea. Most of the investigations dealing with this subject are concerned, however, with the Entomostraca, while the behavior of the higher forms has been less studied. It is apparent, moreover, that the experimental work done has been chiefly upon adults, while little attention has been given to the behavior of the larval forms of those Crustacea, as the macrurous decapods, which undergo an extensive metamorphosis. It is the aim of the present paper to demonstrate certain phases in the reactions of larval and early adolescent stages of the American lobster (Homarus americanus) to light, and to analyze these reactions, so far as possible, into their constituent factors.

In the study of reactions to light it is apparent that the lack of a satisfactory terminology has led to considerable confusion. This is manifest when we attempt to apply the definition of positive or negative phototaxis, as given by LOEB, to the types of behavior which we find, for instance, in the lobster and in the shrimp, Palemonetes (LYON 1907). LOEB (1905, p. 29) states that "positively heliotropic animals are compelled to turn their oral pole toward the source of light and move in the direction of the rays to its source." In the larval lobsters, however, there may be a difference between the signs of body-orientation and what may be called progressive orientation. In body-orientation the animal in question *turns* with reference to the source of light; in progressive orientation it moves toward or from the source of light. Employing these terms, we may say that the body-orientation of the larval lobster under stimulation by light is invariably negative, whereas the progressive orientation may be either positive or negative, as the conditions of the case determine.

Secondly, what do we mean by intensity and by direction of light? Are we justified in assuming that a stimulus such as light can be effective in causing either kind of orientation through its directive quality? The answer to these questions depends largely upon arbitrary definitions. YERKES' (1903) exposition of what constitutes a phototactic reaction as differentiated from a photopathic reaction indicates very nearly the meaning that will be given to these terms in the present paper. Attention may be called to one difference, however. It is inferred by YERKES that the sign of the phototactic response is dependent upon the previously assumed body-orientation of the organism. This is by no means necessarily true, for in the case of the larval lobster, it is clear that the orientation of the body has absolutely nothing to do with the sign of the consequent progressive orientation. For our present purposes we may, therefore, slightly modify the definition of YERKES by describing *a phototactic reaction as one in which the organism tends to place the longitudinal axis of the body parallel to the direction of the rays and to approach or recede from the source of those rays.*

If we so limit the meaning of a phototactic response, what shall we say regarding the nature of the so-called photopathic response? It is entirely possible (and indeed in the case of the larval lobsters, most probable) that again the view of YERKES (1903), that a photopathic reaction is one in which an organism "selects" a particular intensity of light, and confines its movements to the region illuminated by that intensity, is correct. But it is not so certain that the photopathic responses of the lobster larvæ are brought about by means of slight phototactic reactions, as YERKES (1903, p. 1) suggests for Daphnia. Therefore, for present needs, we may conclude that *a photopathic reaction is one in which an organism, without previous assumption of a body-orientation, "selects" regions of optimal light-intensity.* In the following account of experiments and observations, we shall see to what extent the behavior of the lobster larvæ conforms to these definitions of phototactic and photopathic reactions.

The movements of Entomostraca toward or from a source of light, and their reactions to rays of different wave lengths have been made the subject of investigation by many naturalists. In the earlier investigations it was commonly concluded that the intensity of light was the most important factor, and that organisms "chose" an optimal intensity. LUBBOCK (1881) and GRABER (1884) found that Daphnia gather in areas of greater light intensity. SCHOUTEDEN (1902) found that older individuals are negatively phototropic. These experiments, as repeated by DAVENPORT and CANNON (1897), YERKES (1899, 1903), and PARKER (1902), showed that Daphnia also manifests phototactic reaction. It was assumed, therefore, that some organisms may react either phototactically or photopathically. Later work of American investigat.rs has demonstrated that the Crustacea are more influenced by the directive factor of the light rays than by the intensity, and, more recently still, KEEBLE and GAMBLE (1904), in their excellent work on the color physiology of the higher Crustacea, have shown that the nature of the background may be an important factor in determining the reaction of many species.

The Malacostraca have received less attention than have the Entomostraca, and it is only for a comparatively short time that anything has been known concerning the reactions of either the larvæ or the adults of decapod Crustacea. With the adult forms of the decapods results have been readily obtained. HOLMES (1901) found that several species of terrestrial amphipods manifest a strong positive phototactic reaction, while all aquatic species are negatively phototactic. We know further from KEEBLE and GAMBLE (1904) that the adult form of Palemon is negatively phototropic and that Hippolyte is positively phototropic. Hippolyte, according to KEEBLE and GAMBLE, not only moves toward the light, but also "prefers" a white to a black background. Macromysis inermis reacts positively or negatively in accordance with the character of the background or the nature of the physical environment. It is positively phototropic on a white background, and negatively phototropic on a black background. Furthermore, when a choice of background is made possible, Macromysis "selects" the black. In the case of Hippolyte, the larvæ respond positively to light, as do the adults. BELL (1906) states that the adult crayfish is "somewhat negatively phototactic" and that difference in the intensity of light made but slight difference in the reactions. Other investigators have shown that the adults of several species of Crustacea react either positively or negatively to light. Very few investigators, however, have studied systematically the reactions of Crustacea in the larval stages. Among the first, LOEB (1893) reported the reactions to light of Limulus in the "trilobite stage." These larvæ, he said, are at first positive, and later, negative. PEARL (1904), by repeating LOEB's experiments, ascertained that this larval stage of Limulus manifests at first a negative reaction, and that later, a relatively small number of individuals gives a positive reaction. It was learned by KEEBLE and GAMBLE (loc. cit.) that the response of the larvæ of Palemon is the direct opposite of the reaction of the adults. BOHN (1905) discovered that the larvæ of the European lobster (Homarus vulgaris), although at first positive in their reaction to light, may later undergo certain changes. HERRICK (1896) states that larvæ of the American lobster react positively to light. BOHN (1905) learned that the reaction of Artemia salina was similar to that of Homarus vulgaris; and the writer has ascertained that the larvæ of the green crab, Carcinus granulatus, react sometimes positively and sometimes negatively, and behave very much like the larvæ of the lobster. The writer can verify the conclusions published by LYON (1906) that the larvæ of Palemon may react either positively or negatively to light.

The results of the small number of investigations which have been made upon the reactions of Crustacea in the larval stages, indicate the desirability of further systematic study of these reactions. PEARL (1904) has well pointed out the value of studying the "ontogeny of reaction," and of applying the knowledge thereby gained to the investigation of the more complex forms of response exhibited by adult individuals. Although the writer has not yet had an opportunity to study the behavior of the adult lobster, the present work shows that in the larval stages there are found diverse types of reaction, differing from one moment to another, and depending upon conditions which, even in the nicest experiments, are by no means readily discoverable; and, furthermore, that it is only by a systematic study of the reactions *through the developmental stages*, that many contradictory points can be cleared up, and the more complex behavior of the older animals explained.

II. BIOLOGY OF THE LOBSTER.

A brief résumé of the biology of the lobster will facilitate the understanding of later considerations. The life of the lobster consists of a series of stages or stage-periods, each of which represents the span of life between two successive moults, or castings of its shell. Of these stage-periods, the first four are passed through very rapidly, since the young creature usually moults four times in the first twenty days of its existence. These first few quickly passed stages (called the larval stages because they denote the successive emergence of one from another) include the most important changes in form, color, and manner of behavior, that the lobster undergoes. In each successive stage the animal is larger than before. The larvæ grow at the time of moulting, but never between moults. From the fourth stage on, each successive stage-period is of longer duration and the changes which the adolescent lobster thenceforth undergoes are correspondingly less significant, being characterized chiefly by alterations in internal morphology as the adult functional type is gradually approximated. The first three stages of the lobster are free-swimming stages, and the activities are without apparent

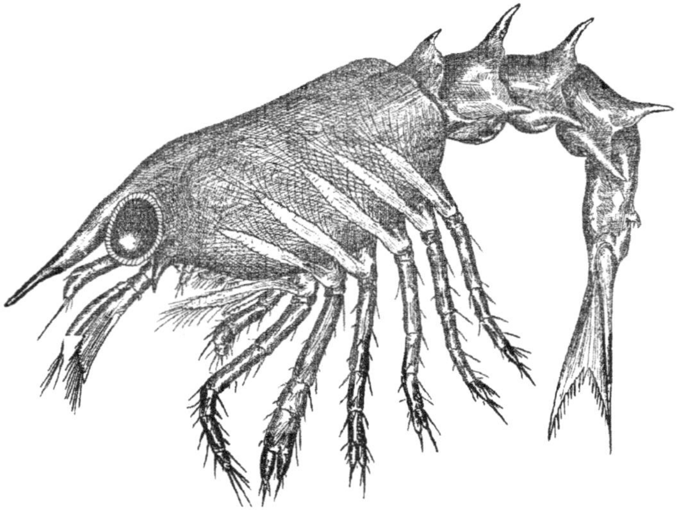

Fig. 1. Showing a young first-stage larval lobster about two days old. The eyes are large and prominent. The exopodites of the thoracic appendages are represented at the beginning of the downward stroke. This figure shows the typical swimming position of larvæ in the first three stages, the plane of the cephalo-thorax bent down at an angle of about 30° from horizontal.

coördination or aim. The larvæ are swept here and there by the tide and possess no power to evade the attacks of numerous enemies.

The swimming of the lobsters of the first three stages is accomplished by means of the feathered exopodites, or outer branches of the thoracic appendages (Fig. 1). These exopodites beat the water with short vibratory strokes, which tend to carry the larva backward or forward or upward as the case may be,[2] and allowing it to sink toward the bottom when their motion ceases. The progressive movement and the body-orientation of the lobster in the first three stages are almost wholly dependent upon the activity of these organs. Occasionally, darting backward movements, caused by the sudden contraction of the abdomen, appear, but these are of slight importance in the reaction to light.

When the lobster moults to the fourth stage, the exopodites are lost. Consequently the forward swimming during and after the fourth stage is dependent upon the action of swimming appendages which after the second stage make their appearance on the under sides of the second, third, fourth and fifth abdominal segments. The fourth-stage lobsters swim with directness and precision, usually near the surface of the water. This surface-swimming may be due to stimulation by light, but, as the writer has suggested elsewhere (1906b), it is not improbable that this form of behavior is due in part to the food-seeking impulse. During the latter part of the fourth stage, contact-irritability begins to play an important rôle in determining the behavior of the young lobster. Now, as in the fifth and later stages, the creature no longer swims at the surface of the water, but seeks the bottom and attempts to burrow in the sand or beneath any object that presents itself. After the fifth stage, the adolescent lobster shows the same type of behavior as during the fifth stage, but with a gradual increase in the tendency to avoid light. Its reactions have now become fixed in every way.

III. APPARATUS AND METHOD OF PROCEDURE.

The manipulation of the various pieces of apparatus here described will be spoken of when the particular experiments in which they are used are mentioned. The room in which the experiments were conducted contained on two of the opposite walls windows 2 feet high and 8 feet long, before which extended work benches or tables. The two windows, which opened respectively to the east and west, were the only source of daylight, and, as occasion required, were heavily screened with black paper or cloth. At appropriate places in these screens were cut openings which could be readily closed. On the table before one of the windows was

[2] For details on method of swimming, see p. 258.

placed a box 2 feet high, 3 feet wide, and 2 feet deep, lined on the inside with black cloth, and containing, on the window side, slits or openings to correspond with the openings in the screen outside the box. On the room side, the box was fitted with a movable black curtain, which permitted the operator to move the jars or other apparatus contained within the box. This arrangement served to control the light falling upon the larvæ, which were put in suitable containers and placed inside the box.

Other pieces of apparatus may be described as follows: *Glass box A*. Of glass boxes two types were used for studying the photopathic and phototactic reactions of the larvæ. One was a rectangular wooden box having glass "windows" in each end and in the bottom. This box, which was 12 inches long, 6 inches wide, and 3 inches deep, was painted dull black on the inside and fitted with a light-tight cover. It was used in experiments which required illumination from the end, from below, or both.

Glass box B—This box was similar in most respects to box *A* (see Fig. 7). It was 12 inches long, 6 inches wide, and 5 inches deep. It had "windows" on each end and along one side. Like box *A*, it was painted black on the inside and was fitted with a light-tight cover. This cover contained three slits so arranged that diaphragms of wood or glass might, in an instant, be slid into place to divide the box transversely into four chambers of equal extent. Then the cover of the box might be removed if desired, leaving the partitions in place. The object of this arrangement was to make it possible to imprison the young lobsters wherever they chanced to be at any given time and so to ascertain, by count, in what manner and in what relative numbers they had responded to certain stimuli.

Of these two boxes, the former, while oftener placed in a level position on a laboratory table, was sometimes used in another way to study the photopathic reaction alone, or the photopathic and the phototactic reactions together. In these cases the box was placed over a light-shaft, which was merely a rectangular tube lined with black cloth, with a height of 18 inches and with a cross section of the same size as the bottom of the box. Over the upper end of this tube or shaft, the glass bottom of box A exactly fitted. At the bottom of the shaft was either a sheet of white paper or a mirror which was so placed as to reflect the rays of light coming from the window up through the shaft to the glass bottom of the over-lying box. The rays that thus passed through the black-lined shaft and entered the box were practically parallel and at right angles to the plane of the bottom of the box. It is clear that, when the water in the box was very shallow, the rays of light passing up the shaft and striking the larvæ could have no directive influence, and that, when they passed through the graded light screens or through plates of colored glass, placed just beneath the glass bottom of the box, they could be effective only through difference in intensity.

Besides these boxes, use was made of certain glass jars, known as museum or brain jars, which were for the most part cylindrical in shape and varied in diameter from 20 to 25 centimeters. For certain experiments these were covered wholly or partially about their circumference with black paper, and the light was made to come from the top, bottom, or through a "window" in the side, as the case might require.

In addition to the apparatus mentioned above, several kinds of glass tubes were employed. Some were ordinary 15 centimeter laboratory test tubes, while others had a length of 40 centimeters and a diameter of 4 centimeters. These tubes were made with rounded ends so that there would be no obstruction to the light striking the tubes even at a slight angle, and the lobsters were introduced through an opening in the top. Another type of tube employed was the Y-tube, constructed of glass tubing, 4 centimeters in diameter as shown in Fig. 5. These proved exceedingly useful in testing the reactions of young lobsters, both to the intensity and the directive influence of the light rays, since the arms of the Y-tube could be readily covered with colored glass plates or fitted with black or white backgrounds, thus producing different conditions of light in each arm of the Y.

In many of the experiments it was desirable to use graded light screens. These were made by adding india ink to a solution of gelatine and allowing this to harden in the form of a wedge. The wedge-shaped screen permitted light to pass through in diminishing amount, from the thin edge to the thick edge, which was quite opaque. Graded light screens of red and blue were also made by adding to the gelatine a solution of eosin or methylene blue. It was by means of these, together with the colored glass plates that differences in the intensity of light were secured.

Since a particular response to light is often interpretable only when the conditions previous to the hour of experimentation are taken into account, it was found desirable to secure such conditions for experiment that all influences which might be instrumental in determining the final reaction of the larvæ either before or during the time of actual experimentation should be clearly recognized. Accordingly, the data to be presented show not only the nature of the reaction of the larval lobsters at a few chosen periods in their life history, but they also make it possible to trace modifications in reaction as the young animals pass on from stage to stage and gradually approach the adult type. Numerical results were usually obtained by counting the larvæ which had been imprisoned in different compartments of the box by the sliding partitions. In other cases a large number of larvæ were put into a glass jar and the reaction of the majority was observed. The separation and selection of larvæ which gave either a positive or a negative reaction to the same stimulus was thus possible, but conclusions have been drawn only after a careful study of the exact accounts of many groups of larvæ. The exact intensity of light used in the experiments was not known, but the experiments were performed on such days and at such times as would make the conditions uniform. Before entering upon a detailed consideration of the experiments as a whole, it will be appropriate to state some ground for assuming that lobster larvæ react both to the intensity and to the directive influence of light. The preliminary experiments which led to this view may be presented as follows:

IV. PRELIMINARY EXPERIMENTS.

Experiment I—Glass tubes 15 centimeters long and 2 centimeters in diameter were filled with salt water and in each were placed six first-stage lobsters two days old. When the tubes were held vertically, there was no tendency shown for the larvæ to gather in any particular region of the tubes. When, however, a strip of black paper was wound in such a manner as to cover the upper half of a tube (Fig. 2) and records were taken every minute, the larvæ became distributed as follows:

Number of larvae present after	Unshaded area.	Area where light and dark meet.	Shaded area.
1 minute	5	1	0
2 minutes	6	0	0
3 minutes	6	0	0
4 minutes	6	0	0
5 minutes	5	1	0
Total	28	2	0

It will be seen that the larvæ "seek" the light area. Next the paper was so arranged that the shaded part was below, as shown in Fig. 3. In this case the larvæ were always found uniformly in the unshaded area. In all cases the body-orientation of the larvæ was determined by the direction of the rays of light, which struck the tube at right angles; but at no time, it would seem, could this directive influence alone have been instrumental in causing the larvæ to remain in the region of greater light-intensity. The same general results were obtained when the tubes were laid horizontally on the table (Fig. 4) but still at right angles to the direction of the incident light rays which came from the side. These tests of the reaction of larvæ in glass tubes appeared at first sight to demonstrate that larvæ of a certain age and stage show a tendency to group themselves in regions of greater illumination, irrespective of the directive influence of light or of reaction to gravity. This, however, is not the only possible conclusion to be drawn from these facts.

To take, for instance, the case of the horizontally lying tube (Fig. 4), in which the larvæ gathered in the illuminated ends (the region of greatest light-intensity), and there oriented to the directive influence of the light. In the darkened area of the tube the larvæ did not undergo body-orientation, but swam about in many directions. When occasionally, they entered the more brightly illuminated end of the tube, they at once oriented to the directive influence of the rays and took the position shown in Fig. 4. Furthermore, the larvæ usually manifested a tendency to retain their body-orientation and thus to remain in the illuminated region when once they had entered it. Here, then, we have a case where the apparent reaction to the intensity of light is, in reality, determined, and maintained, partly at least, by the orienting response to the directive influence of light. In other words, the larvæ did not, in this instance, "select" the region of greater light-intensity because of the intensity *per se*, but because they became imprisoned in it through orientation as a result of the directive stimulus. It is only through rays which strike the larvæ directly from above or from below that an approximately non-directive influence can be obtained.

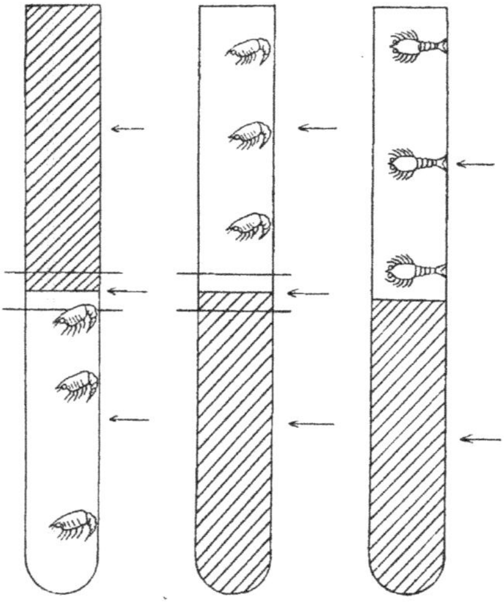

FIG. 2. FIG. 3. FIG. 4.

FIGS. 2 and 3 show the orientation of the larvæ in tubes standing in the vertical position; Fig. 4, in the horizontal. The arrows represent the direction of the light rays striking the tubes from the side. The cross hatching represents the parts of the tubes covered with black paper.

Experiment 2. Reaction to intensity of light—In this experiment use was made of the glass-bottomed box *A* with the light-shaft, and the colored-glass plates or graded light screens. First the glass plates were arranged over the top of the light-shaft in the order blue, green, orange, red. The box was filled with salt water to a depth of 15 mm., and ten first-stage larvæ were placed therein. The light-tight cover was then put in place and the larvæ were allowed five or more minutes to become acquainted with their new environment. The result was as follows:

TIME.	BLUE.	GREEN.	ORANGE.	RED.
After 5 minutes	10	0	0	0
After 10 minutes	10	0	0	0
After 14 minutes	9	1	0	0
After 17 minutes	9	0	0	1
After 19 minutes	9	0	1	0
Totals	47	1	1	1

The results of these tests and others in which the order of the glass plates was changed, demonstrate the tendency of the larvæ to group themselves in the blue, which was the more brightly illumined region. Similar experiments were performed with graded light-screens (strips of paper of different thickness or a gelatine wedge) substituted for the glass plates. The results in every case indicated that here also the larvæ reacted to difference in intensity. These experiments were performed with a belief that it is the refrangible rays of the spectrum alone that are active in determining the phototropic reactions of animals and plants. MINKIEWICZ (1906), however, has found that, although positively heliotropic animals usually react positively to the rays of shortest wave-length (violet or blue), and negatively heliotropic animals usually react to the rays of greatest wave-length (red or yellow), these phenomena of positive phototropism and positive chromotropism are not necessarily found together in the same organism. It is uncertain whether or not the larvæ of Homarus manifest chromotaxis. At the present time it can be said that the observed reactions of the larvæ to colored lights agree so well with the reactions to lights of varying intensity as determined by light screens, that they may fairly be considered responses to difference in intensity of light.

Experiment 3. Reaction to the directive influence of the rays—To demonstrate the response of the larvæ to the directive influence of the light rays the description of a single experiment will suffice. Similar experiments will be recorded later, in another connection. The conditions of this experiment are like those described in Experiment 2, i. e., box *A* was mounted over the light-shaft on the colored glass plates arranged in the order blue, green, orange, red. The ten first-stage larvæ contained in the box were more or less constantly oriented in the region of greatest light intensity, that is over the blue glass. Next, the small window situated at the red extremity of the box was opened to the diffuse light of the room. The result was that the larvæ immediately oriented to the new rays, left the region of greatest light-intensity, the blue area, and moved backward in the direction of the incident light rays toward the source of the weaker light and, at the same time, into a region of lesser light-intensity at the red end of the box.[3] The distribution at the end of 19 minutes was as follows: blue, 4 individuals; green, 1; orange, 1; red, 24. Here it appears that the larvæ, which at the beginning of the experiment were grouped in the area of greatest illumination under the influence of non-directive rays, were forced by the directive influence of the new rays to move from a region of greater into one of diminished light-intensity. As will be observed later, this experiment was tried under a great variety of conditions, and with larvæ of different stages and ages, with uniform results. Whenever the larvæ had an opportunity to move in the direction of the rays, they would do so, notwithstanding the fact that they thus passed from a region of greater to one of less illumination.

In the paragraphs immediately preceding, the purpose has been merely to indicate that in the behavior of the lobster larvæ we may observe reactions both to the intensity of light and to the directive influence of the light rays. The latter depends, first, upon the unequal stimulation of the two eyes, and second, upon the degree of illumination which affects both eyes. The conclusions which have been drawn from these few experiments receive further support from other experiments. But first, it is necessary to know whether there is any form of reaction common to all larval lobsters. To answer this question, which is of primary importance, it will be necessary to report in detail a series of tests, which were made upon many groups of lobsters during different periods of their metamorphosis and under different conditions of stimulation by light.

Experiment 4. Case 1—In several instances larvæ which had been hatched from one-half hour to one hour were put in a glass jar, which was in turn placed in the dark box and submitted to illumination on one side from a narrow window. In every case

[3] If this experiment appears uncritical because of the lack of information regarding the exact intensities of light at the opposite ends of the box, it may be answered that the intensity of light was measured by the only method available. Sensitized paper was placed inside the box, one strip over the end window, the other over the bottom at the blue end. The results showed that the light entering the blue end through the bottom of the box was much stronger than that entering the end window from the room.

the young larvæ at once swam backwards toward the source of light and grouped themselves closely together at the window side of the glass jar. So perfect was this orientation on the part of the newly hatched larvæ that, out of 100 individuals, not one showed a negative reaction.

Case 2—When the same larvæ were put in one of the long 40-centimeter glass tubes so placed in the dark box that the tube was parallel to the direction of the incident rays, the young lobsters in every case swam rapidly to the end of the tube nearest the window and remained there until the tube was reversed, when they again swam toward the window. These reversals might be continued for hours.

Case 3—When the same individuals, or other larvæ of the same group, were placed in box *B*, and this was turned with one end toward the window, the reaction of Case 2 occurred. They swam backward toward the window.

Case 4—Another group of fifty first-stage larvæ three days old was placed in a glass jar in the dark box and illuminated from the small window. All were definitely positive. Next, the circumference of the jar, except a vertical strip three inches wide on the light side, was covered with black paper, and the jar was so placed that direct sunlight had access to the open side. The larvæ immediately gathered on the darker side of the jar and remained there for one and a half minutes, after which they again returned to the sunlit side and remained there in bright sunlight as long as they were observed.

Experiment 6.—July 19, 3:30 p.m. Fifty first-stage larvæ, two days old, were put in a glass jar and this was placed in the dark box. Though the light was not bright at this time in the afternoon all the larvæ gave a positive reaction. The jar now was placed on a black background in the bright sunlight on the west table. Evey lobster moved to the room side of the dish away from the light. Within two minutes, many began to go back to the window side, and this continued until all were again gathered there. After four minutes, however, they again returned to the room side and remained there for ten minutes, at the expiration of which time they were about equally divided between the room side and the window side of the jar. They were now put back in the dark box, and with the slight intensity of light at 7 o'clock in the evening, all were reacting positively. By 8:30 the box was fairly dark and the orientation of the larvæ was indefinite. Suddenly the rays of a powerful acetylene light were thrown upon the jar. Immediately a negative reaction took place and continued for two minutes, when some of the larvæ began to return to the light. At the expiration of four minutes all the larvæ were reacting positively, and this reaction continued for several hours.

Experiment 7—July 22, 9:30 a.m. Forty-four second-stage lobsters, six days old, were placed in the glass jar, in the dark box. Eleven came at once to the room side of the jar. The jar now was moved nearer the small (three by three inch) window. As a result seventeen out of forty-four individuals gathered on the room side, but the definiteness of the positive reaction on the part of the window-side lobsters was lessened by desultory swimming. The jar was next placed on the west table, the room side and top of the jar being shielded by black paper. All the larvæ came to the room side of the jar. When replaced in the dark box (in light of much lesser intensity), the reaction again became uniformly positive.

Experiment 8—July 23, 9 a.m. Forty second-stage larvæ, seven days old, were placed in the glass jar on the east table, and exposed to strong light. All the larvæ at once oriented on the room (darker) side of the jar. These lobsters were next placed on the west table where the negative reaction continued throughout the afternoon. From 6 to 8 o'clock in the evening the light faded gradually. At 7:35 the body-orientation was nearly lost, but the orientation on the room side of the jar with diminishing definiteness remained in effect until 7:50, when the light had faded quite away and the lobsters were scattered throughout the jar.

Experiment 9—July 28, 9:30 a.m. Twenty third-stage lobsters, twelve days old, were placed in the glass jar in the dark box on a white background and submitted to light of slight intensity coming through the small window. All showed a strong positive reaction, and gathered on the window side of the jar. The next day in the afternoon, about fifty third-stage larvæ of the same group, now thirteen days old, were placed in the glass jar in the dark box on white background and submitted to light of medium intensity. Nearly all of the larvæ oriented on the room side of the jar, thus demonstrating a definite negative reaction.

Experiment 10. Case 1—June 26, 9 a.m. Ten fourth-stage lobsters, fifteen days old, were placed in the glass jar in the dark box and submitted to light of medium intensity from the small window. There was no appreciable tendency to undergo either body or progressive orientation. The lobsters were much engaged in eating one another.

Case 2—The fourth-stage lobsters mentioned in the preceding paragraph were fed on chopped clam meat and placed in box *A* with the black interior. Light was admitted through the end window. Records of four tests made at two-minute intervals show that while nine were neutral in reaction, six were positive, and twenty-five were negative. The box was next lined with white paper and the same fourth-stage lobsters were submitted to the same external light conditions. The results show twenty-six positive, twelve neutral, and twelve negative individuals.

Case 3—August 7, 2:30 p.m. When twenty fifth-stage lobsters, twenty-five days old, were put in box *A* and illuminated through the end window, all, without exception, oriented in the dark end of the box.

Conclusions concerning the permanence of these reactions through the stages—In explanation of the ten experiments recorded above, it should be stated that the writer had at his command large numbers of larval lobsters of approximately the same age and stage which had been subjected throughout the whole of their early life to the same conditions of environment. Therefore it was possible to make a detailed systematic study, not of a few isolated individuals alone, but of whole groups. The result of this study is expressed in these experiments.

Whatever else the foregoing facts may demonstrate, the answer to our first question is evident. *There is no constant form of reaction on the part of the larval lobsters to the directive influence of the light rays.* For this reason one has no warrant for saying, without reservation, that the larval lobster is either positively or negatively phototactic. If it had been necessary to depend for material upon a few individuals of uncertain age, and to draw conclusions regarding the general behavior of all the larvæ after observing the behavior of these few individuals, the outcome would of course be far less satisfactory than in the present instance. It is to be regretted, perhaps, that no means were at hand to make a critical determination of the exact intensities of light to which the larval lobsters gave their recorded reactions, but it is apparent that such a refinement of method would not change the general conclusions reached.

With the foregoing facts in mind, it is clear that the problem before us becomes, not, what reactions do the larval lobsters in general give to light, but *how do the lobster larvæ of a certain age react to light under certain known conditions?* To this rather more complex question attention will now be given.

V. SYSTEMATIC ACCOUNT OF THE REACTIONS TO LIGHT OF LOBSTERS IN THE LARVAL STAGES.

What is the nature of the reactions to light through the successive developmental stages, and by what conditions is it determined? Regarding the first of these points, it should be borne in mind that the subject matter concerned cannot be treated concretely, but that it is necessarily scattered through the long series of observations which follows, and that it is only from a consideration of the series *as a whole* that a clear idea of the gradual modifications in the reactions from the first to the sixth stage of the lobster's life can be obtained. As to the second point of inquiry, it is at once perceived that the conditions or factors which we seek to discover are of two sorts:

1. Conditions which are peculiar to a certain definite age or stage in the development of the larva, and which may be designated as physiological conditions.

2. All outside influences, including the intensity and multiplicity of stimuli brought to bear upon the animals.

In the following discussion it will be found of advantage to consider these two kinds of modifying conditions together; for they are found to be very much inter-related when a consideration of their mutual importance in bringing about any orientation of the young lobsters is involved.

It may be appropriate to mention at this point the method of securing the data here presented. The futility of taking young larvæ at random from the hatching bags without knowledge of their age or previous history was recognized early in the course of the investigation. It was considered advisable to work only with those lobsters whose previous history was definitely known. To this end the exact time of hatching of certain groups of larvæ was noted. In the large canvas hatching bags, used at the Wickford Station, hundreds of larvæ hatch in a single hour, and observations were made, as a rule, twice each day (morning and afternoon), upon individuals taken from these groups, whose age was accurately known. During the course of the study, the history and the daily reactions of three groups of larvæ were followed and recorded. For the following account of the reactions of the first-stage larvæ, for instance, the records of these three groups for the first day, the second day, the third day, etc., were used. Only the reactions which appeared to be the most constant and typical have been introduced here. Therefore, although many variations in reactions were found to occur, the following section describes the typical daily reactions of the larval lobsters from the time of hatching through the fifth stage of their existence.

1. First Larval Stage.

As has been shown by preliminary observations and the experiments already mentioned, the lobsters of the first larval stage are usually strongly positive both in their photopathic and in their phototactic reactions. These reactions are manifested strongly in the few hours directly after hatching, when, as we shall presently see, the young lobsters react definitely, and to very slight differences in the intensity of illumination. When half-hour old lobsters were placed in the glass jar, and submitted to any kind or intensity of light (daylight, artificial, or colored), they responded well (especially when the intensity was increased by a white background) to slight differences in illumination; and reacted uniformly and invariably by moving, tail foremost, toward the source of light. In case of two sources of light, on opposite sides of the jar, the larvæ would respond to the rays which were the more intense. If the rays from two sources of light were introduced at right angles to each other, the resultant reaction, as has been shown for other organisms by many investigators, was determined according to the law of the parallelogram of forces.

It would appear that, in the behavior of the first-stage larvæ, we have the most delicate reactions to slight differences in light intensity that occur throughout the life of the lobster. During the early hours of the first larval stage, no individuals reacted negatively to the directive stimulus of the light, while in the later stages, although a majority of the larvæ manifested definitely one reaction or another, there were usually a few individuals which gave responses that were either indefinite or opposite to the rule.

Experiment 11. *Case 1*—Ten first-stage larvæ, five hours old, were placed in a glass tube 40 cm. long, and this was laid on the table at right angles to the plane of the window and parallel to the light rays entering through a narrow slit in the screen. All of the larvæ at once oriented themselves at the window end of the tube. Next, blue, green, and yellow glass plates were placed successively over the end of the tube next the window, leaving the opposite end clear, but none of these changed the definiteness of the positive reaction. When, however, an orange glass was used, the larvæ paused midway in the tube, at the border line of the orange light, and in their final orientation were scattered between this region and the orange end of the tube. When a red glass was superimposed, all the larvæ took a position at the border line of the red and the clear glass, this region representing the junction of the areas of strong and weak illumination.

Experiment 12. *Case 1*—In this experiment the glass bottomed box *A* was set up over the light-shaft with the colored glass plates arranged in the order, red, orange, green, blue, as described on p. 207. The box was filled to a depth of one inch with water and first-stage larvæ, twenty-four hours old, were introduced. Five minutes was allowed for the larvæ to become acquainted with the new environment. Records of four tests then made showed that while thirty-eight larvæ gathered in the blue area, only one was found in the red, one in the orange, and none in the green. Changing the order of the glasses in no way changed the results. This apparently demonstrates that there is a definite tendency on the part of these larvæ to orient themselves over the glass plates which admit the brightest light; and that the precise order of the plates makes no difference in orientation.

Case 2—In this instance the order of the glass plates was red, orange, green, blue. The same larvæ used in the above tests were employed, but the conditions of the experiment were changed. The window in the end of the box corresponding to the red glass was uncovered and the diffuse light from the room was allowed to stream through the box longitudinally. The object of this was to discover whether the larvæ which had previously given so definitely the positive photopathic reaction, could be induced to enter the region of diminished light intensity (at the red end of the box). In other words, whether the phototactic reaction could be made to overcome the photopathic. Between each of the successive tests mentioned below, the light from the room and the light through the shaft were cut off in order that a scattering of the larvæ through the box might occur. In other cases the position of the box was reversed; and in still others both the position of the box and the order of glass slides, changed. The results of four tests are as follows (the arrow represents the direction of the light entering the end window of the box):

Test.	After	Distribution of larvae.			
		→ Red.	Orange.	Green.	Blue.
1	45 minutes	8	0	0	2
2	9 minutes	8	0	0	2
3	14 minutes	8	1	0	1
4	18 minutes	9	0	1	0
Totals		33	1	1	5

Case 3—In this instance the red and orange glass plates were removed and black paper substituted. The photopathic reaction was found to be definitely positive, the young larvæ grouping in the blue area. Now, as before, the window at the end of the box corresponding to that overlying the black paper was opened to the subdued light of the room, while brilliant daylight entered the blue end of the box. As will be observed, the conditions of this experiment are similar to those of Experiment 11, save that, in this instance, a greater difference between the intensity of light at opposite ends of the box existed. Between tests the light from both sources was cut off and the larvæ were allowed to scatter. The results, which may receive the same interpretation as those of Experiment 11, are tabulated below (the arrow indicates the direction of light entering the end window of the box):

Test.	After	Distribution of larvae.		
		→ Black.	Green.	Blue.
1	5 minutes	10	0	0
2	7 minutes	7	3	0
3	10 minutes	10	0	0
Totals		27	3	0

Conclusions from Experiments 11 and 12: In the results of the foregoing experiments, we have further evidence to support the conclusions drawn from Experiment 3. In Experiment 12 the larvæ passed from a region of greater (blue) to one of lesser (the red, or in Case 3, the black) light-intensity in moving toward the source of light in the direction of the incident rays. It must be assumed that in Case 3, there was a much greater difference in the intensity of light at the two ends of the box (overlying the blue glass and the black paper respectively) than in Case 2, or in Experiment 3. These experiments were performed many times, under several different conditions of light, and with larvæ of ages varying from a few hours to two days. The same results were obtained in every case, except that in the older first-stage larvæ the reactions were not so definite (more individual variations) and a stronger light was required to bring about the same responses as were manifested by larvæ under four hours old. In these cases, as also in Experiment 3, rays of lesser intensity (but in a horizontal plane) which struck the larvæ in such a way as to cause a body-orientation in which a normal swimming position was still maintained, were more influential in determining a progressive orientation than were the more intense rays which struck both eyes equally, but which came from below, and had a tendency (as will be shown in detail later) to throw the larvæ out of their normal swimming position. As the writer has shown elsewhere (1907a), galvanotactic reactions in the young lobsters occurred only when the tail or the back was turned wholly or partly toward the anode. Although at first sight it appears that the causes for this condition of reaction can have nothing in common with the causes which determine a progressive orientation to the directive influence of light rays only when the swimming position is favorable, it may not be inappropriate to suggest that here also the direction of the impact of light with reference to the axis of the body of the larva, may have some influence on the reaction.

Experiment 13. Case 1—Ten larvæ, twelve days old, were placed in box *A*, mounted over the light-shaft. When the glass plates were arranged in the order designated below, the photopathic reaction was as follows:

TEST.	AFTER	DISTRIBUTION OF LARVAE.			
		Red.	Orange.	Green.	Blue.
1	5 minutes	0	1	2	7
2	10 minutes	1	1	1	7
3	15 minutes	1	1	2	6
4	20 minutes	2	0	0	8
5	25 minutes	0	1	0	9
Totals......................		4	4	5	37

Case 2—When the order of the glass plates was changed to red, blue, orange, green, the following results were obtained: Red, 3; blue, 31; orange, 3; green, 3.

Case 3—After redistribution of the larvæ had taken place, the small window opening at the green end of the box was uncovered to the diffuse light of the room. The resulting reactions were as follows:

TEST.	AFTER	DISTRIBUTION OF LARVAE.			
		Red.	Blue.	Orange.	Green.
4	2 minutes	0	3	3	4
2	4 minutes	1	3	3	3
3	7 minutes	1	3	1	5
4	10 minutes	1	2	3	4
Totals		3	11	10	16

Case 4—Once more the order of the glass plates was changed to blue, green, orange, red, and the window at the red end was uncovered to the light of the room. The results of the three sets of tests were: Blue, 9; green, 5; orange, 3; red, 13.

Experiment 14—The following observations deal with cases of larvæ suddenly submitted to a light of great intensity, as for instance when they are brought from subdued daylight into full sunlight, or when the brilliant rays from an acetylene lamp fall upon larvæ which had been for sometime in darkness.

Case 1—July 18, 4 p.m. Fifty first-stage larvæ, about thirty hours old, which had been reacting positively in lights of low or medium intensity, were placed (in a glass jar) in the bright sunlight of the west table. Every larva at once moved to the room side of the jar. Within a few minutes, however, all returned to the window side of the jar. Ten minutes later they were divided about equally on each side. Next they were returned to the dark box and submitted to the weak light from the small window. Here they manifested a definite positive reaction which continued until evening. At 8:30 these fifty larvæ were suddenly submitted to the intense rays of an acetylene light. The result was a universal negative reaction. Within two or three minutes, however, a few larvæ began to return toward the light, and within four minutes all had become positive in their reaction.

Case 2—A group of fourth-day first-stage larvæ in the glass jar was subjected to light of low intensity and found to manifest a positive reaction; when subjected to a much stronger light the same larvæ were still universally positive. This reaction, once established, endured through the period of gradually diminishing intensity of light accompanying the coming of night. The next morning these (now fifth-day) larvæ were found to be negative in reaction. It was feared, however, that the manner of reaction might have been changed because of the long period of confinement which they had undergone. For this reason a fresh lot of twenty-five larvæ from the same group (fifth-day, of the first and second stages), was secured. It was observed at this time that about a third of the number of those in the hatching bag had moulted into the second stage, and that the others were very near the moulting-period. When these larvæ were put in the glass jar, placed in the dark box and submitted to subdued light from the small window, six tests showed fifty-five to be negative, and ninety-five positive. When these same larvæ (now thirteen first-stage and twelve second-stage), under the conditions of stimulation stated above, were subjected to light of still greater intensity by placing the jar nearer the small window of the dark box the results showed that fifty-nine were negative and forty-one were positive.

At 3:30 p.m. these same larvæ were removed from the dark box and placed (in the glass jar) on the west table, where they were suddenly subjected to the bright afternoon sunlight. Every larva came to the room side of the jar and remained there so long as observed.

Case 3—The larvæ mentioned above were liberated and another lot of twenty-five (of the same group, but all in the second stage) was secured at 8 o'clock in the evening. The intense rays of the acetylene light were suddenly directed upon one side of the jar.

This resulted in a sudden and universal positive reaction which, however, soon became indefinite. The larvæ gradually returned to the darker side of the jar and, as in the case mentioned above, remained there so long as observed.

Case 4—When, on the other hand, another group of larvæ which was reacting positively to a light of low intensity, was brought by slow degrees into a light of great intensity, there resulted no sudden, temporary change of reaction such as that observed above. The reaction usually remained unmodified, but if it was reversed it remained permanently so. The same statement holds for larvæ which had been reacting negatively to light of low intensity. When they were brought by slow degrees into light of great intensity, seldom did a sudden temporary change in reaction result.

Conclusions from Experiment 14: The stimulation brought about by suddenly submitting larvæ to intense light may cause at least two kinds of response: first, in the case of early first-stage lobsters (about thirty hours old, and manifesting previously a positive reaction), a definite and universal, though temporary, negative response; second, in the case of early second-stage larvæ (about five days old, and giving previously a negative reaction), a definite and universal, though temporary, positive response. From Case 4 it appears that a gradual change of intensity (extending over an equal or even a greater range of intensities) may not bring about a similar result, although a permanent reversion in the reaction may sometimes ensue.

Larvæ which have recently moulted are most susceptible to slight differences in light-intensity; and the reaction of such larvæ is frequently negative, while the reaction of larvæ which are approaching the moulting-period is more often indefinite or positive.

Experiment 15. Case 1—The following experiment involved the use of the Y-tubes described on p. 207. Ten positively reacting lobsters, five hours old, were placed in the tube at the end designated *a* (Fig. 5, *B*). The Y-tube was then placed in position in the dark. Over one arm was laid a red glass, over the other arm an orange glass, and then the screen was drawn from the window to allow the light rays to strike the tube in the direction shown in Fig. *B*. Tests were made about five minutes apart. After each, the return of the lobsters to the (*a*) end of the tube was induced merely by reversing the tube so that the end (*a*) was toward the window; the position of the red and orange glass was also reversed. The distribution at the end of each test was as follows:

Test.	Red arm.	Stem.	Orange arm.
1	0	0	10
2	0	1	9
3	0	1	9
4	0	2	8
Totals	0	4	36

Case 2—Next, green and blue glass plates were substituted for the red and orange, the method of the experiment otherwise remaining the same, and the green and the blue glasses were reversed in position at the end of each test. A series of four tests showed the following results: Green arm, 11; stem, 2; blue arm, 27.

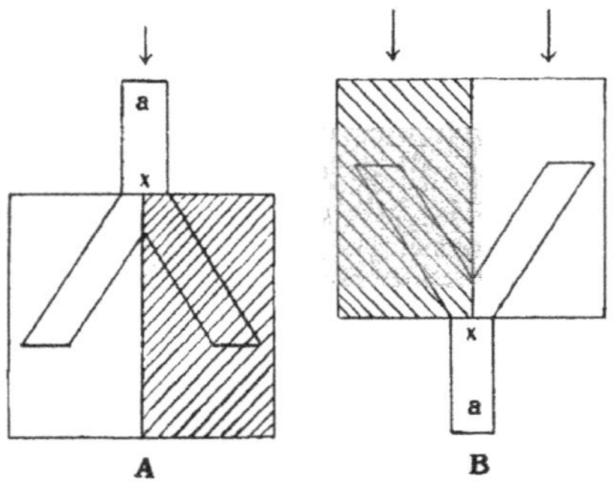

Fig. 5. Showing the Y-tubes as set up for experiment. The arrows indicate the direction of the light rays. The cross-hatched areas represent the glass plates of the darker color laid over the arms of the tubes. The ends designated *a*, represent the starting point for the negatively reacting (A), or the positively reacting (B), larvæ.

Case 3—One more test of the reaction of this group of positive larvæ was made at this time, which was far more delicate than either of the preceding, for the difference in the intensity of the glass plates used was less. In making a selection of glass slides two were chosen which had been purchased for "red glass." On close inspection, however, and by test with sensitized paper, it was observed that one slide was somewhat lighter in color tone than the other. These glasses were used in the next experiment. The darker of them may be designated as red, the lighter as ruby. The results of four tests were as follows: Ruby arm, 21; stem, 10; red arm, 9. In the last experiment, with this group of larvæ, it was found that a great intensity of light, striking the red slides, was required to bring about reaction; and that, even then, several larvæ would remain in the region designated *x* (Fig. 5, *B*), near the junction of light and dark. These experiments were repeated with both black and white backgrounds for the arms of the Y-tube. The results agreed with great uniformity, differing only in the length of time required for the reaction. From these last experiments we may conclude that the first-stage lobsters, at the age of five hours or less, are extremely sensitive to slight differences in the intensity of light, more so in fact than older lobsters of the first and later stages; for it was seldom with these older lobsters that the delicate reaction to the ruby and the red glasses observed in Experiment 15, Case 3, could be induced.

Experiment 16. Twenty first-stage larvæ, slightly over two days old (for which to light of nearly all intensities reactions on the first and second day had been positive), were put in the glass jar, and this in turn was placed in the dark box. They were submitted to light from a small window one inch wide and two inches high, before which the colored glass plates could be placed so as to illuminate one side of the jar with red, blue, green, or orange rays, as the case might be. The reaction in each of these lights was as follows:

LIGHT.	POSITIVE.	NEGATIVE.
Red	20	0
Orange	20	0
Green	19	1
Blue	18	2
White*	15	5
Day	3	17

* Subdued daylight passing through one or two thicknesses of white paper.

Here it is shown that the negative reaction to lights of great intensity, which was first discovered in larvæ thirty hours old (Experiment 14, Case 1), and which, as we shall see, persists for a variable length of time, has become accentuated and remains for the time permanent. The next series of observations were made upon lobster larvæ on the fourth day after hatching. Many of them were earing the moulting-period and preparing to pass into the second stage.

Experiment 17—July 17, 8:30 a.m. About one hundred fourth-day, first-stage lobsters (Group A) were taken from one of the hatching bags and placed in the glass jar in the dark box. The majority reacted positively to daylight through the small window. At 1 o'clock, when examined again, about one-half of them were reacting negatively. The jar was then removed and placed in the light of the west window where the intensity was greater. At once every larva became negative in reaction.

In order to determine whether this mode of reaction was a natural incident in the life of the larvæ of this age, or whether the response had been induced as a result of their having been so long subjected to experimentation, twenty-five first-stage larvæ (Group B) were removed from the same group as that from which the larvæ mentioned above were taken. When these twenty-five were put in a glass jar and placed in the west window beside the group mentioned above, they gave a positive reaction. After five minutes, half were positive and half negative. At 5:30 the sun was low and the light weak, but all the larvæ gave a negative reaction, which persisted, as did the negative response in Group A mentioned above, until far into the twilight.

It may be further noted in this connection, that five of the larvæ which reacted negatively in the afternoon were placed in absolute darkness for four and a half hours. It was believed that the positive reaction might be renewed; but this was not the case when they were again brought into daylight of several intensities.

Experiment 18. Case 1—July 20, 4 p.m. A number of fourth-day, first-stage larvæ were removed from the hatching bag and put in the glass jar. This was placed in the dark box and the larvæ submitted to red light through the three by three inch window. The resulting reaction was positive and remained so even when the intensity was still further diminished by inserting numerous sheets of paper behind the red glass. Finally, a point was reached where the positive orientation was lost and a homogeneous scattering occurred. When the intensity of the light was again increased, the positive orientation returned; but, with a still greater increase in intensity, this response became again less definite, and finally, in the more intense blue and white light, the negative reaction again appeared.

Case 2—In the evening, when other observations were made upon the same group under the influence of the acetylene light, burning dimly, the reaction in the glass jar was positive under all the colored glass plates. When the intensity was increased by substituting a lamp which burned more brightly, the group divided, half going to the positive and half to the negative side. When the intensity was increased still further (reinforced by a brilliant oil burner and reflector) a greater number gave a negative reaction. As it afterward transpired, the larvæ used in these last tests did not moult to the second stage until on or after the fifth day.

Case 3—July 23, 1:20 p.m. Fifty fourth-day, first-stage larvæ were put in the glass jar and placed in the dark box. In the red light the reaction was definitely positive. The reaction under the different intensities obtained by colored glass plates may be tabulated as follows:

COLOR.	POSITIVE.	NEGATIVE.
Red	50	0
Orange	47	3
Green	43	7
Blue	36	14
White	23	27

Case 4—July 31, 10 a.m. Twenty-eight first-stage and second stage larvæ of the fifth day (all nearly ready to moult to the second stage) were put in the glass jar and placed in the dark box. Under lights of different intensities the results were as follows:

LIGHT.	POSITIVE.	NEGATIVE.
Red	28	0
Orange	22	6
Green	18	10
Blue	12	16
White	14	14
Daylight	0	28

In this particular case it was observed that under the orange light the negative larvæ were of the second stage, while those which retained for the longest time the positive reaction (in the case of the blue and white glasses), were the lobsters which were nearest to the moulting-period. When fresh, clean larvæ, which had moulted into the second stage within a very few hours, were selected and submitted to several different intensities of light, they invariably gave the negative reaction.

Case 5—July 23, 1 p.m. Twenty fifth-day, second-stage larvæ were taken from one of the hatching bags and put in the glass jar. This was placed in the dark box and the larvæ were submitted to illumination from the red light. There was some random swimming, but the general reaction was positive, except in white light, in which three were positive and seventeen negative. Next, the jar was removed from the dark box and placed on the west table in subdued sunlight. Here the reaction was definitely negative. At 4:30 when the jar was returned to the box (at this time in the afternoon the light was much less intense than earlier) a positive reaction was obtained in red, orange and green light.

Conclusions from Experiments 16, 17, 18: From the result of the last three experiments the following tentative conclusions may be drawn. The general negative reaction to light of great intensity, begins on about the third day of the first stage, continues for the most part uninterruptedly until the moulting-period is near; just before the moult the. reaction becomes indefinite or, more often, positive; directly after the moult into the second stage (which occurs on the fourth or fifth day of the first-stage-period), the reaction to lights of nearly all intensities again becomes definitely negative.

2. *Second Larval Stage.*

Experiment 19. Case 1—July 19, 8:30 a.m. Observation of a group of sixth-day, recently moulted second-stage larvæ demonstrated that a negative reaction took place when the larvæ were put in the glass jar and placed in the dark box. This was true for daylight coming through the three by three inch window, and in both blue and green light. In the case of orange and yellow light, however, the reaction was similar to that in either yellow or orange, but perhaps less definite. It may be here recorded that a group of first-stage larvæ, about one and a half days old, subjected at the same time to these conditions, gave a positive reaction, not only in orange, but also in blue, and even to white light. These reactions took place on both black and white backgrounds, but they were more definite on white. But when the stimulus of the orange rays was continued for ten minutes or more, in this case also, the negative reaction began to appear again and many larvæ came to the room side of the jar.

Case 2—July 23, 5 p.m. The larvæ used in this case were of the seventh-day group of the second stage, having been taken from the hatching bag at 9 a.m. At 5 p.m. under red, orange, green, blue and white lights, entering through the three by three inch window, all were definitely negative. They had also shown a negative reaction in several intensities of light in the morning. At 7 p.m. further observations were made on the same group of larvæ. The following quotation is from the daily note book.

"July 23, 7 p.m. One of the best demonstrations of the persistency of the negative reaction of these seventh-day larvæ was exhibited this evening. Larvæ taken from the hatching bags at 9 a.m. have reacted negatively at every observation during the day. At 7 p.m. it was observed that this group, which still remained in the glass jar near the west window, continued to present a definite negative reaction. This negative response continued until 7:55 p.m., when the light became too faint to determine either a body or a progressive orientation. Here it is to be observed that the negative reaction on the part of these second-stage larvæ was continued through a long series of gradually diminishing intensities of light. After all signs of body-orientation or progressive orientation had vanished in the case of the group of larvæ mentioned above, the intense light from the acetylene lantern was suddenly thrown open one side of the glass jar. A most definite negative reaction resulted. This response, it will be observed, is different from that recorded in Experiment 14, Case 3, for in the latter case the sudden illumination determined a definite positive reaction."

Experiment 20. Case 1—July 24, 9 a.m. Thirty eight-day, second-stage larvæ were taken from one of the large bags and put in the glass jar in the dark box. The time of moulting into the third stage was near at hand, and many of the individuals were already "fuzzy" and sluggish in their movements. Illumination through the three by three inch window, by the colored lights, gave these reactions:

COLOR.	POSITIVE.	NEGATIVE.
Red	30	0
Orange	27	3
Green	17	13
Blue	13	17
Day	13	17

Before we state the next case, one consideration must be noted. In the previous pages, use has been made of such terms as "third-day," "seventh-day," and "eighth-day" larvæ, to distinguish the age, and roughly the stage, of certain groups of lobsters. Because of the use of these terms, it must not be supposed that there is always a constant relation between the age and the stage of the larvæ. Among the larvæ of a single group which have been hatched and have developed under similar conditions, a fairly constant relation between the age and stage is invariably maintained. But for different groups of larvæ, this correlation does not necessarily exist, for it is entirely possible, and indeed it very frequently happens, that a group of seventh-day larvæ may be in the third stage, while a lot of eight-day individuals are in the second stage. The differences in rate of development are due to such factors as water density, temperature, food-supply, and conditions of light and darkness, which, as the writer has shown (Hadley '06b), may act either directly upon the body processes, or indirectly by favoring or preventing the growth of various body parasites such as diatoms, protozoa, and algæ that naturally develop in profusion on the bodies of the young larvæ. This explanation will perhaps make clear why, in the following case, we apparently retrace our steps to consider the case of seventh-day larvæ. In point of fact, these larvæ were, at the time of experimentation, somewhat further developed than were the eighth-day larvæ mentioned in Case 1.

Case 2—July 20, 9 a.m. Twenty seventh-day larvæ (eight second-stage, twelve third-stage) were removed from the hatching bag, put in the glass jar, placed in the dark box and illuminated by the light through the three by one inch window. After a half hour, observation showed that the larvæ were equally divided between the window side and the room side of the jar. After five minutes' exposure to red light, thirteen larvæ were positive and seven were negative. When, however, the amount of light was increased by opening the large three by three inch window, only three larvæ remained positive while seventeen became negative. This proportionate reaction endured for several hours, or until observation ceased.

Case 3—July 20, 8 p.m. Twenty seventh-day, early third-stage larvæ were taken from one of the hatching bags, placed in the glass jar, and illuminated by an acetylene light. A more or less scattering negative reaction at first resulted. When the amount of light was increased by supplementing the acetylene with a brilliant oil burner the response was more definitely negative.

Case 4—July 21, 9 a.m. Twenty-two eighth-day, early third-stage larvæ were taken from one of the hatching bags and put in the glass jar in the dark box. When subjected to subdued daylight through the three by one inch window, sixteen out of twenty-two gave the negative reaction. In orange light the reaction was seventeen negative, five positive; in red light eighteen negative, four positive. Here attention may be called to the fact that these third-stage larvæ gave a negative reaction to practically the same intensity of light as determined a positive response for larvæ in the late second stage.

Case 5—August 3, 2 p.m. Twenty eighth-day, early third-stage larvæ were taken from the hatching bags and put in the glass jar in the dark box. They were submitted to the colored lights, with results as follows:

COLOR.	POSITIVE.	NEGATIVE.
Red	8	12
Orange	6	14
Green	3	17
Blue	2	18
White	6	14
Day	0	20

Conclusions from Experiments 19 and 20: The conclusions which we draw from the two foregoing experiments support further those formulated for Experiments 16, 17 and 18, on p. 228. In Experiment 20, Case 1, was observed the definite positive response which was manifested toward the end of the second larval stage when the moulting-period was near. In Case 2, where a group of larvæ which included individuals of both the second and third stages was used, it was observed that the reaction was either positive or negative; and that those larvæ which gave the negative reaction most definitely or gave it first were usually the larvæ of the early third stage. In Cases 4 and 5, in which only third-stage larvæ were employed, it was observed that, in general, the reactions to lights of nearly all intensities were negative. As in the case of the first-stage larvæ, it was found that the reaction of second-stage larvæ, just before the moulting-period, usually changed from negative to positive, and again became negative at the beginning of the third larval state.

3. *Third Larval Stage.*

By the ninth day it is only in exceptional cases that the larvæ have not entered the third stage; and it frequently happens that they are nearly ready to enter the fourth. The swimming of the third-stage larvæ is much like that of the earlier stages except that in the third stage there is greater difficulty in using the swimmerets of the thoracic appendages, especially during the last part of the stage. One reason for this is the fact that, as the larvæ grow older and larger, they more often play the host to multitudes of diatoms, algæ and protozoa which gather in such quantities as seriously to interfere with the processes of swimming and eating. In the preparation for the moult from the third to the fourth stage, moreover, occur the most important changes that the young lobster undergoes in the course of its life. These changes appertain not alone to modifications in the external form of the body and to the form and functions of many of the body appendages, but also to points of internal structure. Among the changes during this period of metamorphosis we may enumerate the following as important in connection with our study of behavior: (1) The loss, in the moult from the third stage, of all functional swimming attachments of the thoracic appendages; (2) the great development of both the first and second pairs of antennæ and of the chelipeds; (3) the accession of functioning swimmerets on the under side of the second to sixth abdominal segments; (4) a great change in the form of the body, and a consequent modification of the manner of swimming.

In view of these important changes, which are taking place in the anatomy of the lobsters as they pass from the third into the fourth stage, it does not appear unjustifiable to believe that these processes have an influence on the behavior of the larvæ even before they emerge in approximately the adult structural type, endowed with a new body form, new functional apparatus and new reactions. We shall now undertake a study of the behavior of the third-stage larvæ as they approach and finally pass this most critical period of their life history.

Experiment 21. Case 1—July 22, 9:30 a.m. Thirty ninth-day, third-stage larvæ were removed from the hatching bag, put in the glass jar and placed in the dark box. Under stimulation by the red rays, although there was no definite positive reaction, most of the larvæ swam about at random on the window side of the jar. When orange glass was substituted for red, half of them came to the room side of the jar. In the case of green glass, a few more reacted negatively, and when blue glass was substituted for green, all but five larvæ gave a negative response. These five did not manifest a definite positive reaction, but swam at random on the window side of the jar. When the colored glasses were removed and the larvæ were submitted to the influence of diffuse daylight through the small window, all reacted negatively.

Case 2—August 4, 9 a.m. Twenty ninth-day, third-stage larvæ were taken from one of the hatching bags and placed in the dark box. Stimulation by the colored light resulted as follows:

COLOR.	POSITIVE.	NEGATIVE.
Red	6	14
Orange	3	17
Green	2	18
Blue	2	18
White	0	20
Daylight	0	20

Case 3—In the present case it was attempted to learn whether the sign of the photopathic reaction in the larvæ of this stage corresponds to the sign of their phototactic reaction. To this end, ten ninth-day, third-stage larvæ, fresh from the hatching bag, were placed in the glass-bottomed box *B*, which was set over the light-shaft and mounted upon colored glass plates. After each observation either a period of five minutes was allowed for a uniform distribution of the larvæ to take place, or the box itself was reversed, leaving the glass plates in the same order. In other instances the order of the glass plates was changed. During this experiment the water in the box was eighteen to twenty mm. deep. The results are presented below:

RED.	ORANGE.	GREEN.	BLUE.
4	0	1	5
1	0	2	7
RED.	ORANGE.	BLUE.	GREEN.
2	1	3	4
1	1	4	4
RED.	BLUE.	GREEN.	ORANGE.
1	5	3	1
2	2	3	3
BLUE.	RED.	GREEN.	ORANGE.
6	0	3	1
5	1	2	2

The larvæ which were used as stated above, and which presented a positive photopathic reaction in every instance, were next transferred to the glass jar and placed in the dark box. Here, and in tubes, the assumed phototactic reaction was uniformly and definitely negative; and this was true in the case of lights which were both of greater and of lesser intensity than in the tests above mentioned.

Case 4—To confirm the results obtained in Case 3, similar tests were made with another group of ninth-day, third-stage larvæ, fresh from the hatching bag. Notwithstanding the fact that this series of observations was not started until 5 o'clock in the afternoon when the light was fading, the results were similar to those obtained in Case 3. That is to say, the photopathic reaction was definitely positive, but the phototactic reaction, as shown when the larvæ were transferred to the glass jar in the dark box, was as definitely negative.

Experiment 22—The following experiment and observations concern the tenth-day, third-stage larvæ. Most of these lobsters were well along in the third stage, and many were covered with body parasites.

Case 1—July 23, 9 a.m. Thirty tenth-day, third-stage larvæ were transferred from one of the hatching bags to the glass jar and placed in the dark box. After having been submitted for one-half hour to light coming through the red glass (three by one inch window), the reaction was uniformly negative. In the case of orange, yellow, green, blue and white light the results were the same. In all of these reactions, however, one fact was noticeable, the body-orientation of these larvæ was much less definite than in any previous case of the same or earlier stages.

Case 2—July 26, 9 a.m. A mixed lot of thirty third-stage larvæ, most of which were ten days old, although some were older and some younger, were transferred from the hatching bag to the glass jar. When submitted to the colored lights in the dark box, the following results were obtained:

Color.	Positive.	Negative.
Red	3	27
Orange	13	17
Green	8	22
Blue	13	17
White	0	20

In consideration of the apparent fluctuations in the sign of reaction manifested by the above-mentioned larvæ, it may be noted that these lobsters represented a group in which some were "early," others "advanced," third-stage larvæ. Indeed many were approaching the third moulting-period; the significance of this for the behavior of the larvæ we shall consider in the next few cases.

Case 3—July 27, 2 p.m. Thirty eleventh-day, third-stage larvæ were transferred to the glass jar and placed in the dark box. Under colored lights, although the general reaction was negative, many were positive. Experiments made upon the larvæ in the glass-bottomed box *B* to determine the photopathic reaction at this time, showed that the larvæ gave neither a definitely positive nor a definitely negative reaction. Other tests indicated a definitely positive reaction. When, however, light was admitted to the box through the end window (as well as through the bottom), first from the red end, then from the blue end, of the box, there resulted a definite negative phototactic response. The arrows show the direction in which the light entered the box.

→ RED.	BLUE.	ORANGE.	GREEN.
1	2	1	6
1	1	0	9
0	0	0	10

→ RED.	ORANGE.	GREEN.	BLUE.
0	0	0	10
0	0	1	9

RED.	ORANGE.	GREEN.	BLUE. ←
5	1	2	2
6	0	2	2
9	1	0	0

The foregoing cases demonstrate that these larvæ manifested a definitely negative phototactic reaction under the conditions of illumination described; and that, by those rays which had a directive influence, they could be driven into a region of either greater or lesser light intensity, as represented by the blue and by the red ends of the box, respectively. It might be argued that, so long as the eyes of the larvæ are homolaterally stimulated, variations in intensity can not cause or change the orientation, and that orientation results only from a heterolateral stimulation. But this is by no means true, for it has been noted in the foregoing pages, and it will be further demonstrated, that slight differences in intensity, when coincident with a homolateral stimulation, may even reverse the index of progressive orientation.

Case 4—July 24, 9 a.m. Thirty-five eleventh-day, third-stage larvæ were transferred from the hatching bag to the glass jar and placed in the dark box. The reactions to the colored lights were as follows:

Color.	Positive.	Negative.
Red	15	20
Orange	16	19
Green	8	27
Blue	8	27
White	7	28

Next, the jar was placed in full daylight, on the table before the west window. All larvæ came to the room side. In this case there were seven larvæ which became the special object of observation, since they invariably manifested a positive reaction until they encountered daylight. This group was set aside, and before night four of the seven had moulted into the fourth stage; consequently their exceptional behavior was due to the fact that they were in a different physiological condition than the majority of the group used in Case 4.

Experiment 23. Case 1—In this experiment is continued the examination of the reactions of other twelfth-day larvæ which were approaching the third moulting-period. Twenty-three larvæ were placed in the glass jar and observed under the influence of the cólored lights in the dark box. The results were as follows:

Color.	Positive.	Negative.
Red	6	17
Orange	15	8
Green	6	17
Blue	4	19
White	3	20

Case 2—At 3:30 p.m. Ten larvæ from the above groups were transferred to the glass-bottomed box *B*, which was set up over the light-shaft upon the colored glass plates. The results were as follows: Blue, 19; green, 3; orange, 4; red, 4. During the course of the day, many of these ten larvæ moulted to the fourth-stage.

Case 3—July 29, 9 a.m. By this date there were very few third-stage larvæ left in any of the groups whose actual age was known. Indeed there are few cases in which the development is so slow that the third-stage larvæ endures to the thirteenth or fourteenth day. In this particular instance, twenty larvæ were transferred to the glass jar and placed in the dark box. The resulting reactions to the colored lights were as follows:

COLOR.	POSITIVE.	NEGATIVE.
Red	16	4
Orange	16	4
Green		
Blue	13	7
White	9	11
Day	0	20

It may be observed in the account of the last three experiments how the general reaction of the third-stage larvæ has gradually changed from negative to positive; and how it requires an increasingly greater intensity of light to determine a negative response in the larvæ which are approaching the fourth stage. In the next case the culmination of this gradual change is reached, since the third-stage larvæ almost uniformly manifest a positive reaction which is as definite as that of the newly-hatched larvæ.

Case 4—July 30, 2 p.m. Thirty fourteenth-day, third-stage larvæ secured from a group in which nearly all had entered the fourth stage, were transferred from the hatching bag to the glass jar and placed in the dark box under the influence of colored lights:

COLOR.	POSITIVE.	NEGATIVE.
Red	30	0
Orange	30	0
Green	30	0
Blue	30	0
White	28	2
Day	23	7

It is here observed that, when as indicated above, the jar was removed from the dark box and placed on the west table in daylight, only seven larvæ became negative. All the others remained positive, even in this light of great intensity. This case represents the strongest and most definite maintenance of the positive reaction in late third-stage larvæ ever observed by the writer. These larvæ moulted into the fourth stage very soon after the above observations were made.

Case 5—In the following test other members of the group of larvæ used in the previous case were employed. The aim was to learn whether or not the photopathic reaction in these larvæ was in agreement with the phototactic reaction described in Case

6. The larvæ were placed in the glass-bottomed box over the light-shaft; fifteen minutes was allowed for the first orientation, and five minutes was given for each of the others:

After	Blue.	Green.	Orange.	Red.
15 minutes	6	2	2	0
20 minutes	6	1	1	2
25 minutes	4	1	5	0
30 minutes	4	5	0	1
35 minutes	3	7	0	0
40 minutes	2	7	0	1
Totals	25	23	8	4

It thus appears that the photopathic reaction of the larvæ was definitely positive. After this series of observations the larvæ were returned to the glass jar and placed on the west table. In the faint daylight which remained, the positive reaction was manifested and continued as long as the light lasted.

Conclusions from Experiments 20, 21, 22 and 23: As has been noted, these experiments deal with the reaction of larvæ as they pass through the third and enter the fourth stage. In Experiment 20 (Cases 2, 3, 4 and 5) it was shown that, in general, the majority of early third-stage larvæ reacted negatively, frequently to light of weak intensity, and invariably to light of greater intensity. In Experiment 21 it appears (1) that this negative response was fairly characteristic of the early third-stage larvæ; (2) that, notwithstanding this negative phototactic reaction, the photopathic response might be definitely positive (Experiment 21, Case 3) thus appearing to indicate that, at least at a certain period in the life of the third-stage larvæ, a positive photopathic reaction and a negative phototactic response may be given by the same individual. Experiment 22 demonstrates (1) that, as the third-stage advanced, the positive reaction, was more frequently and more easily determined by light of all intensities (Case 2), and that an increasingly strong illumination was required to bring about a negative reaction (Case 4); (2) that the photopathic response, if anything, remains throughout the stage, positive (Case 3), while the sign of the phototactic response may change with the intensity of the light (Cases 2 and 4).

In Experiment 23 it is observed that the negative reaction to strong light was still prominent in the behavior of the twelfth-day larvæ (Case 1), while on the thirteenth and fourteenth days, as the moulting-period to the fourth stage approached, the negative reaction was less easily determined (Cases 3 and 4). It was observed furthermore, that these larvæ continued to manifest a very definite positive photopathic response (Case 5), and that this was maintained until the end of the stage-period.

General conclusions on the behavior of larvæ of the first three stages—As the larvæ, after the very definite positive photopathic and phototactic reactions characteristic of the first part of the first larval stage, pass on through the first stage-period, lights of low intensity (red, orange, twilight, etc.), gradually lose their efficiency in bringing about a positive phototactic reaction, while, on the other hand, lights of a greater intensity (green, blue, daylight, etc.) determine, more and more easily, a negative response. This negative phototactic response, which may enter on the third day of the first stage-period, changes again to positive as the first-stage larvæ draw near the first moulting-period. At this time, the lights of low intensity are again effective in bringing about a positive reaction, which is maintained until the larvæ have moulted into the second stage.

While the photopathic reaction of newly moulted second-stage larvæ remains positive, the phototactic response is more often negative, and this negative response is commonly maintained until toward the end of the second stage-period. At this time, as was observed in the first stage-period, a positive reaction again becomes manifest as the larvæ approach the period of moulting into the third stage.

While the positive photopathic reaction still obtains, the newly moulted third-stage larvæ commonly manifest a negative phototactic reaction, and this, as was the case with the second-stage larvæ, is retained until the moulting-period into the fourth stage approaches. At this time the reaction again becomes positive, and continues so until the larvæ have entered the fourth stage. These general points in the behavior may be illustrated by the following diagram (Fig. 6).

The foregoing facts serve to emphasize further the statement made on an earlier page, that we can not justly say that the larvæ of Homarus are positive to light or negative to light, or that they react in this way to intensity, and in that way to the directive influence of the light rays. But these observations do show that the larval lobsters manifest a type of behavior which includes widely varying kinds of reaction, even to the same stimulus. The point has been, not to learn what reaction the lobster larvæ give to light, but to ascertain the conditions which so play upon the mechanism of these organisms as to produce the wide range of responses observed. The causes of the daily and the hourly variations in the kinds of reactions manifested by organisms is a field which is, even at the present day, largely given up to speculation, and all sorts of explanations have been brought forward from the view of the rhythmical succession of certain movements resulting from purely internal stimuli, to the view of cycles of change in certain metabolic products under the influence of external stimulation, and their consequent reaction upon the nervous processes of the organism. The fact of variations in the reactions of larvæ of the European lobster (Homarus vulgaris) has been noted by

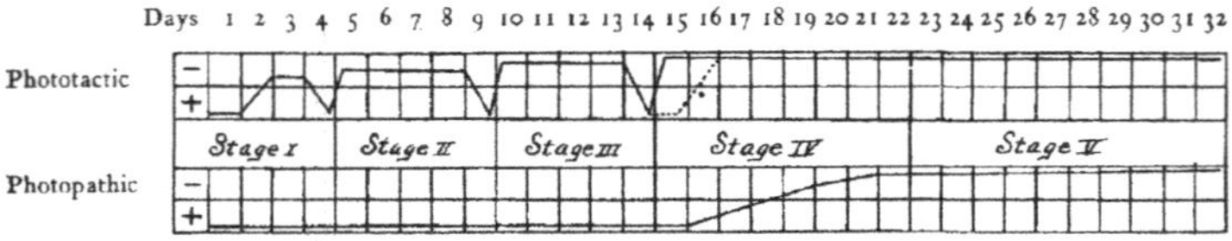

FIG. 6. Diagram recapitulating the nature of the phototactic and photopathic reactions of lobsters in the first five stages. The dotted line in the upper series indicates that the early fourth-stage lobster may give a positive phototactic reaction to light of very great intensity. For further explanation, see General Conclusions on p. 239.

BOHN (1905, p. 10). After having made several observations upon recently hatched larvæ, he comes to the following conclusion. "De ces diverses observations, il semble résulter que *le sens de deplacement des larves de homard subit des variations occellantes de signe*, qui, bien que influencées par l'éclairement actuel sont en relation avec les heures de la journée."

Although certain phases of behavior in some marine animals may be explainable on the ground of rhythmically recurring reactions which bear a certain relation to the hours of the day, the present writer's experience with larvæ of all ages and stages of the American lobster makes it quite impossible to attribute the variations in the reaction of lobster larvæ to such causes. The hours of the day, except as they are accompanied by corresponding differences in the intensity of light, have nothing to do with the form of reaction displayed by the lobster larvæ. This is readily shown, first, by the fact that, at any corresponding time on two successive days (and especially when a moult has intervened), the reactions of the same larvæ may be quite dissimilar. Here the reactions are explainable on the grounds of (1) the stage of the larvæ and their age in the stage-period; and (2) the intensity of the light and of other stimuli which are brought to bear. This conclusion is shown furthermore by the fact that larvæ at corresponding times in the different stage-periods usually manifest similar types of reaction. Thus does it appear that, although the diverse forms of reaction are partly due to the differences of intensity in actual illumination, the underlying cause is some physiological change which the larvæ undergo as they gradually approach and pass the crises of their moulting-periods.

4. Fourth Stage.

We come now to a consideration of the reactions to light of the fourth-stage lobsters. It has been observed in previous pages that the most striking change, not only in body-form but also in life-habits, which takes place in the life of the lobster, occurs during the transition from the third to the fourth stage. It is the aim of the present section to analyze the reactions of the fourth-stage larvæ and to exhibit the conditions which determine or modify these reactions.

In the previous pages the lobsters under consideration have been referred to as the "second-day" or the "fifth-day" larvæ, etc., as the case might be, and this terminology was of advantage, because the first three stage-periods are so brief that changes may occur even in two consecutive days. The fourth stage-period, however, is much longer (usually eight to twelve days) and the differences in reaction on two consecutive days may be slight or inappreciable. For this reason, then, in the following consideration we shall divide the fourth stage-period into three parts, viz: the early, the mid, and the late fourth stage-period.

Experiment 24. Reactions of early fourth-stage lobsters. Case 1. Photopathic reactions—July 28, 10 a.m. Ten early fourth-stage larvæ were put in the glass-bottomed box *B* and this was placed over the light-shaft. The arrangement of the colored plates and the resulting orientations were as follows. The results indicate that the early fourth-stage lobsters show a slight tendency to remain in the more brightly illuminated areas.

Red.	Orange.	Green.	Blue.
1	1	2	6
2	2	1	5
2	3	2	3
5	6	5	14
Red.	**Orange.**	**Blue.**	**Green**
4	1	2	3
2	2	2	4
0	2	5	3
0	1	4	5
6	6	13	15
Green.	**Blue.**	**Orange.**	**Red.**
2	4	1	3
1	4	2	3
3	3	2	2
6	11	5	8

Totals.

Blue.	Green.	Orange.	Red.
38	26	17	19

Case 2. Phototactic reaction—August 7, 5:30 p.m. Ten early fourth-stage larvæ were placed in box *B* and the end window of the box was opened to the diffuse light of the west window. As has been explained, this box was so constructed that in a moment glass plates could be slid through the cover, and into such position that they would divide the floor area of the box between the ends into four equal parts. Beginning with the end toward the light these may be numbered 1, 2, 3, 4, respectively, and the results showing the imprisonment of the larvæ in two instances, may be recorded as follows:

1.	*2.*	*3.*	*4.*
2	2	0	6
7	0	0	3
2	0	2	6
0	0	2	8
11	2	4	23

1.	*2.*	*3.*	*4.*
2	1	2	5
0	1	2	7
0	0	1	9
0	0	3	7
2	2	8	28

In the second of these instances an acetylene light was used, the intensity being diminished by inserting a red glass between the burner and the window in the end of the box toward the light

Case 3. Phototactic reaction—August 9, 3:40 p.m. Ten early fourth-stage lobsters were transferred from the confining bag to box *B*, and this was placed with the end toward the west light. In this case, colored glass plates were placed at intervals in front of the end window to modify the intensity of the light entering the box. A five-minute intermission was allowed between observations. The results, which clearly establish a negative phototactic reaction, are presented in the following table:

COLOR.	*1.*	*2.*	*3.*	*4.*
	2	1	2	5
	1	1	2	6
	2	1	3	4
	0	0	2	8
Orange	5	3	9	23
	0	1	2	7
	0	1	4	5
	1	3	2	4
	0	1	1	8
Red	1	6	9	34
	1	0	1	8
	0	0	0	10
	1	1	0	8
	1	2	2	5
Green	3	3	3	31
Totals	9	12	21	78

Case 4. Phototactic reaction—In the following case, in which the same lobsters were used the source of illumination was the acetylene light and the intensity of the light which entered the end window of box *B* was modified in two ways: (1) by the colored glass plates placed between the light and the window in the end of the box; (2) by the distance of the light from the box. The results, which demonstrate a definite negative phototaxis, were as follows (in all cases the figure *1* indicates the division of the box nearest the light; *4* the division farthest from the light):

Color.	Distance.	*1.*	*2.*	*3.*	*4.*
Red	2 inches	3	1	2	4
		1	0	1	8
		1	2	3	4
		1	1	1	5
	Totals	6	4	9	21
Red	6 inches	2	1	1	6
		1	3	2	4
		3	0	1	6
		1	2	2	5
	Totals	7	6	6	21
Red	12 inches	3	1	2	3
		4	2	1	3
		1	3	2	4
		2	4	4	2
	Totals	10	8	9	12
Blue	12 inches	0	1	4	5
		1	1	3	5
		3	2	0	5
		2	2	0	6
	Totals	6	6	7	21
Blue	2 inches	2	0	1	7
		1	1	2	6
		2	1	2	5
		2	1	3	4
	Totals	7	3	8	22
Totals		36	27	39	97

Case 5—One observation on the behavior of the early fourth-stage lobsters is difficult to harmonize with the reactions mentioned in the previous cases. When at night the rays from an acetylene light were brought to bear upon very early fourth-stage lobsters, swimming in the confinement bags, they would sometimes swim directly toward the light. This reaction was often so strongly manifested that the natural rheotactic response to the influence of the water current circulating in the bags was quite obscured in the areas of greatest illumination, because the young lobsters followed—so to speak—the course of the rays from the acetylene lantern. If this reaction represents a true phototactic response, then it must be said that very early fourth-stage lobsters may, under appropriate conditions of stimulation, respond positively to the directive influence of light, not, as do the earlier stages or the late fourth-stage by turning from the light, but by "heading"

into it. In an earlier paper (HADLEY 1906b), the writer has assumed this to be a true phototactic response. One other instance which appears to support this view may be recorded as follows.

Case 6—Ten sixteenth-day, fourth-stage lobsters were placed in a large slender dish, which was set in the dark box. The larvæ manifested no tendency to undergo either body-orientation or progressive orientation. Next, the same lobsters were placed in the glass-bottomed box, now lined with white paper, which greatly intensified the light within. This box was put with the end window toward the bright sunlight, and the records of five trials (ten larvæ in each) indicated that, when the light was sufficiently intense, the early fourth-stage lobsters might give a positive phototactic reaction. In this instance twenty-six larvæ were positive, twelve negative, and twelve neutral.

When the white paper was removed, and four more tests were made, the results showed that twenty-five larvæ were negative, six positive, and nine neutral.

Experiment 25. Reaction of mid-fourth-stage lobsters. Case 1. Phototactic reaction—August 10, 3:30 p.m. Ten mid-fourth-stage lobsters were transferred from the hatching bags to box *B*, and the experiment was continued in daylight as in Experiment 24, Case 2. The results show a definite negative phototactic reaction, and may be tabulated as follows (similar results were obtained when a white lining in the box was used, though in this case, they showed a less definitely negative reaction):

COLOR.	*1.*	*2.*	*3.*	*4.*
Orange	1	1	1	7
	0	1	2	7
	0	0	1	9
	1	1	2	6
Totals	2	3	6	29
Blue	0	1	2	7
	1	0	1	8
	1	1	2	6
	1	1	1	7
Totals	3	3	6	28
Ruby	0	0	1	9
	0	2	1	7
	1	1	4	4
	1	0	3	5
Totals	2	3	9	25
Grand Totals	7	9	21	82

Case 2. Photopathic reaction—August 9, 3:30 p.m. Ten fourth-stage lobsters were removed from one of the confinement bags and placed in 16 mm. of water in the glass-bottomed box. The glass plates were arranged in the order given below, and tests were made at five-minute intervals. The results, which showed a diminished tendency to remain in the areas of greatest illumina tion, are represented in the following table:

Red.	Orange.	Green.	Blue.
3	2	1	4
4	0	3	3
1	2	1	6
2	2	4	2
2	4	2	2
12	10	11	17

When, some hours later, the same lobsters were tested again the results of five trials were as follows: Blue, 13; green, 9; orange, 7; red, 21; apparently in this instance it can not be said that the mid-fourth-stage lobsters were either positively or negatively photopathic. Yet the last instance shows a tendency toward a negative reaction.

Experiment 26. Reaction of late fourth-stage lobsters. Case 1. Photopathic reaction—August 12, 2 p.m. Ten late fourth-stage lobsters were transferred from one of the confinement bags (where the majority had already entered the fifth stage) to the glass-bottomed box which was placed over the light-shaft in order to test the photopathic reaction. In this case nine consecutive tests were made, three minutes being allowed for each orientation. The results, which are characteristic of all other tests, and which show a tendency on the part of the lobsters to avoid the light, may be recorded as follows:

No.	Blue.	Green.	Orange.	Red.
1	1	2	1	6
2	1	0	3	6
3	3	0	2	5
4	3	4	1	2
5	1	2	1	6
6	3	1	0	6
7	5	3	0	2
8	1	1	1	7
9	4	2	2	2
Totals	22	15	11	42

Case 2. Phototactic reaction—August 10, 3:30 p.m. Ten late fourth-stage lobsters were taken from one of the hatching bags and put in box *B*, which was placed in the dark box so that the end window faced the light, the intensity of light being modified in each case by interposing colored glass plates between the end window and the light. The tests, which were made at three-minute intervals, and which showed a very definite negative reaction, were as follows (in the fourth tests of the first and last sets respectively, one lobster was accidently killed, thus making the totals incomplete):

COLOR.	*1.*	*2.*	*3.*	*4.*
	1	1	1	7
	0	1	2	7
	0	0	1	9
	1	0	2	6
Orange	2	2	6	29
	0	1	2	7
	1	0	1	8
	1	1	2	6
	1	1	1	7
Blue	3	3	6	28
	0	0	1	9
	0	2	1	7
	1	1	4	4
	1	0	3	5
Red	2	3	9	25
Totals	2	8	21	82

Conclusions on the reaction of fourth-stage lobsters—The observations thus far made upon the behavior of fourth-stage lobsters appear to demonstrate the following points: (1) Throughout the entire fourth stage-period (with the exceptions noted under Experiment 24, Cases 5 and 6), the lobsters manifest a negative phototactic reaction, which is accentuated in the latter part of this stage. This behavior is quite different from the positive reaction which supersedes the negative in the case of second and third-stage larvæ just previous to their moult into the third and fourth stages respectively; (2) This type of reaction after the first part of the fourth stage-period, cannot be reversed or modified, as was the case in earlier stages, by using different intensities of light (3) The photopathic reaction, which in the early fourth-stage lobsters is definitely positive, changes by the latter part of the stage to negative in the majority of individuals. Thus it can be observed that, just as the third-stage larvæ might at the same time (or successively) manifest both a negative phototactic and a positive photopathic reaction, so may the lobsters of the fourth stage. Other points regarding the behavior of fourth-stage lobsters will receive consideration in connection with the subject of contact-irritability

5. Fifth Stage.

The body-form of the fifth-stage lobster is similar to that in the fourth-stage, and we might therefore expect to find similar types of reaction. It will be seen, however, that there are many points of difference in behavior which are of such a nature that they can not be attributed, either wholly or in part, to changes in body-form or in the swimming appendages. The changes are doubtless the consequence of modifications which have taken place in the body-processes or in the physiological states of the lobsters themselves, and which have resulted from the cumulative stimulation during the earlier life of the lobsters. Generally speaking, it may be said that the reactions of the fifth-stage lobsters are fairly typical for the adult form, and are especially characterized by the light-shunning tendency. This form of behavior could be observed readily by watching the lobsters in their confinement cars; but, for the sake of certainty, the same experiments, to which the larvæ of earlier stages had been subjected, were repeated with the fifth-stage lobsters. Since the reactions did not appear to undergo any noticeable modification as the lobsters passed through the fifth stage, there is no need for considering the early, mid and late fifth stage-periods separately, as was done for fourth-stage lobsters. The type of reaction presented in the early fifth stage-period differs in no way from the behavior of lobsters in the late fifth stage-period; and both are characteristic of the behavior in all later stages.

Experiment 27. Case 1. Photopathic reaction—In the first instance, ten fifth-stage lobsters were transferred from one of the confinement bags to the glass-bottomed box and this was placed over the light-shaft. The method used was the same as in previous experiments. In the second instance the blue glass was removed, and the space where it had lain was left clear, thus permitting the reflected daylight to enter this area of the bottom of the box. The results of both tests show a negative reaction which was more definite in the second instance.

BLUE.	GREEN.	ORANGE.	RED.
2	1	2	6
2	1	2	5
3	1	1	5
2	3	2	3
8	6	7	19

DAYLIGHT.	GREEN.	ORANGE.	RED.
0	2	2	6
0	3	2	5
1	2	2	5
0	2	2	6
1	1	3	5
0	2	4	4
2	12	15	30

Case 2. Phototactic reaction—Further demonstration of the definitely negative phototactic response of fifth-stage lobsters was given by the experiments on contact-irritability (Exp. 29, p. 256). Here is clearly shown the extreme manifestation of this negative phototactic response, which frequently would have culminated in fatal results by driving the lobsters from deep to shallow water and leaving them stranded where they would certainly have died had they not been returned to the water at the end of the experiments. Here, as has been found in the case of many animals, the total behavior is completely dominated by the light influence. It may be said further that in the case of the fifth-stage lobsters light of different intensities does not cause a change of reaction from positive to negative, or from negative to positive, as was the case in the earlier stages; nor do we ever find the individuals "heading" into the light, as may be the case in the fourth-stage larvæ. For the fifth-stage lobsters any intensity of light which influences their behavior in any degree, determines, under experimental conditions, both a negative body-orientation and a negative progressive orientation.

In the foregoing pages it has been shown that larvæ which were positively photopathic could be made to pass from regions of greater to regions of lesser light intensity by submitting them to the directive influence of light of sufficient strength. In these cases, it was observed that the photopathic reaction was invariably subservient to the phototactic, although the latter was also very dependent upon a certain optimal intensity for bringing about a positive or negative response. In the following instance we shall observe that, although the directive influence of the light rays is capable of modifying the orientations which relative intensities of light have determined, still the directive influence can not quite obliterate the evidence of a photopathic reaction, as was possible in the younger larvæ. In other words *the tendency of the fifth-stage lobster to "select" the darker regions has become almost as firmly fixed as has the tendency to react negatively to the directive influence of the light rays.* In the first larval stages the photopathic response invariably gives way to the phototactic. In the fifth the two tendencies clash; and the resulting orientation of the lobster is determined, not by one, but by both of these factors.

(*A.*) *Case 3. Photopathy versus phototaxis*—Ten fifth-stage lobsters were put in box *B*. This was mounted upon the colored glass plates over the light-shaft as in previous experiments. The preliminary observation showed that there was a definite tendency for the lobsters to congregate at the red end of the series of glass plates, thus demonstrating a negative photopathic reaction. Now the window at the red end was opened to diffuse light. After a period of ten minutes, observations of the position of the lobsters were begun, and continued at five-minute intervals. The following results show that, although the negative phototactic response is still manifested, it has been greatly modified by the tendency on the part of the lobsters to avoid the brightly illumined area at the end of the box:

Daylight.	Green.	Orange.	Red.
1	4	2	3
1	3	3	3
2	3	3	2
1	3	4	2
1	2	4	3
2	3	3	2
8	18	19	15

In the next case, the end of the glass plate series, which in the previous instance admitted reflected daylight, was covered with a blue glass and the illumination of this area thus rendered less intense, while the end window of the box (at the red end) remained open, as in the last experiment. The results, which demonstrate that the phototactic reaction had still further overcome the photopathic, were as follows:

BLUE.	GREEN.	ORANGE.	RED.
2	2	4	2
1	3	2	4
3	2	3	2
3	2	2	3
4	3	2	1
2	3	2	3
15	15	15	15

In the last two instances it becomes apparent that the fifth-stage lobsters, unlike the early-stage larvæ, could not be forced, by the directive influence of the light rays, into an area of greater light-intensity. In other words, the tendency to manifest a negative phototactic reaction was not sufficiently strong to overcome the tendency to give a negative photopathic response.

(*B.*) *Experiment 28. Phototaxis leading to fatal results*—Before bringing to a close this consideration of the reactions to light in lobsters of the fourth and fifth stages, it may be appropriate to introduce the results of some experiments whose aim was to show the extreme nature of some phototactic reactions. In other words, attempt was made to determine whether or not the strong directive influence of the light rays could compel the larvæ so to act that they would do injury to themselves as in the familiar case of the moth that flies into the flame, or of Ranatra, mentioned by HOLMES (1906). The reactions of the fourth-stage and fifth-stage lobsters will be considered together.

Case 1. Fourth-stage lobsters—For this series of experiments box *B* was set up as represented in Fig. 7, being supported at one end so that the bottom of the box made an angle of about fifteen degrees with the table. The box was filled with water so that when it was slanted, the water-line did not quite reach the angle made by the bottom and upper end, *B*. In this way there was created an inclined plane, slanting from the window end, *A*, of the box to the higher end, *B*. The water consequently diminished in depth as the end, *B*, was approached. At this end there was an inch or more of the bottom of the box not covered by water. The light from the window, *L*, was reflected into the box by the mirror,

M, for the purpose of discovering whether the larvæ in presenting their negative phototactic reaction, would allow themselves to be driven into the shallow water. By means of a hole in the bottom of the box, the water could be withdrawn very gradually (a few drops a minute), so that if the larvæ persisted in remaining in the

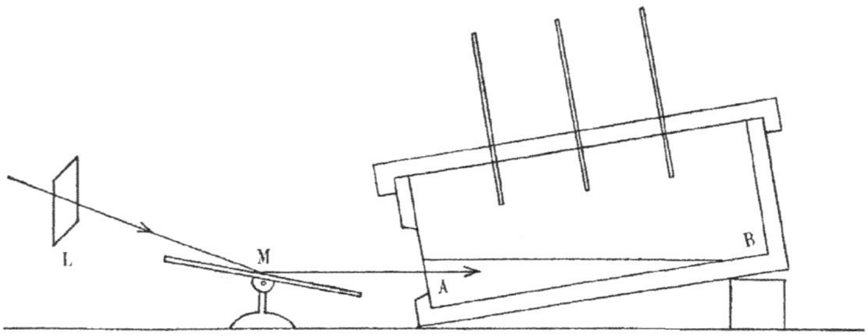

FIG. 7. Diagram of apparatus as set up to test the extreme phototactic reactions, leading, in the case of fourth and fifth-stage lobsters, to fatal results. *L*, source of light; *M*, reflecting mirror; *A*, end of box adjacent to "window;" *B*, end of box not covered with water, where the lobsters were stranded. In the cover of the box are shown the sliding partitions.

shallow area, they would, in the course of a few minutes, be stranded on the dry bottom. Ten fourth-stage lobsters were first used for experiment and the results, ascertained by counts made as in all other cases, were as follows: (The arrow shows the direction of the light coming through the end window of the box, while the numbers at the top of the columns represent the division areas of the box):

TEST.	→ *1.*	*2.*	*3.*	*4.*	NUMBER STRANDED.	TIME (after).
1	1	1	3	5	4	5 minutes
2	1	2	1	6	3	10 minutes
3	0	1	1	8	6	20 minutes
4	1	1	3	5	5	50 minutes
Totals.....	3	5	8	24	18	

The results of this experiment and of several others similar to it, show, that out of a total of twenty-four larvæ which gathered in the area farthest from the light, eighteen allowed themselves to be stranded rather than to retrace their course into deeper water, and in so doing to approach the light.[4]

Case 2. Fifth-stage lobsters—When the same experiment involved the fifth-stage lobsters, the results were similar. The only difference that could be observed was that the intensity of the reaction was greater for the fifth-stage than for the fourth. The result of twelve tests, each with ten lobsters showed the distribution to be as follows: Area 1, nine; area 2, ten; area 3, twenty-one; area 4, eighty, of which seventy were "stranded." These last would have perished, had they not been returned to the water at the end of each successive test.

(*C.*) *Conclusions concerning the reactions to light of fifth-stage lobsters*—The results of the foregoing experiments on the reactions of fifth-stage lobsters, demonstrate the following points: (1) Like the fourth-stage lobsters, the fifth-stage lobsters are negatively phototactic from the beginning of the stage to the end of it, and this holds good for all intensities of light which cause any reaction whatever. (2) Unlike the early fourth-stage but much like the late fourth-stage lobsters, the fifth-stage lobsters are negatively photopathic from the beginning of the stage to the end. (3) This negative photopathic reaction, unlike the photopathic reactions of the earlier stages (in which case the photopathic reaction was entirely subservient to the phototactic), has itself become a well grounded tendency, and, although it can be modified, it can not be entirely obliterated (so far as its value in causing a certain orientation is concerned) by the tendency to react to the directive influence of the light rays. (4) The intensity and energy with which the late fourth-stage, but especially the fifth-stage, lobsters manifest a negative phototactic reaction may lead to results fatal to the lobsters themselves.

(*D.*) *Contact-irritability versus reaction to light*—In the preceding section the phototactic and the photopathic reactions, together with some points of their inter-relation, have been considered. We shall now examine that response of lobsters to solid portions of their immediate physical environment which may be ascribed to contact-irritability or thigmotaxis.

It frequently happens that single types of reaction (phototaxis, chemotaxis, geotaxis, and the like) may be studied to best advan-

[4] It should be noted, however, that the water in no case receded more than 5 to 10 mm. as measured horizontally on the bottom of the box.

tage only when another stimulus of known effect is present and operative. For instance, if the two conditions of stimulation which respectively bring about a photopathic and a phototactic reaction are so arranged as to oppose one another (i. e., by determining opposite reactions in the larvæ), and if the constant effect of one set of conditions is known, then it is possible to form an estimate of the persistency of the reaction determined by the opposed set of conditions. For example, if light rays of low intensity coming through the end of box *B*, resulted in driving the enclosed larvæ, which had just previously given a negative photopathic reaction, to the opposite end of the box, and at the same time forced them from a region of low into a region of high intensity, we should say that the negative photopathic reaction of these larvæ was of slight importance as compared with the phototatic. If, on the other hand, it was learned by experiment that the rays entering the end window of box *B* would not force the negatively photopathic larvæ from the dark into the brightly illuminated end of the box, but resulted in their gathering in the middle of the box (for instance, in the green or orange area) then it might be inferred that the negative photopathic reaction had a greater influence in determining the final reaction of the larvæ, although it was in this case directly and strongly opposed by the tendency to manifest a phototactic reaction. In the following experiments, made to discover the value of contact-irritability in determining the reaction of the larvæ, the principle mentioned above was made use of, and in this instance a combination was made between experimental conditions which would allow the demonstration of contact-irritability, and those which would insure the manifestation of negative phototaxis if no other modifying conditions (such as contact-irritability) were present. But before going farther with the description of the technique of the experiments, a few observations on the behavior of the lobster larvæ under natural circumstances may be considered. This may form a better basis for the consideration of experiments dealing with contact-irritability versus reaction to light under the especially devised conditions to be described.

It might reasonably be imagined that the loss of the swimming branches (exopodites) of the thoracic appendages, which takes place with the entrance to the fourth stage, would at once determine a very radical change in the habits of lobster larvæ. We should surmise that the larvæ would immediately abandon their pelagic manner of existence and enter upon a more sedentary life among the rocks and weeds of the sea bottom. But this is by no means the case, for never in the life history of the lobster do we find surface swimming more strongly manifested than in the fourth stage, and just after the loss of those accessories without which swimming would have been impossible in any of the earlier stages. The energetic surface-swimming of the fourth-stage lobsters was evident from many observations, made under both natural and experimental conditions. It was observable not only in the large hatching bags but also in the quiet water surrounding the bags and hatching apparatus. One case is especially noteworthy. In July a steam launch, of which the captain lost control, rammed one of the floats which suspended six large hatching bags containing lobsters in various stages. As a result many fourth-stage lobsters were suddenly liberated in the water about the hatchery. When order had been restored, an attempt was made to recover the lost lobsters, and over five hundred of the fourth-stage which were swimming actively at the surface of the water were picked up with scrim nets. A far different phenomenon obtains in the behavior of fifth-stage lobsters under natural conditions. This is illustrated by an interesting sequence of changes in the swimming habits. When the majority of the lobsters in the bags were in the fourth-stage, they usually swam near the surface. As the larvæ moulted into the fifth stage, fewer lobsters were to be seen. The reason for this was ascertainable if one poked with a stick about the mass of weeds and algæ adhering to the sides and bottom of the bag. Here could be found, carefully hidden, a large number of fifth-stage lobsters. By the time all the individuals in the bag had passed to the fifth-stage, scarcely one could be discovered swimming freely. Whenever a number of fifth-stage larvæ were liberated in the open water, it was an interesting sight to observe them swim for a moment, then turning head down, disappear for good in the deeper water—a great contrast to the behavior of the fourth-stage lobsters under similar conditions.

Another set of observations refers to the burrowing instinct of the young animals. When early fourth-stage lobsters were transferred to glass dishes, on the bottom of which was a layer of sand, gravel and a few broken shells, they at first paid no heed to these conditions, but for several days continued to swim as persistently as ever. Finally, however (usually within two or three days after having been placed in the dish), the lobsters began to plough through the sand of the bottom, especially near the rim of the container, and to construct burrows beneath shells, stones or other objects in the sand. Yet, even after these burrows were completed, the fourth-stage lobsters seldom remained in them, but came out and crawled rapidly over the bottom or swam more or less actively near the surface of the water. When, on the other hand, fifth-stage lobsters were introduced into the dishes containing sand, gravel, and shells they commenced burrowing at once and when the burrows were completed they showed a much greater tendency to remain therein than did the late fourth-stage larvæ. Although the fifth-stage lobsters came out for food, free swimming was seldom indulged in during such sorties. The question now arises as to what conditions or factors cause the energetic surface-swimming of the early fourth-stage lobsters and the bottom-seeking and burrowing habit of the late fourth and the fifth stage. Are these reactions to be explained as phototropic, geotropic, or thigmotropic reactions? Or do all three of these, and perhaps still other factors, unite in determining the final result? While we are not yet prepared to venture an answer to these queries, the records of a few simple experiments which were undertaken to ascertain the value of the part played by contact-irritability in determining the orientation of the fourth and fifth stage lobsters, under certain known conditions, will be presented.

Experiment 29. Fourth-stage lobsters—The technique employed in the present experiment was as follows: One-half of the bottom of box *B*, was sprinkled with sand to the depth of five mm., the box was filled with salt water to a depth of 3 cm., ten early fourth-stage lobsters were introduced, and the box covered. The aim was to learn whether, in the total absence of light, the larvæ would "choose" either the sanded or the clear area. The result of a typical test is presented below. The readings were taken every five minutes, and after each reading the lobsters were caused to distribute themselves about the box:

Sanded area.		Clear area.	
1.	*2.*	*3.*	*4.*
4	3	2	1
2	2	3	3
3	2	2	3
1	1	4	4
10	8	11	11
18		22	

These and other tests were made, but in no case was it apparent that the early fourth-stage lobsters showed any preference for the sanded area. When, in another series of four trials involving ten lobsters each, the window at the sanded end of the box, was opened so as to allow the rays to stream through, every lobster but one was driven to the compartment farthest from the light. When this experiment was tried with late fourth-stage lobsters, it appeared that a greater number remained on the sanded area, even in the presence of the light conditions mentioned above. The results of a typical experiment of this sort involving five trials of ten lobsters showed that, while thirty were driven to the clear space, ten remained on the sanded area.

Experiment 30. Case 1. Fifth-stage lobsters—In this instance ten fifth-stage lobsters were placed in box *B* as arranged for the previous experiment, no light being admitted at the end of the box. The record of seven trials separated by a period of from five to ten minutes, showed a decided preference for the sanded areas; while forty remained on the sanded region, only twenty gathered on the clear area.

Case 2—In the next instance the end window at the sanded end of the box was opened to the light, but with a red glass so interposed that the intensity of light in this region was not great. A period of from ten to forty-five minutes was allowed for each orientation. Although the influence of the light tended to drive the lobsters off the sanded area the results of six trials (ten lobsters each) showed that thirty-seven fifth-stage lobsters remained in contact with the sand, while twenty-three moved to the clear area.

Case 3—In the next series of six trials (ten lobsters each) the intensity of light was modified by substituting an orange glass before the end window. The results showed twenty-five on the sanded area, thirty-five on the clear.

Case 4—In the last series of six trials (ten lobsters in each) the conditions were still further modified by removing the orange glass and thereby greatly increasing the intensity of the light which entered the end window of the box. This demonstrated that a light of great intensity would drive the fifth-stage lobsters off the sanded area. At the end of the experiment only thirteen lobsters remained on the sanded area, while forty-seven remained in the clear region. Finally, the sand was removed from the box, and the reaction of these lobsters was tested with unobstructed light entering the end window. The resulting reaction was invariably and definitely negative; and this with light of all the intensities used in the previous cases.

Conclusions from experiments on contact-irritability versus reaction to light—Although these experiments can hardly be called critical, they demonstrate that the presence of the sanded area in the box did modify the reactions of the fifth-stage lobster. That there was manifested a tendency to remain in contact with the sand, to burrow in it, and not to be dislodged by such intensities of light as would normally rout the entire group of lobsters and send them to the end of the box farthest from the light. These facts, moreover, cannot be said to hold true for the fourth-stage lobsters that were used in the foregoing experiments, and which showed no well defined preference for the sanded area, at least in the early part of the stage-period.

VI. MECHANICS OF ORIENTATION.

The aim of the present section is to report the results of a series of observations which were made in order to answer the following question: By what movements of the lobster larvæ are the reactions to light accomplished? In our effort to answer this question we shall, for the present, attempt to avoid so far as possible considerations which deal directly with the ultimate causes of orientation; in other words, we shall limit ourselves to the observation of the actual movement of the body, or of certain parts of the body, of individual larvæ; and attempt to show what relation exists between these movements and the external factors which appear to determine them. First, however, it is necessary to establish some points regarding the natural behavior of the larvæ when the influence of external stimuli is at the minimum.

1. The normal behavior of the larvæ—In view of the fact that swimming constitutes the chief activity of the larval lobsters, our question resolves itself into the following: What is the nature of the normal swimming? When one first observes the behavior of individual larvæ amidst the thousands contained in the large hatching bags no difference is evident in the swimming of the first three stages. In all instances the back of the larva is, for the most part, uppermost, the abdomen bent under and downward at an angle of about 60° from the longitudinal axis of the cephalo-thorax, which in turn is inclined about 30° from the horizontal plane. In daylight this position may be maintained without modification for several minutes, but the equilibrium is often interrupted by other body-movements which, upon superficial examination, appear to be of a most diverse and ill-ordered nature. There are leanings, turnings, fallings, somersaults, revolutions and rotations which follow each other in no apparently definite sequence, and which disturb the general equilibrium greatly or slightly as the case may be.

Whether the balanced equilibrium, the devious rotations or other activities are present, the exopodites or swimming attachments of the thoracic appendages beat the water more or less constantly with short vibratory strokes, sometimes lifting the larvæ high toward the surface, and again allowing them to sink to the bottom, where they frequently lie for some moments almost motionless, only again to resume their varied activity. Now they swim forward, now backward, now lurch to the side, now to the rear, always maintaining more or less energetically these apparently aimless movements. Such is the nature of the swimming in daylight or other brilliant illumination; but for our purpose it cannot be called the normal swimming of the lobster larvæ. It is only under special conditions that the latter may be observed; and, in view of the fact that it is the conditions of light which influence more strongly than any other factors the behavior of the larvæ, it is only when they are under certain light-conditions that we may expect to find manifested what we may call the characteristic or normal swimming.

The twilight or nocturnal swimming of the larval lobsters invariably gives us the fairest example of natural behavior. At such times alone (or when the larvæ are submitted to artificially produced twilight) variations in temperature and the multiplicity of conflicting cross-light influences are eliminated. Frequently when the twilight was so dim that observation was rendered difficult, the swimming was delicate and regular, and the young larvæ would mount up, bird-like, to the surface of the water, hover many seconds in a single position, or swim backward or forward with equal ease. In such a case, when a lighted match was brought near the side of the jar in which the larvæ were confined, the same restless and uncertain swimming, characteristic of the diurnal activities, was again manifested, together with the accompanying leanings and rotations. From these facts it may be assumed that the twilight swimming of the larvæ probably represents the natural behavior or at least the behavior that arises purely from the internal states themselves; and that the peculiar antics characteristic of the daylight swimming represent a type of behavior chiefly due to the action of external stimuli.

The question now naturally arises—Do the various turnings, rotations, leanings, and fallings which constitute the apparently haphazard behavior of the larval lobsters when swimming in daylight or other brilliant illumination, give any indication of method? Observations have given a suggestion as to the means whereby we may attempt to ascertain the value of certain light-conditions in determining these peculiar forms of behavior.[5]

If larval lobsters of any of the first three stages are subjected to the influence of light which comes from one direction only, as from the side, the first fact observable is that the larvæ undergo a certain body-orientation; they turn away from the light and place the long axis of the body parallel to the direction of the rays. The second fact which may be noticed is that the larvæ move in the direction of the light rays either toward or from the source of illumination. A third fact, which is of prime importance and which involves those stated above, is that no matter whether the progressive movement of the larvæ be toward or away from the source of light, the orientation of the body (head away from the source of light) remains unchanged. To state the matter briefly we may say that, whatever the nature of the progressive orientation of the larvæ, *the body-orientation is at all times, and under all conditions, negative.* Bohn (1905, p. 8) has clearly pointed out this fact for the larvæ of the European lobster. In this regard he says: "En général, les larves de homard se placent dans le sens négatif; même, dans les premières heures après l'éclosion, alors qu'elles se groupent vis-à-vis des lamps, leur tête se tourne du côté opposé, et les larves s'approchent de la lumière en regardant l'obscurité, c'est-à-dire en reculant. Ainsi, après l'éclosion, l'orientation a lieu dans le sens nègatif, mais le déplacement se fait dans le sens positif. Dans le suite, si le sens de l'orientation

[5] Many of the observations which follow were made previous to the writer's knowledge of the excellent work of Georges Bohn (1905) along similar lines, upon the larvæ of the European lobster, Homarus vulgaris. The writer would acknowledge, however, his great indebtedness to this investigator, whose work has proved suggestive in the highest degree, and whose observations on the mechanics of behavior the writer has been able, in the majority of instances, to verify as well as supplement.

reste le même, le sens du déplacement peut changer." LYON (1906) has recorded a similar observation for several larval stages of Palemon. This condition of affairs is rather at variance with the majority of observations on the phototactic reactions of animals and it is contrary to the condition of body-orientation which we find in the fourth stage of the lobster itself, for in this stage (at least in some of the assumed phototactic reactions) the body-orientation brings the head toward the source of illumination instead of away from it as is invariably the case in the first three stages.

The question has already arisen as to what we may mean by a positive phototactic reaction, for in this case it is clear that we may very frequently have a negative body-orientation coupled with a positive progressive orientation. Until we know more regarding the differences between body-orientation and progressive orientation, it may be considered safe to say that the *direction of the progressive movement*, with respect to the source of illumination, may be held as the surest criterion of the sign of the phototactic response of animals. On the other hand the point has been made clear by some writers, that in the body-orientation of organisms the definite relation of the body-axis to the lines of active force is the primary consideration for all problems of progressive orientation. However this may be, we have before us at least one instance wherein, although the relation of the body-axis to the lines of force is an important consideration, the body-orientation *per se* has little or nothing to do with the question of the positive or negative progressive orientation of the organism; for as we have already observed, conditions which invariably determine a negative body-orientation may determine either a positive or a negative progressive orientation, as other circumstances demand. We may, therefore, first concern ourselves with the *mechanics of progressive orientation* and then turn with better understanding to the *mechanics of body-orientation*, for these two reactions apparently depend upon quite different circumstances.

2. *The mechanics of progressive orientation*—The only means of locomotion possessed by the larvæ of the first three stages are the exopodites of the thoracic appendages and the strong, flexible abdomen with its broad terminal fan (Fig. 1). It is but seldom, however, that the latter is used, and never when it is a question of progressive orientation to light. We are then confronted with the problem: How, by the motion of the thoracic exopodites

alone, is the larval lobster able to execute those movements which determine his progress either toward the source of illumination or away from it?

If the larval lobsters in any of the first three stages be put in a glass jar which is surrounded by black paper and placed in subdued daylight, the short vibratory strokes of the exopodites can be readily observed. At one time, certain individuals may be seen to swim rapidly backward, and again forward, with no apparent change in the position of the body or in the direction of the stroke of the exopodites. If, however, the thoracic appendages themselves be carefully watched, one can observe that, from time to time, these limbs undergo either a forward shifting (extension) as shown in Fig. 8, or a backward shifting (contraction) as shown in Fig. 9. This change from the "anterior" position to the "posterior" position may occur at short intervals, each position may persist for some seconds, or there may be a successive alteration with periods of longer duration in either one position or the other. It may be observed further, that when the thoracic appendages take the "anterior" position, the direction of the strokes of the exopodites becomes somewhat forward as well as downward, and the *resulting motion of the larvæ becomes backward and upward.* When, on the other hand, the thoracic appendages assume the "posterior" position, the stroke of the exopodites becomes backward and downward; and *the resulting motion of the larvæ becomes forward and upward.* During a great part of the time, the upward movement of the larvæ, as a result of the outward and downward stroke of the exopodites, does little more than compensate for the natural tendency to sink toward the bottom. For this reason the

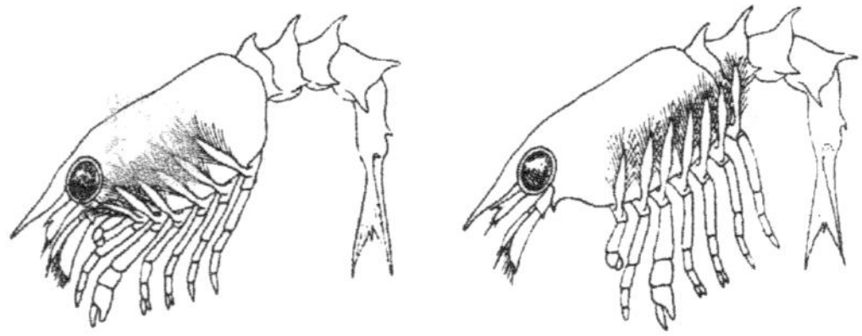

FIG. 8. FIG. 9.

FIG. 8 shows a larval lobster with the thoracic appendages in the extended or 'anterior" position; the resulting movement is forward and upward.

FIG. 9 represents the appendages in the contracted or "posterior" position; the resulting movement is backward and upward.

progress of the larvæ may often be directly forward or directly backward with but slight deviation from the horizontal plane; while at other times, when the stroke of the exopodites is directly outward and downward (exclusive of either the "forward" or "backward" factor), the larvæ may mount to the surface in nearly vertical lines.

It thus becomes evident that the progression of the larvæ, backward or forward, upward or downward, is largely determined by the position (state of extension or contraction) of the thoracic appendages. In other words, if for the greater part of the time these appendages are in the "anterior" position the phototactic reaction of the larva is positive; but on the contrary, if the thoracic appendages are more frequently in the "posterior" position, then the consequent reaction of the larvæ is negative. Naturally the next important question which arises is: What conditions determine the "anterior" or the "posterior" position of the thoracic appendages? It cannot be questioned that these changes are directly due to certain variations in the intensity of the illumination and are modified by the "physiological state" of the larvæ themselves; and that, furthermore, the state of extension or contraction of the thoracic appendages, and the stroke of the exopodites, are regulated to a great degree through the mediation of the eyes and the nervous system of the larvæ. But further consideration of this subject must be postponed until later. In the meantime we may turn our attention to the mechanics of body-orientation.

3. The mechanics of body-orientation—Under the present heading we shall consider the nature of those peculiar movements which the lobster larvæ undergo when they are under diverse and changing conditions of stimulation, in order to explain the cause of these actions and to show their relation to certain definite laws which may be said to regulate to a great degree the body-orientation of the larvæ. As we have observed, it is the influence of light which is most active in determining the behavior of the larvæ; furthermore, it is in the absence of such influences as diverse and changing conditions of illumination afford that the most realistic picture of the normal behavior of the larvæ is obtained. It will then prove the most practical method of approaching this problem, first, to obtain conditions of light which allow natural behavior (normal swimming); and then, by gradually modifying these conditions, to observe the effects upon the behavior of the larvæ.

A. THE EFFECTS OF DIRECT LIGHTING AND SHADING. *Technique and Methods of Observation*—This section deals more especially with the directive influence of light rays so introduced as to strike the larvæ from different directions; from before, from behind, from the side, from above, from below, or obliquely to the body-axis. These conditions were obtained, for the most part, in two ways. The larvæ were placed either in a cylindrical glass jar, or in an especially constructed rectangular glass box (similar, perhaps, to the révélateur used by BOHN), three inches wide, six inches long, and two and a half inches deep, all sides and the bottom being of glass. Either of these receptacles might be placed in the dark box already described. To regulate the intensity, slides of colored glass were used as in the earlier experiments,

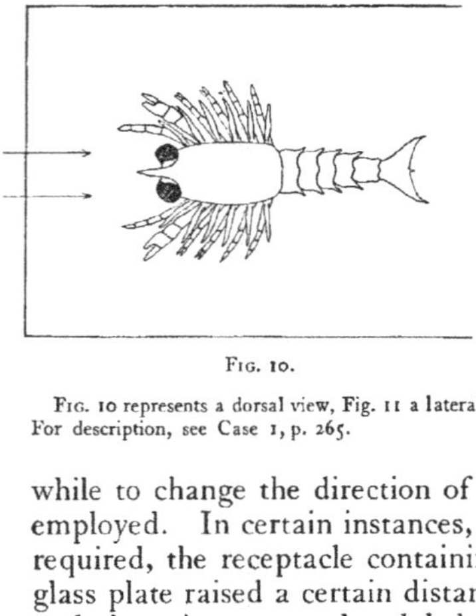

FIG. 10.

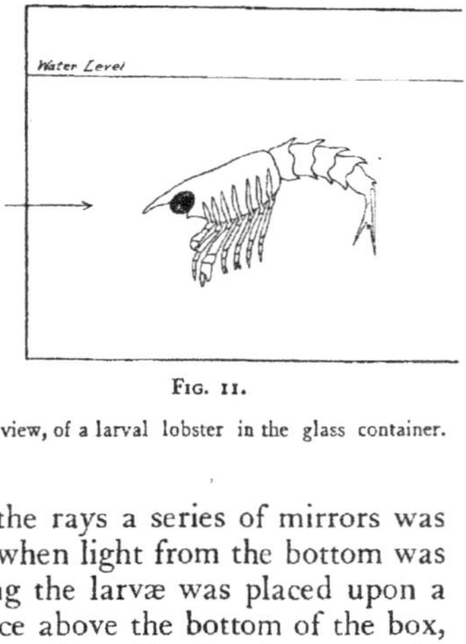

FIG. 11.

FIG. 10 represents a dorsal view, Fig. 11 a lateral view, of a larval lobster in the glass container. For description, see Case 1, p. 265.

while to change the direction of the rays a series of mirrors was employed. In certain instances, when light from the bottom was required, the receptacle containing the larvæ was placed upon a glass plate raised a certain distance above the bottom of the box, and the mirror was placed below. In still other instances the direction or the intensity of the light was modified by the use of light-absorbing (black) or light-scattering (white) backgrounds. These were used more frequently when the observations were made in diffuse daylight, and the subdued light came to the glass containers from several different directions. From the experiments it appears very probable that in determining the orientation of the organisms, the backgrounds were instrumental only in regulating the amount and the general direction of the light which they reflected or absorbed. First, however, we shall consider the effects of suddenly throwing the light from a certain direction upon larvæ oriented in various positions.

Case 1. Illumination from before—In the first instance the behavior of a single larva was studied (the stage does not matter). It was oriented in the rectangular container, in the dark box with its head toward the three by one inch window, which was closed (Fig. 10), but in such relation to the glass box that its longitudinal axis was parallel to the direction of the rays of light coming from this window when it was opened. While the larva was so oriented, the screen was drawn aside and light from the small window was allowed to strike the larva "head-on." Under these conditions, one of two reactions resulted. The larva underwent either a forward or a backward somersault, or rotation, which brought the back below with the head directed away from the source of illumination. Whether the rotation was backward or forward made no difference in the resulting orientation and which one occurred depended upon the direction of the rays of light which struck the eyes of the larva. In normal swimming the body of the larva in any of the first three stages is bent about 30° from the horizontal. Now if the rays of light had the direction of *A* or *B* (Fig. 12) the rotation was usually forward, while if the light came from below, direction *C*, the rotation often was backward. After this first orientation the larva (position *B'*) frequently performed a rotation on its long axis, either to the left or right, which brought the back again uppermost, and it then progressed in the direction of the rays, either toward or away from the source of illumination.

Corollary 1—If the rays striking the eyes of the larva had the slightly oblique direction shown in Fig. 13, *a* or *c*, but were in direction or plane *B* (Fig. 12), then the larva pivoted at the middle of its own longitudinal axis and swung to one side or the other, always keeping the back uppermost.

If the rays of light took the direction designated a^1 — a^4 or c^1 — c^4, the result was the same; the larva swung until the longitudinal body-axis was parallel with the incident rays, and the head was directed away from the source of illumination.

Corollary 2—If the rays striking the eyes of the larva had the oblique direction, a^1 — a^4 or c^1 — c^4 (Fig. 13) in plane *A* of Fig. 12, then the resulting movement was a combination of the forward rotation and the side swing (Cor. 1). In other words, the larva performed a side-somersault, and ended with the back directed below and to the side. Whether it turned to the left or to the right depended upon the direction of the rays in either the *a* or the *c* series. At the end of this reaction the larva usually became righted again with the back above and the head away from the light, and continued its progressive orientation in one direction or the other according as the reaction was positive or negative.

Corollary 3—If the rays striking the eyes of the larva had the oblique direction $a^1 - a^4$ or $c^1 - c^4$, and were in plane *C* of Fig. 12, the resulting reaction was a combination of the backward rota-

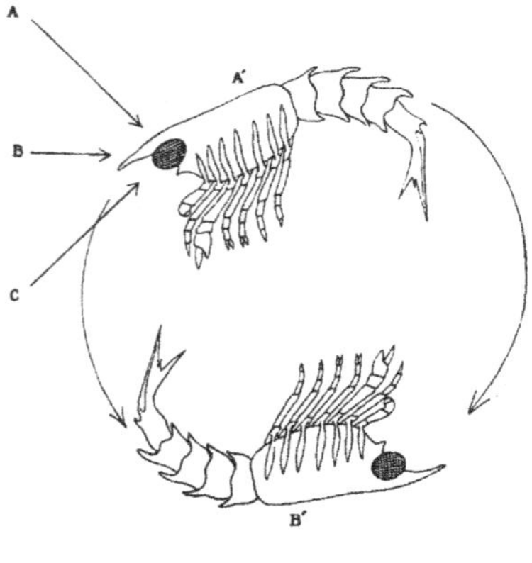

FIG. 12. For description, see Case 1, Cor. 1.

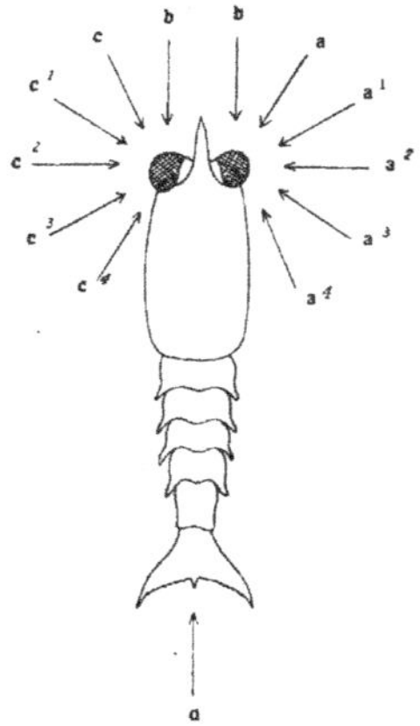

FIG. 13. For description see Case 1, Cor. 2.

tion and the side swing (Cor. 1). That is to say, the larva performed a backward side-somersault, became oriented as in Corollary 1 and 2, again turned the back uppermost, with the eyes directed away from the source of light, and continued its progressive orientation, in one sense or the other.

Case 2. Larva lying with back downward; head toward light—In these instances, the larva was oriented head toward the (closed) window, and back downward. The rays were introduced from before, as in Case 1. It may be said that this orientation was difficult to obtain. Often it was necessary to wait fifteen minutes or more before it occurred, then at the proper moment the light was admitted and the consequent reaction observed. On the other hand, it was common to find the larvæ on their backs and oriented obliquely to the rays of light. When the larva was oriented in this manner and the light was admitted, there usually occurred either a forward or a backward rotation (Fig. 14), but the forward rotation was most common. Whichever one occurred, however, the final orientation was the same: the back of the larva was again brought uppermost, and the head was directed away from the source of light.

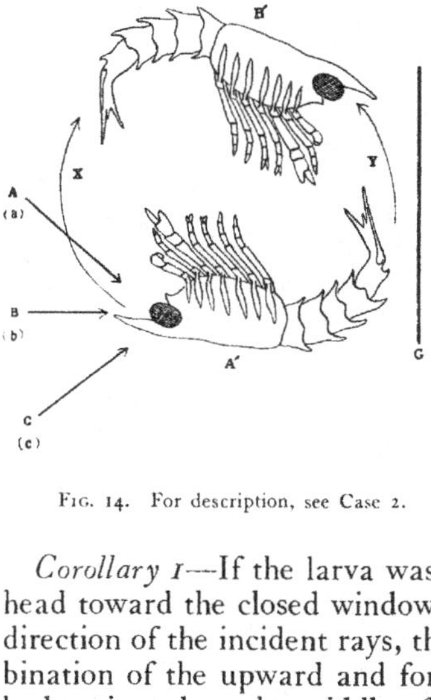

FIG. 14. For description, see Case 2.

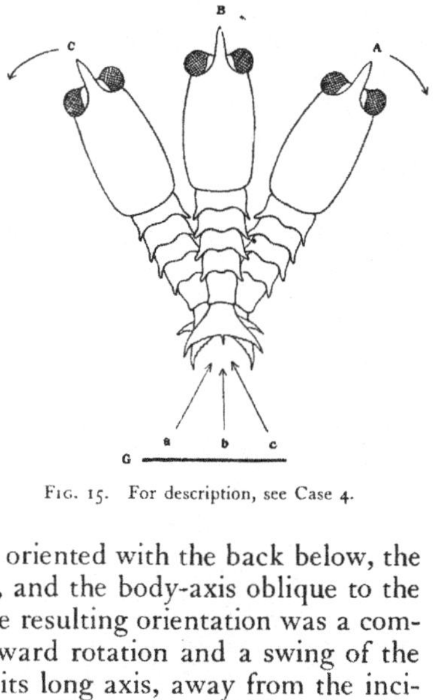

FIG. 15. For description, see Case 4.

Corollary 1—If the larva was oriented with the back below, the head toward the closed window, and the body-axis oblique to the direction of the incident rays, the resulting orientation was a combination of the upward and forward rotation and a swing of the body, pivoted on the middle of its long axis, away from the incident rays (this last reaction was similar to Case 1, Cor. 1, except that in the former instance the larva oriented back below). The final orientation was as in Case 2 (Fig. 14, *B'*). Whether the incident rays were in plane *A*, *B*, or *C* did not appear to make as much difference in the manner of orientation when the lobster was lying back below. It was observed that rays coming from above (plane *A*) more frequently determined the backward rotation; and that rays coming from below (plane *C*) more often determined a forward rotation.

Case 3. Larva lying with the side downward; head toward light—In this case, the larva was oriented with one side uppermost and the head turned toward the source of light. The conditions may be represented by Fig. 14, if it be imagined that for the present case the larvæ are lying in a horizontal plane rather than in the vertical as originally intended in this figure. The arrows *A*, *B* and *C* represent rays in the same vertical plane, while (*a*), (*b*) and (*c*) represent them in a horizontal plane. When the light was admitted to a larva so oriented, the reaction was similar to that described under Case 2. In the present instance, however, when the rays had the direction (*a*), the backward rotation was more likely to occur than when the rays had the direction *A* as in Case 2. Rays in the direction (*b*) or (*c*) almost invariably determined a forward rotation, in which, if the larva was fatigued, it would merely turn through 180° in the same plane, and become oriented, still lying on the side, but with its head away from the source of light. If, however, the larva was fresh and active at the end of the rotation of 180° in the arc of a circle (*A′*), it would rotate through 90° on its longitudinal axis and come into the normal swimming position with the back uppermost and the head directed away from the source of light.

Case 4. Larva oriented with back above; head directed away from the source of light—When the larva was thus oriented and the light was so introduced that the rays streamed in a direction parallel to the longitudinal axis of the larva, no change in the body-orientation took place. The progressive orientation, however, might continue as either positive or negative. In case, however, the light came from the sides *a* or *c* (Fig. 15) the larva reacted by swinging (pivoted on the middle or end of its longitudinal axis) to either one side or the other, and it might then undergo positive or negative progressive orientation. If the direction of the rays changed through the seriés, *a*, *b*. *c*, the larva could likewise be made to swing as regularly as a pendulum and for long periods of time, according as the light came from one side or the other. Indeed the animal was quite at the mercy of the influence of light.

In case the light came somewhat from above as shown in Fig. 16, *A*, the larva would incline itself farther forward, the number of degrees of rotation depending upon the degree of the angle formed by *A* with the horizontal. When the angle was slight the forward rotation of the larva was but a few degrees, and it continued to swim in this body-position, and might undergo a positive or negative progressive orientation, as ordinarily. When, however, the angle between *A* and the horizontal was greater, the degree of rotation of the larva was proportionately greater, and in certain cases it might undergo a rotation of 180° and fall to the bottom.

When, on the other hand, the incident rays struck the larva in the direction of *C* (Fig. 16), then the larva underwent a backward rotation whose degree was dependent upon the breadth of the

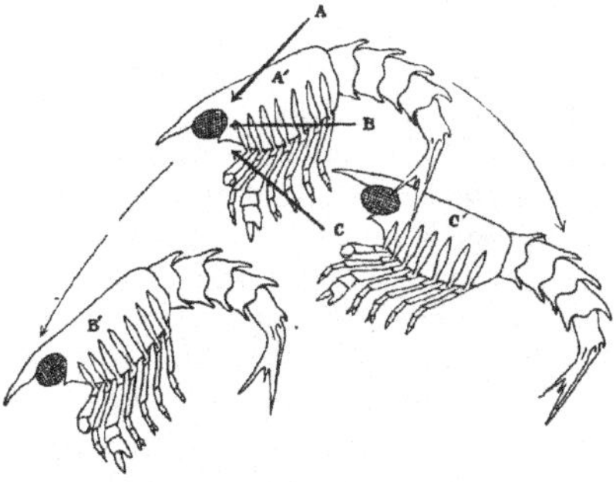

FIG. 16. For description, see Case 4.

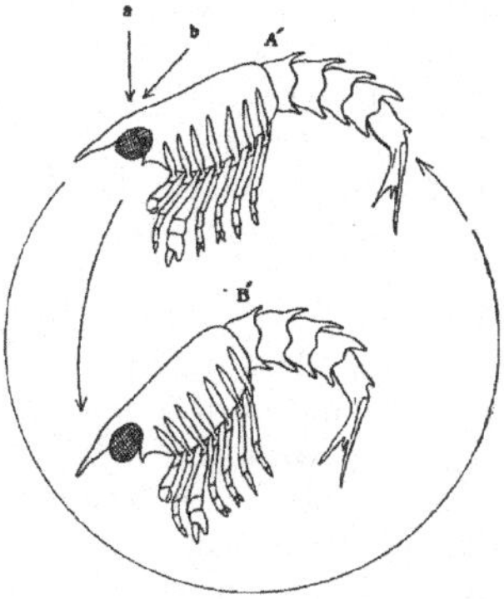

FIG. 17. For description, see Case 6.

angle between *C* and the horizontal. If the angle thus formed was slight, the backward rotation of the larva was correspondingly slight, and it would continue to swim in the position designated *C* (Fig. 16), undergoing positive or negative progressive orientation as other conditions of light might determine. If the angle formed between *C* and the horizontal was great, the degree of backward rotation of the larva was proportionately greater, and a fall to the bottom, tail downward, might result.

Corollary 1—When the direction of the rays was determined by compounding the vertical series of light factors (*A*, *B*, *C*, Fig. 12) with the horizontal series (*a*, *b*, *c*, Fig. 15), the resulting reaction was a combination of the two types of behavior described above.

Case 5. Larva oriented with back above and longitudinal body-axis at right angles to direction of light rays—When the larva was oriented as above and the rays were introduced at right angles to the longitudinal axis (Fig. 13, a^2, c^2) the behavior was similar to some phases of Case 1, Cor. 1. The larva swung directly away from the source of light until its longitudinal axis was parallel to the light rays, with the head directed away from the source of light. Obviously the swing might cover from 1° to 90° and either positive or negative progressive orientation might follow.

If the larva was lying with the back below, but otherwise oriented as in the previous instance to the directive influence of the rays, the reaction was the same; namely, a swing to one side. This resulted in placing the longitudinal axis parallel to the rays of light. Frequently, in such case, the larva would undergo a rotation on its own axis, so that it assumed a position with the back uppermost and the head directed away from the source of light. Whether or not this "righting reaction" occurred, appeared to depend largely upon the degree of freshness. Individuals which had undergone fatigue more frequently refused to rise from the bottom. It was at no time possible, however, to fatigue the larvæ to such an extent that they would not give the "swinging-reaction" into line with the light rays. By alternately changing through an arc of 30° the direction of the light which struck the larvæ from behind (Fig. 15, *a*, *b*, *c*), they could be made to swing, pivoted on the middle or end of their longitudinal axis, in an arc of equal degree. This pendulum-like activity in answer to the change in direction of the light-stimulus was extremely constant and in no case was it observable that the reaction was diminished by fatigue in spite of long periods of such alternate directive stimulation. It may be added here that prolonged direct stimulation from behind never produced a change in the body-orientation of the larva. The progressive orientation, however, might take place in either the positive or the negative sense.

Case 6. Larva oriented with back above; light enters from above—Under the conditions mentioned above, the larva was forced to give one or two reactions, depending upon the degree of intensity and the suddenness of introduction of the light:

(1) In some instances (especially when the light had the direction, *b*, Fig. 17), the larva first rotated through an arc of greater or less curvature and finally assumed a new swimming position

with the longitudinal axis of the body bent at a greater angle from the horizontal plane (Fig. 17, *B′*). This new swimming position was usually maintained so long as the conditions of light remained the same, but was sometimes replaced by the second form of reaction, which usually occurred when the light had the direction *a*, and which was merely an exaggerated form of the first.

(2) In this second type of reaction the rotation of the larvæ was not limited to an arc of a few degrees, but was extended into a forward "somersault." This in turn took place in one of two ways: (*a*) the larva might accomplish a rotation of 360° and return to its original position with the back above, but since the stimulation from above remained the same, it would not rest in this position, but would continue for a time to perform complete rotations without pause, after which it would come to rest as shown in Fig. 17, *B′*. This new swimming position was sometimes maintained as long as the conditions of light remained unchanged, though it might give place to further rotations; (*b*) the larva might, as a result of the forward rotation, come to rest with the back directed below, but this orientation was only momentary, because the influence of the light from above immediately determined a backward rotation. This last reaction might culminate when the larva had gained the new position shown in Fig. 9, *B′*, or it might be continued into one or more backward rotations through 360° and culminate after a greater or less number of such rotations, by coming into the new swimming position mentioned above. This orientation would be maintained as long as the same conditions of light were in effect; or it might be interrupted from time to time by rotations in arcs of varying degrees, and in either of the directions mentioned above.

Corollary 1—If, when the larva was oriented as in Case 6, the light was introduced from both sides and above, the resulting reaction was a combination of the forward rotation and the side swing. If the light came from above and behind (Fig. 17, *b*, then the direct assumption of the new swimming position *B′* more frequently resulted without the variable number of rotations through 180° or 360°.

Case 7. Larva oriented with back below; light enters from above—Under the above conditions of orientation (Fig. 18) there was usually one constant form of reaction. The larva would undergo a backward rotation through about 120°, and come into a new swimming position with the axis of the body bent downward several degrees from the normal swimming position (perhaps 45° from the horizontal), the exact amount appearing to be dependent upon the intensity of the light. This new swimming position was usually maintained as long as the conditions of light remained unchanged. It might sometimes be interrupted by backward rotations through 360°. These rotations invariably culminated in the assumption of the new swimming position (Fig. 18, *B*). In case the direction of the rays was both from the side and from above the resultant reaction was a combination of the reaction described above and the direct side swing.

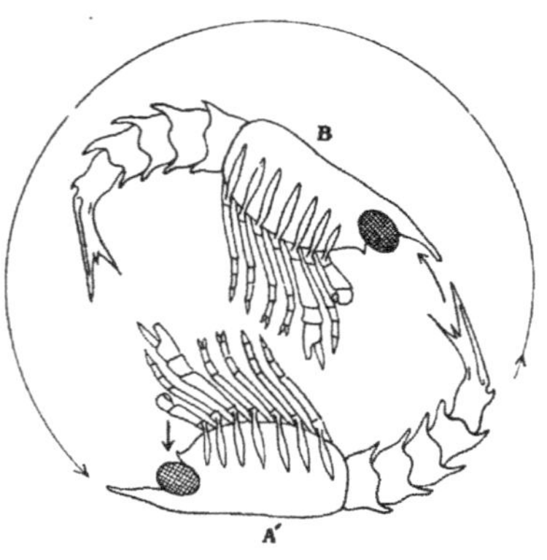

FIG. 18. For description, see Case 7.

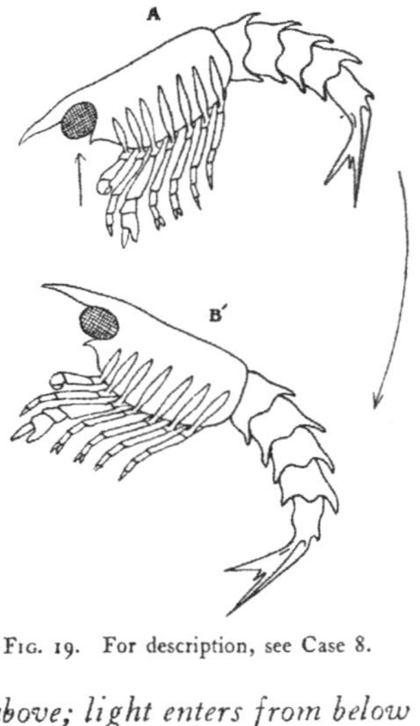

FIG. 19. For description, see Case 8.

Case 8. Larva oriented with back above; light enters from below—Under these conditions of orientation, the nature of the reaction was similar to that described in Case 6. Usually there resulted a direct backward rotation through a few degrees, which produced a new swimming position, Fig. 19, *B'*. This was usually constant while the conditions of light remained the same, but it was sometimes interrupted by backward rotations through an arc of greater extent, or even by a variable number of complete backward rotations through 360°. At the end of these, however, the new swimming position *B'* was invariably assumed. Combinations of the directions of the light (as both from the side and from above) produced modification in the reaction, but these could at any time be predicted if the individual constituents of the light were known.

Case 9. Larva oriented with back below; light enters from below —Under the conditions of orientation stated above the resulting reaction was similar to that described under Cases 6 and 8, but reversed. As in these instances, one of two results usually occurred: (1) The larva would undergo a forward rotation through a variable number of degrees, and assume directly a new "swimming-position" as shown in Fig. 20, *B′*. It was readily observed that the head was directed upward and away from the

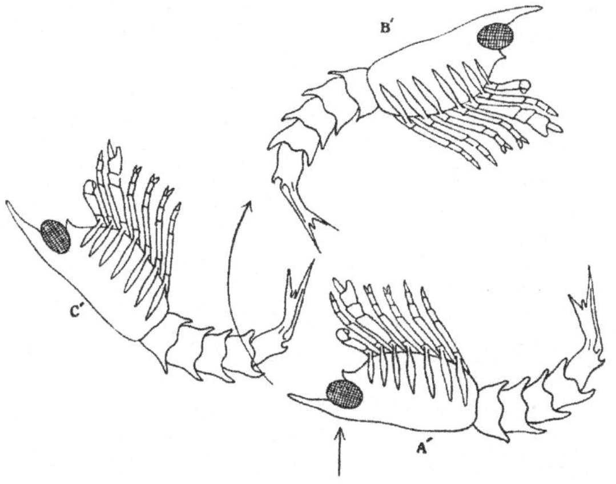

FIG. 20. For description, see Case 9.

light, not downward at an angle of about 30° from the horizontal, as in the normal swimming position; (2) it might happen, however, that instead of assuming this orientation the larva would merely come to an orientation with the back below and with the head directed upward as a slight angle as shown in Fig. 20, *C′*. It might, again, undergo one or more complete rotations forward through 360° and then assume the new position shown in Fig. 20, *B′*, which position might be retained as long as the light conditions remained unchanged. The definiteness in these two reactions could be modified, as a result of changing slightly the direction of the light.

In addition to the facts regarding the effect of direct lighting upon body-orientation, which have been presented in the form of these nine cases, several other conditions might be mentioned:

1. If the longitudinal axis of the larva was parallel to the direction of the incident light rays, and the head away from the light, then the introduction of light produced no change in the body-orientation, but it might cause a positive or a negative progressive orientation.

2. In order that the unmodified forward or backward rotation might occur, it was learned that the light rays must strike both eyes with equal intensity, and consequently in a direction exactly perpendicular to any transverse body-axis of the larva.

3. In case the incident rays came from a direction that was not exactly perpendicular to the transverse axis of the larva, be the angle of difference ever so slight, the perfect backward and forward rotation would not occur, but would be greatly modified by swingings of, and revolutions on, the longitudinal axis of the body.

4. This type of behavior could not be observed unless the conditions of light were reduced to a single directive influence, and this factor handled with very great precision.

The effect of blocking the illumination—In the previous section we have examined the reactions which were brought about by suddenly introducing rays of light in directions which maintained a certain definite and specified relation to the longitudinal or transverse axis of the larval lobsters. In the present instance, however, we are to consider the nature of the reactions which are produced as a result of suddenly excluding or blocking the principal source of light by which the larvæ have just previously been stimulated. The "cut-off" was made by closing the window through which the light came, and thus leaving the larvæ in the subdued and diffuse light which entered the dark box from the room. Since the body-orientation of the larvæ to the directive influence of the light is always the same, obviously there could not be many different varieties of orientation caused by the change in the conditions of light. Such as were possible, however, may be described as follows:

Case 10. Larva oriented with the back above and the longitudinal body-axis exactly parallel to the incident rays—In case the larva was oriented as described above, when the light was shut off there usually resulted a forward rotation through 180°. This reaction caused the larva to become oriented (often on the bottom) with the back below and the head toward the previously existing source of light. This position was not maintained, but was succeeded by a "righting reaction," usually a revolution on the longitudinal axis, which brought the back again uppermost. After this response the larva might swim in diverse directions.

Case 11. Larva oriented with the back above and the head away from the light, which comes slightly from the side—If, when the larva was oriented as described above, the light was suddenly cut off, there resulted a swing of the long body-axis so that the larva was brought more or less nearly to face in the opposite direction; i. e., in the direction from which the light had previously come. This orientation, however, was not permanent, but other consequent reactions occurred and the larva might swim in one of several directions.

Case 12. Larva oriented as in Fig. 17, B′—If the direction of the light was from above, and the orientation of the larvæ as in Fig. 17, *B′*, when the light was cut off, the head of the larvæ would swing upward to face the direction from which the rays had previously come. Consequently, however, the orientation became that of the normal swimming position.

Case 13. Larva oriented as in Fig. 20, B′—When, as the result of light stimulation from below (as in Case 9), the larva was oriented with the head directed upward, and the illumination was suddenly cut off, the head of the larva would swing downward to face the direction from which the light had previously come; sometimes the larva would perform a rotation in an arc of greater or less extent and fall to the bottom. The body-orientation with head downward was not maintained, however, but was at once superseded by the normal swimming position.

It thus appears from these cases that there was usually an excessive movement to produce the new body-orientation; but that these movements invariably ended in the assumption of the normal swimming position.

Résumé of experiments on the effects of direct lighting and shading—(A) The effect of suddenly submitting the larval lobsters to a light which has a directive influence is to cause the larvæ to orient themselves in such a manner that the longitudinal axis of the body finally assumes a definite relation to the direction of the light rays. This orientation is a position with the long axis of the body parallel to the light rays, and with the head turned away from the source of light. (B) The effect of suddenly blocking the light to which the larvæ are reacting phototactically is to cause a new body-orientation by which the head is usually brought to face the direction from which the light had previously come. In either of the cases mentioned above the body-orientation is brought about by a single motor reflex or by a longer or shorter series of motor reflexes, some of which are "over-produced" movements.

These movements include the following types:

1. *Forward or backward rotations,*[6] *or somersaults*—These were rotations in an arc, of a few degrees, which directly determined a new swimming position with the head raised or lowered, depending upon the direction from which the light or shadow had been introduced. In other cases these rotations took the form of a variable number of complete rotations through 360°, either backward or forward, in which the body of the larva formed a constant part of the circumference.

2. *Revolutions on the longitudinal axis of the body or rollings*—The revolutions or rollings took place either to the right or left, but usually in such direction that the back of the larva became directed more or less toward the light. They might be through a few degrees, or they might exceed 90°, in which case the larva fell to the bottom. In the case of larvæ one of whose eyes had been injured this revolution took place very rapidly, oftne at the rate of one hundred and fifty per minute, and always in a determined direction, the normal eye over, the injured eye under (HADLEY 1907b).

3. *Swingings of the longitudinal axis of the body*—These reactions were swingings in such a direction that the head was brought by the shortest path to face the dark, and the tail to point toward the light.

[6]Three similar types of movement are described by BOHN (1905, p. 4) as follows:

1° *Mouvement de manège*—l'animal décrit un cercle de plus ou moins grand rayon, l'axe du corps, courbé en arc, faisant partie constamment de la circonférence; la rotation se fait tantôt dans le sens des aiguilles d'une montre, tantôt dans le sens inverse. Parfois, au lieu de décrire un mouvement de manége pur, l'animal décrit des courbes de rayon variable qui constituent une sorte de spirale.

2° *Mouvement de rotation en rayon de roue*—l'axe du corps ne dévie pas; il est une des parties d'un des rayons du cercle décrit, et non une partie de la circumférence du cercle: la tête peut se trouver a la circonférence ou au centre.

3° *Mouvement de rotation sur l'axe, ou roulement:* l'animal tourne autour d'un axe longitudinal qui traverserait le corps dans sa longeur; la rotation commence par une inclinaison de l'animal d'un côté, et le sens de la rotation se trouve ainsi determiné. Le roulement peut s'accompagner d'un mouvement de translation et devient un mouvement en pas de vis.

4. *Rotations in the radii of a circle*—In these the longitudinal axis of the larva formed a radius, and with either the head or the tail at the center the animal rotated about a fixed point. These reactions were uncommon and, as yet, unexplained.

These four types of movement seldom occurred separately, except under especially devised experimental conditions. Under natural conditions, they were usually combined to form a composite action. To the previously mentioned simple components, however, all the more complex movements of the larval lobsters could be reduced.

B. THE EFFECT OF SCREENS AND BACKGROUNDS—It is probable that the reactions which are brought about through the use of backgrounds, are, generally speaking, dependent upon the same factors and conditions of illumination which are effective when light-absorbing or light-scattering screens are used. The term "screening" has been employed by BOHN (1905) to designate his method of submitting organisms to the influence of surfaces of light and shade. This investigator made use of screens of black and white of such size that he could readily bring them close to the sides of the glass containers in which the organisms under observation were placed. He has made a special study of the reactions of Crustacea to the influence of such screens, and in several instances the observations of the writer upon the larvæ of Homarus americanus merely confirm certain points in BOHN's earlier work. In many instances, however, new facts have been added.

The influence of white screens—The lobster larvæ were confined in a cylindrical jar, crystallization dishes, or in a rectangular glass container. The latter was used most frequently. The larvæ were then placed in the dark box and this was illuminated in such a manner that a general twilight was produced and the directive influence of light was at a minimum. While making observations it was even found necessary that the writer should wear a black mask over his face and collar, and, often, darken his hands in order not to modify the uniform light. For white screens pieces of white cardboard were employed, and brought over, under, or beside the receptacle containing the larvæ, as the case might require. Sometimes the screen was brought gradually toward the container, sometimes abruptly; but in all cases the results were definite and agreed with great uniformity. In order to secure the best results with the white screen, it was found best to reduce the intensity of light within the dark box below the degree used in the case of the black screens. The results of the series of experiments with white screens may be summarized as follows:

Case 14—When the larva was oriented with back above and the screen, held vertically, was so introduced from before that its plane was at right angles to the longitudinal axis, and parallel to any transverse axis of the larva, there resulted a rotation through 180° with, perhaps, a fall to the bottom. After this, and as a result of a revolution on the body-axis, a "righting reaction" usually occurred and the back would again be brought above. Now, with the head directed away from the white screen, the larva might either approach or depart from it, according as the progressive orientation was positive or negative. Sometimes, instead of producing a rotation through an arc of 180°, the larva underwent a series of rotations, its body forming a constant part of the circumference. The final orientation mentioned above would, however, invariably succeed. In case the screen was not held squarely before the larva, but somewhat at an angle to any transverse axis, the consequent reaction was a direct side swing away from the screen in order to place the longitudinal body-axis perpendicular to, and the head away from, the screen. In other cases there resulted a combination of the side swing and the forward rotation, so that the larva performed a sort of "half-somersault," and eventually assumed the normal swimming position, with the head directed away from the screen, as pointed out above.

Case 15—When the larva was oriented with the back above and the screen, held vertically, was made to approach the posterior end of the larva, no change in the body-orientation resulted. There might occur, however, either a positive or a negative progressive orientation.

Case 16—In this case the larvæ were swimming promiscuously about the container. When the screen was made to approach larvæ which held a position with the back above and one side turned toward the screen, these larvæ experienced a swing of their longitudinal axis so that the head came to be directed away from the screen and the longitudinal body-axis at right angles to the plane of the screen.

Case 17—When the screen, held horizontally, was made to approach, from below, a larva which held the normal swimming position, one of two reactions (which probably represent different degrees of the same reaction) resulted: (1) The larva would swing the head upward as shown in Fig. 20, *B'*, and maintain this swimming position so long as the light condition remained unchanged, or (2) it might, on the other hand, experience this same reaction in an exaggerated form, i. e., there might result a backward rotation through 180°, which reaction would cause the larva to fall to the bottom and to assume a position with the back below and with the head directed upward at a slight angle as shown in Fig. 20, *C'*, Usually, however, this form of orientation resulted only when the light was of greater intensity, such as that secured in cases of direct illumination.

Case 18—When the larva was oriented in the normal swimming position and the white screen was made to approach from above, the reaction was similar to that described for Case 6, p. 270. The one difference was that while the direct lighting often caused a number of complete rotations through 360° before the final body-orientation was assumed, the white screen, on the other hand, usually acted by changing the swimming position directly to that of Fig. 17, *B'*. This difference in response was probably due to the difference in the intensity of light (direct or reflected) coming from above.

The black screen—The method of conducting the experiments with the black screen was almost the same as that for the white screen. There was one point of difference. It was found that, in order that the black screen should determine any reaction of the larvæ, it was necessary to have a slightly greater illumination within the dark box. The following report of cases shows the result of making the screen to approach, from various directions, the larvæ diversely oriented.

Case 19—When the larva was in the normal swimming position and the back screen was presented opposite the head, and at right angles to the longitudinal axis, the orientation was not changed, but was retained constantly so long as the screen remained in position.

Case 20—When the larva was in the normal swimming position and the screen was made to approach from behind, so that its plane was parallel to a vertical plane passing through both eyes of the larva, there usually resulted a forward rotation of 180° in the arc of a circle. This reaction brought the back of the larva below, and the head toward the black screen. This position was at once further modified by a revolution of 160° on the long body-axis, either to the left or right (determined by the nature of the lateral or secondary illumination), and the larva again assumed the normal swimming position, but with the head directed toward the black screen. In case the plane of the screen was not exactly parallel with the vertical plane passing through the eyes of the larva, the reaction was not represented by the simple forward rotation, but was modified by side movements.

Case 21—When the larva was in the normal swimming position and the black screen approached from the side, several reactions might occur. Most commonly the larva underwent a swing of its longitudinal axis so that the head was brought to face the screen. Another reaction sometimes observed was a rolling, or revolution, on the long body-axis, in such a manner that the back moved away from the screen. At the same time there occurred a swing of the longitudinal axis which caused the head to be directed toward the screen. These two reactions might occur simultaneously, and the resulting reaction be a blending of the two components mentioned above. The rolling on the longitudinal body-axis was seldom over 90° from normal (back above), usually less. Yet in cases where the illumination in the dark box was greater, or when the screen was introduced suddenly, the rolling motion might exceed 90°, and the larva fall to the bottom of the container.

Case 22—In this instance the larva oriented in the normal swimming position and the screen was made to approach from above. This combination produced several forms of reaction. In cases where the general illumination in the container was not great, the larva merely experienced a slight change in the direction of the longitudinal body-axis; the head assumed a superior position, so that the long axis of the body was nearly horizontal, or even directed upward at a small angle, rather than bent downward at an angle of 30° from horizontal, as in the normal swimming position. On the other hand, if the illumination was greater, the larva might undergo a rotation on its own longitudinal axis through 180° and fall, back downward, to the bottom. Whatever reaction occurred, it could be explained as an effort of the larva to turn the head toward the black screen, and the degree to which this was attained depended very much upon the intensity of illumination throughout the container. The type of reaction mentioned above was demonstrated to better advantage in the following experiment. A large tube containing a number of larvæ was placed in an upright position on the laboratory table, and the upper half covered with a roll of black paper. The larvæ gathered in the more brightly illumined end of the tube, which was below. So long as they swam in the lower part of the illuminated area, they assumed the normal swimming position, but whenever they came into the upper regions, and approached the edge of the black paper, the direction of the longitudinal body-axis was changed from 30° below horizontal to 30° or even more above the horizontal plane.

Case 23—In the following case the larva was oriented in the normal swimming position and the screen was made to approach from below. As a result the larva usually reacted by a slight forward rotation, the head passing through an arc of a few degrees, and producing a still greater angle between the longitudinal axis and the horizontal plane. This new swimming position was seldom subject to further modifications so long as the light conditions remained unchanged. Regarding the reactions of the larvæ of Homarus vulgaris under similar experimental conditions, BOHN (1905, p. 11) remarks: "Si la larve nage le dos dirigé le haut, il y a roulement de 90° ou de 180°, la par suite la larve dévie lateralement ou tombe."

Such a result as the above was not observed by the writer. On the other hand, it was observed that, whatever the body-orientation of a group of larvæ might be previous to the approach of the black screen from below, its presence usually determined a rise of the larvæ from the bottom of the container to the upper waters, where normal swimming was manifested so long as the screen beneath remained in place. When it was removed, however, or replaced by a white screen, the consequent reaction was, as we have already seen, characterized by rotations and revolutions through 90° or 180°. These reactions in turn resulted in bringing the larvæ again toward the bottom, and in determining a consequent absence of larvæ in the regions near the surface of the water.

Case 24—In this instance the larvæ were oriented with back below, and the black screen was made to approach from behind in such a manner that the plane of the screen was parallel with a vertical plane passing through both eyes of the larva. Under these conditions (see Fig. 14, *A'*) the reactions were as follows. When the black screen, *G*, was introduced, the larva, *A'*, underwent a forward rotation through an arc of 180°, and assumed the normal swimming position, *B′*, with the back uppermost and the head facing the screen. This orientation was maintained with a greater or less degree of constancy so long as the conditions of light remained the same. If, on the other hand, the screen was so placed, or the larva had such a position, that the plane of the screen was not exactly parallel to the vertical plane passing through the two eyes of the larva, a different reaction was experienced. In this instance the first response was a revolution on the longitudinal axis, usually through 180°. This resulted in bringing the back of the larva uppermost, and was usually followed by a swinging of the longitudinal axis, which brought the head to face the screen. The direction of this side swing (to the left or the right) was determined by the angle which the longitudinal axis of the larva made with the screen. For instance in Fig. 15, the larva designated *A′* would swing to the right, while the larva designated *C′* would swing to the left, each in the direction indicated by the arrows. In other words we may say that the larva would swing in that direction which brought the head, by the shortest course, to face the screen. But the two reactions mentioned above might, as in previous cases, be blended to form a composite reaction, which differed from either of its simple components.

Case 25—In the present instance the larva was oriented lying on its back and the screen was introduced from before. Under these conditions, as in Case 19, there was no modification in the body-position. In certain instances the larva underwent a revolution through 180° on its longitudinal axis and assumed a position with back above and head still directed toward the black screen; but in the great number of cases the orientation remained unchanged.

Case 26—In case the larva was oriented with the back below and the screen was made to approach from the side the reactions were as follows. The larva experienced a rolling or revolution on its longitudinal axis, in consequence of which the back moved away from the screen through an angle of 90°, occasionally more. At the same time there was a swinging of the longitudinal axis, itself, so that the larva came face to face with the screen, eventually with the back uppermost. During this reaction the larva often departed from the screen. As in Case 21, mentioned above, these two reactions might occur at the same time, and then the resulting reaction was a composite.

Case 27—In the present case the larva was oriented with back below and the black screen was introduced from above. Under these conditions it usually underwent a slight forward rotation with a consequent rise from the bottom, and came into a new swimming position with the longitudinal axis directed somewhat upward as shown in Fig. 20, *B′*.

Case 28—In this instance the larva was oriented with back below and the black screen was introduced from beneath. The reactions were usually as follows. The larvæ underwent a revolution of about 180° on its longitudinal axis, and assumed practically the normal swimming position, with the back uppermost and the head bent downward at an angle of about 30°. In other cases, however, this new position was brought about by a different sort of reaction; namely, a backward rotation through an arc of 180°. This resulted in throwing the larva again into the normal swimming position.

Generally speaking, we may say that, when black or white screens were made to approach larvæ of any one of the first three stages, diversely oriented, the larvæ manifested two forms of response. First, a motor reflex, which tended to place the longitudinal axis in a certain relation to the plane of the screen; secondly, and subsequent to the first response, a progressive orientation, toward or away from the screen, as the luminosity of the screen, the physiological state of the larvæ, and other conditions of the case, determined. When the white screen was used, the larvæ commonly became oriented with the head directed away from the screen. In the case of black screens, on the contrary, the head was directed toward the screen and the back more or less away. These reactions occurred whether the screens were made to approach from above, below, behind, or the side. After body-orientation had taken place, the larvæ might approach or recede from the black or the white screen, according as they were reacting positively or negatively.

The mechanics of reaction upon which orientation to the screens was found to depend, agree, for the greater part, with the types of reaction to black screens reported by BOHN (1905), who has made a careful study of the effects of causing a black screen to approach the larvæ of Homarus vulgaris, diversely oriented. There are, however, certain disagreements. First, it is certainly true that bringing the black screen parallel to the longitudinal axis of the larva frequently determined a rolling of the larva on its own longitudinal axis, whatever the original orientation may have been. But in Case 21, certain orientations of the larva were noted in which these rollings did not occur. It is true, moreover, that the progressive orientation often took place in that direction in which the back was directed. But several instances were observed wherein the orientation to the black screen resulted merely from a swinging of the longitudinal axis of the larvæ so that the head was directed toward the screen and where consequent progressive orientation was either a movement backward or forward, head foremost or tail foremost, as in positive or negative phototaxis.

We have now examined somewhat in detail the effects of sudden illumination and of sudden shading, the effects of white screens and of black. If we now compare the detailed results of these studies, we note that the effects produced by introducing a white screen are comparable with those obtained by suddenly admitting illumination, while the results brought about by black screens are comparable to those determined by suddenly cutting off the light. In other words, the larvæ appear to respond to the influence of screens of black and white by reactions which are dependent upon the same simple forms of response observed under the conditions of direct lighting and shading.

In view of this correspondence in the nature of reaction to direct lighting and to screens of black and white, it may be considered probable that the screens and backgrounds are instrumental in determining the behavior of the larvæ, only in so far as they are themselves the source of (reflected) illumination. Thus, when the black background causes a swing of the larva, as a result of which it comes to face the screen, we cannot say that the primary factor is the blackness of the screen; but rather that the small amount of light reflected from the screen permits rays of light from other directions to become effective, The larva "heads" to the black screen because his eyes encounter no light rays coming from this direction; and he turns away from the white screen because his eyes encounter stronger reflected light from this than from any other direction.

The effect of backgrounds—The question of the influence of backgrounds in determining the orientation of crustacean larvæ has been brought forward by KEEBLE and GAMBLE (1904). Aside from the effects of screening, the more general problem of backgrounds did not receive especial attention in the course of the present investigation, but, as we shall see, the question of screening which we have discussed in the preceding section is probably only a single phase of the problem of backgrounds. The following experiments which were performed more or less at random in connection with other experiments, but which deal with the question of backgrounds, may, however, be presented.

By the term background, as it is used in the present case, is meant the permanent color-tone of the surrounding walls (as a whole or in part) which confined the young larvæ. This condition was somewhat different from that determined by the use of screens which were movable and could be placed at any angle with reference to the body-axis of the larvæ. Backgrounds were employed in several different ways. They were sometimes represented by the black or white lining of the reaction boxes; again, by the ground upon which the glass dishes or tubes rested, and in still other cases by the outer covering of these dishes, or tubes. The subject may be considered under two heads: (1) the effect of backgrounds in connection with the purely photopathic response; (2) their effect in determining the "choice" of a particular region of light-intensity when phototaxis also is operative. In view of the fact that the investigation of the first phase of this problem was not undertaken in the present work, we may pass directly to the consideration of the second point stated above.

The effects of backgrounds in connection with both the phototactic and the photopathic response—Under this head we may consider those conditions of experiment, which, although they be chiefly productive of reactions to the directive influence of the light, nevertheless were modified by response to the intensity of the light. These conditions were secured by the use of Y-tubes. The following experiments serve to show why, in the case of the larval lobsters, the tendency to gather in the brighter areas (assumed positive photopathy) is often associated with positive phototaxis; and why a tendency to gather in the darker areas (assumed negative photopathy) may be associated with negative phototaxis. In the diagrams of Fig. 21 are represented the Y-tubes as set up for experiment. Those whose arms are above were arranged for experiment with larvæ having positive phototactic reaction; those whose arms are at the bottom, for larvæ having a negative phtototactic reaction. In tubes *A* and *B* one side of one arm was fitted with a band of black paper which extended half over the circumference of the arm and a very short distance down each stem. In tubes *C* and *D* the same arrangement existed, save that white instead of black paper was used. In every case the light rays came from the window in the direction of the arrows. In all cases of larvæ manifesting a negative reaction, the start was made at the end of the tube (lying horizontally on the table) nearer the window. In the case of positively reacting larvæ, the start was made from the end of the tube farthest from the window. The end marked *a* in every instance was the end *from which* the larvæ moved, the purpose of the test being to determine in which arm of the tube the larvæ would eventually gather.

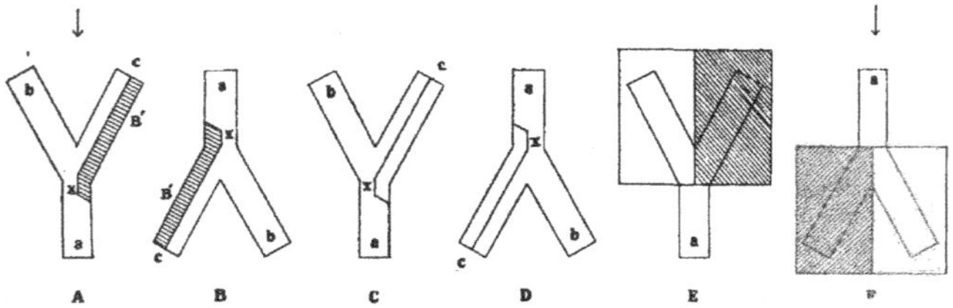

FIG. 21. Showing the Y-tubes set up for experiment. In every case the light came from above in the direction of the arrows. The tubes whose arms are above were set up for positively reacting lobsters; those whose stems are above, for negatively reacting lobsters. In tubes *A* and *B* the cross-hatched areas represent the part covered with black paper. In tubes *C* and *D* the clear area was covered with white paper. Tubes *E* and *F* are shown equipped with the glass plates placed over the arm. In every instance the larvæ were started from the end of the tube designated *a*. For further explanation, see Cases 29–33 inc., pp. 286-289.

Case 29—The tube was arranged as in Fig. 21, *A*. Ten positively-reacting, first-stage larvæ were placed in the Y-tube, and, by certain manipulations of the light and by virtue of their positive reaction, they were made to congregate in the stem end. Then suddenly, the direction of the light was changed so as to come in the direction of the arrows. Immediately the larvæ oriented with their heads toward the end *a*, and passed through the tube toward the light. As soon as they approached the region marked *x* they came under the influence of the dark background bounding the side of the tube. Immediately, as we have seen to be the case in previous instances, the longitudinal body-axis swung so that the head came to face, more or less obliquely, the dark background, *B′*. The directive influence of the rays, however, continued to draw the larvæ on, but since they must travel in the direction in which the tail pointed, they entered the arm *b*, and passing close to the inside continued until further progress was prevented by the end of the arm. Space will not be taken to show the numerical results of this and similar experiments. Suffice it to state that nearly all of the positively reacting larvæ, of whatever stage or age, when submitted to these conditions of experiment, reacted as has been described above. This experiment was modified by placing the Y-tube so that the uncovered arm of the tube rested upon a piece of black paper. The results were invariably the same; the majority of the larvæ progressed to the arm of the tube not overlying the black ground.

Case 30—In this case the conditions of the experiments were further modified by reversing the Y-tube so that the arms pointed away from the window. In this instance larvæ which were manifesting a negative reaction were employed, and were first placed in the end (*a*), nearer the window. When the light was admitted the larvæ at once oriented with their heads directed away from the light and began to move away from the window. When they had reached the point designated *x*, they immediately underwent a swing of the longitudinal axis, as in previous cases, so that the head was directed toward the black ground, bounding the outer surface of the arm *c*. Thus they would continue, passing close to the inner wall of the tube until the majority had gathered in this arm. In this instance, however, the larvæ would usually rest between *x* and *c*, instead of moving to the end of the arm.

Case 31—Here the black background bounding the outer side of one arm was exchanged for a white ground of the same size and having the position shown in Fig. 21 *C*. Third-stage larvæ giving a positive reaction were employed for the experiment. They were started in the end *a*. When the light was admitted, the usual body-orientation resulted, and the larvæ began their progression through the tube toward the window. When they had arrived at *x* they came under the influence of the white ground and turned their heads away from this side. Progressive orientation then continued and the larvæ eventually became grouped in arm *c*. Similar results were obtained when half of this arm of the tube was laid over a sheet of white paper.

Case 32—The previous experiment was further modified by reversing the Y-tube so that the arms were directed away from the window (Fig. 21, *D*). Larvæ which were giving a negative reaction were employed. They were placed in the end *a*, and the light was admitted. After the usual body-orientation had taken place, the progression away from the window began. When the larvæ reached the point *x*, and had come under the influence of the white ground bounding one side of the tube they would swing their heads toward the right and continue their progress until all were gathered in arm *c*. This was somewhat unexpected. It eventually transpired, however, that the white ground bordering the outer surface of the tube did not act as a reflector or intensifier of the light rays, but as an opaque shield, cutting off the rays which would otherwise have entered the arm *c*. Thus, as in Case 30, the negatively reacting larvæ had merely grouped themselves in the arm where the light was least bright. When the Y-tube was so placed that half of arm *c* rested upon a sheet of white paper the result was different. The larvæ congregated in arm *b*, which was, under these conditions, the region of least light intensity.

Case 33—The four cases mentioned above were supplemented by other experiments involving the use of colored glass plates. As described in Experiment 15, these plates were so placed over the arms of the Y-tube that a difference in the intensity of light striking one arm was caused by interposing a red, orange or yellow glass plate between that arm and the source of illumination. In these cases the positively reacting larvæ gathered in the arm where the light-intensity was the greater, while the negatively reacting larvæ grouped themselves in the arm where the light was least bright. As a rule, the larvæ of earlier stages seemed to be more susceptible than the others to slight differences in the intensity of light at the entrance to the arms.

Thus is explained the tendency for positively reacting larvæ to gather in regions of greater light-intensity, and on the other hand, the tendency of negatively reacting larvæ to congregate in regions of lesser light-intensity. This condition of affairs has, no doubt, given many investigators reason to believe that such reactions are but manifestations of a positive photopathy; and that photopathy and phototaxis are fundamentally the same. We now know, however, that the reaction just described in Case 5 is due to the combined effects of two tendencies; the one to turn the head toward the dark areas (areas of non-stimulation); the other to move in the direction of the longitudinal axis of the body either toward or from the source of light. Were we dependent upon such experiments as these for our belief in the existence of a separate response to light-intensity, regardless of directive influence of light, we might well say that the photopathic and phototactic responses are, in the end, one and the same. But the writer has adduced in the previous section other data which separate more clearly these two types of reaction.

VII. ANALYSIS.

It has for some time been the custom to state that certain organisms are positively phototactic or positively photopathic, and that other organisms are negatively so. The index of reaction for several crustaceans has been so recorded, but the observations are usually incomplete, often uncritical, and sometimes of questionable significance. It is true that, in a very general way, organisms react positively or negatively to light. For instance, it may be said that the lobster shuns the light, that Palemonetes is attracted by the light, and that the larvæ of Limulus avoid the light. The definite statement, however, that the larvæ of Limulus are negatively heliotropic, or that Palemonetes and larvæ of Homarus are positively phototactic, is as inadequate as would be a biography written on the basis of a single day's association with a human individual. It may be true that by the time the adult stage is reached, the reactions of many animals have become more or less stereotyped, so that reactions like those of the moth to the flame, are easily predictable. In the larval and adolescent stages, on the other hand, the reactions are frequently more variable. To say that the lobster of the second larval stage is positively phototactic or positively photopathic is, as has been demonstrated, by no means a correct interpretation of the facts of the case, for slight changes in the conditions of stimulation may be sufficient to reverse the index of reaction. This variability doubtless occurs in many arthropods. It thus becomes evident that, although the young lobsters may be regarded as machines upon which many different external forces act and cause certain reactions, still (except for the definite body-orientations which are invariably determined by the directive influence of the light rays) they are machines the nature of whose operations can seldom be predicted unless the age, the stage, the kind and degree of the stimulus, are accurately known. These conditions of reaction indicate the extent to which the behavior of young lobsters is determined by their physiological states; and the foregoing experiments show in what way these physiological states change, not only from one stage-period to another, but even during the same stage-period, through the influences of metabolism, development, and perhaps still other factors. The extent to which the natural behavior of animals in their natural environment can be explained on the basis of the results of laboratory experiments depends largely upon the animal and the kind of reactions involved. It is quite probable that some of the characteristics of reaction, which have been described in the present paper, determine in a large measure, the daily behavior of the larval and early adolescent lobsters when they are in their natural environment. Unfortunately, however, we know too little regarding the behavior of lobsters under natural conditions, to attach great importance to far-reaching explanations of their daily activities on the basis of laboratory experiments. A few points, however, may be noted. The reports of biological surveys make it clear that, at the surface of the ocean or of bays in which lobsters are known to live and breed, the stage most often taken in the tow-nets is the fourth; the larval stages are much less frequently found, the fifth stage seldom, and later stages never. Observations which were made on lobsters of different stages taken from the Wickford hatchery and liberated in the surrounding waters of Narragansett Bay yield similar evidence regarding the immediate natural distribution. In these cases the lobsters of the larval stages were found to swim for a brief time, then gradually disappear from the surface; the fourth stage lobsters swam actively at the surface so long as they were observed; while the fifth and all later stages plunged at once into the deeper water and were immediately lost to sight.

As the writer has already suggested, it is impracticable to attempt to explain the natural behavior of larvæ of the first three stages, on the basis of the reactions which have been discussed at some length in the present paper. The light (depending upon its intensity and directive influence; and upon the age, stage, and previous condition of the larvæ) may determine at one time a positive, at another a negative, response, so that the general reaction of groups of lobster larvæ can in no way be readily predicted. One exception to this may be stated. The first-stage larvæ, directly after hatching, would be strongly drawn to the surface of the water by virtue of both their photopathic and of their phototactic response. After the first day or two, however, begins that modification and variation in the phototactic action which, for groups of uncertain age and condition, makes any accurate prediction of their movements quite impossible.

In the case of the fourth-stage lobsters there is a better basis for the correlation of the natural and experimental types of behavior. We know that, under experimental conditions, hungry fourth-stage larvæ, when submitted to food stimuli, will rise immediately to the surface of the water and swim about excitedly for some moments; we know also that the early fourth-stage larvæ, under certain experimental conditions will leave a region of low light intensity and remain in regions of greater light intensity. We have learned, moreover, that the same fourth-stage larvæ, under different experimental conditions, will usually shun the light when it has a single directive influence, and travel in the direction of the rays away from their source. Finally, we have observed that the fourth-stage lobsters, except in the latter part of the stage-period, show a definite tendency to remain at the surface of the water.

The question now arises: What is the cause of this surface-swimming? Is it a response to the intensity of light, to the directive influence of light, to hunger, or to gravity? Although we know something of the effects of several of these factors when they act separately, it is difficult to ascertain their individual influence when they work in combination. If, however, we can discover any parallel between a certain type of reaction under experimental conditions and a certain mode of behavior under natural conditions, and find that as one is modified or lost the other is also, then, and then only, are we justified in believing that we know the determining cause of the particular type of natural behavior in question. We have such a parallel between the photopathic (and occasionally the phototactic) reactions and the surface-swimming tendency of the fourth-stage lobsters. As the former becomes modified and is eventually replaced by the negative reaction, so the latter is changed and finally gives way to the bottom-seeking tendency as the lobsters pass on through the fourth stage-period. With such a parallel before us, it cannot be doubted that there exists a certain causal relation between the positive photopathic reaction and the surface-swimming tendency on the one hand, and the negative photopathic reaction and the bottom-seeking tendency on the other. But the photopathic reaction may not alone be responsible for the surface-swimming tendency on the part of the fourth-stage lobsters. The presence of food particles in the water excites them strongly, and causes them, when in the glass jars, to swim excitedly at the surface of the water. It therefore appears quite within the bounds of possibility that chemotropism may also play a part in determining the surface-swimming of the fourth-stage lobsters.

The explanation of the behavior of the fifth and all later stages, in the light of the foregoing experiments, rests upon a more certain basis. We have observed that the fifth-stage lobsters invariably manifest both a negative phototactic and a negative photopathic reaction. In general this may be said to explain the fact that lobsters in the fifth and all later stages shun the light at all times. Little work was done on the behavior of the older lobsters, and it is hoped that future investigations may continue along this line.

In connection with the mechanics of orientation, the writer has shown that the reaction of larval lobsters to light is made up of two components—body-orientation and progressive orientation; and that the former is primary while the latter is secondary. In the earlier pages of this paper it was demonstrated that the *progressive orientation* is dependent upon a great number of conditions, and that the orientation responses are relatively complex reactions which are dependent in great measure upon the obscure, changing, internal conditions which are embraced under the general term, "physiological states." In later pages, on the other hand, attention has been directed to those conditions of light which determine the *body-orientation* alone; and the results recorded have made it clear that the movements producing the body-orientations are types of action which simulate more closely pure reflexes, direct, constant, and invariable.

As Bohn (1905a) has well said, it is impossible to take definite account of the complicated series of phenomena which take place in the nervous system of animals even as low as the arthropods, for these are dependent not alone upon complicated connections between neurons, but also upon their variable states. Yet it is apparent that this difficulty applies rather (at least in the reactions of the larval lobsters) to those movements which determine the *progressive orientation* to light, than to those which determine *body-orientation*. Even in the latter somewhat less complicated and more easily explained phenomena, however, we are still far from recognizing the underlying causes.

It is true that we can understand in a way why the "posterior position" of the thoracic appendages determines a negative response, while the "anterior position" determines a positive response. We can, moreover, understand why a more intense illumination of the eye on one side causes a greater activity of the swimmerets on that side, and a consequent swing of the larva away from that side. This phenomenon was well shown by experi-

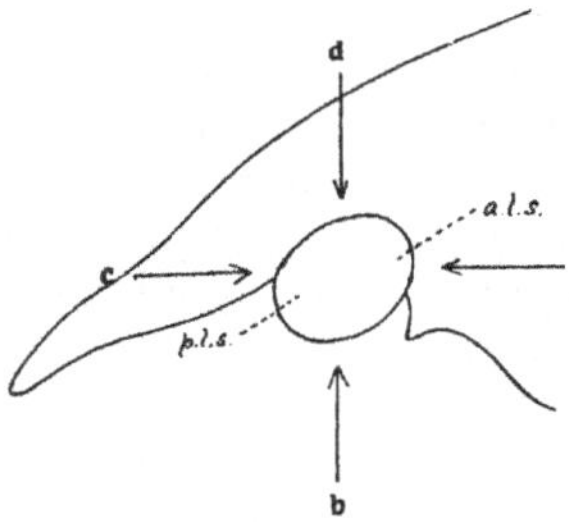

FIG. 22. Diagram showing the rostrum and one eye of a larval lobster; *a, b, c, d* represent direction of light striking the eye from behind, below, in front and above; *a.l.s.* represents posterior lateral surface; *p.l.s.* represents anterior lateral surface. For further explanation, see p. 297.

ments which the writer performed upon larvæ with blinded eyes (HADLEY 1908). These experiments demonstrated that, when the right eye was blinded, the direction of forward swimming was invariably to the right; in other words, the exopodites beat more vigorously upon that side of the body whose eye was most stimulated, and the larva was, in consequence, "pulled around" like a boat. These reactions are explainable on the grounds of a heterolateral stimulation and a consequent unequal action of the muscles on the two sides of the body. But we do not understand as clearly how or why the action of the light striking with equal intensity the corresponding areas of the posterior surface of the eyes (Fig. 22), for instance, brings about these "anterior" or "posterior" positions of the thoracic appendages, and the consequent positive or negative reactions. Nor do we understand why, when the larva is in one "physiological state," a certain intensity of light (striking equally the posterior lateral surface of the two eyes) causes a positive reaction, while if the same larva is in another "physiological state," the same light (striking with the same intensity the same parts of the eye-surfaces) causes the opposite reaction; or again, why when the larva is in the *same* "physiological state," one intensity of light causes a positive reaction, while light of slightly less intensity determines a negative reaction. No more do we know why the illumination of the upper surface of the eyes (Fig. 22, *d*) causes a forward rotation; or the illumination of the lower surfaces (*b*), a backward rotation; or the illumination of the anterior surface (*c*), a forward or a backward rotation. These as yet unexplainable conditions of reaction may well convince us that, however simple and mechanical some of these reactions appear to be, many of them are extremely complex, and indicate a very complex relation between the different regions of the eyes and the nervous centers. Yet, as has been stated, to such a degree as any of these reactions can be explained, those which are concerned in the processes of body-orientation are more easily interpretable on the "simple-reflex" hypothesis. In view of this fact the writer would differ from the conclusion reached by BOHN (loc. cit., p. 41): "Tous ces phénomènes (the reactions of larvæ of Homarus vulgaris) sont en relation avec des états physiologiques particuliers. Sous l'influence de l'éclairement, l'état physiologique des larves de homard ne tarde pas à changer, et les tropismes aussi." The present writer would limit the application of this theory to those reactions of the larval lobsters which are concerned with progressive orientation, excluding body-orientation.

Regarding the relation of the type of reaction found in the larval lobsters to the tropism theories, inference has already been made in the preceding paragraphs. First, to what extent does the behavior found in the larval lobsters agree with the local action theory of tropism? The primary demand of this theory is that the body of the organism should become so oriented with respect to the source of illumination that the anterior end is made to point either toward or from the source. Under these conditions the index of reaction is said to be positive or negative, according as the organism moves toward or from the light. "This orientation is produced, according to this tropism theory, by the direct action of the stimulating agent on the motor organs of that side of the body on which it impinges. A stimulus striking one side of the body causes the motor organs of that side to contract or extend or to move more or less strongly. This, of course, turns the body till the stimulus affects both sides equally; then there is no occasion for further turning and the animal is oriented" (JENNINGS 1906a, p. 266). This is also brought out by HOLT and LEE (1901, p. 479), "The light operates, naturally, on the part of the animal which it reaches." Thus, this tropism theory requires that, in order to determine the direction of movement, the stimulus must act more strongly on one side of the body than on the other. It is needless to say also that in the majority of cases the same conditions of stimulus which cause an animal to direct the head away from the source of the stimulus, also determine a movement in the same direction. Therefore, if we separate, as has been done in this paper, *body-orientation* from *progressive orientation*, we can say that, in most organisms, the index of body-orientation agrees with that of progressive orientation; the conditions of stimulation which cause the one likewise determine the other. Let us now see to what extent the behavior of the larval lobsters agrees with these requirements of the local action theory of the tropisms. In order to treat the matter concretely we must consider it under two heads. First, body-orientation; then, progressive orientation.

It has been shown in the previous pages that, whatever the sign of progressive orientation may be, the *body-orientation is invariably negative*; and that this body-position is produced as a result of diverse reactions which are attributable to the relative intensities of light which strikes the eyes of the larvæ. This body-orientation, moreover, is consta t; it is not dependent upon the age, stage, previous st mulation, hunger, "physiological state," or upon any modifications of the external stimulus, such as changes in intensity, duration of stimulation, etc. The orienting reaction always comes about in the same way, so that we here have a case where the "same-stimulus-same-reaction" principle invariably holds. In other words, the reactions by which the larval lobsters secure the characteristic body-orientation are typical and invariable motor-reflexes.

Beyond producing the body-orientation, the direct motor-reflex ceases to influence the behavior of the larval lobsters. From this moment on, a multitude of conditions appear to be brought to bear to determine the consequent *progressive orientation* of the young animals in one sense or the other. No longer can we say, "same stimulus, same reaction" (SPAULDING 1904), for there is now no constant form of reaction even to the same stimulus. The reactions appear to be no longer so dependent upon the nature of the *external* stimulus, but are more largely regulated by the "physiological states." This we might consider as the cumulative result of a long series of previously acting stimuli, to which others are constantly being added with two effects; first, of bringing about a definite reaction determined by the nature of the stimulus and by the present physiological state; second, of further modifying the physiological state itself, so that even the reapplication of the same stimulus might provoke a quite different reaction. It can not be doubted that the series of changes, which occur in the behavior of the lobster larvæ as they pass through the successive stages, is largely due to this gradual modification of the physiological condition—the cumulative effect of a long series of antecedent stimuli.

We may sum up the preceding paragraphs by saying; (1) The reactions by which the *body-orientation* of larval lobsters is produced are invariable motor reflexes, and the method of such orientation is, therefore, quite in accord with the requirements of the local action theory of tropisms. (2) The reactions by which the *progressive* orientation is produced, although appearing to be simple reflexes, are not invariable but are dependent upon many conditions of stimulation, and especially upon the physiological states.

In view of these facts, it appears that, while the body-orientation of the larval lobsters is not of primary importance in determining the index of the progressive response to the directive influence of the light rays (since the body-orientation and the progressive orientation are dependent upon quite different factors), still it is of primary importance in determining the general line along which the movement shall take place, either toward or from the source of light. It is shown by these points that this type of response is not in agreement with JENNING'S theory (1906b), in which the process of orientation is of secondary importance, for neither the immediate nor the final body-orientation of lobster larvæ to light can be characterized as a "selection from among the conditions produced by varied movements" (JENNINGS 1906b, p. 452). Indeed there are no "varied movements" in the reactions by which the body-orientation to light is brought about. The only way in which the term "random movements" can be applied to the orientation of the larval lobsters is in its relation to the variable extent of the revolutions or rotations. It cannot be denied that this *degree* may be dependent upon the physiological states of the larvæ (for instance, fatigue or freshness), but, after all, this point is irrelevant to the present discussion, since it is the direction of the immediate turning and not the extent of it, which is the important consideration.

The foregoing experiments throw but little light upon the question of intensity of light versus direction of light. Indeed it is probable that the latter phase of the problem is not of great importance except in cases where the light rays are effective by passing through the body as in the case of the electric current, which, as the writer has shown elsewhere (HADLEY 1907a) causes reaction only when the direction of the current holds a certain relation to the longitudinal axis of the larvæ. It is clear, however, that the direction of the light rays does modify the reactions of the larval in two ways: (1) By determining *which of the two eyes shall be most stimulated*, thus causing a body-orientation in which the longitudinal body-axis is thrown into line with the direction of the light rays, so that the eyes shall be equally stimulated; (2) by determining *what parts of the surfaces of the two eyes shall be stimulated equally*, and thus producing a body-orientation in which the posterior lateral surface (Fig. 22, *a.l.s.*) of the eyes receives the strongest stimulation, and the anterior lateral surface (*p.l.s.*) the least. These reactions, and the consequent progressive orientations of the larvæ, the writer has called reactions to the directive influence of the light. That there may be, in addition to these responses, reactions to the intensity of light as HOLMES (1901) and others have considered possible, it is still permissible to believe, and in the earlier pages of this paper the writer has pointed out some reactions of larval lobsters, which, although not perfectly understood, may be included under the head of photopathic response.

The foregoing experiments were carried on at the Experiment Station of the Rhode Island Commission of Inland Fisheries at

Wickford, Rhode Island, where exceptional facilities were found for obtaining material of all ages and stages. The writer's thanks are especially due to Prof. A. D. MEAD of Brown University for making possible an opportunity for this line of inquiry and for material assistance; to Dr. R. M. YERKES of Harvard University, and to Dr. H. E. WALTER of Brown University for friendly criticism during the preparation of the paper; also to Mr. E. W. BARNES, Superintendent of the Wickford hatchery, for many kindnesses.

VIII. SUMMARY.

1. Larval and early adolescent lobsters present both phototactic and photopathic reactions as these responses are defined on p. 201.

2. There is no constant type of response for all larval lobsters, but a modification of reaction occurs through the metamorphosis of the larvæ.

a. First-stage larvæ, directly after hatching, give definitely positive phototactic and photopathic reactions which endure for about two days, after which the phototactic reactions change to negative, becoming positive again shortly before moulting into the second stage.

b. Both early second-stage and early third-stage larvæ manifest a negative phototactic reaction, which usually becomes positive shortly before moulting into the third and fourth stages, respectively.

c. The photopathic reaction of the first three larval stages is commonly positive from the beginning to the end of the stage.

d. The phototactic reaction of the fourth-stage lobsters is usually (i. e., except in cases where intense light is used in connection with early fourth-stage lobsters) negative throughout the stage-period, and the photopathic reaction, positive during the early fourth stage-period, eventually becomes negative.

e. During the fifth stage-period, and in all later stages, both the phototactic and the photopathic reactions are strongly negative.

3. While the photopathic reaction of the larval lobsters remains constant, the phototactic reactions are subject to modification as a result of changes in the intensity or in the direction of light.

a. During the early first stage-period no intensity of light used changes the index of the phototactic or of the photopathic response, but later an intense light may reverse the index of the phototactic reaction.

b. Throughout the second and third stage-periods, the index of the photopathic reaction is not reversible, but during the early part of these periods the negative phototactic reaction, and during the latter part the positive phototactic response, may be reversed temporarily by using light of great intensity (suddenly introduced).

c. During the fourth stage-period the negative phototactic response can not be reversed (except in such instances as are noted in Exp. 24, Cases 5 and 6), but the positive photopathic reaction of the early fourth stage-period may be reversed temporarily by using light of very great intensity.

d. None of the negative responses of the fifth-stage lobsters can be reversed by using light of any intensity whatsoever.

e. Submitting larvæ to darkness for periods of 2 to 12 hours does not change the index of reaction.

4. The reactions to light can be modified by other factors; contact-irritability is first manifested in the middle or later part of the fourth stage-period, and henceforth determines (about equally with light) the behavior of early adolescent lobsters.

5. Laboratory experiments explain some of the aspects of the behavior of the young lobsters under natural conditions of environment: (1) The positive photopathic reaction, and the positive phototactic reaction (to lights of very great intensity) together with the response to food stimuli may unite in determining the surface-swimming of the early fourth-stage lobsters. (2) The negative photopathic reaction, the negative phototactic reaction together with the response to contact-stimuli may unite in causing the late fourth, fifth and all later-stage lobsters to leave the surface water, and to burrow at the bottom of the sea.

6. The reaction of larval lobsters to light depends upon two factors; body-orientation and progressive orientation.

7. The body-orientation is invariably negative and is due to the difference in illumination of the two eyes of the larva. It is brought about by invariable reflex movements which tend to bring the longitudinal axis of the body parallel to the rays of light, with the head away from their source.

8. The progressive orientation may be either positive or nega-

tive, and is due to the position (extension or contraction) of the thoracic appendages. If these have the "anterior position," the reaction is positive; if they have the "posterior position," the reaction is negative. These positions appear to depend upon the intensity of light which strikes the posterior lateral surface of the eyes equally.

9. The larvæ orient to screens and backgrounds of black and of white by reflex movements identical with those by which they react to direct illumination and shading.

10. The reactions by which the body-orientation to light is produced, are invariable motor-reflexes, quite in accord with the local action theory of tropisms. The reactions by which the progressive orientation to light is produced, although appearing to be simple reflexes, are not invariable or constant, but dependent upon "physiological states."

11. In all the reactions to light (except the photopathic) the body-orientation is of primary importance, since progressive orientation cannot occur until the body-orientation has been established.

12. None of the reactions to light can be interpreted as "a selection from among the conditions produced by varied movements." They are not trial (and error) reactions, in the sense in which this expression is used by JENNINGS and HOLMES.

IX. LIST OF REFERENCES.

BELL, J. C.

1906a. The reactions of the crayfish to chemical stimuli. *Journ. of Comp. Neurol. and Psychol.*, vol. 16, p. 299.

1906b. Reactions of the crayfish. *Harvard Psychological Studies*, vol. 2, p. 615.

BOHN, GEORGES.

1905a. Impulsions motrices d'origine oculaire chez les Crustacés. *Bul. institut gèn. psychol.*, no. 6, p. 1–42.

1905b. Attractions et ocillations des animaux marins sous l'influence de la lumière. *Institut gèn. psychol. Memoirs*, i, p. 108.

GRABER.

1884. Grundlinien zur Erforschung des Helligkeits- und Farben-sinnes der Thiere. *Prag und Leipsig.*

HADLEY, P. B.

1906a. The relation of optical stimuli to rheotaxis in the American lobster. *Am. Journ. of Physiol.*, vol. 17, pp. 326–343.

1906b. Annual report of the Rhode Island Commission of Inland Fisheries for 1905, pp. 237-257.

1907a. Galvanotaxis in larvæ of the American lobster. *Am. Journ. of Physiol.*, vol. 19, pp. 39–51.

1907b. Annual report of the Rhode Island Commission of Inland Fisheries for 1906, pp. 181-216.

1908. Reaction of blinded lobsters to light. *Am. Journ. of Physiol.*, vol. xxi, pp. 180–199.

HERRICK, F. H.

1896. The American lobster. *U. S. Fish Commission Bull.*, vol. 15, pp. 1–252.

HOLMES, S. J.
1901. Phototaxis in Amphipods. *Am. Journ. of Physiol.*, vol. 5, p. 211.
1905. The selection of random movements as a factor in phototaxis. *Journ. of Comp. Neurol and Psychol.*, vol. 15, p. 98.

HOLT, E. B. AND LEE, F. S.
1901. The theory of phototactic response. *Am. Journ. of Physiol.*, vol. 4, p. 460.

JENNINGS, H. S.
1906a. Behavior of the lower organisms. *The Macmillan Co., New York.*
1906b. Modifiability in behavior. II. Factors determining direction and character of movement in the earthworm. *Journ. Exper. Zoöl.*, vol. 3, pp. 435–455.

KEEBLE, F., AND GAMBLE, F. W.
1904. The color-physiology of higher Crustacea. *Philosophical Transactions of the Royal Society of London*, Series B, vol. 196, pp. 295–388.

LOEB, J.
1893. Ueber künstliche Umwandlung positiv heliotropischer Thiere in negativ und umgekehrt. *Arch. d. ges. Physiol.*, vol. 56, p. 247.
1905. Studies in General Physiology. *Decennial Publications of the University of Chicago*

LUBBOCK, SIR JOHN.
1881. On the sense of color among some of the lower animals. *Journ. Linn. Soc.*, vol. 16 p. 121.

LYON, E. P.
1906. Note on the heliotropism of Palæmonetes larvæ. *Biol. Bull.*, vol. 12, p. 21.

MINKIEWICZ, R.
1906. Sur le chromotropisme et son inversion artificielle. *Comptes Rendus de l'académie des Sciences, Paris*, Nov. 19, 1906.

PARKER, G. H.
1902. The reactions of copepods to various stimuli and the bearing of this on daily depth migrations. *Bull. of the U. S. Fish Comm. for 1901*, pp. 103–123.

PEARL, R.
1904. On the behavior and reactions of Limulus in early stages of its development. *Journ of Comp. Neurol. and Psychol.*, vol. 14, p. 138.

SCHOUTEDEN, H.
1902. Le phototropisme de Daphnia magna. *Annales de Société entomologique de Belgique* vol. 46, pp. 352–362.

SPAULDING, E. G.
1904. An establishment of association in hermit crabs. *Journ. of Comp. Neurol and Psychol.*, vol. 14, pp. 49–61.

YERKES, R. M.
1899. Reactions of entomostraca to stimulation by light. *Am. Journ. of Physiol.*, vol. 3 pp. 157–182.
1903. Reactions of Daphnia pulex to light and heat. *Mark Anniversary Volume*, pp. 361–377.

THE REACTION TO LIGHT OF THE DECAPITATED YOUNG NECTURUS.

BY

ALBERT C. EYCLESHYMER.

(From the Anatomical Laboratory of St. Louis University.)

During the summer of 1904 a large number of young Necturi (15–18 mm.) were decapitated by pinching with fine forceps. The heads were cut off, at slightly different levels, at about the exit of the common trunk of the seventh and eighth nerves. Although the percentage of fatalities ran high, many of the larvæ lived until the yolk was absorbed, usually about three months. The larvæ used in the following experiments were decapitated on July 10 and in early September they had grown to a length of 30 mm.

That the young and old Necturi are negatively phototropic is a matter of everyday observation both in the natural environment and in the aquarium.

In testing the effects of various kinds of light on the normal and decapitated larvæ they were placed in a small glass aquarium about 60 cm. long, 30 cm. deep and 25 cm. wide. One-half of this aquarium was then painted black and the top covered with a black board. The larvæ, both normal and decapitated, were then subjected to sunlight of varying degrees of intensity. The rays were condensed by a hand glass and by concave mirrors, and were also passed through ground glass. The light of the room was controlled by an opaque curtain so that varying degrees of intensity could be obtained.

In order to test the effects of artificial light, the normal and decapitated larvæ were taken into a photographic dark-room and the aquarium was placed in such a position that a sixteen candle-power electric light illuminated one-half of the aquarium. In the same manner one-half of the aquarium was exposed to the light from an arc lamp. Further experiments were made by controlling these lights with condensers and mirrors. In all cases both the normal and the decapitated animals, when together or separate, react in the same manner as they do in diffuse daylight and direct sunlight. They are negatively phototropic.

In case the normal larvæ are unable to escape a bright light they almost invariably orient themselves in such a position that the light falls with equal intensity upon the two sides of the body. It was also noted that in the great majority of cases the heads were turned toward the light. The decapitated individuals showed the same orientation, except that the heads were about as frequently turned from the light as toward it.

A sharp pencil of rays of either sunlight or electric light when thrown on the tail of the normal animal causes a quick response. This indicates that the tail is especially sensitive, which is in agreement with the observations of DUBOIS on Proteus. In the same manner the decapitated animals respond more readily when the rays are concentrated upon the tail than when they are concentrated on other parts of the body.

During the summer of 1902 larvæ were reared in glass aquaria, beneath which were placed pieces of black, white, red, yellow, green and blue paper. Although a large number of counts were made to determine the percentage over the different colors, at successive intervals, there seemed to be no decided preference for one color over another. A second set of observations, the following year, seemed to show that by far the highest percentage of larvæ were found over the green, whether this was placed on the side of greatest or least diffuse daylight.

In 1905 the same experiment was repeated with the decapitated larvæ, but fifty-two counts showed nothing definite beyond the fact that the larvæ were most frequently found on the colors in the half of the spectrum toward the violet end.

It is of interest here to recall that DUBOIS ('90, p. 356) says: "I have observed that Proteus, under the same condition as the blinded Triton, shows a preference for the following colors in a decreasing series: first, dark, then red, yellow, green, violet, blue and white light." In the Proteus with normal eyes DUBOIS found the reaction towards the various colors was in the following decreasing series: first, dark, then yellow, then green, red, blue, violet. It should be added that these results were not obtained with monochromatic light.

Concerning the reactions of Amphibia to light, there is some difference of opinion. The earlier observations of GRABER ('84, p. 121) seemed to show that Rana esculenta is negatively phototropic and LOEB considered this probable. PLATEAU ('89, p. 82), however, found that R. temporaria is positively phototropic. PARKER ('03, p. 30) also found that R. pipens is positively phototropic, not only in the normal condition, but also when the eyes are removed. Later Miss TORELLE ('03, p. 487) discovered that Rana virescens and R. clamata are positively phototactic at ordinary temperatures, but that raising the temperature to 30° C. accelerates the rate of positive response, while a lowering of the temperature to 10° C. produces movements away from the light. KORANYI ('93, p. 6) says that microscopical changes in the retina of Rana may be effected by the exposure of the skin, as well as the eye itself, to light.

The results of experiments on Urodeles seem to be more uniform than those of experiments on the Anura. CONFIGLIACHI and RUSCONI were probably the first to point out that some of the Urodeles are negatively phototropic. They noted that Proteus always retreats towards darkness. These investigators thought the effect upon the skin, rather than upon the eyes, caused the animals to seek darkness. GRABER'S ('84, p. 96) experiments on the young of Triton, in which he found them negatively phototropic, even when their eyes had been removed and their heads covered with black wax, led to the assumption that the skin can be stimulated by light. DUBOIS ('90, p. 356) who covered the eyes with gelatine and lampblack, concludes that Proteus distinguishes light from obscurity both by the eyes and skin, but that the dermatopteric sensibility is far less powerful than the ocular sensibility. WHITMAN ('98, p. 302) says of the young Necturus: "It is interesting to see how little the eyes are depended upon in finding a piece of meat. A bit dropped in front of a young Necturus receives no attention after it reaches the bottom. An object must be in motion in order to excite attention, and it is not generally the moving form that is directly perceived, but the movements of the water, traveling from the object to the sensory hairs, are felt, and in such a way as to give the direction of the disturbing center with most surprising accuracy. If a bit of beef is taken up adhering to the point of a needle and held in the water, the vibrations imparted to the needle by the most steady hand will be sufficient to give the animal the direction. If the meat falls to the bottom, and the needle is held in place, the animal approaches the needle and tries to capture it without paying the slightest attention to the meat lying directly below. If, after the meat has fallen, the needle is withdrawn and touched to the surface of the water behind or at one side of Necturus, it turns instantly in the direction of the needle not because it sees, but because it feels wave motions coming from that direction. Long experience with Necturus, and with many of its nearer állies, enables me to speak very positively on this point. When it is remembered that in the higher animals the direction of sound waves is given by the auditory sense organs, which are primarily surface sensillæ homologous with those in the skin of Necturus, it may not seem so strange that the animal directs its movements in the way described. Necturus can see, but it can feel (perhaps we should say hear) so much more efficiently that its small eyes seem almost superfluous."

All the facts thus recorded seem to show that the eyes of the young Necturus, as well as those of many other Urodeles, are not highly functional structures, and that when the animal is deprived of their use the dermatopteric sense adequately compensates for the loss.

As PARKER ('05, p. 418) has well said, "The ability of the spinal nerve terminals to be stimulated by light may now be said to be established for certain fishes, amphibians and reptiles; and this fact is not without interest in connection with the theories of the origin of the vertebrate retina."

The many attempts to explain the inverted position of the vertebrate retina early led to hypotheses by LANKESTER ('80), BALFOUR ('85) and BEARD ('88) that the eyes are structures which have been evolved from light perceiving organs which were at one time located in the unclosed neural plate. BISCHOFF, KÖLLIKER, HIS, VAN BENEDEN and others long since observed in mammals a very early appearance of the optic vesicles. HEAPE ('84) observed the optic vesicles in the mole when the neural folds were widely open in the head region. KEIBEL ('89) later observed that in the guinea pig a like early differentiation of the optic vesicles occurs. WHITMAN ('89) discovered that in Necturus there is a very early appearance of the eye, "its basis being discernible as a circular area long before the closure of the neural folds of the brain."

No one, however, had ever shown that the optic vesicles were present in the neural plate at the time the neural folds first appear, until the writer ('93) showed that in Rana palustris the *Anlagen* of the optic vesicles not only appear as a pair of pigmented areas, but that these areas are made up of pigmented columnar cells so different from the cells in the remainder of the neural plate that there could be no reasonable doubt of their being specially differentiated areas. By following these areas step by step during the period of closure of the neural folds it was definitely established that these areas formed the bases of the future retinæ.

Shortly after the publication of the writer's observations LOCY ('93) found a series of depressions in the unclosed neural plate of certain Elasmobranchs which he thought represented paired sensory structures, probably, of a visual character.

In a word, it may be said that the evidence has been slowly accumulating from the morphological side in support of the hypothesis that the retina belongs to the cutaneous sensory system.

The evidence from the physiological side is equally confirmatory. PARKER ('05, p. 419) who has recently carefully reviewed the literature states that "This sensitiveness of the vertebrate skin to light is probably a remnant of that primitive condition from which the lateral retinas were derived, and possibly served as a basis from which the temperature terminals of the skin in the higher vertebrates developed."

In conclusion, then, one may say all the evidence goes to show, as JOHNSTON ('05, p. 241) has well stated, that "the retina belongs morphologically, as well as physiologically, to the cutaneous sensory system."

BIBLIOGRAPHY.

BALFOUR, F. M.
'85. A Treatise on Comparative Embryology. *London, Macmillan & Co.*

BEARD, J.
'88. The Old Mouth and the New. *Anat. Anzeiger*, Bd. 3, pp. 15–23.

DUBOIS, R.
'90. Sur la perception des radiations lumineuses par la peau, chez les Protées aveugles des grottes de la Carniole. *Comp. rend. Acad. Sci., Paris*, Tom. 110, pp. 358–361.

EYCLESHYMER, A. C.
'93. The Development of the Optic Vesicles in Amphibia. *Journ. Morphol.*, vol. 8, pp. 189–194.

GRABER.
'84. Grundlinien zur Erforschung des Helligkeits und Farbensinnes der Thiere, *Prag*, pp. 1–322.

HEAPE, W.
'84. The Development of the Mole. *Stud. Morp. Lab., Cambridge, Eng.* Vol. 2, pp. 30–69.

JOHNSTON, J. B.
'05. The Morphology of the Vertebrate Head from the Viewpoint of the Functional Divisions of the Nervous System. *Jour. Comp. Neurol. and Psychol.*, vol. 15, no. 3, pp. 176-253.
KEIBEL, F.
Zur Entwicklungsgeschichte der Chorda bei Säugern. *Arch. f. Anat. u. Physiol.*, pp. 329–338.
KORANYI, A. v.
'93. Ueber die Reizbarkeit der Froschhaut gegen Licht und Wärme. *Centralb. f. Physiol.*, Bd. 6. pp. 6–8.
LANKESTER, E. R.
'80. Degeneration. *London.*
LOCY, WM. A.
'93. The Derivation of the Pineal Eye. *Anat. Anzeiger*, Bd. 9, pp. 169–180.
PARKER, G. H.
'03. The Skin and Eyes as Receptive Organs in the Reaction of Frogs to Light. *Amer. Jour. Physiol.*, vol. 10, pp. 28–36.
'05. The Stimulation of the Integumentary Nerves of Fishes by Light. *Amer. Jour. Physiol.* vol. 14, pp. 413–419.
PLATEAU, F. I.
'89. Recherches expérimentale sur la vision chez les arthropodes. *Mémoires couromés de l'Academe royale des sciences, des lettres et des beaux arts, de Belgique.* Tom. 43, pp. 1–91.
TORELLE, E.
'03. The Response of the Frog to Light. *Amer. Jour. Physiol.*, vol. 9, pp. 466–488.
WHITMAN, C. O.
'88. Some New Facts about the Hirudinea. *Jour. Morph.*, vol. 2, pp. 585–599.
'98. Animal Behavior. Biological Lectures. *Ginn & Co., Boston, Mass.*

RECENT STUDIES UPON THE LOCOMOTOR RESPONSES OF ANIMALS TO WHITE LIGHT.

BY

E. D. CONGDON.

During the last few years attention has been given to the light reactions of nearly all the large groups of invertebrates. The sudden appearance of data upon the photic responses of animals differing greatly in habits and in mechanism of locomotion has naturally resulted in a variety of opinions as to the proper classification of their orientations. The wide latitude as to precision of light control, amount of quantitative experiments, emphasis laid upon the mechanism of locomotion, and the like, exhibited by various investigators, has increased this diversity. Nevertheless recent discussions make good the claims of trial[1] and phototaxis[2] as two mutually exclusive but closely associated categories within which most features of animal light response may find a place. Papers not concerned with these subjects may in most cases be best considered in relation to the animal group to which they refer. The period to be given attention extends from the year 1900 to 1907 inclusive.

PHOTOTAXIS.

Although some of the postulates of the mechanical phototactic theory of a few years ago have not survived, there can be no doubt that most of the animals to which it was applied have one characteristic in common. They align with the light by a movement whose direction has a definite relation to a localized photic stimulus. Some recent papers may help us to determine the accuracy and speed with which they align and the relation of bilateral symmetry to the procedure.

HARPER's accounts ('05, '07) of the behavior both of the earthworm Perichæta and the larva of the insect Corethra have an important bearing upon the questions just suggested. The earthworm is found to react by the trial method if the light be of low intensity. Under greater illumination exploring movements in the direction not aiding orientation gradually disappear. The animal then aligns itself with the light by a few quick turns. This procedure illustrates the fact that the turning provoked by localized stimulus may consume an appreciable amount of time and may consist of a series of movements. It might be mentioned here that PARKER found Planaria to orient phototactically by a curved course of some length.

The larval Corethra has a discontinuous jerky locomotion. The successive advances are invariably in a path bending towards the source of light. Neverthe-

[1] The expression "trial and error" may be shortened to "trial" because the second term is implied in the first.

[2] The view of RADL is here adopted that the older term phototropism should be applied to all motor responses of animals which are produced by light as does geotropism to those produced by gravity, chemotropism to those produced by chemical stimuli, etc.

less they do not produce alignment because the animal always curves too far. In spite of its zig-zag path the larva always responds to the greater illumination of one side by turning in that direction. Alignment is defeated by a peculiarity in the method of locomotion.

HARPER points out the inconsistency of applying to Corethra the mechanical theory of phototaxis as it was stated by DAVENPORT for the earthworm. It was the suggestion of that author that orientation would result in a simple and mechanical way if we suppose, first, that light directly modifies the tonus of the muscle and, second, that optimal illumination gives the highest tonus. Under these conditions the side of the animal towards the more nearly optimal light would contract the most and the animal thus turn towards optimum. In spite of the fact the Corethra is a worm-like larva, the theory cannot apply to it because it contracts on the side away from the optimum.

RADL ('03) is the author of the most complete account of phototaxis that has yet appeared. A considerable part of his monograph is occupied by his own study upon insects and other arthropods. A variety of interesting facts have been brought to light. Butterflies of various families may be found toward sunset perching upon flowers with body pointed away from the sun, wings outspread and head raised or depressed so as to bring the back of the wing as nearly perpendicular as possible to the sun's rays. In the middle of the day certain species close their wings and align with the light. In general RADL says: "Some butterflies so orient with the sun's rays that in weak sunlight they expose the greatest possible surface, in strong sunlight the smallest surface of the wings." He leaves the explanation for later investigators. BOHN has described similar orientation in butterflies. Certain dragon flies persistently orient with the right side to the sun. Midges have a curious way, little understood, of flying in a circle or spiral within a small area at some point near a light. This place may be forsaken and a new position taken up, only to repeat the previous behavior. Actively moving Cladocera are found in dense swarms within the free spaces among the algal clumps in fresh water ponds. There is a clear band several centimeters wide between them and the shore. The face of the moving mass is clean cut and follows every irregularity in the lateral surface of the algal mass. A similar condition of things is not obtained in the laboratory and an explanation has not yet been discovered.

RADL also found that many aquatic arthropods show an orientation to the light divorced from their locomotor response. Daphnia, for example, regularly orients with its back to diffuse or direct sunlight, while at the same time moving about in a non-directive way. It will turn its back upward if the light be made to come from above or downward if the direction be reversed. Animals were kept in an inverted position for two weeks in this way with no diminution in the precision of the response. He also obtained that locomotor reaction described by DAVENPORT, YERKES and others, for Crustacea, which is characterized by rapid orientation to light and a symmetrical arrangement of the body in relation to the direction of the rays. The conditions which determine whether orientation with back towards the light or with orientation with long axis in line with the light and accompanied by locomotion, shall control the animal have not been determined.

Other peculiarities regarding the relation of body to locomotion have been recently described for arthropods. Pycnogonids (COLE '01) go towards the light with the head leading provided they are crawling, but swim toward it with the abdomen in advance. Palæmonetes larvæ swim with abdomen toward the light. LYON ('07) caused them to move head first towards the dark by diluting the seawater.

A considerable part of RADL's investigation, like that of LOEB and LYON at an earlier period, relates to the movements of insects and other arthropods placed upon disks rotating in various planes. Some animals remain standing upon disks and automatically turn their heads to maintain their orientation to the light. Some, if upon a slowly moving horizontal disk, keep in such a position that they do not lose their orientation to light or to the surrounding, fixed environment. The relation of these reactions to the explanation of light responses can best be made clear after recalling a view expressed by LOEB ('07, etc.). Binocular vision, he believes, is phototactic because a pair of eyes are always placed symmetrically in respect to the center of the field of vision by virtue of their adjustment so that it will fall upon the middle points of their retinas. RADL conceives phototaxis as a response to localized stimulus resulting in symmetrical adjustment. He believes binocular vision is phototactic in his sense of the term phototaxis. At the same time he acknowledges that such phototaxis has two novel elements: namely, the substitution of the varied field for a simple source of light, and the orientation of organs instead of the whole body. In regard to the former, he admits that there has not as yet been brought forward any series of phototactic reactions to fields of gradually increasing complexity as a proof that orienting to them is essentially the same as orienting to a simple source of light. RADL together with LOEB and LYON have found that orientation of eyes as well as head are shown by compensatory movements to be very common in insects. So frequently does it occur that RADL is led to say that the essential of arthropod phototaxis is the orientation of the eyes, and that the adjustment of the body follows only at times, and is of secondary importance.

HADLEY ('06, '06a) has recently shown that young lobsters keep a constant position relative to the bottom while in moving water. This is partly due to orientation toward the fixed field about them. The optical portion of the process may be directly compared with compensatory locomotion upon a revolving disk. Mechanical compensatory movements due primarily to light, and resembling those of the young lobster, are described by LYON ('04) in a study of the rheotaxis of certain fish. LOEB ('07a) finds that marked compensatory head movements are made by the reptile Phrynosoma upon the revolving disk. His experiments show them to be in part due to optical reflexes. It is evident that the compensatory movements of vertebrates, in so far as they are optical in origin, have in them the qualities of similar arthropod movements. If the term phototaxis may be applied to the binocular vision of arthropods, it also can rightly be used for vertebrates. Such a statement needs the corollary that phototaxis probably expresses only a tithe of the nervous activity involved in the binocular vision of the vertebrate.

There has been presented a fair illustration in the variety of locomotion which may result from the response to localized stimulus as described by recent authors. We are therefore in a position to consider whether these different procedures have anything further in common. Opposite conditions in the complexity of aligning movements are illustrated by the earthworm and flatworm as compared with certain crustaceans, as Daphnia. Some animals also stand in contrast with Daphnia because of the greater irregularity of their course to or from the light. Thus in Corethra peculiarities of locomotor mechanism produce a zigzag course. In spite of the great variety which these animals just mentioned show in their accuracy of orientation because of differences of locomotor mechanism and other factors, they have in common, that they align with the light more or less accurately as a result of its differential effect upon the opposite sides of a bilateral symmetrical body. There is thus a response to localized stimulus. The available evidence goes to show that animals responding to localized light stimulus have in general this same character. Even the bell-shaped jelly-fish and a spherical form such as Volvox come within the category inasmuch as radially symmetrical animals must be also bilaterally symmetrical. In the further use of the term phototaxis we shall therefore imply alignment by the differential effect of light upon the sides of bilaterally symmetrical organisms.

JENNINGS attaches little value to this view of phototaxis because it does not pretend to seek a full explanation of things as did the old mechanical theory. He says, "In order to retain any of its value for explaining movements of organisms, it would have to hold at least that the connections between the sense organs and the motor organs are of a perfectly definite character so that when a certain sense organ is stimulated a certain motor organ moves in a certain way." It is to be granted that there is little of an explanatory character in phototaxis as defined above. Nevertheless it has the value which attaches to all categories. It represents a certain stage in the classification of facts, and is a unit of behavior which will simplify further attempts at analysis in the same direction.

LOEB and RADL rightly claim that there is a graded series between such a locomotor response as we have just defined and the training of the two eyes of a vertebrate upon any object. The comparative anatomy of various types of eyes, as well as those experiments upon light response which bear upon the subject, strongly indicate that there is also a graded series between orientations to a single source and those to a varied field. The question of the practicability of applying the term phototaxis, which originally referred to locomotor responses of lower animals alone, to a series including the orientation of eyeless animals on the one hand and of the vertebrate eyes upon the other, is simply one of convenience in terminology. In this paper it will be used in the wider sense.

Perhaps no contribution has appeared which shows more clearly the relation between phototaxis and the general nervous activity of an animal than does the study by HOLMES ('05a, '07) of the reactions of the insect Ranatra to light. The behavior of the animal is dominated to a surprising degree by photic stimuli. It is marked not only by phototaxis of the body but its eyes and breathing tubes sway towards an alignment with the light even when the animal is not engaged in locomotion. If various parts of the eyes be blackened there results the phototactic response which we would expect if the part of the environment dark to the animal were really devoid of light. HOLMES points out that this slavish and mechanical response is probably due to simple reflexes.

But the behavior of Ranatra also reveals more complex nervous processes existing side by side with phototaxis. Hemisection of the brain destroys light response almost completely. Therefore it is probable that the crossing optic fibers in the brain are part of the reflex arc. A number of stereotyped procedures such as hunting food and cleaning the body may inhibit phototaxis. Of especial interest is the result of blackening all but a small posterior portion of one eye. There is a marked disturbance of orientation as one would expect. In spite of this fact, the animal in time learns to move towards the light quite accurately. HOLMES argues that no simple reflex can explain orientation under these conditions.

The first experiment definitely directed to determining the relation of phototaxis and the image-forming power of the eye is described by Parker ('03a). He made use of the positively phototactic butterfly, Vanessa. The animal was placed in such a position between a window and a candle that the intensities of light from the two sources were equal where they fell upon the animal's body. Under these conditions Vanessa flew towards the window, thus demonstrating that it can distinguish between the size of luminous fields. Phototaxis is preceded by a choice of the field to which it orients. The experiment, as PARKER points out, furnishes an answer to the query why positively phototactic winged insects do not fly towards the sun. They seek instead the larger mildly illuminated patches upon the earth's surface.

COLE ('07) employed PARKER's test upon a number of terrestrial animals and thus increased our knowledge as to the relation of phototaxis and the power of forming primitive images. The animals were placed perpendicularly to the line joining two parallel screens and equidistant from them. The light given off by the screens per unit surface was inversely proportional to their size. Therefore the total light intensities of their surfaces were equal. The dung worm Allolobophora, the insect larva Tenebrio, the cockroach Periplaneta, the European garden snail Helix, and the blinded cricket frog Acris did not give a greater number of turnings to one field than to the other. On the other hand the flatworm Bipalium, and the small crustacean Oniscus showed some little power of discrimination. Vanessa, Ranatra, and two species of frogs with eyes intact distinguished readily between the screens and always oriented to a particular one of them.

A discussion of the relation between perception of detailed images and phototaxis appears in a recent work upon vision by NUEL ('04).

TRIAL.

JENNINGS was the first to apply the idea of trial, long recognized for vertebrates, to invertebrates as well. We shall consider the papers on this subject relating to the earthworm by PARKER and ARKIN, SMITH, ADAMS, HOLMES and HARPER before turning to the protozoan studies of JENNINGS.

The methods used by SMITH ('02) in studying the earthworm are valuable in giving, as it were, a birdseye view of its activities in light of rather weak intensity. She devised a means of plotting upon paper the path of the worm for a considerable distance. Exploring movements were shown by spurs upon the line indicating the animal's course. When worms are started with their bodies perpendicular to horizontal light, they go in various directions, varying from directly toward to directly away from the light. In the great majority of cases the course is obliquely from the light. Exploring movements are especially common when the anterior end of the worm encounters stronger illumination or an unfavorable surface. Often they are preceded by a recoil. Although the fact is not emphasized by the writer, her diagrams show that exploring movements toward the light are not followed up so frequently as those away from it.

We must turn to HOLMES ('05) for the application of the trial idea to the worm. He makes the statement that the first effect of moderate light upon the earthworm is the production of exploring movements, of the anterior end, haphazard as to direction, with possibly a few more away from the light than toward it. The second effect is to check the movements toward the light. As a result the animal becomes roughly oriented negatively to the light. He calls the process "the selection of random movements," and points out that it resembles the trial method of higher animals, with the reservation that there is here no learning by experience.

HARPER ('05) gives us a very reasonable explanation as to the mechanism of the exploring movements. He finds that the extension of the anterior segments of the worm presents more fully to light certain cells, probably photoreceptive, which lie near the dissepiments. An animal must extend its anterior end well out in a certain direction, therefore, before light can produce inhibition of further movement.

PARKER and ARKIN ('01) had published, previously to the appearance of the papers by SMITH and HOLMES, an account of the orientation of the earthworm Allolobophora. Their method of procedure was to tabulate the movements of the anterior end in a large number of trials made upon individuals placed transversely to the direction of the light. There were 66 per cent of movements straight ahead, 4 per cent toward the light, and 30 per cent away from it. The view was taken that the 4 per cent toward the light indicate disturbing influences of other stimuli, and so that it is probable that 4 per cent of those away from the light have a like cause. The remaining 26 per cent of those away from the light indicate a tendency of the animals to orient to the stimulation of light in the phototactic way.

Another test of photic response was devised which gave very suggestive results. Light was thrown perpendicularly at different times upon the anterior, middle, or posterior thirds of the body. The percentages indicating the orienting effects are 10.2, 2.4 and 1 respectively as compared with 26 per cent of turns from the light when the entire body was illuminated. It is evident that the condition of the trial reaction as described by HOLMES is present when only the anterior end is illuminated. Yet if the rest of the body be also exposed to light the orienting response more than doubles in amount. The experiment suggests the unreasonableness of thinking that this elongated animal, sensitive to light along its whole length, should make no use, in its orientation, of that wide difference of intensity which must often exist between its opposite ends.

A recent experiment by COLE ('07) suggests that the importance of antero-posterior differences of intensity could be found if a partial shadow were cast upon the earthworm's anterior end when in a field of horizontal light perpendicular to the long axis. A difference of illumination of the two sides of the anterior end would exist such as would fulfill the conditions for a turning by the trial and error method. At the same time if the difference in intensity of the anterior as compared with the posterior end of the animal were effective we should expect a movement straight into the shadow.

ADAMS ('03) applied the methods of PARKER and ARKIN to Allolobophora with the intention of determining the effect of twelve different intensities of light ranging from 192 candlemeters to .012 candlemeters. At 192 candlemeters there were 41.5 per cent negative movements which showed the orienting influence of light. At 8 candlemeters there was an increase to a maximum of 59 per cent of negative reactions. The percentage decreased gradually to 3 per cent at .012 candlemeters. The very low intensity of .0011 candlemeters was found to produce a preponderance of positive movements. This increasing proportion of precise movements away from the light tallies in a general way with the behavior of Perichæta when it forsakes all non-orienting movements in strong illumination. But Allolobophora shows a slight falling off of direct reactions from 8 candlemeters up to 192 candlemeters instead of the uniform increase seen in Perichæta. Of course it cannot be expected that the different genera of worms used by the two experimenters should agree in the details of their reactions.

The trial method as described by HOLMES with its production and checking of varied movements is confirmed by HARPER as far as Perichæta is concerned, and it is hinted at by SMITH. The observations of PARKER and ARKIN do not invalidate its occurrence, because we do not know that they attempted to record a checking of exploring movements. There is, therefore, little doubt that there are both phototactic and trial phases in the behavior of worms, as well as that dependent upon the relation of the stimuli anterio-posteriorly along the animal.

Only those parts of JENNINGS' study ('00, '04, '05, '06, '06a) of Protozoa need be considered here which refer to the method of orientation to light and to his conception of the trial reaction. He found that alignment takes place by a swinging of the anterior end of the animal away from a structurally defined side due to an unfavorable change of intensity. This he terms an avoiding reaction. In case it is initiated by an abrupt entry into a field of perpendicular light of unfavorable intensity there is usually a quick return to the ordinary spiral course. The turn often serves to take the animal out of the unfavorable field. If it does not accomplish that end the process will be repeated until it gets out or becomes acclimated to the new conditions. If at any time it blunders into a field of favorable intensity it is evident that it will be held there as in a trap. A second variety of the reaction usually occurs if the animal be moving at an angle with horizontal light. The beat of the cilia which produced the swing is then likely to continue longer and the anterior end move around a larger circumference than usual. If forward motion be entirely stopped it may describe the surface of a cone or disk by whirling on its posterior end. Some part of the curve which is traversed by the anterior end of necessity leads into increasingly favorable light intensity and the stimulus for the swinging, which was an unfavorable change of illumination, is thus removed. The ordinary spiral course is resumed but the direction is now more nearly in line with the light. By a series of such turns, often very close together the protozoan soon becomes directed as nearly toward the light as its spiral motion will permit.

The following definition of trial is given by JENNINGS ('06) which he applies to the protozoan methods and to that of the earthworm as well. "The organism performs varied movements, some features of which are not determined by the localization of the stimulus but by other factors; it then continues those movements which bring it into or toward a certain condition; this condition usually being a greater or less action of the stimulating agent as the case may be."

This statement of trial differs from that of HOLMES in two of its features. JENNINGS does not confine varied movements to such as are produced by an unfavorable change of illumination. A comparison of earthworm and protozoan varied movements will show whether the latter may be considered due to a change of light intensity. The exploring movements of the earthworm constitute its varied movements. They may be clearly distinguished from the movements which carry the animal along because they are confined to the anterior end of the body. The avoiding reaction of a protozoan, upon the contrary, may consist of ordinary locomotor movements modified by a swing from the structurally defined side due to an unfavorable change of illumination. It is the avoiding reaction which constitutes the varied movement of the protozoan. All components of its motion are not evoked by change of light intensity. Then why refer in a definition of trial to the method by which the varied movement is produced? HARPER ('07) in a recent paper gives a reason. There are a great number of irregular movements, especially among lower animals, which by carrying their possessor into a large number of regions help them better to test the surroundings. Such, among others, are the spiral movements of the protozoan not involved in the avoiding reaction, and certain writhings of insect larvæ. These do not aid in orientation to light and in most cases do not result from unfavorable change of illumination. The trial reaction is therefore to be considered as resulting from varied movements produced in whole or in part by change of light intensity.

JENNINGS does not make the checking of some varied movements an essential part of trial. He says merely: "Movements are continued which bring the animal into or toward the favorable condition." A moment's consideration of the method of the trial procedure in Protozoa makes the reason for his attitude clear. There is first an increase of the ciliary stroke producing the movement from a structurally defined side. When the anterior end of the animal in pursuing the enlarged spiral is brought into more favorable light intensity the increased vigor of stroke disappears. The process is not a checking of any movement by unfavorable illumination.

The orientation of protozoan and earthworm plainly have some differences. Yet the two have sufficient in common to warrant their inclusion within a single category. Orientation by trial then consists in the production of varied movements which are at least in part produced by an unfavorable change of illumination, and the following up of those leading towards favorable illumination.

JENNINGS' account of protozoan behavior to light was soon followed by a paper from MAST ('06) upon the protozoan, Stentor. As the latter says: "JENNINGS laid particular stress on the detailed movements of the individuals while I directed most careful attention to the regulation of the stimulus."

One carefully planned device which he employed gave him a graded field of vertical light. When subjected to it, Stentor, which is negatively phototropic, becomes directed in some path which does not lead into greater intensity. That is to say, it becomes oriented to such an extent that its head points within 90° to one side or other of the line which would carry it most directly toward the dark.

MAST raises a doubt as to whether the avoiding reaction of Protozoa is a trial response at all. His examination of the threshold of light stimulation for different parts of a Stentor shows that the peristomal region is probably much more sensitive than the rest of the surface. Therefore light stimulus is most likely to be received in this region. Inasmuch as the animal turns from the peristomal side in an avoiding reaction it is giving a definite response to a localized stimulus just as truly as is an insect which upon one eye being blinded turns away from the remaining one. The localization of light sensitiveness and the turning from the localized area cannot as yet be considered as an established fact.

Does the protozoan show an alignment of a bilaterally symmetrical body to the light? It does so only imperfectly because of its spiral movement. Inasmuch as many phototactic animals only roughly approximate a straight course because of peculiarities in their locomotor mechanism, such a condition would not prevent the regarding of protozoan orientation as phototactic. There is a consideration, however, that would do so, even though it were proven that a response to localized stimulus occurs. There are no paired bilaterally symmetrical sensory areas through whose unsymmetrical stimulation orientation is accomplished.

DIRECTION VS. INTENSITY.

RADL ('03) makes the statement that from a physio-chemical point of view there can be no question as to whether intensity or direction is the primary factor in the action of light upon an animal. The amount of change produced in the protoplasm by light is due to the amount of energy given up by the light, and that in turn is a function of intensity, not of direction. That this view does not exhaust the question is shown by a test which HOLT and LEE ('01) applied to the protozoan Lyncæus, in imitation of the earlier experiments by COHN. Their apparatus consists of a wedge-shaped tank containing dilute india ink suspended over an aquarium. Light from above produces in an imperfect way a field graded in intensity from one end of the aquarium to the other. By varying the angle of incidence upon the prism the light is given an oblique direction within the aquarium and the gradation of the field is little changed. Lyncæus aligns itself with the light, and goes slavishly into either greater or less intensity according as the rays slant in the one or the other direction. This kind of reaction had been previously used as an argument that direction is the essential factor of light response. HOLT and LEE applied in explanation of the reaction VERWORN's suggestion that if an animal always turned toward the shaded side of its own body it would of necessity align with the light. Thus Lyncæus is forced into either light or dark areas while responding in an unvarying way to the difference of intensity upon the sides of its body. Although HOLT and LEE have thus explained the behavior of the animal satisfactorily by means of intensity changes, they did not settle the question of the relative merits of intensity and direction.

We owe to MAST ('07) the first conclusive proof that orientation is due primarily to intensity. He used for his purpose the colonial protophyte Volvox. An individual was first illuminated by two like pencils of light. As a result it took a course intermediate between them. Then without changing the direction of either beam one of them was modified in intensity. The organism now changed its orientation, bending its course somewhat toward the beam that had become relatively stronger. COLE ('07) illustrated the same point in another way upon the two worms Allolobophora and Bipalium. A partial shadow was cast upon the anterior end of a worm which has been pointed toward the nearly horizontal light. The creature in spite of the fact that it is negatively phototropic went into the shadow, thus moving almost directly toward the source of light. Serpulid larvæ, though phototactic, were found by ZELENY ('05) to go into greater light intensity whether it led them in the direction of the light or not.

It is not necessary to seek for further evidence that light produces stimulation through variations of intensity. Direction plainly affects the intensity of light upon the body or the retina by the casting of shadows or by the complications introduced through eyes of varying position and visual angle. TORREY ('07) has recently recalled to mind the view that light may possibly show an orienting effect dependent upon direction in a way analogous to the action of an electric current. Such a theory would not explain the orientations of Volvox and Allolobophora.

PROTOPHYTA.

MAST ('07) has produced a well rounded and thorough piece of work upon the photic reactions of Volvox. Although the form is classified as a plant its locomotion is of a protozoan type and so is of interest here. The first part of his report gives an analysis of its curious method of locomotion. His apparatus is carefully planned and the methods applied to determining its behavior are various. Equal attention is given to the reactions of segregated individuals and of large numbers taken together. Volvox is found to orient by phototaxis, although through a peculiarity of its locomotion its path is at a slight angle with the light. The light response is analyzed into a series of avoiding reactions of the individuals comprising the colony. Various factors such as previous condition of illumination, the stage of development of the individual, etc., are described as modifying the light reaction.

COELENTERATA.

Gonionemus is an interesting object of study as a type of the primitive and unique group of the jelly fishes. YERKES ('02, '03a, '04, '06) contributed a paper upon its light reactions in a series treating of different phases of its nerve physiology. MORSE ('06, '07) has also devoted some attention to the subject.

According to YERKES ('03a) Gonionemus is decidedly phototactic under certain conditions of illumination. The response to localized stimulus can be readily seen if the individual in a negative condition happens into a band of light of graded intensity such as may occur at the edge of a shadow. The side of the bell toward the light which is most intensely illuminated contracts most strongly and the animal thus turns back into the shadow. The juxtaposition of contracting and stimulated regions results in a localized response reduced very nearly to its simplest terms. MORSE has confirmed YERKES in the occurrence of directive response by observing single medusæ in various conditions of illumination.

A marked photokinetic effect occurs. As LOEB earlier found for Planaria the medusæ will collect in a shadow because as soon as their active movements bring them there they come to rest. This method of non-orienting response to intensity of illumination has been termed negative photokinesis. It has been already stated that a graded field at the edge of a shadow may produce a phototactic orientation of stragglers which directs them back into the shadow. Thus non-orienting light response and phototaxis coöperate.

PLATHELMYNTHES.

Three investigations have been recently published upon the flatworms by PARKER and BURNETT, GAMBLE and KEEBLE and WALTER.

PARKER and BURNETT ('00) so planned their experiments upon the negative Planaria as to determine the importance of the eyes in orientation, and to show the relative importance of light as compared with other factors in locomotion. Single animals were placed at the center of the horizontal surface marked with a circular scale and directed toward the zero point. The angle at which they emerged was recorded, as well as the time consumed in the trip. If individuals with eyes were started toward the light it was found that they would, on the average, bend 78°

toward the right or left. Animals in a healthy condition, but with head cut off, showed a directive effect by an average bending of 57°. In case the initial direction coincided with that of the light the deviation from a straight line dropped to 24° for a normal animal and 35° for those without eyes. The conclusion is that phototaxis is only in part due to the eyes.

The amount of non-directive wandering which takes place was learned from the angle obtained under vertical light. Animals with eyes wander on the average 27°. The bending of 78° by animals headed toward the light must be discounted by this much to obtain the directive effect of light upon them. The course away from the light with 24° of wandering has a turning of only 3°, due to the directive action of light. The results of the comparison of the undirected with the directed course of Planaria are in point with the criticism made by TORREY ('07) upon JENNINGS' view of the manner in which animals may move forward after they have once become oriented. The latter believes that a straight course may be regarded as due in part at least to a lack of any stimulus of light or any other agent which would tend to turn it. TORREY takes the position that the straight course may be due to balanced rather than to non-stimulation. As a matter of fact, we have seen that Planaria's course from the light is influenced only slightly by the light.

Convoluta, to which GAMBLE and KEEBLE ('03, '03a) gave their attention, is a sedentary planarian containing a large amount of chlorophyl. The animal gives a positive response in strong illumination which is markedly greater or less depending upon whether the bottom of the aquarium is white or black. The conditions of tonus found in this creature are peculiar, very likely because of the presence of chlorophyl in its body. If kept in darkness for a while its muscles become contracted and its movements sluggish. Very strong light produces a similar effect except that the animal is now unusually susceptible to being broken to pieces if handled. Convoluta lives within the tide lines and periodically moves to the surface of the sand. The changes in tonus give GAMBLE and KEEBLE an explanation of this procedure.

The study of planarian light reactions by WALTER ('07) is one of the most extensive and many-sided contributions that have as yet appeared upon the light-reactions of any group.

A comparison is made between representatives of several genera in regard to nine different varieties of response which he distinguishes in the animals. Diagrams are given of typical paths followed by the various species if allowed to roam in an aquarium until they come to rest. WALTER says of these, "It may be affirmed that the generic differences are so pronounced that one could take a miscellaneous unidentified assortment of such records and correctly assign the great majority of them to the proper genera." Two species of one genus show a nearer relationship in behavior than do the different genera.

Among the conditions of illumination which were applied to the worm are a series of intensities of non-directive vertical light, including zero intensity, changes in the strength of the entire field of non-directive light, two adjacent non-directive fields of differing intensity, directive light of constant and of varying intensities. Animals in the dark make many double turns which are termed "indefinite" as they evidently are not of orienting value. In non-directive light they are replaced by single turns. The stimulating effect of simultaneous change of intensity over the entire animal varies with the rapidity of change. Decrease of illumination is more of a stimulus than an increase.

A large number of observations upon the effects of other tropisms, physiological states and various internal factors are recorded. For example, some individuals were found to change from the usual response of the species for a time and then to return to it.

By a certain arrangement of conditions it is shown that negative photokinesis is overcome by phototaxis. The tendency to wander may result in many excursions contrary to the phototactic influence. This increases the effectiveness of negative photokinesis by bringing the animals into dark regions to which they would not come through phototaxis.

CRUSTACEA.

The photic reactions of crustaceans have received much attention. Their phototaxis is characterized by quickness of alignment with the light and straightness of course. Photokinesis is often strikingly marked. One is especially impressed in looking over the papers upon the group by the variety of ways in which a reversal of phototaxis has been produced.

TOWLE ('00) found that the positive Cypridopsis could be readily made negative by squirting it through a pipette. Negative animals could less readily be turned to positive by the same means.

In a series of papers upon entomostracan light reactions by YERKES ('00, '03) a similar condition of things is described for Cypris and Daphnia. Cypris is made positive in this way, and Daphnia faintly negative. In the latter animal it could not be determined whether the opposite effect could be produced, for the negative condition was at the best very weak and transitory. YERKES believed that in general the reversal most readily effected is from the less to the more common reaction for each species. There is some probability that the stimulus producing the change is thigmotactic inasmuch as according to PARKER's observation, the crustacean Labidocera though affected in like manner to these others when squirted through a pipette, is not influenced as to its light response by vigorous shaking.

YERKES attempted to find whether increase of intensity calls forth greater accuracy of phototactic response. The question was answered by observing the duration of trips of constant length made by Daphnia and Cypris under various conditions of illumination. There occurred a marked shortening of the period occupied by a trip if the illumination was increased. YERKES believes this partly due to a straightening of the course and therefore to more accurate orientation. The difference between its phototaxis and that of Corixa, therefore, consists only to the degree of accuracy of the orientation. Daphnia was found to recoil and turn back into the light if its head came into a shadow somewhat as in the avoiding reaction of Protozoa.

YERKES has been able to obtain a physically perfect graded field of vertical light by means of a lens consisting of the segment of a cylinder. All light which passes through the bottom of the aquarium is deflected away by a mirror, thus avoiding reflected light. A Daphnia placed in the apparatus goes obliquely upward toward the lighter end at an angle of 45°; it is evident that this is partly due to an attempt at phototactic alignment with the vertical light. The significance of the horizontal component of this motion is not stated.

Daphnia does not seek an optimum but moves unhesitatingly into the most intense illumination it can reach. The harmful effects of strong light are shown by jerky and disorganized movements. PEARL and COLE ('01) have described a like photokinetic effect in a variety of animals which they subjected to the light of a projection lantern. A leech, a nemertean worm and a small crustacean are rendered especially active by strong light until they show exhaustion by sluggishness and insensibility to tactile stimuli. HOLMES ('05a) finds that Ranatra acts like Daphnia in strong light, yet when it has been in the dark for a time it is not only sluggish but negative. The amphipod Orchestia which lives under drift seaweed is negative for a time when exposed to daylight, but turns positive much as does Ranatra. The beach flea Talorchestia, though a nocturnal animal, gives a positive response as strong as any that has been recorded. Ranatra also responds in a positive way with great vigor. The positivity of these dark-loving animals requires explanation.

Increase of temperature hastens a change of Ranatra in a positive direction and accentuates the positive response when it is already present. Dipping into water gives a negative reaction which is probably a contact effect. HOLMES ('01) discovered that the immersion of certain terrestrial amphipods will also effect a reversal.

Labidocera was found by PARKER to behave toward light differently from Daphnia and Ranatra. It reacts positively in diffuse light, but turns strongly negative in direct sunlight. He cites a number of similar cases. The possible adaptive value of the reaction does not need to be pointed out.

SMITH ('05) brings forward a reasonable explanation of the gradual change of sense of response in a number of crustaceans, when subjected to a marked increase or decrease of illumination. It depends upon the fact that in Gammarus annulatus, as in many other crustaceans, the retinal pigment of the individual put from the dark into the light migrates distally at a rapid rate for about fifteen minutes, then moves more slowly for the remainder of an hour. This mechanism protects the more sensitive parts of the eye from over illumination. A large part of a group of animals subjected to strong light change their response within fifteen minutes. At the end of an hour nearly all will be positive. A possible relation between pigment migration and photic response is evident.

DIURNAL MIGRATION.

PARKER and ESTERLY contribute to the explanation of the movements of plankton crustacea and HARPER of insect larvæ from their nocturnal position at the surface of the water to greater depths during the daytime. PARKER ('02) concerned himself with Labidocera æstiva, a typical marine plankton crustacean. He first made sure that geotropism could not account for the migration by any reversal through the agencies of temperature and density. Weak illumination gave a positive response; daylight produced a negative reaction sufficiently strong to overcome the negative geotropism. He thus explains the migration: "Females rise to the surface with the setting of the sun because they are positively phototactic to faint light and negatively geotropic; they descend into deep water at the rising of the sun because they are negatively phototactic to strong light, their negative geotropism being overcome by their negative photopism. The males follow the females in migration because they are probably positively chemotropic toward the females."

A peculiarity of the method used by ESTERLY ('07) upon Cyclops consisted in subjecting animals after a long period in the dark to a series of intensities in various orders of succession. By this means his records show the reaction of the animal to each intensity after exposure to each other intensity. This arrangement suggests a labor saving way of studying various kinds of previous stimuli upon light response. He finds that Cyclops is neutral to artificial lights of low intensity and negative to those of high intensity if it be subjected to them after confinement in darkness. Exposure for some time to light of any intensity makes it negative in all kinds of illumination. Especially interesting is the influence of light upon the geotropic response. Under illumination so diffuse as to be non-directive the animals are strongly positive in their geotropism. If light be removed they become negative. ESTERLY concludes that phototropism is of little importance in the diurnal migration in a direct way. Light, however, probably produces some photo-chemical change in the animal, as a result of which positive geotropism occurs during the day.

HARPER's ('07) work has to do with the insect larva Corethra. The animal is positively geotropic in strong light whether it come from above or below, and negatively geotropic in dim light. It is also distinctly phototactic. It is evident that the chief effect of the light upon migration must be due to its action upon geotropism. HARPER thinks that it is likely that while the animal would go down in the day and come to the surface at night obedient to geotropism it would also respond phototropically by collecting in well illuminated areas at whatever level it happened to be swimming.

There are several extremely interesting and rather lengthy quantitative studies upon the distribution of plankton in various American and European lakes of which a recent example has appeared in the work of JUDAY ('04). It is not advisable to discuss them here as their interest is chiefly ecological. It may be said, however, of JUDAY's work that it shows conclusively that light is a very important though by no means the only factor in diurnal migration. It brings to light the fact that the downward migration of plankton begins long before sunrise if not before midnight. The reason of this early departure from the surface is not forthcoming.

INSECTA.

In his paper on the light reactions of the pomace fly Drosophila, CARPENTER ('05) gives us one of the first general analyses, by laboratory methods, of the light responses of a winged insect. Like many Crustacea, Drosophila is strongly phototactic as well as photokinetic. Like Daphnia it will go into any strength of light without reversal. Very great intensity produces not only rapid movement but apparent loss of coördination. By shaking the jar containing the flies they are rendered negatively geotropic. It is a suggestive fact that as in the case of plankton crustaceans light may affect the geotropism while not acting directively itself. CARPENTER believes its effect is produced through a stimulating action much like that of the mechanical shaking.

MOLLUSCA.

FRANDSEN ('01), MITSUKURI ('01), WALTER ('06) and BOHN have given more or less attention to the behavior of gasteropod mollusks toward light within the period of this review. It would not be profitable to consider their papers since not only has there been a lack of an extended recent study of the group but there is also investigation under way upon their light reactions.

PISCES AND AMPHIBIA.

Certain work upon the compensatory movements of vertebrates has already been referred to while discussing existence of phototaxis in binocular vision. The attention of several workers has also been recently given to locomotor light responses of fishes and amphibians, in particular those dependent upon the light sensitiveness of the skin. Previously a few amphibians were known to possess this dermal function and blind newts of the genus Triton had been found to collect in the shade.

Two primitive vertebrates, Amphioxus and the larval lamprey, have been examined by PARKER ('05, '06) with the resulting discovery that both can perceive light through the skin. Amphioxus is phototactic and negatively photokinetic. The larval lamprey has these same characteristics. It is somewhat startling to learn that the latter animal orients even when the head is removed. The especial sensitiveness of the tail to light is correlated with a habit of burrowing head first until the rest of the body is covered. The earliest study of the orientation of blind fishes in response to light was made by EIGENMANN ('00) upon Ambylopsis. PAYNE ('07) has within the year repeated and extended EIGENMANN'S observations. The animals gather in shade by some other process than direct orientation. They give a stronger response under vertical than under horizontal illumination. A photokinetic effect seems to be present along with movements suggesting discomfort. The experiments employed are simple and not devised to carry very far the analysis of the complex nervous activities of the animals.

PARKER ('03) demonstrates the occurrence of phototaxis for blind amphibians by means of frogs whose optic nerves have been severed. The animals can orient promptly, but only occasionally do they move toward the light. If the skin of a normal animal be covered orientation takes place readily by means of the eyes.

TORELLE ('03) concerns herself with the behavior of the frog without reference to the relative activities of eyes and skin. Orientation is tested in certain experiments by placing the frog in a box 12 in. long with a glass window 9 in. wide and 5 in. high. It may be objected to this procedure that the animal was presented with a pattern of light and shade whose image possibly covered only a portion of the retina at once. The only type of phototaxis which could be inferred from orientation toward the window was that variety present in all binocular vision. The other methods employed by TORELLE are not thus open to criticism, and her conclusions are well established. She, as well as PARKER, finds that orientation to light frequently takes place without locomotion. If one eye be covered the animal orients obliquely to the light when at rest. In spite of this fact it goes directly toward the light as did Ranatra when so blinded. When an individual is put in direct sunlight it will do one of two things. If there is a possibility of its walking into the shadow without losing its orientation to the sun it is likely to do so, but frequently it varies the process by hopping into the shadow and then turning around so as to come to rest with its head toward the light. The field toward which the animal will orient for the time being is thus chosen in the same manner as COLE describes for certain insects. Some features of the frog's response are more mechanical than those we have just been considering. While sitting facing the light it may be made to raise or depress its head in an effort to keep its alignment with the rays if their angle of incidence upon it be changed. The behavior of the animal at low temperatures is probably related to its hibernating habit. Below 10° C. in air it becomes negatively phototactic and crouches down making feeling motions with its head. A very similar condition obtains when it is in water.

THE REVERSAL OF PHOTOTROPISM BY MEANS OF CHEMICALS.

A few scattered cases have been known where some substance in solution has changed the light response of an aquatic animal supposedly through its chemical effect. LOEB ('06) alone has attempted a duplication of this process experimentally. He has succeeded in making a number of organisms positive by adding a trace of acid to the water containing them. Fresh water Copepods, Daphnia, Gammarus, and Balanus larvæ, as well as Volvox, have been made positive in this way. LOEB thinks it probable that the hydrogen ion is the active factor, because normal salts of effective acids produce no change. Alkalies affect the phototropism by destroying the activity of the acid. Volvox differs from the crustaceans in its behavior toward alkalies in that they act directly upon it to make it positive. Hydrochloric, oxalic, and acetic acids reverse, but less quickly, than carbon dioxide. Low temperature has the same effect as acid upon most of the animals mentioned. It may also be made to reinforce the effect of acid.

A hypothetical substance within the animals which is affected by the temperature and chemical condition of the water is invoked by LOEB to explain these reversals. His theory is also extended to cover changes due to temporary physiological states, as for instance the reversal of the Porthesia larva upon becoming well fed. He argues that acid can not favor the production of a chemical compound producing a positive condition because less acid is required to produce the positive state at 10° C. than at 20° while the velocity of a chemical reaction is more rapid at the higher than at the lower temperature. Therefore a substance favoring negative reaction is built up by the protoplasm and its formation or activity is hindered by acid. Or possibly a compound favoring positive reaction and situated in the body may have its activity checked by a different one in the retina. Acids by hindering the formation of this last would produce positive phototropism. An increase in temperature would augment the velocity of its formation and so produce negative response. LOEB did not mention that a rise in temperature makes some animals positive; this fact makes necessary a more general form of the theory. /

MAST ('07) in his paper on Volvox makes use of the principle of reversible chemical reactions in much the same way as LOEB. He further takes into consideration the significance of the substances on the opposite sides of the equation and recognizes that any theory must explain reversal in either direction for each kind of stimulus affecting the light reaction. He has especially in mind the reversal due to change of light intensity but applies it to other kinds as well. His reasoning is as follows for the case in which great intensity changes the usual positive reaction to negative. Let X stand for a substance, on the one side of the equation and Y for one upon the other. Suppose an increase of X beyond a definite amount to produce a positive reaction and of Y a negative one. When X and Y are equal the animal is neutral. Since change of temperature produces a new equilibrium in any reversible chemical reaction, thus altering the relative amounts of the substances on opposite sides of the equation, it is reasonable to suppose that light changes can do the same, inasmuch as they also affect the amount of energy involved. If we suppose intense light to fall upon an animal in which the substances are of the proportion $X = Y$

we would soon get a decrease of X and an increase of Y. The equation would read $X - = Y +$. Y being increased from the amount giving neutrality the animal is made negative. A reversal of this process would occur in sufficiently weak light. Acclimatization of the animal consists in changing the proportions of X and Y which give the neutral reaction.

The correlation by LOEB and MAST of reversal of light response and reversal of chemical reaction is suggestive and tempting. Unfortunately it is difficult to test its validity by experiment. Also, while it is beyond question that light may cause chemical change it is doubtful whether it can produce a reversal of reaction.

A considerable number of agencies have been referred to by which reversal may be brought about. Among them may be named heat, mechanical and chemical stimuli, the various tropisms, condition of development of the individual, temporary physiological states, such as hunger and sexual activity, previous stimuli, and certain stereotyped procedures. Inasmuch as two fairly definite types of light response have by this time become distinguished, there is encouragement to study the effect upon them of the various agencies which have just been named. LOEB and some others have already given a certain amount of attention in this direction. It is one of the tasks which lie next at hand in the comparative psychology of lower animals.

BIBLIOGRAPHY.

ADAMS, GEORGE P.

1903. On the Negative and Positive Phototropism of the Earthworm Allolobophora fœtida (Sav.) as determined by Light of Different Intensities. *Amer. Jour. Physiol.*, vol. 9, pp. 26–34.

BELL, J. CARLETON.

1906. Reactions of the Crayfish. *Harvard Psychological Studies*, vol. 2, pp. 615–644.

CARPENTER, FREDERICK W.

1905. The Reactions of the Pomace Fly (Drosophila ampelophila Loew) to Light, Gravity and Mechanical Stimulation. *Amer. Nat.*, vol. 39, pp. 157–171.

CHIMELEVSKY, V.

1904. Ueber Phototaxis und die physikalischen Eigenschaften der Kultur tropfen. *Beihefte, Bot. Cent.*, vol. 16, pp. 53–66.

COLE, L. J.

1901. Notes on the Habits of Pycnogonids. *Biol. Bull.*, vol. 2, pp. 195–207.

1907. The Influence of Direction vs. Intensity of Light in Determining the Phototropic Responses of Organisms. *Science* (n. s.), vol. 26, p. 784.

1907a. Experimental Study of the Image Forming Powers of Various Types of Eyes. *Proc. Amer. Acad. Arts and Sci.*, vol. 17, pp. 355–417.

DEARBORN, G. V. N.

1900. Notes on the Individual Psycho-physiology of the Crayfish. *Amer. Jour. Physiol.*, vol. 3, pp. 404–433.

DREYER, G.

1903. Influence de la lumière sur les Aimbes et leurs Kystes. *Ov. Danske Selsk.*, p. 399–421.

DUBOIS, RAPHAEL.

1905. Apropos l'heliotropisme. *C. R. Soc. Biol.*, vol. 58, p. 299.

EIGENMANN, C. H.

1900. The Blind Fishes. *Biol. Lect. M. B. L., Woods Holl, for 1899*, pp. 113–126.

ESTERLY, C. O.

1907. The Reaction of Cyclops to Light and Gravity. *Amer. Jour. Physiol.*, vol. 18, pp. 47–54.

FRANDSEN, P.

1901. Studies on the Reaction of Limax maximus to Directive Stimuli. *Proc. Amer. Acad. Arts and Sci.*, vol. 37, pp. 185–229.

GAMBLE, F. W. AND KEEBLE, FREDERICK.
1903. The Bionomics of Convoluta roscoffensis, with Special Reference to its Green Cells. *Proc. Royal Soc. Lond.*, vol. 72 pp., 93–98.
1903a. The Bionomics of Convoluta roscoffensis with Special Reference to its Green cells. *Quart. Journ. Mic. Sci.*, vol. 47, pp. 363–431.
GARREY, W. E.
1904. A Sight Reflex shown by Sticklebacks. *Biol. Bull.*, vol. 7, pp. 79–84.
HADLEY, PHILIP B.
1906. Observations on some Influences of Light upon the Larval and Early Adolescent Stages of the American Lobster. Preliminary Report. *36th Annual Rept. Comm'r Inland Fisheries, Rhode Island*, pp. 237–257.
1906a. The Relation of Optical Stimuli to Rheotaxis in the American Lobster, Homarus americanus. *Amer. Jour. Physiol.*, vol. 17, pp. 326–343.
HARPER, E. H.
1905. Reactions to Light and Mechanical Stimuli in the Earthworm Perichæta bermudensis, Beddard. *Biol. Bull.*, vol. 10, pp. 17–34.
1907. The Behavior of the Phantom Larvæ of Corethra plumicornis Fabricius. *Jour. Comp. Neurol. Psychol.*, vol. 18, pp. 435–456.
HERMES, WILLIAM B.
1907. An Ecological and Experimental Study of Sarcophagidæ with Relation to Lake Beach Débris. *Jour. Exp. Zoöl.*, vol. 4, pp. 45–83.
HOLMES, S. J.
1901. Phototaxis in the Amphipoda. *Amer. Jour. Physiol.*, vol. 5, pp. 211–234.
1903. Phototaxis in Volvox. *Biol. Bull.*, vol. 4, pp. 319–326.
1905. The Selection of Random Movements as a Factor in Phototaxis. *Jour. Comp. Neurol. Psychol.*, vol. 15, pp. 78–112.
1905a. The Reactions of Ranatra to Light. *Jour. Comp. Neurol. Psychol.*, vol. 15, pp. 305–349.
1907. Observations of the Young of Ranatra quadridentata, Stal. *Biol. Bull.*, vol. 12, pp. 158–164.
HOLT, E. B. AND LEE, F. S.
1901. The Theory of Phototactic Response. *Amer. Jour. Physiol.*, vol. 4, pp. 460–481.
JENNINGS, H. S.
1900. Studies in the Reactions to Stimuli in Unicellular Organisms. V. On the Movements and Reactions of the Flagellata and Ciliata. *Amer. Jour. Physiol.*, vol. 3, pp. 229–260
1904. Contributions to the Study of the Lower Organisms. *Carnegie Institute of Washington, Publication 16.*
1905. The Method of Regulation in Behavior and in other Fields. *Jour. Exper. Zoöl.*, vol. 2, pp. 473–494.
1906. Modifiability in Behavior. II. Factors Determining the Direction and Character of Movements in the Earthworm. *Jour. Exper. Zoöl.*, vol. 3, pp. 435–455.
1906a. Behavior of Lower Organisms. *New York.*
1907. Behavior of the Starfish Asterias farreri de Loriol. *Univ. Cal. Pubs.*, vol. 4, pp. 53–185.
JENNINGS, H. S. AND CROSBY, J. H.
1901. Studies, etc. VIII. The Manner in which Bacteria react to Stimuli, especially to Chemical Stimuli. *Amer. Jour. Physiol.*, vol. 6, pp. 31–37.
JUDAY, C.
1904. The Diurnal Movements of Plankton Crustacea. *Trans. Wisc. Acad. Sci. Arts Let.*, vol. 14, pp. 534–568.
JENSEN, PAUL.
1904. Die physiologischen Workengen des Lichtes. *Verh. Ges. d. Naturf. Leipzig, for 1903*, vol. 75, pp. 240–254.
LOEB, J.
1900. Comparative Physiology of the Brain and Comparative Psychology. *New York.*
1904. The Control of Heliotropic Reaction in Fresh Water Crustaceans by Chemicals, especially CO_2. *Univ. of Cal. Pub., Physiol.*, vol. 2, pp. 1–3.
1905. Studies in General Physiology. *Chicago.*
1906. Ueber die Erregung von positivem Heliotropismus durch Saure inbesondere Kohlensaure, und von negativem Heliotropismus durch ultraviolette Strahlen. *Arch. f. ges. Physiol.*, vol. 115, pp. 564–581.

LOEB, J.
1907. Concerning the Theory of Tropisms. *Jour. Exper. Zoöl.*, vol. 4, pp. 151–156.
1907a. Ueber die Summation heliotropischen und geotropischen Wirkungen bei den auf der Dreischiebe ausgelosten compensatorischen Kopfbewegungen. *Arch. f. ges. Physiol.*, vol. 116, pp. 368–374.
LONGSTAFF, G. B.
1905. Heliotropism in Parage and Pyrameis. *Trans. Entom. Soc. Lond. for 1905*, pp. 28–29.
LYON, E. P.
1904. On Rheotropism. I. Rheotropism of Fishes. *Amer. Jour. Physiol.*, vol. 12, pp. 144–161.
1907. Note on the Heliotropism of Palaemonetes Larvæ. *Biol. Bull.*, vol. 12, pp. 23–25.
MAST, S. O.
1906. Light Reactions in Lower Organisms. I. Stentor cœruleus. *Jour. Exper. Zoöl.*, vol. 3, pp. 354–394.
1907. Light Reactions, etc. II. Volvox globator. *Jour. Comp. Neurol. Psychol.*, vol. 17, pp. 99–180.
MITSUKURI, K.
1901. Negative Phototaxis and other Properties of Littorina as Factors in Determining its Habitat. *Annotations Zoologicæ Japoneses*, vol. 4, pp. 1–19.
MORSE, MAX.
1906. Notes on the Behavior of Gonionemus. *Jour. Comp. Neurol. Psychol.*, vol. 16, pp. 450–456.
1907. Further Notes on the Behavior of Gonionemus. *Amer. Nat.*, vol. 41, pp. 683–688.
NUEL, J. P.
1904. La Vision, *Paris.*
PARKER, G. H.
1902. The Reactions of Copepods to Various Stimuli and the Bearing of this on Daily Depth Migrations. *Bull. U. S. Fish Comm. for 1901*, pp. 103–123.
1903. The Skin and the Eyes as Receptive Organs in the Reactions of Frogs to Light. *Amer. Jour. Physiol.*, vol. 10, pp. 28–36.
1903a. The Phototropism of the Mourning-Cloak Butterfly, Vanessa antiopa Linn. *Mark Anniversary Volume*, pp. 453–567.
1905. The Stimulation of the Integumentary Nerves of Fishes by Light. *Amer. Jour. Physiol.*, vol. 14, pp. 413–420.
1906. The Reactions of Amphioxus to Light. *Proceed. Soc. Exper. Biol. Med.*, vol. 3, pp. 61–62 (213–232).
1907. The Interrelation of Sensory Stimulations in Amphioxus. *Science* (n. s.), vol. 25, pp. 724–725.
PARKER, G. H. AND ARKIN, L.
1901. The Directive Influence of Light on the Earthworm Allolobophora fœtida (Sav.). *Amer. Jour. Physiol.*, vol. 5, pp. 151–157.
PARKER, G. H. AND BURNETT, F. L.
1900. The Reactions of Planarians with and without Eyes to Light. *Amer. Jour. Physiol.*, vol. 4, pp. 373–385.
PAYNE, FERNANDO.
1907. Reactions of the Blind-fish Ambylopsis to Light. *Biol. Bull.*, vol. 13, p. 317.
PEARL, RAYMOND AND COLE, L. J.
1901. The Effect of Very Intense Light on Organisms. *3d Rept. Mich. Acad. of Science*, pp. 77–78.
RADL, EM.
1901. Ueber die Phototropismus einiger Arthropoden. *Biol. Cent.*, vol. 21, pp. 75–86.
1903. Untersuchungen über die Phototropismus der Tiere, *Leipzig.*
1906. Einige Bemerkungen und Beobachtungen über den Phototropismus der Tiere. *Biol. Cent.*, vol. 26, pp. 677–690.
REESE, A. M.
1906. Observations on the Reactions of Cryptobranchus and Necturus to Light and Heat. *Biol. Bull.*, vol. 11, pp. 93–99.
ROTHERT, W.
1903. Ueber die Wirkung des Aethers und Chloroform und die Reizbewegungen der Mikroorganismen. *Jahrb. f. wiss. Bot.*, vol. 39, pp. 1–70.

SCHOENICHEN, WALTER.
1904. Die Empfindlichkeit der Nacht-Schmetterlinge gegen Licht-Strahlen, *Prometheus*, vol. 16, pp. 29–30.
SCHOENTEDEN, H.
1902. Le Phototropisme de Daphnia magna Straus (Crust.). *Ann. Soc. Ent. Belgique*, vol. 46, pp. 352–362.
SMITH, AMELIA C.
1902. The Influence of Temperature, Odors, Light and Contact on the Movements of the Earthworm. *Amer. Jour. Physiol.*, vol. 6, pp. 459–486.
SMITH, GRANT.
1905. The Effects of Pigment Migration on the Phototropism of Gammarus annulatus. *Amer. Jour. Physiol.*, vol. 13, pp. 205–216.
TORELLE, E.
1903. The Response of the Frog to Light. *Amer. Jour. Physiol.*, vol. 9, pp. 466–488.
TORREY, HARRY BEAL.
1907. The Method of Trial and the Tropism Hypothesis. *Science* (n. s.), vol. 26, pp. 313–323.
TOWLE, ELIZABETH W.
1900. A Study on the Heliotropism of Cypridopis. *Amer. Jour. Physiol.*, vol. 3, pp. 345–365.
WALTER, HERBERT EUGENE.
1906. The Behavior of the Pond Snail Lymnæus elodes Say. *Cold Spring Harbor Monograph No.* 6.
1907. The Reactions of Planarians to Light. *Jour. Exp. Zoöl.*, vol. 5, pp. 35–162.
YERKES, ROBERT MEARNS.
1900. The Reaction of Entomostraca to Stimulation by Light. II. Reactions of Daphnia and Cypris. *Amer. Jour. Physiol.*, vol. 4, pp. 405–423.
1902. A Contribution to the Physiology of the Nervous System of the Medusa Gonionemus murbachii. Part I. The Sensory Reactions of Gonionemus. *Amer. Jour. Physiol.*, vol. 6, pp. 434–449.
1903. The Reactions of Daphnia pulex to Light and Heat. *Mark Anniversary Volume*, pp. 359–377.
1903a. A Study of the Reaction and Reaction-time of the Medusa Gonionema murbachii to Photic Stimuli. *Amer. Jour. Physiol.*, vol. 9, pp. 279–307.
1904. The Reaction-time of Gonionemus murbachii to Electric and Photic Stimuli. *Biol. Bull.*, vol. 6, pp. 84–95.
1906. Concerning the Behavior of Gonionemus. *Jour. Comp. Neurol. Psychol.*, vol. 16, pp. 457–463.
ZELENY, C.
1905. The Rearing of Serpulid Larvæ with Notes on the Behavior of the Young Animals. *Biol. Bull.*, vol. 8, p. 308.

LITERARY NOTICES.

Pfungst, Oskar. Das Pferd des Herrn von Osten (Der kluge Hans). Ein Beitrag zur experimentellen Tier- und Menschen-Psychologie mit einer Einleitung von Prof. Dr. C. Stumpf sowie einer Abbildung und fünfzehn Figuren. Pp. 193. *Leipzig: Verlag von Ambrosius Barth.* 1907.

Clever Hans has been in the public eye for four or five years and many charming magazine stories have been written about his wonderful and supermundane powers. Hans and his master, Herr VON OSTEN, apparently were first brought to scientific and world-wide popular notice by the zoölogist, SCHILLINGS. So wonderful were the attainments of the horse that a "Hans commission" of thirteen men was chosen from widely different scientific fields and asked to solve the question as to whether there was any secret means of rapport between horse and master. The commission reported that Herr VON OSTEN did not, at least consciously, control the responses of the animal by means of signals.

STUMPF's investigations of the behavior of Hans began on the thirteenth of October and were continued until November 29, 1904. Herr O. PFUNGST and Dr. E. v. HORNBOSTEL were present during these observations. The main conclusion reached was to the effect that visual signs of one kind or another were utilized by the horse in making the proper responses.

O. PFUNGST then continued the work in two ways. First, he made a thorough test of the various acts of Hans, then determined the sensory cues to which the horse reacted: Second, he substituted human subjects for Hans, who were required to answer questions (similar to those put to Hans) by utilizing the same kind of data which Hans employed.

The experimental work was conducted partly in an open court and partly in a large, white tent. Carrots, sugar and bread were the rewards for correct answers. All questions asked were put in such a form that the answers could be given by tapping a certain number of times with the foot.

1. Can Hans read numbers? Printed or written numbers were placed on cards and exhibited to Hans. Hans was supposed to tap the appropriate number of times. Two methods were tried. First, the questioner himself was ignorant of the number displayed; second, the questioner knew the correct answer. When the questioner was ignorant of the answer, only 8 per cent of correct responses was returned. On the other hand, when the questioner knew the answer, 98 per cent of correct answers was returned.

2. Can Hans read words? Such words as "Hans," "Stall," etc., were printed on placards and arranged in a numbered series on a board. The horse was asked to indicate by tapping on which placard any chosen word lay. When the word chosen was unknown to the experimenter, no correct answers were returned, when known to the experimenter, 100 per cent of correct answers was given.

3. Can Hans spell? The letters of the alphabet were arranged in horizontal rows on a board. Hans had to indicate first the row, and then the position in the row, of each letter called for in the word. The experimenter did not know the positions of any of the letters of the alphabet except *s* and *a* (the positions of these were purposely ascertained). Hans was asked to spell such words as "Schirm," "Arm," "Rom" and "Hans." Under these conditions, Hans was a complete failure. Afterwards, when the questioner knew the positions of all the letters, the horse not only could "spell," but also could answer questions involving several long words.

4. Can Hans make arithmetical calculations? The method adopted in this test was as follows: Herr VON OSTEN would whisper a number into the ear of the horse which was unknown to the rest of the observers. PFUNGST would then give another number in the same way and then the horse was asked to add the two numbers. The answer, of course, was unknown to all. In 31 tests of the above type, the horse returned correct answers in three cases. In 31 cases where the questioner knew the answer, 29 correct responses were made.

5. Can Hans even count? The Russian kindergarten counting device (abacus) was used in this experiment. First, the questioner turned his back upon the machine and then shoved forward a certain number of balls. (The questioner in no case knew the number of balls which he had actually pushed forward.) The horse was then asked to indicate the number of balls which had been advanced. No correct answers were given. On the other hand when the questioner knew the answer, Hans in all cases responded correctly.

6. Memory tests. In the absence of the experimenter, a number, or the day of the week, was mentioned to the horse which he was to indicate to the experimenter when the latter returned. In ten trials, only two correct answers were returned. One of the two correct answers was the number three which Hans always "played" when in doubt.

7. Musical memory. A little one octave harmonica was operated in an adjoining room. Hans was asked to indicate whether the first, second, or third, etc., tone had been played. When not attended by the experimenter, the horse always failed. When the questioner could be observed by the horse, all the answers were correct.

In summarizing the results of these experiments, we find that when the questioner knew the answer to the proposed query, from 90 to 100 per cent of the horse's responses were correct. On the other hand, when the answer was unknown to the questioner, the highest percentage of correct answers was 10. According to the author, these latter correct answers must be ascribed to accidents. PFUNGST concludes "that Hans can neither read, count nor perform calculations with numbers. He can distinguish neither coins nor cards. He is not acquainted with the calendar nor with our system of time. He cannot even recall a number given him but a moment before. Finally, there is no trace of a musical ear. From all this, we must conclude that the horse is unable to work independently, but is dependent upon his environment for particular stimulations" (free translation).

After the above data had been obtained, the author tested very carefully the means by which Hans gets his cue. Without going into detail in this part of the work, it may be said at once that if visual stimulation were cut off by means of blinders (the horse has a wonderfully wide field of vision) the horse could no longer give the correct responses. In making his responses, it was observed that Hans never looked at the objects to which he was supposed to react, but always at his questioner.

The sensory stimulations from which Hans took his cue consisted of *certain slight movements of his questioner's body*. After Herr VON OSTEN had stated the problem, he tended always *to bend the head and trunk slightly forward;* whereupon Hans extended his right foot and began to tap without putting his foot back after each successive tap. When the desired number of taps was reached, Herr VON OSTEN would give *a slight upward jerk of the head*. At this second signal, the horse would retract the foot to its normal position (this last movement was never counted). Now when the horse had ceased to tap, the questioner would raise the head and trunk to an upright position. This second and more extensive movement (that is, more extensive than the slight upward jerk of the head) cannot be regarded as the signal for the retraction of the foot. If the larger movement, however, did not follow the slight upward jerk of the head, the horse would give a single vigorous tap with the *left* foot, without however first extending it. Horizontal movements were without effect in eliciting responses from Hans. All downward motions of the body, eyebrows, nostrils, arms, etc., were signs to begin the tapping movement, whereas the raising of these parts was a signal to cease tapping.

The results of the experiments conducted in the laboratory upon human subjects in the rôle of Hans form an interesting contribution to the study of the psychology of involuntary movements.

Surely, this careful and painstaking work of PFUNGST may be prescribed as an antidote henceforth and forever to those untrained but enthusiastic observers who may be filled with the desire to describe the doings of pet animals in glowing anthropomorphic terms.

J. B. W.

Yale Psychological Studies. Edited by CHARLES H. JUDD, n. s. vol. 1, no. 2. *Psychological Review Monograph Supplements*, vol. 8, no. 3, pp. 227–423. 1907.

The second half of the first volume of the new Yale Studies contains five papers, at least four of which show that perfection of experimental technique which is characteristic of Professor JUDD's laboratory. In Tonal Reactions, Dr. E. H. CAMERON gives an interesting study of tones as produced by the human voice, both with and without distractions. "The attempt to sing a uniformly sustained tone is not successful. The beginning of the tone is markedly irregular and there is a tendency to raise the pitch towards the end of the tone. There is usually a harmonious relation between the sung tone and the distracting tone." Mr. F. N. FREEMAN contributes some Preliminary Experiments on Writing Reactions which connect interestingly with the Analysis of Reaction Movements contained in the first half of the volume. Both of these papers have important bearings on the mechanism of consciousness, of which the full significance, in the reviewer's opinion, will appear only later. Mr. H. N. LOOMIS reports on Reactions to Equal Weights of Unequal Size. He finds that the weight of smaller size is usually raised later than the larger weight, and that the hand which lifts this latter has a much greater muscular tension. With practice the illusion tends to disappear, and so too these differences in the manner of lifting. In Studies in Perceptual Development Professors JUDD and D. J. COWLING describe experiments in which subjects learned to draw complex figures. Each figure was shown repeatedly for ten seconds, and after each view the subject reproduced once as well as he could what he had seen. The issue concludes with some remarkable Photographic Records of Convergence and Divergence, including a theoretical discussion of the mechanism of perceptual unity, by Professor JUDD. In general, lateral movements of both eyes in the same direction are a more thoroughly established form of coördination than are movements of convergence and divergence; but many particular facts of movement are brought out which are well worth studying in detail. As is well known, the author denies the significance ordinarily attributed to sensations of movement in the formation of spatial and other percepts, and his view is expressed in the very pregnant and, in the reviewer's opinion, just proposition: "The only concept which is of any value in the clear explanation of perceptual unity is the concept of coördination," i. e., the coördination of motor response. E. B. H.

The Archives of Psychology. Edited by R. S. WOODWORTH. *New York, The Science Press.*

The *Archives of Psychology* is a continuation of the psychological part of the *Archives of Philosophy, Psychology, and Scientific Methods*, of which one volume, consisting of the following monographs, was published.

Measurements of Twins. By EDWARD L. THORNDIKE.
Avenarius and the Standpoint of Pure Experience. By WENDELL T. BUSH.
The Psychology of Association. By FELIX ARNOLD.
The Psychology of Reading. By WALTER FENNO DEARBORN.
The Measurement of Variable Quantities. By FRANZ BOAS.
Linguistic Lapses. By FREDERIC LYMAN WELLS.
The Diurnal Course of Efficiency. By HOWARD D. MARSH.
The Time of Perception as a Measure of Differences in Sensations.
By VIVIAN ALLEN CHARLES HENMON.

Below we give review notices of the numbers of the *Archives of Psychology* which have appeared.

Norsworthy, Naomi. The Psychology of Mentally Deficient Children. *Archives of Psychology*, no. 1, pp. iii + 111. $1.00. 1906.

The author presents the results of an experimental study of groups of defective children, and discusses the scientific and practical significance of the facts which she has discovered. She briefly summarizes what little experimental work had been done in this field, previous to her own investigation, and insists upon the need of the exact measurement of a number of the important characteristics of normal and defective children.

"I have sought to determine," she writes, "(1) whether the mental defects of idiots are equaled by the bodily, (2) whether idiots form a separate species or not, and (3) whether the entire mental growth is retarded, that is, whether there is a lack of mental capacity all around." In order to get data for the solution of these problems she made measurements of the following traits. Mental Traits: Efficiency of perception; memory of unrelated ideas; ability in the formation of abstract ideas; ability to appreciate relationships and to control associations; perception of weight; and motor control. Physical Traits: Height; weight; pulse and temperature.

The measurements which were made led the author to conclude: (1) That there is a decided difference between bodily and mental deficiency (p. 69); (2) that idiots seem not to form a special class or species, at least as far as intellectual traits are concerned, but that they are included as part of a large distribution (p. 77); and (3) that there is not among idiots an equal lack of mental capacity in all lines (p. 82).

Of special interest to educators, and to others who are interested primarily in applications, is the discussion of the education of defectives. It is the author's belief that the difference between the defective and the normal child is one of degree and not of kind, and that for this reason the educational methods applied to the former should not differ in principle from those which are used for the latter.

R. M. Y.

Franz, Shepherd Ivory. On the Functions of the Cerebrum: The Frontal Lobes. *Archives of Psychology*, no. 2, pp. 64. 50c. 1907.

This monograph is one of a projected series on the functions of the cerebrum, particularly on those of the so-called association areas. The first section (pp. 5–11) is introductory and historical; the second (pp. 12–28) summarizes and criticises previous studies on the frontal lobes; the third (pp. 29–34) gives the author's methods and the fourth (pp. 35–62) his results.

In criticising previous results the author concurs with others in giving slight weight to MUNK's statement that dogs show motor disturbances of the trunk muscles after extirpation of the frontal lobes. He noticed no such disturbances, at any rate, in the cats and monkeys on which he operated. Nor do his experiments lend any support to FERRIER's conclusion from experiments on monkeys (confirmed in part by GRÜNBAUM and SHERRINGTON for the chimpanzee) that "frontal centers are concerned with the movements of the head and eyes" (p. 14). A somewhat detailed review of the evidence adduced by many to show that the frontal lobes are the centers of inhibition, attention or of higher mental processes leaves the impression that this evidence is either inconclusive or is actually opposed to these inferences. His general criticism of such work is that the observations are too casual and that the accounts show too clearly the lack of "a careful analysis of the mental condition" (p. 24) to be taken as univocal proof.

FRANZ attempts, in his own experiments, neither primarily to discover possible motor centers nor a connection between the frontal lobes and so-called higher mental processes, but "to determine whether or not animals with frontal lobe destruction retained simple associations or could form associations" (p. 30). The animal (a cat or a monkey) was placed in a box from which, by pulling a string or turning a button, it could escape and obtain food. For monkeys an intricate "hurdle" was sometimes used. The habit was thoroughly formed while the animal was still in a normal condition and its retention tested after the lapse of some weeks or after severance of the frontal lobes, which were left in situ, from the rest of the cerebrum. In some cases the operation was first performed and then the attempt made to form the habit. "If the time for the performance of the act of turning the button or pulling the string, etc. (after the operation), remained the same as when the earlier experiments were made, we are warranted in saying that the association is retained and the nervous connections for the performance of the habit have not, in the case of extirpation, been interfered with" (p. 34). In cats, the attempt was made to extirpate always in front of the crucial sulcus and often the section was made in the immediate neighborhood of the supraorbital fissure. For monkeys the endeavor was to limit the lesions to that portion of the cerebrum anterior to the precentral fissure. From neither of these extirpated regions did stimulation give any constant motor response. After the formation of the habit, but before the operation, the animals were put away for a week or more with no practice, when their retention of the habit was again tested.

The results on four cats are first reported, all of which were practiced until they could release themselves from the box in from 1 to 6 seconds. Both frontals were then removed. During periods varying from seven to fourteen days after the operation these animals were tested and were found to have lost the habits that they had formed. In these, and in the other cases as well, two minutes were given the animal in which to open the box. On ten monkeys the same operation (severance of both frontals), after the required habit had been formed, resulted in the loss of the habit by six of the animals and its retention by four, although the time of performance, in the latter cases, was somewhat longer than before the operation.

As the result of the experiments designed to test the ability of the animals to learn a habit for the first time after the operation (on both frontals) had been performed, or to relearn it after it had once been lost as a result of the operation, it was found that two cats easily acquired the box habit after both frontals had been severed from the rest of the cerebrum and that two of the cats and two of the monkeys could relearn the habit lost after operation. In the case of one of the cats that learned the habit after operation the author remarks that "the curve of learning in this animal was about the same as that in an animal before the removal of the frontals" (p. 57).

That the loss of newly acquired habits after the removal of the frontal lobes, which is the more frequent result in these experiments, is not due to surgical shock and does not result after the removal of other parts of the cerebrum is shown by the cases in which habits were retained even after the frontal lobes had been severed and by others in which the parietal lobes were removed without detriment.

The inference that the author draws from his results and from those of other investigators is that "the frontal lobes are concerned in normal and daily associational processes and that through them we are enabled to form habits and, in general, to learn" (p. 64). Four cases were mentioned, however, in which the formed habit was retained after cutting away both frontals, two in which the habit was learned for the first time after the operation and four in which it was relearned. The author suggests, in explanation, since the learning period, in those cases in which the habit was retained even after operation, was longer than in other cases, that it had become a reflex and was therefore due to the functioning of the lower centers, identifying these cases with others in which he observed that habits of long standing, such as coming on call or jumping on the experimenter's shoulder, were also retained after the operation. The instances of learning and of relearning, after the removal of the frontal lobes, he supposes to have been due to the activity of the still uninjured parts of the cerebrum, in particular of the remnants of the frontal lobes still left intact after operation. Both suggestions are interesting but, as the author himself feels, further and more crucial experiments are needed to define the exact difference between a *new* habit (not retained after operation) and an *old* habit (retained). It is unfortunate, in this connection, that no exact record is given of the length of the whole training period for each animal, of the intervals between tests and of the number of trials at each test. YERKES[1] has recently shown the importance of such records in determining the efficacy of training. Further, if those portions of the frontal lobes which were severed from the rest of the brain are *normally* concerned in the formation of habits, ought habits to be learned

[1] YERKES, R. M. The Dancing Mouse. *New York.* 1907.

as quickly (as one case, at least, indicates) by the parts still left intact after the operation? In short, before the general inference from the experiments can be looked on as more than highly probable, further investigation and more exhaustive records, of the kind just indicated, are much to be desired. FRANZ himself intends to experiment further. ROSWELL P. ANGIER.

Thorndike, Edward L. Empirical Studies in the Theory of Measurement. *Archives of Psychology,* no. 3, pp. 45. 50c. 1907.

A discussion of statistical methods, in the light of the author's experience. Measurements of type and variability, and measurements of relationships are considered with a view to convenience, economy, and directness as well as to precision.

Lapisky, Abram. Rhythm as a Distinguishing Characteristic of Prose Style. *Archives of Psychology,* no. 4, Pp. iii+44. 50c. 1907.

Ruediger, W. C. The Field of Distinct Vision, with Special Reference to Individual Differences and their Correlations. *Archives of Psychology,* no. 4, pp. 68. 1907.

The author mapped out the field of acute vision in eighteen subjects, with a view to finding "characteristic individual differences" and to ascertaining whether the size of this field is correlated with "reading rate, the color zones, visual acuity, retinal inertia, and other phenomena of vision." The field of acute vision is defined as the area within which the letters *n* and *u* (of a certain font of type) are discriminated 75 (and again 90) per cent of the cases, when exposed for a fixed length of time (less than the reaction-time of the eye).

The shape of this field is found to vary "in different individuals from a 'square-oval,' about twice as long horizontally as wide vertically, to a circle;" and "the size of the field varies approximately as 2–1 in the horizontal diameter, as 1.5–1 in the vertical diameter, and as 2–1 in area." There is some correlation between the size of this field and the acuteness of vision itself, which amounts, if the latter is determined by GALTON's test, to nearly +.69 (PEARSON coefficient). In this correlation not eighteen but twelve subjects are used, and this value of the coefficient should not be further employed without a careful reading of the author's text (pp. 46–49). "There is little or no correlation between the horizontal extent of distinct vision and the 'A' test, the number of lines that can be seen simultaneously, reading rate, and the number of pauses per line. Reading rate apparently does not correlate with any of the attributes of vision, but it correlates highly with the smallness of the number of reading pauses per line."

A simple method of Professor WOODWORTH's is described, for measuring correlation (pp. 37–39). It can be used wherever the individuals compared with regard to a character, can be ranked in ordinal series, and it takes into account this order of the individuals, but not the amounts by which they differ in regard to this character from one another. E. B. H.

Jones, E. E. The Influence of Bodily Posture on Mental Activities. *Archives of Psychology,* no. 6, pp. 61, 50c. 1907.

The author briefly sums up the chief results of his work as follows: "Pitch is discriminated better (with the body) in the vertical than in the horizontal position;

tactile discrimination is slightly more acute in the horizontal than in the vertical; visual memory is both more rapid and subject to fewer errors in the horizontal than in the vertical position; auditory memory shows the same result as the visual memory; adding can be done more rapidly and with greater precision in the horizontal posture; subjects show greater signs of fatigue in the horizontal than in the vertical posture; a greater number of taps per minute can be made in the vertical than in the horizontal position; and the vertical position is favorable to the strength of grip."

Wells, F. L. A Statistical Study of Literary Merit with Remarks on Some New Phases of the Method. *Archives of Psychology*, no. 7, pp. 30. 30c. 1907.

Yerkes, Robert M. The Dancing Mouse: A Study in Animal Behavior. Animal Behavior Series, vol. 1, pp. xxi + 290. *New York: The Macmillan Company.* 1907.

It is not putting the matter too strongly to say that Dr. YERKES has in this book given us the most valuable contribution that has yet been made to the study of animal behavior. Having become interested about four years ago in some specimens of the curious Japanese dancing mouse, and finding them readily tamed, easy to care for, and comparatively quick to learn, he undertook a thorough investigation of their sensory equipment and intelligence. The results are stated clearly and concisely in the present volume. Every step in the methods used, every stage in the reasoning processes by which the author's conclusions were reached, is given, so that the book is a real text-book in experimental method.

The dancing mouse, as is well known, gets its name from the fact that when placed in an open space it makes peculiar whirling or circling movements. These movements have been thought to be due to a malformation of the equilibrium apparatus in the ear, and support seems to be given to the theory by the fact that the mice have defective hearing. Yet the statements of various experimenters who have examined the ear are so conflicting that no definite inference can be drawn from them. Dr. YERKES concludes that to explain the peculiar movements of the dancer "the structure of the entire organism will have to be taken into account," and at the same time he finds no "satisfactory ground for considering the dancer as either abnormal or pathological"—an assertion the truth of which would seem to depend upon the meaning assigned to the term "abnormal." To account for the disagreement among different observers of the behavior of the animal, some of whom say that the mouse is markedly deficient in balancing power, while others find no striking defect in this respect, YERKES adopts the suggestion of CYON that there are at least two varieties of the dancing mouse, and has observed evidences of the existence of two different strains among the specimens examined by himself.

The net results of the author's work with some four hundred individuals may be grouped under four heads: those concerning the mouse's power of sense discrimination; those concerning its learning capacity; those bearing on questions of experimental method, and those of interest to students of biogenetics.

1. Three classes of sensory discriminations: auditory, brightness, and color, were investigated. The mice of one of the two lines of descent represented were, with the exception of one litter, throughout their lives insensitive to sounds. Those of the other line showed sensitiveness for a day or two during their third week

The mice displayed ability, varying with the individual, to discriminate different shades of gray paper, though their capacity in this direction was less than that of a human being. One mouse was subjected to tests of the validity of WEBER'S Law, in discriminations of the degree of illumination in different compartments; and the law was found to hold, the proportion $\frac{DR}{R}$ lying between 1–10 and 1–15. These discriminations were greatly improved by practice.

As regards color discrimination, much of the behavior superficially to be classed under this head appeared on further investigation to be really based on the brightness value of the colors. This value is apparently quite different in the case of the mouse from what it is in the human subject. The red end of the spectrum is much darker to the mouse, being indistinguishable from black and darker than any green or blue. There is some evidence that the mice can discriminate green and red "by some other factor than brightness," but on the whole the problem of their color vision is not solved. The dancer was found to be incapable of distinguishing between two equal illuminated areas of equal brightness but different form. Mice that have learned a labyrinth path are little disturbed in traversing it by being made to do so in darkness, by washing the labyrinth so as to destroy smell clues, or by moving the labyrinth to one side so that the former track is removed. They were a good deal disturbed when the floor was covered with smoked paper, but Dr. YERKES thinks this disturbance was a general one and not the result of loss of a clue. The rôle of the senses of sight, smell, and touch in the learning of a labyrinth path, quite a different problem from the effect of eliminating a sense after the animal has learned the path, was not experimentally tested; observation of the general behavior of the animals led to the conclusion that these senses are all used, but in different degrees by different individuals.

2. The results that bear especially on the mouse's learning capacity are as follows. The dancer is capable of forming habits that involve turning in one direction or another (labyrinth habits), and habits that require in addition visual discrimination, but the former are acquired much more rapidly than the latter. A regular labyrinth, involving turns alternately in two directions, is learned with especial speed. Useless habits occasionally persist for some time, a fact which several other investigators have noted. The mice did not learn by imitating each other. Putting the animals through one of the reactions which they learned, that of climbing a ladder, did aid the learning process. Dr. YERKES's conclusions regarding sex differences in learning capacity are less convincing than his other inferences from results. He says, "In labyrinth tests the female is as much superior to the male as the male is to the female in discrimination tests." Yet in the tables upon which the first part of this statement is based, although the average number of tests which had to be given before the labyrinth was perfectly learned was 18.7 for the males and 13.8 for the females, five of the ten males learned with fewer tests than did five of the females. "Of the five pairs of individuals whose records in white-black training appear in Table 43," says the author, "not one contradicts" the statement that the males are superior to the females in discrimination experiments. This table contains the results of tests of black-white discrimination made at the rate of ten per day; but Dr. YERKES himself points out that the females did better than the males when twenty tests per day were given.

The experiments on the relation of docility to age were not completed. As regards the persistence of the habits acquired, the white-black discrimination habit it is concluded, "may persist during an interval of from two to eight weeks of disuse," but "is seldom perfect after more than four weeks." The color discrimination habits never persisted more than two weeks. The white-black discrimination was re-learned, after all traces of it had disappeared, in a shorter time than had been required for its original acquisition, thus suggesting "the existence of two kinds or aspects of organic modification in connection with training; those which constitute the basis of a definite form of motor activity, and those which constitute bases or dispositions for the acquirement of certain types of behavior." As ROUSE found to be the case with the pigeon, experience with one form of labyrinth made the learning of another form easier.

3. The suggestions on method which the book contains are among its most valuable features. In the first place, Dr. YERKES finds that the best motive to employ in studying the learning processes of mammals is not hunger, which is variable in intensity, unfavorable, in its extreme degrees, to the exercise of the animal's full powers, and inhumane; but the punishment of mistakes by slight electric shocks, given through wires on the floor of the discrimination box or labyrinth. So far as my recollection serves, the author first used this method in his experiments on the frog. It has certainly a decided advantage over any method where the motive is constituted by a continuous state in the animal, such as hunger or the desire to escape from confinement. Any continuous state is likely to vary in intensity, and discomfort under confinement is likely to diminish as the animal becomes used to its surroundings. An intermittent stimulus, given when mistakes are made, has a much better chance of producing a constant effect.

Another interesting point in method concerns the evidence which was obtained that apparent color discrimination was really, in some measure at least, brightness discrimination. This evidence consisted in the fact that mice which had been trained to choose a compartment illuminated by green light in preference to one illuminated by red light, on being offered a choice between black and white compartments, chose the white, although before the green-red training they had shown no such preference. Thus it looked as if they had been choosing the green partly, at least, as the lighter of the two visual impressions. This method might well be given a wider scope in the study of the sensory discriminations of animals. It is puzzling, by the way, in view of the evidence that red is much darker to the mouse than to the human observer, to find that a tendency to choose red rather than yellow in tests with colored cards is explained as the result of previous training to go to the brighter compartment in white-black tests.

The results of training tests are throughout stated in terms not of the time required to perform the act, but of either the number of errors made or the number of tests required before the formation of a perfect habit. The time required in traversing a labyrinth, for example, is in the author's opinion a poor index of the perfection of the habit. In this respect his position is the opposite of that taken by WATSON, who states the results of his tests of the white rat wholly in terms of time.

On the whole, the most favorable number of tests per day in the discrimination series, taking into account economy of time and fatigue for both experimenter and animal, was found to be ten.

For labyrinth experiments, the author recommends the use of a standard maze in which "errors by turning to the right, to the left, and by moving forward should occur with equal frequency and in such order that no particular kind of error occurs repeatedly in succession."

4. Finally, Dr. YERKES made some studies on the phenomena of heredity in the dancing mouse. The subjects of his experiments belonged to two separate lines of descent, which presented certain characteristic differences, the individuals of one line being more like ordinary mice than those of the other line. Observations on several generations indicated a certain inheritance in the latter line of descent of a tendency to whirl to the left in dancing, while those of the former line showed no such tendency. Four generations, a male and a female in each, were tested to see if the training of the parents in white-black discrimination facilitated the learning process in the offspring. The results showed no evidence of the inheritance of this acquired character.

From this superficial survey it will be seen how rich both in result and in suggestion the book is. No less admirable is the spirit in which the work has been carried on; a spirit of scientific conscientiousness, of modesty, and of humane sympathy, untinged with sentimentality, for the animals experimented on. Of comparative psychology, in the sense of an attempt to interpret the mental states of the subjects, there is very little in the book, and that little does not strike one as its most successful feature. For instance, when in labyrinth tests the procedure was adopted of allowing a mouse, as a preliminary, to traverse the maze and escape without getting an electric shock, it is said that this was for the purpose of allowing the animal "to discover that escape from the maze was possible," but there is no discussion of the terms in which such a possibility may have functioned in the mouse's consciousness in subsequent tests, whether as a memory image or merely as an increased tendency to movement. Again, when one method of reaction in the discrimination experiments is designated "choice by comparison," one is left with the interesting problem as to what sort of process the comparison of two stimuli in the mouse's mind may be, and it might perhaps have been better to use a term that would have had a less decided psychic implication. However, if there is but little attempt at interpretation of the mental aspect of the facts observed, the book is almost an ideal example of the kind of work which promises to put comparative psychology on a firm scientific basis.

MARGARET FLOY WASHBURN.

Davis, H. B. The Raccoon: A Study in Animal Intelligence. *American Journal of Psychology,* vol. 18, pp. 447-489. 1907.

This paper describes the habits and instincts observable in adult raccoons in captivity, and it presents the results of experiments to test learning, color perception, and imitation in these animals, together with comparisons of these results with those obtained by other investigators. In this review only the descriptions of experiments and the discussion and interpretation of results will be examined. These may be taken up under the general headings learning, color perception and imitation, and comparisons.

Learning. In the author's experimental study of learning the raccoons were allowed to unfasten the door of a box, reach into it, and get food. Single fastenings, a group of two and a group of three latches, and finally, two combination-locks, each composed of four of the previously learned single fastenings, were used in the tests. The combination locks demanded that their elements be operated in a fixed order.

We are told that in these experiments each animal at first attacked the box with indiscriminate clawing, but finally settled down to a single habitual method of operating the latch or latches. The formation of this habit was due to "the omission of unnecessary movements and the combination of those required," exactly as described by THORNDIKE in the case of cats. "The steps by which perfection is reached are *very short* and *blindly* taken"[1] (p. 468). Yet despite this gradualness and blindness of the learning process Mr. DAVIS's adult raccoons showed a "*nearly equal facility*" with monkeys in learning to undo fastenings (p. 487), and their curve of learning follows closely the type of those for the higher animals and *for man* (p. 477).

Mr. DAVIS states, "Experience with former fastenings holds over in the case of new ones leading the animal, at least in certain cases, to begin his attack at the place on the surface of the food-box where he has been accustomed to work. (This has been found by THORNDIKE in the case of cats and denied by COLE in the case of raccoons)." This very important conclusion is based, so far as can be judged from a very obscure statement, on only eight reactions of a single animal. During four of these trials the animal stood on his head and clawed where the latch had been. Presumably the vividness of this experience and COLE's remark concerning an easily discriminated fastening have led Mr. DAVIS to say that such performances are denied in COLE's paper, yet on p. 218 of that paper it is distinctly stated that one of COLE's raccoons clawed twice, another four times *at the side of the door where a latch had been in the preceding test.* Nevertheless COLE's results are characterized as "exceptional." Are we to infer that his animals should have clawed eight times instead of six, or that they ought to have stood on their heads while doing so? Notwithstanding the tremendous weight given by Mr. DAVIS to this performance of a single individual, we are told that the raccoons seemed to reach a sort of "generalized manner of procedure" which enabled them to deal more promptly with any new fastening. This half-subjective, half-objective term, "generalized procedure" is vague in the extreme. Does it mean, as pointed out by COLE (p. 218) that "in future new boxes the animals seemed to pick out the new latch and work directly at that as if experience led them to attack movable objects within the box, or else objects which gave a click or other sound when operated?" COLE continues: "These facts, with others to be mentioned, indicate, I think, that the raccoon's learning to operate a latch includes something more than a mere mechanical coupling up of a certain instinctive act with a given situation." KINNAMAN (p. 122) says more boldly, "It looks very much like the possession of a general notion fairly well represented by projecting-thing-has-something-to-do-with-it, and so they attacked the projecting thing and not something else." If the conclusion as to "generalized procedure" does mean this, it seems to contradict point blank the conclusion based on the special procedure of the single animal which clawed eight times at the side of the door where a bolt had been in the preceding test. Probably THORNDIKE would be first to protest against confirmation by the exceptional behavior, super- or sub-normal of a single animal, when this behavior was contradicted by other records.

[1] Italics are the reviewer's.

According to the author, "Perception of the essential relations, if present at all, is dull and stupid in the last degree" (p. 468). Yet "there is an evident ability to respond *to small differences in complex relations*. How far the perception of such relations really enters is, however, at present in doubt" (p. 477). Practical denial is weakened to doubt within a few pages, because of the contrast between what one raccoon did and what they all did. However, the study of the perception of *relations* in animals is a new field. It is enough for most workers at present if they can *prove* the perception of an easily discriminated object and say with certainty whether it is a visual, olfactory, or tactile perception. Perhaps, however, Mr. DAVIS means that the raccoons did not perceive the fastenings, though they responded to small differences in them, for his confirmation of THORNDIKE commits him to the latter's statement (p. 80) that "the loop is to the cat what the ocean is to a man when thrown into it, when half asleep."

This apparent conflict between conclusions is evident throughout the paper and it seems to be due to the old discrepancy between isolated observations and the final and impersonal result of systematic records, a conflict which THORNDIKE tried hard to terminate.

In working the combination locks the animals learned "order" and "amount of effort" at somewhat different rates. "The table seems to show that the memory of the order is more readily perfected than that of the muscular adjustment required for each particular locking device" (p. 469). Does this mean two types of "memory," as is indicated by this quotation, or merely two rates of habit forming, or a type of memory and a case of habit forming, which we should expect to develop at two different rates? Varying the locking devices would doubtless have explained the phenomenon or analyzed it for us. If it is due to mere defect in method, varying the device would have eliminated it. So the observation seems significant for future tests. If we can find two widely divergent rates, we may find a distinction between habit and association. KINNAMAN, who tested monkeys with combinations very similar to those used by Mr. DAVIS, does not call our attention to any difference in the rate of learning "order" and "amount of effort." This adds interest to Mr. DAVIS's observation.

An excellent table is given of the first forty trials of raccoon No. 1 with the single fastenings and groups. The generalized curves are too greatly reduced to be of much value. A curve for KINNAMAN's monkey's is given, but KINNAMAN counted the entire time "no matter whether the monkey was before or behind the box, whether prancing around it or jumping up and down on top of it, so long as he was trying to open it. Some of these efforts were in nowise directed toward the latch" (KINNAMAN, p. 115). DAVIS, on the contrary recorded the time during which the animal was *in contact* with the locking device (DAVIS, p. 465). Surely this difference must be taken into account in valuing his conclusions that the raccoons show a nearly equal facility with KINNAMAN's monkeys in learning to undo fastenings, and that "the monkeys would seem to be a little less clever at the start" (p. 476). COLE had previously concluded that "in the rapidity with which it forms associations the raccoon seems to stand almost midway between the monkey and the cat, as shown by the numerical records for those animals. In the complexity of the associations it is able to form it stands nearer the monkey" (COLE, p. 261). His method of timing agreed with KINNAMAN's and with THORNDIKE's, though his animals were young and exhibited "play trials." Mr. DAVIS's method of timing and his failure to mark intermissions in practice precluded the presentation of records of "play trials."

Although Mr. DAVIS finds marked differences in learning due to practice effects, we are nowhere told in what order he used the several fastenings. This is a most serious omission and greatly impairs the value of his paper for comparative purposes. For example, in one case he tested a raccoon on fastening No. 3, then passed to No. 9, and in the latter case finds remarkable stupidity. Now if fastenings No. 9 were really the ninth fastening this animal had tried, the result is new and unusual, but if it is the fourth, third, or second fastening, the animal's behavior agrees closely with that of other raccoons, and entirely loses its marvelous character.

Further, we read that while all the raccoons were "fully grown" when received, yet distinct differences in learning were found between the younger and the older animals. We should expect, therefore, a statement of the approximate age of each animal when received. Instead we are offered what seems to be the most useless possible statement, namely, "the approximate age of each animal" *at death or escape or for the summer of 1907*, some months after his work was completed (cf. pp. 462, 448). This defect greatly lessens the value of the experimental records. Fortunately reparation can be made by the publication of the date at which each animal was received, the length of time during which it was tested, and the order in which the tests were given. This addendum ought certainly to appear.

Finally, so far as we can learn from the paper, only one box was used and the latches were all so fastened to it as to be most inconspicuous. This, of course, tends to stamp in the box feature of the situation, and, therefore to make the animal more dependent on kinæsthetic sensations. Varying the position and size of boxes and doors gives control experiments which rarely fail to modify an investigator's first conclusions.

The conflict between the author's several conclusions leaves one in doubt whether the raccoons are, in intelligence, nearer the cats, which possibly have "no images or memories at all," or nearer the monkeys which exhibit even "a low form of general notion."

Color Perception. The color perception of the raccoons was tested, and it is concluded that they do not discriminate colors as such, but depend on differences in brightness alone for their successful reactions. While the tables seem to show this they may prove merely that discrimination of brightness is easier for the animals to make than discrimination of color, for the method employed is very defective.

With the first piece of apparatus used the raccoons could both look into and reach into the vessel which contained the food, and into the five similar vessels which were empty. The experimenter must have felt this disadvantage, for the second piece of apparatus "*did not allow the animal to look into the container in which the food was placed*" (p. 479), but the food could be obtained by reaching through an opening 2 by 1½ inches, in the vertical slides. The food was placed back of one color and when this was moved every other color was given a new position also. Thus with the second device each container could be explored by touch. If the animal reached into a no-food vessel an error of color discrimination was recorded against him. There seems thus no means of distinguishing true errors of color discrimination from the cases in which the animal paid no attention to the colors, except that brightness tests gave better results than color tests. "The two pieces of apparatus were used indifferently." Under these conditions there were 52 per cent of right choices in brightness tests, and 24 per cent in color tests. One of the four animals made 40.7 per cent of right choices in the color tests. This fact is ascribed to brightness differences in the colors used. It must be remembered that by chance alone the animals would have made 17 per cent of right colors. It seems possible, therefore, that both averages are too low, due to the steady pressure of an instinctive impulse, for the reviewer has used the first piece of apparatus and found that *apparently the raccoons could not pass a single food container without both reaching into it and looking into it.* Instead, the animal would go to one end of the row of vessels, explore the first one carefully both by touch and sight, then the next, and so on until the vessel with food in it was found; then it would go on in the same way to the end of the row, and back again, rarely skipping a single vessel. The raccoon has, then, a very strong instinctive impulse to reach into and to look into all sorts of openings.

Imitation. No certain cases of imitation were discovered by the author.

Comparisons. Mr. DAVIS "correlates" his results with those of BERRY on the white rat and of COLE on the raccoon. BERRY very properly compares the behavior of a rat which learns by trial and error with one given an opportunity to learn by imitation and concludes, "It seems to me that we ought to be able to say *a priori*, in the light of these facts, that no ordinary rat would be able to open a door by pulling a string, simply from having seen another do it, without first making a number of random movements." To this Mr. DAVIS replies, "It is upon such a slender basis that Mr. BERRY infers imitation" (DAVIS, p. 483). This seems quite unfair. It is upon no such basis that BERRY infers imitation, but upon repeated experiments in which the imitator developed a tendency (not present before) to pull a knot after seeing another rat pull it many times. On such experiments as a "basis" with most carefully arranged control tests, which proved that the tendency was due to example alone, Mr. BERRY makes the very conservative observation quoted, and first rate confirmation of its truth has been forthcoming. WATSON has demonstrated the immense rôle that kinæsthetic sensations play in the life of the rat, and BERRY has found an advance upon this grade of imitation in other animals. Why invert BERRY's argument to the neglect of his recorded facts?

So BERRY is said to "beg the whole question" because he "lays great stress on the visual sensation as the chief factor in what he calls the final imitative act." Here again it would seem as if BERRY were, instead only allowing proper importance to kinæsthetic sensations, since the rat which learns by trial and error has them, while the imitator must depend first of all on sight. The reviewer is unfamiliar with the behavior of rats, but he can say of raccoons that the experimenter had better lay *very great stress* on their visual sensations of movements for they are almost as skillful muscle readers as the trained dog, "Roger," of recent fame. The danger is that the experimenter will not ascertain until too late how delicate are the movements which the animals can detect by sight.

In a second correlation, what COLE described as due either to imitation or to the presence of visual images in raccoons has, what seems to Mr. DAVIS, a third explanation. As this additional explanation offered is based on a complete misunderstanding of the conditions of the experiment we may pass it by with the remark that perhaps COLE was not clear in his description.

A third correlation with COLE's work, and one already referred to, is of so vital importance in interpreting the behavior of raccoons that all the relevant statements must here be brought together. (1) THORNDIKE (p. 80) is authority for the statement that "cats would claw at the loop or button when the door was open," and "at the place where the loop has been though none was there." (2) COLE, pp. 218, 253 found that raccoons did neither of these things, though they were given opportunities to do both. (3) When the latch is fastened to the side of the box and on the opposite side of the door from that of the immediately preceding test the raccoons did claw a few times (COLE records six times by two young animals, in their earlier trials; DAVIS, apparently, eight times, one adult animal, whether earlier or later trials is not stated) at the place where the latch had been. If Mr. DAVIS had used an easily discriminated fastening like a loop or platform his results would have been more easily comparable with THORNDIKE's and COLE's. He does not say that the raccoon ever clawed at a fastening when the door was open. Surely we must not expect the raccoon, in his early trials, to limit his efforts to projecting objects as he will do in his later trials. Do Mr. DAVIS's records, then, really agree with THORNDIKE's statement that "the loop is to the cat what the ocean is to a man when thrown into it when half asleep" (THORNDIKE, p. 80)? This is a phrase meant to describe about as near a total lack of discrimination as "thought can pump out of itself." Is it consistent with the raccoon's responding to small differences in complex relations? Truly the small number of cases must be recorded, but they must not be overloaded with conclusions.

It is, indeed, an ungracious task to find defects in another's work and in a previous résumé the reviewer largely refrained from it. Yet the first step in assigning the true value to the record of an investigation is to compare interpretations in the same paper. From such comparisons emerges a truth, well known to most investigators. In our early experiments with an animal his behavior suggests ambiguous or contradictory conclusions. This is a hint from the animal that our apparatus or method or both need modification. Watch the animal closely and the direction the modification should take will be suggested. Refuse to modify the method or cling to apparatus already used with some other animal and you remain in the first stages of your work with contradictory or doubtful conclusions on every side. Care in varying the conditions seems to show, however, fairly consistent behavior in any one type of animal.

Finally it appears that cats and monkeys are so widely different in intelligence that it is very difficult to interpret the behavior of raccoons as agreeing with that of both the other animals.

L. W. COLE.

The Journal of
Comparative Neurology and Psychology

VOLUME XVIII | OCTOBER, 1908 | NUMBER 4

A COMPARISON OF THE ALBINO RAT WITH MAN IN RESPECT TO THE GROWTH OF THE BRAIN AND OF THE SPINAL CORD.

BY

HENRY H. DONALDSON,

Professor of Neurology at the Wistar Institute.

WITH PLATES II AND III AND ONE CHART IN THE TEXT.

In this paper it is proposed to present data illustrating the growth of the brain and spinal cord of the albino rat, and also to compare their growth in this animal with that in man.

As a preliminary to this study, it was necessary to determine for the rat the growth curve of the entire body. The observations on this point were published in 1906 under the title "A comparison of the white rat with man in respect to the growth of the entire body" (DONALDSON '06). In that paper it was shown that the growth curve of the rat exhibited all the phases found in the human growth curve, and, further, that the curves for the two sexes were similarly related in both the forms examined. In the present study, therefore, we shall have the advantage of examining the growth of the nervous system in an animal, the general growth curve of which is similar to that of man, and this fact should enhance the significance of the results.

The observations to be presented are unique, as the literature contains no extended record of the growth of the brain and of the spinal cord in any mammal below man. Moreover, the observations on man are open to a good many qualifying criticisms, and it will be most advantageous, therefore, to postpone comment on them until the data from the rat have been presented.

This study of the rat was begun thirteen years ago, and during the interval the records have been accumulating. Throughout this period the rat colony has been composed always of the albino variety of Mus norvegicus (HATAI '07), although occasionally, of course, the colony has been recruited from outside sources. By thus extending the observations over a long period the material has lost perhaps a shade in homogeneity, but on the other hand, something has been gained for the general value of the results.

In collecting the data, I have been assisted by my students and members of the laboratory staff, and I desire on this occasion to acknowledge my indebtedness to those who have worked with me. I am indebted chiefly to my colleague, Dr. HATAI, who has been of the greatest aid in the mathematical treatment of the observations, since without this assistance, the publication of the results must have been delayed still longer.

In presenting the observations, the effort has been made to condense them as much as possible, while at the same time furnishing all the facts which would enable other observers to control the conclusions. To this end, there is printed a complete table of the individual observations. (See General Table at the end of this paper.) All the formulæ and the descriptions of the methods by which the data have been treated are of course given, and in addition, the results have been condensed as usual in the form of curves or tables. The formulæ are given only once in each instance, and then referred to by number when they reappear.

It has not been deemed necessary, however, in view of the general table, to print at the same time correlation tables or the intermediate calculations.

TECHNIQUE.

It is to be expected that during so long a period the methods of observation should have changed somewhat and also should have been improved. In giving the technique for removing the brain and spinal cord and for making the other measurements the methods described are those now used, it being understood that if, in any instance, there was previously a deviation from the procedure which might modify the results, this fact has been taken into account.

The procedure was as follows: Just before feeding time, i.e., when the stomach is comparatively empty, the rat was chloroformed and notes made on the age, sex and any important conditions which might have modified the development of the animal. It was then weighed to the tenth of a gram, and the body length taken with calipers from the tip of the nose to the anus, the animal lying on its side, and being gently extended to its full length.

The measurement was recorded in millimeters as the "body length." From the anus to the tip of the tail, a second measurement was taken, which gives the length of the tail, and this was recorded as "tail length." The animal was then eviscerated.

The spinal cord was next exposed, gently raised by the filum terminale, and the nerve roots clipped away (caudo-cephalad) close to the cord. The division between the brain and the cord was made at the tip of calamus scriptorius or just caudad to it. The skull was then opened from the dorsal side, and the brain removed.

Immediately after removal, the brain was put in one closed weighing bottle, the cord in another, and each weighed separately. The meninges of both brain and cord were left intact. Such blood as they contained, was therefore included in the weight.

After the first weighing, the brain and cord were dried at a temperature between 90° and 95° C. for a week or more, then reweighed, and the percentage of water determined.

In the following pages we shall discuss only the weights of the body, brain and cord, and their relations to one another, leaving for later consideration, the data on body length and on the percentage of water in the brain and the spinal cord.

The observations on the growth of the brain will be presented first.

GROWTH OF THE RAT'S BRAIN.

Table 1 contains 680 records (462 male, 218 female) of the weight of the rat's brain. The changes in the weight of the brain are most readily appreciated when the records are arranged in relation to the increase in the total body weight. Such an arrangement is made in chart 1, plate ii, on which all the individual records that could be entered without confusion are shown. To avoid confusion, however, it was necessary to omit a total of 37 records (26 males, 11 females). The impression given by this chart is therefore somewhat less strong than that warranted by the observations. As can be seen by inspection, the "scatter" of the individual entries is not very great.

The entries on chart 1 suggest that the weight of the brain in the male rats is heavier than in the female. To test this animals of like weight must be compared, and since the females run to only 255 gms., the numbers available for comparison are somewhat reduced (424 males, 218 females). When the data are tabulated, the values given in table 2 show that the weight of the male brain exceeds that of the female in 84 per cent of the groups and is on the average 1.5 per cent greater.

TABLE 1.

Giving the mean observed and calculated weights of the brain and of the spinal cord in the albino rat. Sexes not distinguished. Brain 680 cases; spinal cord 647 cases.

Body weight (in gms.).	Brain weight (in gms.)				Spinal cord weight (in gms.)			
	No. of cases.	Observed.	Calculated by formula [1].	Calculated on 7th root	No. of cases.	Observed.	Calculated by formula [3].	Calculated on 2.7 root.
A	*B*	*C*	*D*	*E*	*F.*	*G.*	*H.*	*I.*
5	58	0.333	0.231		58	0.036	0.033	
15	60	0.977	1.009		58	0.103	0.115	
25	52	1.285	1.244		47	0.180	0.178	
35	53	1.367	1.362		47	0.227	0.228	
45	42	1.441	1.442		42	0.254	0.269	
55	43	1.473	1.502		41	0.283	0.305	
65	32	1.488	1.550		32	0.309	0.337	
75	34	1.559	1.590		32	0.333	0.365	
85	21	1.588	1.625		20	0.362	0.390	
95	22	1.618	1.656		21	0.392	0.413	
105	21	1.674	1.683	1.683	21	0.419	0.434	
115	24	1.683	1.707	1.705	24	0.432	0.453	
125	18	1.706	1.729	1.725	16	0.465	0.471	
135	15	1.763	1.750	1.744	14	0.481	0.488	
145	19	1.718	1.769	1.762	17	0.489	0.504	
155	25	1.754	1.786	1.779	21	0.506	0.519	
165	19	1.771	1.802	1.795	19	0.542	0.533	
175	13	1.827	1.818	1.810	13	0.556	0.546	
185	16	1.781	1.833	1.824	14	0.530	0.559	
195	15	1.803	1.846	1.838	15	0.598	0.571	
205	17	1.809	1.859	1.851	15	0.582	0.582	
215	8	1.873	1.871	1.864	8	0.605	0.593	0.593
225	8	1.813	1.883	1.876	8	0.595	0.604	0.603
235	10	1.890	1.894	1.888	9	0.626	0.614	0.613
245	6	1.900	1.905	1.899	6	0.637	0.624	0.622
255	4	1.900	1.915	1.910	4	0.620	0.633	0.632
265	7	1.921	1.925	1.920	7	0.653	0.642	0.641
275	6	1.983	1.934	1.931	6	0.690	0.651	0.649
285	3	1.950	1.943	1.941	3	0.710	0.660	0.658
295	3	1.950	1.952	1.950	3	0.683	0.667	0.667
305	3	2.117	1.960	1.960	3	0.683	0.675	0.675
315	3	2.083	1.969		3	0.737	0.683	

TABLE 2.

Showing the mean brain weight according to sex. 424 males, 218 females.

Body weight (in gms.).	BRAIN WEIGHT OBSERVED (IN GMS.)				Percentage difference between female brain weight and that of male taken as the standard.	
	No. of cases.	Males.	No. of cases.	Females.	−	+
5	37	0.356	16	0.324	8.9	
15	47	0.971	13	0.998		2.7
25	39	1.306	13	1.222	6.4	
35	39	1.370	14	1.358	0.9	
45	20	1.475	22	1.410	4.4	
55	31	1.489	12	1.432	3.8	
65	26	1.489	6	1.483	0.4	
75	28	1.568	6	1.517	3.2	
85	17	1.591	4	1.575	1.0	
95	9	1.650	13	1.595	3.3	
105	10	1.680	11	1.668	0.7	
115	14	1.664	10	1.710		2.7
125	11	1.750	7	1.636	6.5	
135	12	1.767	3	1.750	0.9	
145	11	1.722	8	1.712	0.5	
155	17	1.768	8	1.725	2.4	
165	9	1.783	10	1.760	1.3	
175	7	1.821	6	1.833		0.6
185	9	1.783	7	1.780	0.2	
195	6	1.833	9	1.783	2.6	
205	11	1.814	6	1.800	0.7	
215	5	1.830	3	1.943		6.2
225			No	females		
235	5	1.930	5	1.850	4.1	
245	2	1.900	4	1.900	0.0	
255	2	1.900	2	1.900	0.0	

Average percentage deficiency in the weight of the female brain 1.5 per cent.

Although the absolute value here given is somewhat greater, this result accords with that of HATAI ('07A) who found the cranial capacity in the male greater by about 0.43 per cent.

BOYCOTT and DAMANT ('08) have found the fatty acids in the male rat to be on the average 4.4 per cent of the entire body weight, and in the female 5.6 per cent. This datum, when applied as a correction to the body weight, would tend to reduce the difference between the brain weights of the sexes. It is further not improbable that the thoracic and abdominal viscera are also proportionally different in the two sexes, and that as a consequence, there is a characteristic sex relation between the weight and length of the body, a condition which would also modify the results which we have obtained by using the crude body weights alone.

In view of these circumstances, it seems permissible in most instances to treat the records for both sexes together, and so the statements which follow are based on the total series of records without distinction of sex, except where such distinction is specially noted.

The theoretical curve, about which the observations cluster, is represented by the continuous line in charts 1 and 3, plates ii and iii, and was found by means of the logarithmic formula [1]

$$y = .569 \log. (x - 8.7) + .554$$

in which y is the weight of the brain in gms. and x the weight of the body in grms. This formula has already been published by HATAI ('08). The values obtained are given in column D of table 1.

The formula [1] just given, was derived in the following manner. Assuming that the weight of the central nervous system is a function of the body weight, we obtain at once the following general expression

$$y = \phi(x)$$

An inspection of the curve of the brain weights, as plotted on the body weights, shows that the rate of growth of the nervous system decreases as the body weight increases. This relation is expressed by the following formula

$$\frac{dy}{dx} = \frac{1}{x} C$$

where C is a constant.
Hence we have

$$dy = \frac{1}{x} C, dx$$

and

$$y = C \int \frac{1}{x} dx = C \log x + A$$

The two constants C and A were determined by the method of the least squares.

When the foregoing formula is applied, the theoretical curve gives a very good graduation of the brain and cord weights for the larger values of x, but fails to adequately represent them for the smaller values of x.

The values obtained by the formula are too high for the brain weight, and too low for the spinal cord weight. In order to meet this difficulty, the constant β empirically determined, has been introduced, and the resulting formula becomes

$$y = C \log (x + \beta) + A$$

in which β is the new constant.

This is the general formula which we have employed for the present work, and it has been found very satisfactory, as will be seen from the tables and charts.

Arranging the rats examined in groups differing by ten grams in body weight, and calculating the mean values of the observed weights of the brain for the mid value of each of these groups, we obtain the curve which is given in chart 3. The mean values (M = the broken line) obtained by so treating the observations, are given in column C of table 1. The table and chart show that the curve based on the means, fits closely with the theoretical logarithmic curve (C = continuous line).

The coefficient of correlation between brain weight and body weight in the case of the 680 records, was determined accordingly to the formula [2]

$$r = \left(\frac{\Sigma (x' y')}{n} - v' v'' \right) \frac{1}{\sigma_1 \sigma_2}$$

(DAVENPORT '04) and is high, being .7639 ±.0108.

For comparison with this result, it may be noted that PEARL ('05) in the case of the total series of Bavarian brains, weighed by BISCHOFF ('80), found the coefficient of correlation between brain weight and body weight to be as follows:

Male	0.1671	±0.0343
Female	0.2260	±0.0412

In the case of Worcester school children 6 to 17 years of age, in which the measurements are more accurate than they could possibly be in the case of BISCHOFF's series, BOAS ('05) found for the following coefficients of correlation between body weight and head measurements:

	Length of head.	Width of head.
Boys	0.43	0.32
Girls	0.41	0.33

Thus in both these series from man, the correlation is less perfect than in the albino rat. However, it must be remembered that the determination of the true body weight, especially when it must be taken postmortem, is much more difficult to make in man than in the rat.

In 535 records (357 male, 178 female) the age of the rat is known, and a similar calculation of the coefficient of correlation between age and brain weight in the male, gives a much smaller value, 0.5177 ± 0.0261, a result which might have been expected from the fact that the body weight of the rat is so easily modified by food and other external conditions. In this case also the coefficient of correlation for man is much less than for the rat.

An examination of either of the charts (1 and 3) shows that between the body weights of 50 to 100 gms. the observations tend to sag below the theoretical curve. For this "sag" no explanation has yet been found. There is of course no cogent reason for expecting that the increase in the weight of the brain must conform to a simple formula, yet it does conform to such a formula, except at the body weights of 50 to 100 gms, and we are therefore justified in expecting that this deviation may sometime be explained.

In order to distinguish between the period of early rapid growth, and the later period of slow growth of the brain, a determination has been made of the limits within which the mature brain changes in weight in a simple relation to the body weight.

Taking as a standard the theoretical brain weight of the heaviest group (315 gms.), as given in column *D*, table 1, and calculating the values for each successive group below this, it is found that as far as the group with a body weight of 105 gms. the brain weight diminishes nearly in proportion to the 7th root of the body weight. The calculated values based on the 7th root of the body weight, are given in column *E* of table 1.

For this distance the straight line formed by the 7th roots of the body weight runs as a chord, of which the logarithmic curve forms the arc. At 105 gms. the chord and arc intersect and a limit is obtained. This point of intersection is arbitrarily chosen to indicate that at which the rapid growth of the brain ceases. Within the limits taken, the maximum deviation of the values obtained by the 7th root of the body weight is 0.5 per cent, the values on the logarithmic curve being considered as the standard. (Compare table 1, columns *D* and *E*, for the body weight group, 185 gms.).

Using the formula of DUBOIS ('98)

$$E : E' :: S^x : S'^x$$

where E and E' are two different encephalic weights, related as a given power of S and S', the corresponding body weights, it appears that the value of x ("the exponent of relation") taken as the 7th root, is in the present instance 0.143. LAPICQUE ('08) has endeavored to show that where individuals of the same species but of different body weights are compared, we should expect the value of x to be 0.25, equivalent to the 4th root of the body weight. To explain why my results do not accord with those obtained by LAPICQUE would require a long critique of his studies on this point. I prefer however to leave this till another occasion, as the introduction of it here would obscure the main point of the present paper.

To explain the essential differences between the rapid and the slow growth of the brain thus indicated, it will be necessary for us first to have information touching the changes in the percentage of water, the chemical composition, the ether-alcohol extract, the degree of medullation and the other histological modifications occurring during growth, so that it is hardly worth while to discuss this question now.

Before leaving the subject of the brain weight, there is still one point more to be presented. It is a familiar fact that rats, even of the same litter and reared together, grow very differently, and therefore at the same age may have widely different body weights. Moreover, either by underfeeding, or by the use of a monotonous and comparatively innutritious diet, animals otherwise normal, may be stunted in their growth.

In the class first mentioned, we have designated those which grew to unusual size as "giants," and those which remained small as "dwarfs." In addition also, we have records on rats experimentally stunted (HATAI '04, '07B and '08).

In the accompanying table 3, there is given a summary of the observed and calculated weights of the brain and spinal cord in these three groups. The calculations are based on the weight of the body at the time of killing, and were made by the use of formula [1] for the brain, and formula [3] for the spinal cord. The individual records used in forming this table 3 do not appear in the general table.

TABLE 3.

Data on special groups; condensed statement; all the measurements are averages.

Group.	No. of cases.		Body weight	Average age in days.	Brain weight. Observed	Brain weight. Calculated.	Difference.	Spinal cord weight. Observed	Spinal cord weight. Calculated.	Difference.
			grams		*gram.*	*gram.*	*per ct.*	*gram.*	*gram.*	*per ct.*
Giants	Males	38	179.8	79	1.728	1.755	−1.5	0.489	0.500	−2.3
	Females	7								
Dwarfs	Males	32	47.2	98	1.333	1.366	−2.5	0.258	0.252	+2.3
	Females	14								
Experimentally stunted	Males	14	92.5	203	1.622	1.620	+0.1	0.401	0.406	−1.2
	Females	12								

On looking at the columns giving the observed brain weights, and comparing these with those calculated, it appears that in the case of the "giants" there is a difference of .027 gm., or 1.5 per cent, in favor of the calculated weight. In the case of the "dwarfs," a difference of .033 gm., or 2.5 per cent, in favor of the calculated weight, and in the case of the rats experimentally stunted, a difference of .002 gm., or 0.1 per cent, in favor of the observed weight. Within the same range of body weights (47.2 to 179.8 gms.), as shown in table 1 and chart 3, the calculated values are on the average 1.6 per cent above the general observed means, so that the special groups in question show on the whole no greater deviation than that found in the larger series. From this it follows that the relations of the brain weight to the body weight are not modified by either excessive or deficient growth under the usual conditions, nor by the deficient growth which may be experimentally induced.

From the foregoing observations on the albino rat we conclude:

1. That for albino rats between 5 and 315 gms. in body weight the mean weight of the brain as observed increases from .333 gm. to 2.083 gms., or 6.2 times, and as calculated, from .231 gm. to 1.969 gm. or 8.5 times.

2. That from birth up to a body weight of about 105 gms. the brain grows rapidly, and after that, more slowly, increasing in the phase of slow growth very nearly as the 7th root of the body weight.

3. That the weight of the brain is closely correlated with the body weight, the coefficient of correlation being 0.7639 ± 0.0108, but less closely correlated with the age, the coefficient of correlation being 0.5177 ± 0.0261.

4. That the relation of the brain weight to the body weight is not essentially modified in either "dwarf" or "giant" individuals, nor in those experimentally stunted.

5. That in these various relations there is no marked distinction between the sexes, although on the average for animals of the same crude body weight, the male has a brain weight 1.5 per cent heavier than that of the female.

The bearing of these results on the corresponding relations as recorded for man will be considered farther on.

We pass next to the observations on the growth of the spinal cord.

GROWTH OF THE SPINAL CORD.

The general table contains 647 records (429 male, 218 female) of the weight of the spinal cord. Chart 2 shows how the individual observations are distributed when these are entered in relation to the body weight in the same manner as in the case of the brain. It has been possible to record clearly on the chart only a fraction of the total records, and so 65 males and 21 females have been omitted.

The determinations of the values according to sex are given in table 4, and show a distinct tendency for the female to have a heavier spinal cord, as the cord is greater in weight in 68 per cent of the groups, and on the average exceeds that of the male by about 2.0 per cent. Although, of course, the absolute differences are here very small, the indications of a difference according to sex are unmistakable.

TABLE 4.

Showing the mean spinal cord weight according to sex.

Body weight gms.	No. of cases.	SPINAL CORD WEIGHT OBSERVED (IN GMS.)			Percentage difference between female spinal cord weight and that of the male taken as the standard.	
		Males.	No. of cases.	Females.	−	+
5	37	0.035	21	0.036		3.0
15	47	0.103	11	0.103		
25	33	0.178	14	0.184		3.3
35	33	0.223	14	0.235		5.4
45	20	0.251	22	0.256		2.0
55	27	0.283	12	0.281	0.7	
65	25	0.307	6	0.315		2.6
75	25	0.331	6	0.338		2.1
85	16	0.360	4	0.370		2.7
95	8	0.395	13	0.390	1.2	
105	10	0.426	11	0.412	3.0	
115	14	0.435	9	0.425	2.3	
125	8	0.473	7	0.447	5.5	
135	10	0.486	3	0.470	3.3	
145	11	0.477	6	0.510		6.9
155	11	0.495	8	0.518		4.6
165	9	0.550	10	0.534	2.9	
175	7	0.521	6	0.593		13.8
185	7	0.504	7	0.556		10.3
195	6	0.590	8	0.605		2.5
205	9	0.592	6	0.567		4.2
215	5	0.590	3	0.630		6.7
225			No	females		
235	4	0.620	5	0.630		1.6
245	2	0.630	4	0.640		1.6
255	2	0.630	2	0.610	3.2	

Average percentage excess in the weight of the female spinal cord, 2.0 per cent.

To discuss this result further observations are required, but pending a well grounded explanation, it must be remembered that WATSON ('05) has shown that the bearing of young has the effect of increasing slightly the weight of the spinal cord in the female, and as many of the females recorded in table 4 had borne young, this is probably one factor in producing the result as it appears in females at or beyond the bearing age. The excess is found, nevertheless, even before puberty.

As in the case of the brain, however, it seems justifiable to treat the sexes together. When so treated, the theoretical curve as shown by the continuous line (*C*) in chart 3 is found by the formula [3]

$$y = .585 (x + 21) - 0.795$$

in which *y* is the weight of the spinal cord and *x* the body weight This formula [3] was derived in the same manner as formula [1].

The means for the weight of the spinal cord, determined as in the case of the brain, follow this curve closely (see chart 3). The

CORRECTION.

On page 357 of THE JOURNAL OF COMPARATIVE NEUROLOGY AND PSYCHOLOGY, Vol. XVIII, No. 4, 1908, Formula (3) is erroneously printed

$$y = .585 \ (x + 21) - 0.795.$$

The correct form is

$$y = .585 \ \text{Log} \ (x + 21) - 0.795.$$

n *H.*
pinal
.8564
f the
body
opos
f the
hole
rela-
ance
apid
in a
far
viest
the
sive
the

values on the logarithmic curve and those determined by the 2.7th root of the body weight become identical. As in the case of the brain, we consider this point of intersection of the two lines to mark the cessation of rapid growth. As far down as the 205 gms. group, then, the weight of the spinal cord is in a simple relation to that of the body weight. Using this fact as a criterion, we may look upon the earlier growth of the spinal cord up to the 205 gms. group as rapid, while after that it is slow.

As in the case of the brain, so in the spinal cord, the variations in the growth of the body which produce "giants" or "dwarfs," or the stunting which may be brought about experiabout 2.0 per cent. Although, of course, the absolute differences are here very small, the indications of a difference according to sex are unmistakable.

TABLE 4.

Showing the mean spinal cord weight according to sex.

Body weight gms.	No. of cases.	Spinal cord weight observed (in gms.)			Percentage difference between female spinal cord weight and that of the male taken as the standard.
		Males.	No. of cases.	Females.	
5					
15					
25					
35					
45					
55					
65					
75					
85					
95					
105					
115					
125					
135					
145					
155					
165					
175					
185					
195					
205					
215					
225					
235					
245					
255					[illegible]

Average percentage excess in the weight of the female spinal cord, 2.0 per cent.

To discuss this result further observations are required, but pending a well grounded explanation, it must be remembered that WATSON ('05) has shown that the bearing of young has the effect of increasing slightly the weight of the spinal cord in the female, and as many of the females recorded in table 4 had borne young, this is probably one factor in producing the result as it appears in females at or beyond the bearing age. The excess is found, nevertheless, even before puberty.

As in the case of the brain, however, it seems justifiable to treat the sexes together. When so treated, the theoretical curve as shown by the continuous line (C) in chart 3 is found by the formula [3]

$$y = .585\,(x + 21) - 0.795$$

in which y is the weight of the spinal cord and x the body weight. This formula [3] was derived in the same manner as formula [1].

The means for the weight of the spinal cord, determined as in the case of the brain, follow this curve closely (see chart 3). The numerical values for the means are given in table 1, column H.

The coefficient of correlation between body weight and spinal cord weight is still higher than that for the brain, being 0.8564 ± 0.0071. As in the case of the brain, there is a "sag" of the observed means below the theoretical curve, between the body weights of 50 and 100 gms. and what has been stated apropos of this on p. 352 applies to the cord also.

A moment's inspection of chart 3 shows that the growth of the spinal cord differs from that of the brain in being on the whole more rapid, and also longer continued. The details of the relations will be taken up later, but the point of importance at this moment is that from the longer continued rapid growth it follows that the increase in the weight of the cord in a simple fixed relation to the body weight does not extend as far down the curve as in the case of the brain. From the heaviest group (315 gms.), the mean cord weight of which is taken as the standard, the weight of the cord diminishes in each successive group according to the 2.7th root of the body weight, until the 205 gms. group is reached, when the values on the logarithmic curve and those determined by the 2.7th root of the body weight become identical. As in the case of the brain, we consider this point of intersection of the two lines to mark the cessation of rapid growth. As far down as the 205 gms. group, then, the weight of the spinal cord is in a simple relation to that of the body weight. Using this fact as a criterion, we may look upon the earlier growth of the spinal cord up to the 205 gms. group as rapid, while after that it is slow.

As in the case of the brain, so in the spinal cord, the variations in the growth of the body which produce "giants" or "dwarfs," or the stunting which may be brought about experimentally, do not modify essentially the relations of the spinal cord to the body, so that the weight of the cord as calculated by the formula [3] corresponds closely with that observed (see table 3).

From the foregoing observations we conclude therefore:

1. That for albino rats between 5 and 315 gms. in body weight, the mean weight of the spinal cord as observed, increases from .036 gm. to .737 gm, or 20.4 times, and as calculated from 0.33 gm. to .683 gm. or 20.6 times.

2. That from birth to a body weight of about 205 gms. the spinal cord grows rapidly, and after that more slowly, increasing in this phase of slow growth nearly as the 2.7th root of the body weight.

3. That the weight of the spinal cord is closely correlated with the body weight, the coefficient of correlation being 0.8564 ± 0.0071.

4. That the relation of the spinal cord weight to the body weight, is not essentially modified in either "dwarf" or "giant" individuals, nor in those experimentally stunted.

5. That in these various relations there is no marked distinction between the sexes, although on the average, the female spinal cord is about 2 per cent heavier than that of the male. This difference probably depends in part on the effect of the bearing of young.

THE ENTIRE CENTRAL NERVOUS SYSTEM.

While a detailed discussion of the weight relations of the entire central nervous system of the albino rat is hardly necessary, in view of what has already been presented concerning the brain and the spinal cord, nevertheless one or two points call for consideration.

The values for the entire central nervous system are entered in table 5, in which the sum of the values for the brain and the spinal cord are given both as observed and as calculated by the formulæ [1] and [3]. The totals for the entire series of groups agree closely, the observed being 0.2 per cent less than that calculated by the formulæ. By dealing with the entire system, we avoid any error which might depend on variations in the point of separation between the brain and the spinal cord.

On determining the period of rapid growth for the entire nervous system and using the same general procedure as before (see pp.

TABLE 5.

Weight of the central nervous system in the albino rat, given in mean values. Calculations according to the formulæ [1] and [3] and the 5th root of the body weight. The heaviest group, 315 gms., is taken as the standard for the calculation according to the 5th root, and at 135 gms., the values by the 5th root and the logarithmic curve coincide.

Body weight. (gms.)	No. of cases.		Weight of the central nervous system. (in gms.)		
				Calculations	
	Br.	Cd.	Observed.	By formulæ [1] and [3].	By $\sqrt[5]{\ }$.
A.	*B.*	*C.*	*D.*	*E.*	*F.*
5	58	58	0.369	0.264	
15	60	58	1.080	1.124	
25	52	47	1.465	1.421	
35	53	47	1.594	1.590	
45	42	42	1.695	1.711	
55	43	41	1.756	1.807	
65	32	32	1.797	1.887	
75	34	32	1.892	1.955	
85	21	20	1.950	2.015	
95	22	21	2.010	2.068	
105	21	21	2.093	2.116	
115	24	24	2.115	2.160	
125	18	16	2.171	2.201	
135	15	14	2.244	2.237	2.238
145	19	17	2.207	2.272	2.270
155	25	21	2.260	2.305	2.301
165	19	19	2.313	2.335	2.329
175	13	13	2.383	2.364	2.357
185	16	14	2.311	2.392	2.384
195	15	15	2.401	2.416	2.409
205	17	15	2.391	2.441	2.433
215	8	8	2.478	2.464	2.457
225	8	8	2.408	2.486	2.479
235	10	9	2.516	2.508	2.501
245	6	6	2.537	2.528	2.521
255	4	4	2.520	2.548	2.542
265	7	7	2.574	2.567	2.561
275	6	6	2.673	2.585	2.581
285	3	3	2.660	2.602	2.599
295	3	3	2.633	2.620	2.617
305	3	3	2.800	2.636	2.635
315	3	3	2.820	2.652	——

352 and 357), it appears that the weight of the central nervous system diminishes in proportion to the 5th root of the body weight, as far as the 135 gms. group. The rapid growth of the entire central nervous system ceases then according to this criterion, at 135 gms. of body weight. The sum of the values determined in

accordance with the 5th root of the body weight (i.e., from the body weights of 135 to 305 gms.) is found to be 0.2 per cent less than the sum of the corresponding values determined by the formulæ, all of which indicates substantial agreement between the three series.

The following table 6 gives the weight of the central nervous system according to sex. In 68 per cent of the groups the male is the heavier, and the values for the male exceed those for the female by 0.8 per cent. The difference is slight, but as already pointed out it seems probable that it is real.

TABLE 6.

Weight of the central nervous system according to sex. Those male groups which are heavier are marked with a star (*).

Body weight.	Weight of central nervous system observed.	
	Male.	Female.
grams.	*grams.*	*grams.*
5	0.391*	0.360
15	1.074	1.103
25	1.484*	1.406
35	1.593*	1.593
45	1.726*	1.666
55	1.772*	1.713
65	1.796	1.798
75	1.899*	1.855
85	1.951*	1.945
95	2.045*	1.985
105	2.106*	2.070
115	2.099	2.135
125	2.223*	2.083
135	2.253*	2.220
145	2.199	2.222
155	2.263*	2.243
165	2.333*	2.204
175	2.342	2.426
185	2.287	2.336
195	2.423*	2.388
205	2.406*	2.367
215	2.420	2.573
225		No females
235	2.550*	2.480
245	2.530	2.540
255	2.530*	2.510

WEIGHT RELATION OF THE BRAIN TO THE SPINAL CORD.

The weight relations of the brain and spinal cord change with age. Using the calculated values, it appears that for a very short period after birth the brain grows more rapidly than the spinal cord (see body weight 15 gms., table 7) but at about the body weight of 15 to 25 gms., the cord begins to grow more rapidly than the brain, and from that time on the ratio of the brain weight

TABLE 7.

Showing the ratio of the weight of the spinal cord to that of the brain in the albino rat.

Body weight.	Calculated by formulae [3] and [1].		Ratio of spinal cord weight to brain weight.
	Cord weight.	Brain weight.	
grams.	*gram.*	*gram.*	
5	0.033	0.231	7.05
15	0.115	1.009	8.74
25	0.178	1.244	6.99
35	0.228	1.362	6.01
45	0.269	1.442	5.35
55	0.305	1.502	4.92
65	0.337	1.550	4.60
75	0.365	1.590	4.36
85	0.390	1.625	4.17
95	0.413	1.656	4.01
105	0.434	1.683	3.88
115	0.453	1.707	3.77
125	0.471	1.729	3.67
135	0.488	1.750	3.59
145	0.504	1.769	3.51
155	0.519	1.786	3.44
165	0.533	1.802	3.38
175	0.546	1.818	3.33
185	0.559	1.833	3.28
195	0.571	1.846	3.23
205	0.582	1.859	3.19
215	0.593	1.871	3.15
225	0.604	1.883	3.12
235	0.614	1.894	3.09
245	0.624	1.905	3.05
255	0.633	1.915	3.03
265	0.642	1.925	3.00
275	0.651	1.934	2.97
285	0.660	1.943	2.94
295	0.667	1.952	2.89
305	0.675	1.960	2.87
315	0.683	1.969	2.85

diminishes. The observed values indicate the same relation, although in a less marked degree. The accompanying table 7 shows this ratio, determined for each of the several body weight groups.

The phase in which the brain grows relatively more rapidly than the spinal cord is found also in man, but so far as the scant human records go it appears to pass over into the phase of the less rapid relative growth of the brain some time before birth (MIES '93). Owing to the immaturity of the rat at birth, however, this earlier phase is just recognizable as a post-natal phenomenon in that animal.

The mean values for the weight of the spinal cord at given brain weights are represented by the dots in chart 4, and are given under "observed" in table 8. In this same table under "calculated" are given also the values for the weights of the spinal cord as determined by calculation. These latter values were obtained in the following manner. Transposing formula [1] to the form

$$\text{Log.}\ (x - 8.7) = \frac{y - .554}{.569} \quad \text{or} \quad x = 8.7 + 10^{\frac{y - .554}{.569}}$$

it was possible to calculate the body weights which belonged to the brain weight values used in this table. From the body weights thus obtained the corresponding weights for the spinal cord were calculated by formula [3]. The continuous line in chart 4, p. 363, is the curve based on these values, and the inspection of the chart shows that the observed values and those calculated agree very closely.

On correlating the brain weight with the observed spinal cord weight, using weight groups for the brain differing by 0.1 gms., the coefficient of correlation is found to be 0.8787 ± 0.006, which is higher than for any relation which we have had occasion to determine. PFISTER ('03), in his studies on the spinal cord and brain in children, has also noted the close correlation between these two portions of the central nervous system.

From the foregoing study of the weight of the entire central nervous system and of the relation of the brain weight to that of the spinal cord, we conclude:

1. That from the 5 to the 315 gms. weight group, the entire central nervous system as observed, increases in weight from

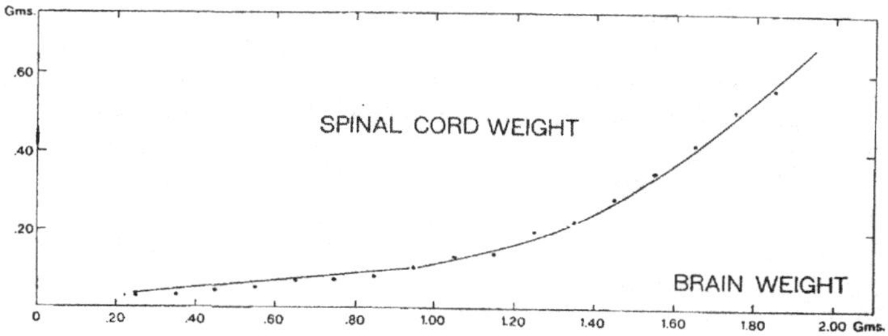

CHART 4. The base line represents the brain weights from 0.20 to 2.00 grams. The corresponding values for the spinal cord weights are measured on the ordinates, and shown by a theoretical curve (continuous line) determined by formulæ [1] and [3] and also shown by a series of dots indicating the observed mean values (see table 8).

TABLE 8.

Showing the mean weight of the spinal cord both observed and calculated through the range of brain weights in groups differing in brain weight by 0.1 gm.

Brain weight.	SPINAL CORD WEIGHT.	
	Observed.	Calculated.
grams.	*gram.*	*gram.*
0.25	0.030	0.036
0.35	0.030	0.046
0.45	0.043	0.053
0.55	0.050	0.067
0.65	0.070	0.079
0.75	0.070	0.085
0.85	0.076	0.094
0.95	0.103	0.106
1.05	0.131	0.124
1.15	0.139	0.148
1.25	0.199	0.180
1.35	0.223	0.222
1.45	0.282	0.274
1.55	0.345	0.336
1.65	0.422	0.408
1.75	0.505	0.488
1.85	0.561	0.575
1.95	0.614	0.666
2.05	0.710	
2.15	0.700	

0.369 gm. to 2.820 gms., or 7.9 times, and as calculated, from 0.264 gm. to 2.652 gms., or 10 times.

2. That for the central nervous system, the period of rapid growth extends up to the 135 gms. group, after which the system increases nearly regularly in proportion to the 5th root of the body weight.

3. When the mean weights of the central nervous system are determined according to sex, it appears that in 68 per cent of the records the weight in the male exceeds that in the female, but on the average the difference is small, amounting to only 0.8 per cent.

4. From birth to the 15 to 25 gm. group, the brain grows more rapidly than the spinal cord, but after that the spinal cord grows faster, so that using the calculated values, the ratio drops from 8.74 at 15 gms. to 2.85 at 315 gms.

5. The coefficient of correlation between the weight of the brain and the weight of the spinal cord is very high, being 0.8787 ± 0.006.

COMPARISON OF THE GROWTH OF THE BRAIN AND OF THE SPINAL CORD IN THE ALBINO RAT, WITH THEIR GROWTH IN MAN.

With the foregoing data in hand, it is possible to make some comparisons between man and the albino rat in respect to the growth of the brain and of the spinal cord.

Five points will be examined:

1. The form of the growth curve according to age.
2. The relative increase in the weight of the brain during the phase of rapid growth.
3. The time taken for this increase.
4. The cessation of the rapid growth of the brain in relation to puberty.
5. Weight of the brain and the spinal cord as modified by sex.

To determine in the albino rat the growth of the brain with age, the calculation according to the formula [1] was made for the brain weight of each of the age groups used by me in constructing the growth curve for the entire body (DONALDSON '06). The results obtained are given in the following table 9 and are plotted in chart 5, plate iii.

TABLE 9.

Showing the weight of the brain at different ages in both sexes of the albino rat. Data on body weight taken from DONALDSON '06, and brain weight determined by formula [1].

Age in days.	MALES.		FEMALES.	
	Body weight.	Brain weight.	Body weight.	Brain weight.
	grams.	*calculated.*	*grams.*	*calculated.*
0	5.4	0.2589	5.2	0.2444
1	5.6	0.2744	5.5	0.2665
2	5.8	0.2909	5.7	0.2825
3	6.3	0.3367	6.2	0.3276
4	6.9	0.4087	6.5	0.3592
5	8.3	0.5197	7.7	0.4971
6	9.1	0.5887	8.5	0.5369
7	9.2	0.5938	8.7	0.5540
8	10.4	0.6851	10.6	0.7126
9	11.3	0.7902	11.1	0.7703
10	12.2	0.8636	12.1	0.8564
11	13;3	0.9311	12.8	0.9027
12	14.8	1.0008	15.1	1.0128
13	15.3	1.0203	15.1	1.0128
14	15.2	1.0186	15.6	1.0313
15	16.5	1.0616	17.7	1.0970
17	17.8	1.0906	19.2	1.1351
19	19.5	1.1420	20.6	1.1660
21	21.2	1.1782	22.6	1.2044
23	22.9	1.2097	24.9	1.2422
25	25.3	1.2483	27.4	1.2776
27	27.4	1.2776	30.0	1.3099
29	29.5	1.3040	31.4	1.3256
31	31.8	1.3299	32.9	1.3414
34	34.9	1.3611	35.7	1.3685
37	37.8	1.3870	39.5	1.4010
40	42.2	1.4218	43.7	1.4326
43	46.3	1.4503	47.9	1.4542
46	50.5	1.4765	52.0	1.4852
49	56.7	1.5106	57.7	1.5157
52	62.5	1.5388	62.9	1.5407
55	68.5	1.5149	68.4	1.5646
58	73.9	1.5863	74.6	1.5890
61	81.7	1.6142	78.4	1.6028
64	89.1	1.6381	85.8	1.6278
67	99.3	1.6676	96.0	1.6588
70	106.6	1.6868	99.8	1.6690
73	113.8	1.7066	105.6	1.6842
76	121.3	1.7213	110.4	1.6858
79	128.2	1.7360	118.8	1.7158
82	135.0	1.7497	124.7	1.7287
85	143.8	1.7664	131.5	1.7428
88	148.4	1.7747	136.0	1.7516
92	152.3	1.7815	139.8	1.7589
97	160.0	1.7943	146.3	1.7709
102	168.8	1.8083	153.1	1.7828

TABLE 9—Continued.

Age. in days.	Males.		Females.	
	Body weight.	Brain weight.	Body weight.	Brain weight.
	grams.	*calculated.*	*grams.*	*calculated.*
107	177.6	1.8215	155.8	1.7874
112	183.8	1.8305	161.4	1.7966
117	191.4	1.8409	168.0	1.8069
124	197.3	1.8488	172.6	1.8141
131	202.5	1.8555	181.0	1.8265
138	209.7	1.8645	185.0	1.8322
143	218.3	1.8749	186.6	1.8344
150	225.4	1.8831	188.2	1.8366
157	227.0	1.8849	188.0	1.8363
164	231.4	1.8899	189.5	1.8383
171	235.8	1.8947	192.2	1.8420
178	239.4	1.8986	197.0	1.8484
185	239.8	1.8990	200.0	1.8523
192	No	Males	202.2	1.8551
216	252.9	1.9126	No	Females
256	265.4	1.9250	No	Females
365	279.0	1.9377	226.4	1.8843
730	308.5	1.9633		

As the calculated values of the weight of the brain vary in the same sense as the body weights, it appears that from birth to maturity the curves for the brain weight in the two sexes are related to each other as are the body weights, and thus the brain weights in the females between the ages of 14 and 52 days are heavier than those in the males. This relation should be confirmed by direct observation before any value is attached to it. On the other hand, after the period of most rapid growth the brain weight of the male is always the heavier, because at like ages the male body weight exceeds that of the female. This portion of the curves is therefore like that in man, and for the same reason.

As previously pointed out, we consider the period of the relatively rapid growth in the brain to cease when it reaches the point where the further increase in weight is approximately in proportion to the 7th root of the body weight. This occurs at about 70 days in the males, and 73 days in the females.

In order to compare the amount of increase in the weight of the brain between birth and maturity in man with that in the albino rat, it is necessary to bring the data on man into the same form as that for the rat. Taking as a basis the data compiled by VIER-

ORDT ('90), we give in the following table his observed values, and also the values obtained from a smoothed curve based on these data. The curve represented by VIERORDT's observations is given in "The growth of the brain" (DONALDSON '95) and also in the American Text-Book of Physiology (DONALDSON '01). The smoothed curve which passed with the least deviation through the rough curve has been drawn, and then the values given by the smoothed curve were determined for each year. These values are entered in table 10 under "calculated" in columns *D* and *F*, and from these, of course, the smoothed curves can be reconstructed.

To so reduce the values of the human records as to make them comparable when plotted with those from the rat, it was necessary to divide them by 700. In chart 5 the weight of the human brain thus reduced is compared with that of the rat, the span of human life being taken as thirty times that of the rat, and the time intervals entered accordingly. When thus plotted, it is seen that the two curves are similar in form. Moreover, if we determine the age at which the rapid growth of the brain ceases in the rat, which is at a body weight of 105 gms. (see p. 352), it is found to fall at about 70 days in the male, and 73 days in the female and it is evident that the average date, 72 days, corresponds very nearly with six years in man.

Between birth and 72 days the rat brain has increased in weight (mean of both sexes combined) from .02517 gm. to 1.6823 gm. or 6.3 times, while in the corresponding interval, the human brain has increased in weight (mean of both sexes combined) from 383 to 1215 gms., or 3.2 times.

We know, however, that the rat is born relatively much less mature than the child. The comparison as it stands, is therefore hardly fair. If we determine for the rat the initial brain weight, which at 72 days would give an increase similar to that observed in man (3.2 times) we find the required weight to be .525 gm., or approximately the weight of the rat's brain between five and six days. Therefore, between the age of five and six days—at which time the rat's brain is certainly more comparable with the human brain than at birth—and 72 days, the brain of the rat increases in weight in the same proportion as does the human brain between birth and six years.

This relation, although derived from a treatment of the data which is admittedly rough, is very suggestive, but it will be hardly

TABLE 10.

To show the increase in the brain weight of man with age. Encephalon weighed entire with pia. (Compiled by VIERORDT).

Age.	Males.			Females.		
	No. of cases.	Brain in gms.		Brain in gms.		No. of cases.
		Observed.	Calculated. (Donaldson)	Observed.	Calculated. (Donaldson)	
A	*B*	*C*	*D*	*E*	*F*	*G*
0 months	36	381	381	384*	384	38
1 year	17	945	945	872	850	11
2	27	1025	1085	961	950	28
3	19	1108	1175	1040	1060	23
4	19	1330	1225	1139	1140	13
5	16	1263	1290	1221	1180	19
6	10	1359	1325	1265	1205	10
7	14	1348	1360	1296	1220	8
8	4	1377	1380	1150	1235	9
9	3	1425	1390	1243	1245	1
10	8	1408	1400	1284	1250	4
11	7	1360	1410	1238	1255	1
12	5	1416	1415	1245	1257	2
13	8	1487	1415	1256	1259	3
14	12	1289	1415	1345	1260	5
15	3	1490	1415	1238	1260	8
16	7	1435	1415	1273	1260	15
17	15	1409	1414	1237	1258	18
18	18	1421	1413	1325	1255	21
19	21	1397	1412	1234	1253	15
20	14	1445	1410	1228	1251	33
21	29	1412	1408	1320	1249	31
22	26	1348	1404	1283	1247	16
23	22	1397	1400	1278	1245	26
24	30	1424	1397	1249	1243	33
25	25	1431	1395	1224	1240	33
Total no. of cases........................ 415				Total no. of cases.............. 424		

*It appears probable that the weight here given in table 10 for the female brain at birth is too high (Handmann '06, S. 35); but it seemed best to hold to one table in this instance, and not attempt to revise any single entry in it. Any lowering of the human brain weight at birth would tend to make the weight relations in man even more similar to those found in the rat than the calculations given farther on show them to be.

profitable to discuss it until we learn both for man and the rat, at what time cell-division in the brain ceases, and so can determine when the increase in weight becomes the expression of simple enlargement alone. Nevertheless, it is of interest to note that while nearly the same fraction of the span of life is used for the rapid growth process in both forms, the actual period required by man is thirty times that for the rat.

The observation as it stands, represents a special instance of the phenomenon already observed by BUNGE ('02) and RUBNER ('08 and '08A), that during the phase of rapid growth, immediately following birth, the smaller mammals double their body weight in a much shorter period of time than does man. The present observation has moreover the interest of applying to an organ in which it is probable that cell division has nearly ceased, so that the increase in weight during this period, is due almost entirely to the mere enlargement of the elements which are for the most part neurones.

It might be urged that to complete the demonstration, it should be shown that during this interval, the same percentage of the limiting brain weight had been attained by both forms. The facts are these. The brain of the rat has a weight (calculated by formula [1]) at 72 days of 1.6823 gm., and at 303 days (corresponding to 25 years in man) a weight of 1.9020 gm., so that at 72 days it has attained approximately 88.4 per cent of its limiting weight.

On the other hand, the human brain (mean of both sexes) has at six years a weight of approximately 1215 gms., which according to the value in table 10, is 92.2 per cent of its limiting weight at 25 years, and 90.8 per cent of its calculated maximum weight at 16 years.[1] Thus the human brain has attained a greater fraction of both its limiting and its maximal weight. The discrepancy seems to depend mainly on the fact that while the early phases of body growth in the rat are similar to those in man, yet the rat continues to grow for a relatively longer period after maturity than man does, and at the same time, the weight of the brain and spinal cord continues to increase with that of the body. This difference in the later phase of body growth therefore is a point which needs to be investigated. At the same time although the early attainment of the maximum weight in man followed by a slow decline in weight through later life, as brought out by several investigators and specially studied by PEARL ('05), may be a normal biological phenomenon, yet it must be frankly admitted that the human records as they stand, are distinctly influenced by the factors represented by the peculiarities of the "hospital population" on the one hand, and the effect of disease, especially

[1] It may be noted in passing that HANDMANN ('06, S. 14-17) finds the maximum brain weights in both sexes between 15 to 17 years.

chronic disease, on the other (GREENWOOD '05, GLADSTONE '05 and BLAKEMAN '05).

Turning next to the relations of puberty to the completion of the rapid growth of the brain, it is worthy of note that the completion of rapid growth in the albino rat at about 72 days, coincides in this animal with puberty, which appears at 65 to 75 days. In man, on the other hand, it precedes puberty from 6 to 9 years. Any interpretation of this difference must await a determination of the finer anatomy of the brain in the two forms, at the time of puberty.

Passing to the spinal cord, much less can be done in the way of comparison owing to the small amount of data on the spinal cord of man. The human spinal cord at birth has a mean weight of about 3.2 gms. (MIES '93) and at maturity of 27 to 28 gms. (ZIEHEN '99). The body weight, length of trunk and sex probably all have an influence on the weight of the cord, but we do not know how much (PFISTER '03).

Using the foregoing values (3.2 gms. and 27.5 gms.) it appears that between birth and maturity the human spinal cord increases in weight about 8.6 times. Taking the calculated weight of the spinal cord in the rat (mean of both sexes) as 0.589 gm. at 303 days (equal to 25 years of human life), we find that the weight which would give an increase of 8.6 times is 0.068 gm. This corresponds to the average weight of the cord between 7 and 8 days, which is nearly the same as the age (5 to 6 days) found for the brain by a like calculation. From this it follows that the cell elements in the spinal cord of the rat enlarge in the same proportion as do those in man, and that these two divisions of the central nervous system in the rat are similarly related to the corresponding parts in man. Calculation shows that the amount of enlargement between birth and maturity is in both forms very nearly 2.5 times greater in the case of the spinal cord than it is in the case of the brain. Expressing this result in terms of neurones, it would mean that the average bulk attained by the neurones of the spinal cord was 2.5 greater than that attained by the nuerones of the brain. The relatively greater weight of the cord of the rat, as compared with the brain, depends of course, on the initial plan of the central nervous system peculiar to that animal.

With regard to the weight of the brain and of the spinal cord as modified by sex, a few words are in place. In the human records, so largely has age been made the basis for the comparison of brain weights, that we have most of us fallen into the habit of thinking of them always in that relation. I wish therefore to emphasize the fact that it is my purpose here to consider the possible influence of *sex* on the weight of the brain and of the spinal cord in animals of *like body weights*, age not being considered.

As has already been shown in the case of the albino rat (pp. 349 and 355) when males and females of like body weight are compared, it is found that the weight of the brain is about 1.5 per cent heavier in the males, and the weight of the spinal cord about 2 per cent heavier in the females. HATAI'S ('07A) studies on the cranial capacities of the male and female rats show even less difference than we have found in the case of the brain itself.

As already pointed out, it seems probable that when the crude body weights of the females are corrected for the excess of fat, and for the relation of stature to body weight, even this difference in the weight of the brain and cranial capacity will be further diminished.

It should be noted moreover that the corrections which would tend to make the brain weights in the two sexes more nearly similar, would also tend to increases the weight of the spinal cord in the female. Such being the relations in the case of the rat, it is of interest to inquire how these matters stand in the case of man. Touching the weight of the brain as correlated with the weight of the body and the body measurements, I will cite only two recent investigations.

BLAKEMAN ('05) on making the necessary calculations, finds that "The English man of the same age, stature, and diametral product as the mean woman, has 1235 gms. brain weight, or only 10 gms. more than the average woman" (1224.90 gms.) and further that "The English woman of the same age, stature and diametral product as the mean man, has 1315 gms. brain weight, or only 13 gms. less than the average man" (1327.69 gms.).

The comparison is far from perfect, and other corrections, the need for which is recognized, would probably further reduce even this small difference.

By a very different procedure LAPICQUE ('08) reaches a conclusion which is quite similar. Taking the values in the following table as given by him,

	BODY WEIGHT.	BRAIN WEIGHT.
Man	66,000 gms.	1360 gms.
Woman	55,000 gms.	1220 gms.

he finds that the 0.56 power of the body weights, gives very nearly the brain weights as observed. The 0.56 power of the body weights represents the relation of the brain weights found by DUBOIS ('98) to subsist between animals of like form, but different species. Leaving aside at this time any discussion of LAPICQUE's general result, I wish merely to point out that the brain weights according to sex, as shown by these data of LAPICQUE, are so related that when the body weight of the female is raised to that of the male, it calls for approximately the same brain weight as is found in the male.

We may conclude, therefore, that in both rat and man the brain weight is nearly the same in both sexes, when the body weights are the same, such small difference as is still found being in favor of the male, but at the same time probably open to further reduction.

If we turn now to the spinal cord, a direct comparison of the weights according to sex is blocked in man by the absence of sufficient data. Some light, however, can be obtained by examining in the case of man the ratio

$$\frac{\text{Brain weight}}{\text{Spinal cord weight}}$$

MIES ('93) gives the following:

Age.	Male.		Female.	
	No. of cases.	Ratio $\frac{\text{B.W.}}{\text{S.C.W.}}$	No. of cases.	Ratio $\frac{\text{B. W.}}{\text{S.C.W.}}$
Birth	10	116.42	11	113.11
Maturity	10	51.33	4	49.47

which shows that in proportion to the brain weight both at birth and at maturity the weight of the spinal cord in the male is less, i.e., gives a higher ratio than in the female.

In a series of eight comparisons, extending in age from one month to 6½ years, and based on 35 males and 38 females, PFISTER ('03) finds the proportional value of the spinal cord weight in each of the eight comparisons, to be less in the male, indicating according to the average of the ratios, about 4 per cent of difference in favor of the spinal cord in the female. From what has been said concerning the weight of the brain and of the spinal cord in the rat when the sexes are compared, it follows that similar relations are found in that animal and the calculations show them.

In view of these facts, and in view of the preceding determinations, that for like body weights, the human male and female have approximately the same weight of the brain, it necessarily follows that where the body weights are alike, the spinal cord in the woman is heavier than in the man; a conclusion which I believe has not been heretofore explicitly stated. It thus appears that in both sexes of man and the albino rat, the relations of the weight of the brain and the spinal cord to that of the body are similar.

From the observations presented in the later portion of this paper, we conclude that man and the rat are similar in the weight relations of their brain and spinal cord, in the form of the growth curves for the brain, in the fraction of the span of life taken for the rapid growth of the brain, and in the proportional development of the brain and cord during this phase.

They differ, however, in the intensity of the general growth processes, which are some thirty times more rapid in the rat than in man, in the relation of the completion of the phase of rapid growth to the appearance of puberty and in the longer continuance of the phase of slow growth in the rat.

Nevertheless, in view of the similarities above named, it appears that by the study of the nervous system of the albino rat, it will be possible to obtain information bearing on certain growth phenomena in man, the direct study of which in the human nervous system is at present impracticable.

GENERAL TABLE.

Mus norvegicus var. albus (albino). The records are grouped according to sex, and in the case of each sex, every record carries its own serial number. In the present table, the records are arranged according to body-weight in two series, Series 1, normal, 458 males, 215 females, Series 2, injected with lecithin (HATAI '03), 4 males, 3 females. Both series were used in forming the special tables. If new observations are published in the future, each new record will bear its own serial number. In case series are to be formed for any purpose which involves the use of new records combined with those previously published, the latter will always bear the serial number given them when they were first printed. By this device, it is hoped that confusion between the new and old records may be avoided.

Series 1. No.	Sex.	Age in days	Weight in gms.			Remarks.
			Body.	Brain.	Cord.	
1	M.	1	3.8	0.2523	0.0242	
2	M.		4.1	0.2092	0.0286	At birth
3	M.		4.3	0.2348	0.0350	At birth
4	M.		4.3	0.2400	0.0310	At birth
5	M.		4.4	0.2092	0.0312	At birth
6	M.		4.5	0.2040	0.0309	At birth
7	M.		4.6	0.2179	0.0330	At birth
8	M.	1	4.6	0.2749	0.0306	
9	M.		4.6	0.2240	0.0270	36 hours
10	M.		4.6	0.2096	0.0342	At birth
11	M.		4.7	0.2150	0.0280	36 hours
12	M.		4.7	0.2310	0.0300	36 hours
13	M.		4.9	0.2052	0.0304	At birth
14	M.		5.0	0.2286	0.0336	At birth
15	M.	1	5.0	0.3111	0.0335	
16	M.		5.1	0.2088	0.0318	At birth
17	M.		5.2	0.2324	0.0316	At birth
18	M.		5.2	0.2336	0.0324	At birth
19	M.		5.3	0 2220	0.0328	At birth
20	M.		5.3	0.2326	0.0328	At birth
21	M.		5.4	0.2034	0.0338	At birth
22	M.		5.4	0.2750	0.0300	38 hours
23	M.		5.5	0.2406	0.0320	At birth
24	M.		5.6	0.2320	0.0350	At birth
25	M.		5.8	0.2578	0.0342	14 hours
26	M.	2	5.8	0.2340	0.0330	
27	M.		5.9	0.2752	0.0342	At birth
28	M.		6.0	0.2710	0.0352	At birth
29	M.		6.1	0.2722	0.0370	At birth
30	M.	2	6.2	0.2340	0.0320	
31	M.		6.4	0.2758	0.0386	At birth
32	M.		6.6	0.2975	0.0409	20 hours
33	M.	10	7.3	0.6982	0.0637	
34	M.	10	7.9	0.7087	0.0658	
35	M.	10	8.2	0.7430	0.0734	
36	M.	5	8.3	0.5050	0.0540	
37	M.	10	8.5	0.7498	0.0750	
38	M.	9	10.7	0.7080	0.0764	

GENERAL TABLE—Continued.

Series 1. No.	Sex.	Age in days.	Weight in gms.			Remarks.
			Body.	Brain.	Cord.	
39	M.	10	10.8	0.8010	0.0744	
40	M.	10	10.9	0.7998	0.0685	
41	M.	10	11.2	0.8494	0.0818	
42	M.	9	11.3	0.7450	0.0730	
43	M.	9	11.7	0.7404	0.0752	
44	M.	10	11.8	0.8560	0.0890	
45	M.	10	11.9	0.8410	0.0701	
46	M.	10	11.9	0.8082	0.0750	
47	M.	10	12.0	0.8600	0.0960	
48	M.	9	12.1	0.7758	0.0784	
49	M.	10	12.3	0.8415	0.0781	
50	M.	10	12.6	0.8271	0.0733	
51	M.	10	12.7	0.8992	0.0975	
52	M.	10	13.4	0.8656	0.0947	
53	M.	9	13.5	0.8010	0.0800	
54	M.	19	13.5	1.0840	0.1250	
55	M.	10	13.5	0.8883	0.0898	
56	M.	10	13.8	0.8978	0.0895	
57	M.	10	13.8	0.8495	0.0825	
58	M.	14	13.8	1.1153	0.1062	
59	M.	14	13.9	1.1099	0.0962	
60	M.	40	14.3	0.9927		
61	M.	19	14.3	1.1580	0.1262	
62	M.	10	14.4	0.9010	0.0895	
63	M.	10	14.4	0.9550	0.1000	
64	M.	19	14.5	1.1812	0.1391	
65	M.	6	14.5	0.8180	0.0820	
66	M.	6	14.8	0.8380	0.0840	
67	M.	12	15.0	1.1020	0.1240	
68	M.	10	15.5	1.0270	0.1060	
69	M.	17	15.8	1.1494	0.1218	
70	M.	65	16.0	1.0389	0.1236	
71	M.	21	16.4	1.0190	0.1132	
72	M.	19	16.5	1.2431	0.1439	
73	M.	38	17.0	1.0990	0.1453	
74	M.	40	17.0	1.1081		
75	M.	41	17.0	0.9695	0.1400	
76	M.	11	17.2	0.9968	0.1010	
77	M.	11	17.7	0.9870	0.0961	
78	M.	15	17.8	1.1200	0.1180	
79	M.	10	18.0	1.1017	0.0962	
80	M.	53	18.0	1.0367	0.1476	
81	M.	58	19.0	1.2039	0.1843	
82	M.	11	19.1	1.0438	0.1050	
83	M.	21	19.5	1.3000	0.1718	
84	M.	19	20.0	1.1184	0.1300	
85	M.	40	21.0	1.2369	0.1745	
86	M.	65	21.0	1.2444	0.1603	
87	M.	45	21.2	1.1661	0.1802	
88	M.	15	21.6	1.2339	0.1158	
89	M.	20	21.9	1.2284	0.1504	
90	M.		22.3	1.2420		

GENERAL TABLE—Continued.

Series 1. No.	Sex.	Age in days.	Weight in gms.			Remarks.
			Body.	Brain.	Cord.	
91	M.	50	24.0	1.2149	0.1881	
92	M.	50	24.0	1.2219	0.1939	
93	M.	20	24.1	1.2764	0.1534	
94	M.		24.6	1.2448	0.1751	
95	M.	20	24.8	1.3336	0.1578	
96	M.	22	25.2	1.2060	0.1690	
97	M.		25.3	1.2440		
98	M.	20	25.5	1.3338	0.1605	
99	M.	21	25.6	1.3790	0.1712	
100	M.	20	25.9	1.3118	0.1694	
101	M.	20	26.0	1.3361	0.1626	
102	M.	86	26.0	1.3734	0.2495	
103	M.	22	26.2	1.3796	0.1806	
104	M.	59	26.4	1.1532	0.1737	
105	M.	21	26.7	1.3330	0.1760	
106	M.	21	27.0	1.3160	0.1700	
107	M.	35	27.0	1.3047	0.1948	
108	M.	45	27.1	1.3083	0.1938	
109	M.	21	27.1	1.3878	0.1734	
110	M.		27.2	1.2940		
111	M.	22	27.5	1.3100	0.1726	
112	M.	20	27.6	1.3156	0.1742	
113	M.	21	28.3	1.4206	0.1856	
114	M.	21	28.4	1.4458	0.1726	
115	M.	20	28.4	1.3404	0.1690	
116	M.		28.5	1.2850	0.2000	
117	M.	45	28.5	1.3462	0.2135	
118	M.	20	28.5	1.3348	0.1662	
119	M.	22	28.7	1.3802	0.1732	
120	M.		28.9	1.2840		
121	M.	50	29.0	1.2879	0.2122	
122	M.		29.6	1.2380		
123	M.	21	29.9	1.4130	0.1792	
124	M.	30	30.1	1.2956	0.2004	
125	M.		30.2	1.4070		
126	M.	30	30.6	1.3035	0.2774	
127	M.	48	31.0	1.3000	0.2392	
128	M.	30	31.9	1.2803	0.1909	
129	M.	38	32.0	1.3300	0.2560	
130	M.	30	32.1	1.3198	0.1983	
131	M.	26	32.4	1.3169	0.1589	
132	M.	22	32.5	1.3650	0.2040	
133	M.	33	32.7	1.3470	0.2300	
134	M.	30	32.8	1.3791	0.2087	
135	M.		32.8	1.3130		
136	M.	35	33.5	1.3450	0.2380	
137	M.	42	34.0	1.4459	0.2426	
138	M.		34.6	1.3120	0.2020	
139	M.	27	35.0	1.3020	0.2030	
140	M.	42	35.0	1.4183	0.2427	
141	M.	81	35.0	1.3890	0.2457	
142	M.	27	35.2	1.3480	0.1970	

GENERAL TABLE—Continued.

Series 1. No.	Sex.	Age in days.	Weight in gms.			Remarks.
			Body.	Brain.	Cord.	
143	M.	30	35.3	1.3292	0.2246	
144	M.		35.7	1.3920		
145	M.	39	35.7	1.3580	0.2430	
146	M.	43	36.0	1.4249	0.2492	
147	M.	31	36.1	1.3236	0.2158	
148	M.	75	36.3	1.2951	0.2453	
149	M.	35	36.9	1.4350	0.2640	
150	M.	50	37.0	1.3407	0.2397	
151	M.	39	37.1	1.4015		
152	M.	30	37.2	1.4090	0.2332	
153	M.	31	37.7	1.4107	0.2122	
154	M.	31	38.0	1.4172	0.2299	
155	M.		38.1	1.3820		
156	M.		38.5	1.3280	0.2400	
157	M.	39	38.5	1.4450	0.2300	
158	M.	76	39.3	1.2875	0.2424	
159	M.	96	39.4	1.3654	0.2742	
160	M.		39.5	1.3410		
161	M.	30	39.6	1.5016	0.2322	
162	M.	36	39.8	1.4060	0.2500	
163	M.	30	40.0	1.4182	0.2228	
164	M.	38	40.8	1.5143	0.2247	
165	M.	51	41.0	1.5261	0.2902	
166	M.	30	42.2	1.5286	0.2434	
167	M.	26	43.2	1.4750	0.2150	
168	M.		43.3	1.3130	0.2490	
169	M.	31	43.6	1.4580	0.2429	
170	M.	30	43.9	1.4140	0.2330	
171	M.	30	44.7	1.5416	0.2324	
172	M.	30	45.5	1.5770	0.2274	
173	M.	32	45.9	1.5550	0.2750	
174	M.	30	46.2	1.5446	0.2448	
175	M.	40	46.8	1.4220	0.2652	
176	M.	25	47.0	1.3930	0.2970	
177	M.	51	47.1	1.4555	0.2623	
178	M.		47.5	1.4650	0.2870	
179	M.	88	47.8	1.4726	0.2850	
180	M.	40	48.6	1.4699	0.2800	
181	M.	70	49.0	1.5383	0.2623	
182	M.	88	49.4	1.3610	0.2637	
183	M.		50.7	1.4300	0.2930	
184	M.	30	50.8	1.5256	0.2602	
185	M.	29	51.0	1.3800	0.2900	
186	M.		51.2	1.5120		
187	M.	88	51.6	1.4309	0.2805	
188	M.	30	51.8	1.5006	0.2580	
189	M.	30	53.1	1.5222	0.2612	
190	M.	30	53.1	1.5244	0.2652	
191	M.	60	53.4	1.5445	0.3001	
192	M.	30	53.7	1.4706	0.2832	
193	M.	92	53.9	1.3675	0.2584	
194	M.	31	54.0	1.6051	0.2680	

GENERAL TABLE—Continued.

Series 1. No.	Sex.	Age in days.	Weight in gms. Body.	Brain.	Cord.	Remarks.
195	M.	100	54.3	1.5056	0.2882	
196	M.		55.6	1.5220		
197	M.	40	55.9	1.5338	0.2995	
198	M.	40	56.4	1.4466	0.2560	
199	M.	50	57.0	1.4412	0.2797	
200	M.	30	57.2	1.5670	0.2804	
201	M.	30	57.2	1.5918	0.2888	
202	M.	40	57.3	1.5072	0.2474	
203	M.	30	57.6	1.5706	0.2804	
204	M.		57.7	1.4200	0.3350	
205	M.		57.9	1.4430	0.3286	
206	M.	50	58.0	1.4452	0.2774	
207	M.	96	58.4	1.4308	0.3207	
208	M.	97	59.0	1.4382	0.2740	
209	M.		59.6	1.4090	0.3278	
210	M.	41	59.9	1.4424	0.2950	
211	M.	96	59.9	1.4979	0.3150	
212	M.		60.9	1.3800	0.2870	
213	M.	50	61.0	1.4296	0.2756	
214	M.	50	62.0	1.6122	0.3015	
215	M.	92	62.6	1.5273	0.3105	
216	M.	40	62.9	1.5854	0.3304	
217	M.	97	63.0	1.4288	0.2911	
218	M.	40	63.0	1.4023	0.2700	
219	M.	89	63.2	1.5823	0.2986	
220	M.		63.4	1.2850	0.2620	
221	M.	41	63.5	1.4695	0.2923	
222	M.	92	64.0	1.5338	0.3072	
223	M.	78	64.1	1.6439	0.3169	
224	M.	50	64.2	1.4891	0.2950	
225	M.	78	64.3	1.4174	0.3205	
226	M.		65.0	1.4480	0.3220	
227	M.	38	65.0	1.4270	0.3000	
228	M.		65.1	1.5525	0.3353	
229	M.	97	65.2	1.4659	0.2885	
230	M.	60	65.6	1.4744	0.2869	
231	M.	40	65.7	1.5039	0.3009	
232	M.	40	67.2	1.4518	0.3048	
233	M.	41	67.8	1.4764	0.3139	
234	M.	100	69.0	1.6307	0.3104	
235	M.	40	69.1	1.5738	0.3334	
236	M.	96	70.0	1.5397	0.3380	
237	M.	40	70.3	1.5188	0.3148	
238	M.		70.4	1.6010		
239	M.	97	70.4	1.4237	0.3067	
240	M.	78	70.6	1.5201	0.3180	
241	M.		70.7	1.4200	0.3480	
242	M.	50	71.0	1.6326	0.3338	
243	M.	60	71.8	1.5414	0.3083	
244	M.	60	72.9	1.5244	0.3037	
245	M.	40	73.1	1.5984	0.3274	
246	M.	51	73.2	1.3864	0.2976	

GENERAL TABLE—Continued.

Series 1. No.	Sex.	Age in days.	Weight in gms.			Remarks.
			Body.	Brain.	Cord.	
247	M.	40	73.9	1.5960	0.3298	
248	M.	115	74.0	1.6012	0.3953	
249	M.	78	74.1	1.5592	0.3188	
250	M.	40	74.4	1.5264	0.3166	
251	M.	40	75.5	1.5958	0.3410	
252	M.	58	75.5	1.6013	0.3328	
253	M.	50	76.0	1.6272	0.3204	
254	M.	40	76.1	1.6800	0.3460	
255	M.	50	76.1	1.6813	0.3821	
256	M.	97	76.3	1.7749		
257	M.		76.6	1.4400	0.3280	
258	M.	60	78.1	1.5315	0.3015	
259	M.	51	78.4	1.4472	0.3060	
260	M.	57	78.6	1.5407	0.3682	
261	M.	125	79.0	1.6388	0.3574	
262	M.		79.2	1.5520	0.3420	
263	M.	48	79.8	1.6400	0.3320	
264	M.	58	80.0	1.5817	0.3168	
265	M.	40	81.3	1.6674	0.3308	
266	M.		82.0	1.6540		
267	M.	116	82.0	1.4691	0.3355	
268	M.	121	83.0	1.6656	0.4106	
269	M.	57	83.3	1.6045	0.3833	
270	M.	51	83.4	1.5864	0.3440	
271	M.	51	83.6	1.5254	0.3402	
272	M.	51	84.9	1.5090	0.3266	
273	M.	59	85.1	1.5003	0.3741	
274	M.	96	85.5	1.5952	0.3733	
275	M.	58	87.1	1.6381	0.3299	
276	M.	52	87.4	1.5824	0.3574	
277	M.		87.4	1.7156	0.3579	
278	M.	115	88.3	1.6955	0.3828	
279	M.	58	89.1	1.5482	0.3354	
280	M.	115	89.3	1.5668	0.3820	
281	M.	51	92.6	1.5614	0.3532	
282	M.	131	94.0	1.6737	0.3700	
283	M.	127	96.0	1.7341	0.4293	
284	M.		96.4	1.5400		
285	M.	50	96.9	1.7744	0.4031	
286	M.	115	97.3	1.6957	0.4136	
287	M.	96	99.3	1.5921	0.3778	
288	M.		99.7	1.7029	0.4059	
289	M.	130	99.8	1.6296	0.4119	
290	M.	52	100.4	1.6774	0.4162	
291	M.	158	101.7	1.7087	0.4388	
292	M.	131	102.0	1.7388	0.4141	
293	M.	161	104.2	1.6268	0.4414	
294	M.	115	105.8	1.6594	0.4012	
295	M.	170	106.0	1.5837	0.3781	
296	M.	300	106.9	1.6755	0.4762	
297	M.	115	107.0	1.7742	0.4274	
298	M.		107.6	1.7790	0.4138	

GENERAL TABLE—Continued.

Series 1. No.	Sex.	Age in days.	Weight in gms.			Remarks.
			Body.	Brain.	Cord.	
299	M.	300	109.8	1.6445	0.4380	
300	M.	121	110.9	1.6934	0.4611	
301	M.	221	110.9	1.7765	0.4809	
302	M.	72	112.6	1.4940	0.3648	
303	M.	115	113.1	1.7444	0.4074	
304	M.	96	113.6	1.6784	0.3686	
305	M.	115	114.7	1.7646	0.4119	
306	M.	57	116.0	1.5226	0.3810	
307	M.	116	116.0	1.7271	0.4104	
308	M.	222	116.2	1.5944	0.4745	
309	M.	221	116.6	1.8536	0.5033	
310	M.	72	117.7	1.5661	0.3792	
311	M.	161	117.8	1.7033	0.4057	
312	M.	116	117.9	1.6680	0.3794	
313	M.	164	119.1	1.6671	0.4524	
314	M.		119.7	1.4621	0.4184	
315	M.	115	120.0	1.7656	0.4213	
316	M.	186	122.2	1.8046	0.5122	
317	M.	155	122.8	1.7301	0.4544	
318	M.	159	125.1	1.7795	0.4437	
319	M.	57	125.2	1.6400	0.4120	
320	M.	158	125.9	1.8213	0.4488	
321	M.	167	125.9	1.5677	0.4654	
322	M.		126.0	1.7660		
323	M.	221	126.3	1.8255	0.5162	
324	M.		128.4	1.6950		
325	M.	222	128.7	1.9465	0.5007	
326	M.	159	130.6	1.8041	0.4751	
327	M.	158	131.0	1.6976	0.4753	
328	M.	121	131.4	1.8024	0.4630	
329	M.	155	131.7	1.7024	0.4565	
330	M.	121	135.5	1.8644	0.5043	
331	M.	222	136.7	1.8831	0.4844	
332	M.	221	137.1	1.9649	0.5381	
333	M.	222	137.2	1.7730	0.4802	
334	M.	128	138.0	1.7383	0.4767	
335	M.	221	138.2	1.7667	0.4923	
336	M.		138.9	1.6390		
337	M.		141.0	1.6950	0.4750	
338	M.	128	141.4	1.7823	0.4392	
339	M.		141.6	1.8470	0.5670	
340	M.	131	141.9	1.7549		
341	M.		142.2	1.6150	0.4940	
342	M.		142.5	1.5990	0.4560	
343	M.	159	142.7	1.7673	0.5120	
344	M.		144.1	1.7590		
345	M.	128	144.2	1.7333	0.4300	
346	M.	57	144.7	1.6656	0.4352	
347	M.	222	147.1	1.7003	0.4855	
348	M.		148.9	1.8975	0.4896	
349	M.	128	150.0	1.8165	0.4915	
350	M.	128	150.0	1.7271	0.5007	

GENERAL TABLE—Continued.

Series 1. No.	Sex.	Age in days.	Weight in gms.			Remarks.
			Body.	Brain.	Cord.	
351	M.	79	150.3	1.6657	0.4547	
352	M.	183	150.5	1.6952		
353	M.	128	151.2	1.7604	0.4339	
354	M.		152.2	1.7280	0.5260	
355	M.		152.4	1.7430		
356	M.	126	153.0	1.7297	0.4713	
357	M.		153.7	1.7420		
358	M.	222	153.7	1.8276	0.4856	
359	M.	193	153.8	1.5666		
360	M.	234	154.0	1.8167	0.5488	
361	M.	222	155.7	1.8441	0.5307	
362	M.	126	156.0	1.7551	0.4853	
363	M.	158	158.0	1.8245	0.5232	
364	M.	221	159.0	1.8364	0.5186	
365	M.	72	159.6	1.8459	0.4984	
366	M.	222	160.1	1.9153	0.5062	
367	M.		161.6	1.7770	0.5890	
368	M.	222	162.2	1.8040	0.4740	
369	M.		163.0	1.7250	0.6150	
370	M.	234	164.3	1.8072	0.5424	
371	M.		166.5	1.7925	0.5355	Over 6 months
372	M.		167.2	1.5924	0.5844	
373	M.	222	167.9	1.8717	0.5164	
374	M.		169.0	1.7963	0.5769	
375	M.	85	169.4	1.7800	0.5240	
376	M.		171.0	1.8291	0.6199	
377	M.		172.8	1.8322	0.4652	
378	M.		176.6	1.8166	0.5232	Over 6 months
379	M.	206	176.7	1.8321	0.5213	
380	M.	80	176.9	1.7290	0.4760	
381	M.	159	177.2	1.8590	0.5564	
382	M.	221	177.6	1.8080	0.5166	
383	M.	80	177.7	1.7320	0.5000	
384	M.	128	180.6	1.9520	0.5017	
385	M.		182.5	1.6420	0.5130	
386	M.		182.5	1.6600	0.4800	
387	M.	155	182.5	1.7519	0.5166	
388	M.		185.0	1.8050		
389	M.	182	186.7	1.6578	0.4862	
390	M.	90	186.9	1.7940	0.5510	
391	M.	221	187.4	1.8000	0.5036	
392	M.		188.0	2.0780		
393	M.		190.5	1.9120	0.6380	
394	M.		190.5	1.8400	0.6920	
395	M.	202	192.1	1.7083	0.5327	
396	M.		195.5	1.7693	0.4857	
397	M.	90	196.0	1.6530	0.5840	
398	M.		198.6	1.7250	0.5810	
399	M.	186	199.0	1.9781	0.5829	
400	M.		200.9	1.8340		
401	M.		201.2	1.8140	0.6570	
402	M.		201.4	1.8840		

GENERAL TABLE—Continued.

Series 1. No.	Sex.	Age in days.	Weight in gms. Body	Brain.	Cord.	Remarks.
403	M.		201.9	1.5220	0.5285	
404	M.	100	202.1	1.8600	0.5800	
405	M.		203.6	1.8714	0.7106	
406	M.	155	204.2	1.8442	0.5646	
407	M.	158	206.8	1.9538	0.5444	
408	M.	165	207.7	1.7490	0.5490	
409	M.		208.6	1.7698	0.5928	
410	M.	107	209.6	1.8442	0.5165	
411	M.		216.2	1.7790	0.6110	
412	M.	158	217.0	1.9337	0.5423	
413	M.		218.0	1.8370	0.5600	
414	M.	165	218.8	1.8050	0.6280	
415	M.		219.9	1.7144	0.6241	
416	M.		221.6	1.8124	0.5774	
417	M.		223.2	1.8496	0.6289	
418	M.	216	223.7	1.7645	0.5564	
419	M.		223.9	1.7988	0.5976	
420	M.	155	224.7	1.9342	0.6034	
421	M.		226.6	1.8900	0.6122	
422	M.	107	229.2	1.9389	0.5888	
423	M.		229.5	1.5766	0.5753	
424	M.		234.8	2.1380		
425	M.	113	236.2	1.8102	0.6108	
426	M.	113	236.5	1.9594	0.6476	
427	M.		238.1	1.9300	0.6180	Over 10 months
428	M.		238.6	1.7298	0.5963	
429	M.	120	241.5	1.8849	0.6503	
430	M.	159	242.1	1.9967	0.5948	
431	M.		250.3	1.9230	0.5810	Over 10 months
432	M.	216	251.0	1.9107	0.5883	
433	M.		260.0	1.8740	0.6110	
434	M.	120	260.0	1.8039	0.6503	
435	M.		261.0	1.7661	0.6447	Over 10 months
436	M.	106	262.8	1.9057	0.5990	
437	M.	113	264.5	1.8214	0.6290	
438	M.		267.0	1.9391	0.6693	Over 10 months
439	M.		267.0	2.0856	0.7056	Over 10 months
440	M.		267.4	1.9700	0.6750	
441	M.	120	269.1	1.9389	0.6793	
442	M.	120	272.5	1.9444	0.6855	
443	M.	140	275.3	1.9389	0.6484	
444	M.		278.0	1.9780	0.6700	
445	M.	140	278.2	1.9173	0.7102	
446	M.	140	279.1	2.0713	0.7633	
447	M.		282.7	1.8200	0.6562	Over 10 months
448	M.		285.0	1.9000	0.7380	Over 10 months and fat
449	M.		285.0	2.0954	0.7154	Over 10 months
450	M.		291.0	1.9000	0.6910	
451	M.		294.5	1.9450	0.6380	
452	M.		297.0	2.0599	0.7147	Over 10 months
453	M.		305.0	2.1037	0.7012	Over 10 months

GENERAL TABLE—Continued.

Series 1. No.	Sex.	Age in days.	Weight in gms.			Remarks.
			Body.	Brain.	Cord.	
454	M.		308.0	2.1791	0.6630	Over 10 months
455	M.		310 .0	2.0610	0.6639	Over 10 months
456	M.		310.2	1.9690	0.7550	Over 10 months
457	M.		316.0	2.1213	0.7626	Over 10 months
458	M.		320.0	2.1527	0.6948	Over 10 months

Series 2. No.	Sex.	Age in days.	Weight in gms.			Remarks.
			Body.	Brain.	Cord.	
459	M.		50.6	1.4016	0.3340	Injected with lecithin, HATAI, '03.
460	M.		55.3	1.3850	0.3050	" "
461	M.		63.3	1.4500	0.3500	" "
462	M.		65.3	1.4690	0.3470	" "

Series 1. No.	Sex.	Age in days.	Weight in gms.			Remarks.
			Body.	Brain.	Cord.	
1	F.		3.9	0.2300	0.0330	At birth
2	F.		4.2	0.2070	0.0260	At birth
3	F.	1	4.5	0.3023	0.0290	
4	F.		4.6	0.2280	0.0300	36 hours
5	F.		4.6	0.2110	0.0270	38 hours
6	F.		4.9	0.2210	0.0210	36 hours
7	F.		5.0	0.2550	0.0280	38 hours
8	F.	2	5.2	0.2260	0.0300	
9	F.	2	5.5	0.2260	0.0300	
10	F.		5.7	0.2830	0.0350	38 hours
11	F.		5.7	0.3320	0.0340	36 hours
12	F.	5	7.0	0.4520	0.0510	
13	F.	5	7.6	0.4610	0.0480	
14	F.	5	7.7	0.4680	0.0450	
15	F.	10	9.0	0.8101	0.0746	
16	F.	10	10.0	0.8312	0.0709	
17	F.	10	11.3	0.8470	0.0860	
18	F.	10	11.6	0.8280	0.0810	
19	F.	9	11.9	0.7120	0.0720	
20	F.	16	13.5	1.0530	0.1200	
21	F.	16	13.5	1.0870	0.1210	
22	F.	9	14.0	0.9109	0.0864	
23	F.	12	15.5	1.0550	0.1150	
24	F.	16	15.5	1.0109	0.1028	
25	F.	16	17.9	1.1480	0.1110	

GENERAL TABLE—Continued.

Series I. No.	Sex.	Age in days.	Weight in gms. Body.	Brain.	Cord.	Remarks.
26	F.	40	18.0	1.0479	0.1561	
27	F.	52	18.0	1.0580	0.1423	
28	F.	40	18.5	1.0637	0.1502	
29	F.	75	21.3	1.0941	0.1722	
30	F.	41	21.5	1.0832	0.1657	
31	F.	46	23.0	1.1977	0.1741	
32	F.	50	23.0	1.1420	0.1770	
33	F.		23.0	1.2205	0.1551	
34	F.		23.9	1.2850	0.1620	
35	F.	21	24.9	1.2940	0.1730	
36	F.	21	27.0	1.3765	0.1606	
37	F.	21	27.9	1.3130	0.1890	
38	F.	50	29.0	1.2251	0.2092	
39	F.	33	29.2	1.3550	0.2220	
40	F.	50	31.0	1.2591	0.2123	
41	F.	63	31.0	1.4015	0.2484	
42	F.	50	35.0	1.3944	0.2313	
43	F.	26	35.4	1.3470	0.2110	
44	F.	50	37.0	1.4924	0.2589	
45	F.	75	37.9	1.2841	0.2372	
46	F.	48	38.0	1.4266	0.2408	
47	F.	49	38.0	1.3338	0.2380	
48	F.	54	39.0	1.4311	0.2444	
49	F.	75	39.0	1.3467	0.2464	
50	F.	51	40.0	1.4330	0.2627	
51	F.	49	41.0	1.3722	0.2445	
52	F.	76	41.8	1.3594	0.2346	
53	F.	49	42.0	1.4042	0.2552	
54	F.	51	42.0	1.4100	0.2630	
55	F.	26	42.5	1.3930	0.2310	
56	F.	76	43.3	1.3501	0.2415	
57	F.		45.0	1.3680	0.2220	
58	F.	30	45.0	1.4420	0.2300	
59	F.	43	45.0	1.3703	0.2515	
60	F.	104	45.0	1.4055	0.2792	
61	F.	124	45.0	1.4247	0.2904	
62	F.	88	45.9	1.2625	0.2480	
63	F.	32	46.0	1.6100	0.2460	
64	F.	78	46.0	1.5850	0.2845	
65	F.	83	46.0	1.3943	0.2389	
66	F.	30	47.9	1.4440	0.2480	
67	F.	32	48.0	1.6070	0.2540	
68	F.	78	48.6	1.2810	0.2632	
69	F.	88	48.8	1.2800	0.2610	
70	F.	60	48.9	1.4028	0.2786	
71	F.	49	49.0	1.3772	0.2392	
72	F.	75	49.8	1.4693	0.2861	
73	F.	97	50.4	1.4194	0.2588	
74	F.	97	50.5	1.3587	0.2716	
75	F.		50.8	1.3770	0.2950	
76	F.	50	52.0	1.3787	0.2533	
77	F.	88	52.0	1.3838	0.2846	

GENERAL TABLE—Continued.

Series I. No.	Sex.	Age in days.	Weight in gms.			Remarks.
			Body.	Brain.	Cord.	
78	F.	78	53.1	1.4486	0.2809	
79	F.	50	54.0	1.3517	0.2602	
80	F.	130	54.2	1.4869	0.3224	
81	F.	78	55.5	1.4829	0.2746	
82	F.	60	55.8	1.4103	0.2834	
83	F.	30	57.4	1.5040	0.2820	
84	F.	60	58.4	1.4213	0.2802	
85	F.	95	59.0	1.5580	0.3042	
86	F.	97	60.0	1.4600	0.2930	
87	F.		61.7	1.4880	0.2975	
88	F.	167	63.0	1.3814	0.2837	
89	F.	78	63.1	1.5745	0.3116	
90	F.		64.6	1.5367	0.3062	
91	F.	130	64.8	1.4536	0.3387	
92	F.	97	68.9	1.4429	0.3211	
93	F.		69.6	1.4950	0.3470	
94	F.	88	71.3	1.6030	0.3179	
95	F.	97	73.0	1.4795	0.3127	
96	F.	48	73.5	1.6550	0.3350	
97	F.		73.7	1.4550	0.3430	
98	F.	116	74.0	1.5665	0.3727	
99	F.		76.4	1.3700	0.3260	
100	F,		77.2	1.5110	0.3359	
101	F.	96	80.5	1.5189	0.3542	
102	F.	116	82.8	1.5950	0.3846	
103	F.	116	88.6	1.6002	0.4010	
104	F.	167	89.1	1.5513	0.3528	
105	F.	96	91.5	1.5325	0.3541	
106	F.	145	91.7	1.5374	0.4307	
107	F.	135	92.0	1.6823	0.3436	
108	F.	116	95.0	1.6614	0.3967	
109	F.		95.4	1.5970	0.4280	
110	F.	151	96.1	1.4999	0.3989	
111	F.	156	97.7	1.7433	0.4246	
112	F.	131	98.0	1.6037	0.3421	
113	F.		98.0	1.5640	0.3710	
114	F.	130	98.6	1.5646	0.3939	
115	F.	116	99.1	1.5897	0.3775	
116	F.	158	99.1	1.6155	0.3730	
117	F.	130	99.8	1.6124	0.4075	
118	F.	158	102.5	1.8228	0.4444	
119	F.		102.5	1.5780	0.4190	
120	F.	135	104.1	1.7545	0.4012	
121	F.	159	104.3	1.8782	0.4294	
122	F.		105.0	1.6336	0.4035	
123	F.	166	106.5	1.6457	0.4451	
124	F.	116	106.7	1.7353	0.4523	
125	F.	131	107.0	1.6747	0.3678	
126	F.	160	107.1	1.6549	0.4003	
127	F.	130	107.8	1.4902	0.3712	
128	F.	176	108.1	1.5381	0.4164	
129	F.	126	112.0	1.7040	0.4440	

GENERAL TABLE—Continued.

Series 1. No.	Sex.	Age in days.	Weight in gms.			Remarks
			Body	Brain.	Cord.	
130	F.	127	113.2	1.7698	0.4459	
131	F.	128	113.2	1.6488	0.3950	
132	F.		114.7	1.7050	0.5340	
133	F.	130	115.8	1.6299	0.4167	
134	F.		116.1	1.7605	0.4123	
135	F.	128	117.7	1.7193	0.4114	
136	F.		118.5	1.4520	0.4120	
137	F.	166	119.0	1.8465	0.4919	
138	F.	130	121.8	1.6970	0.4397	
139	F.	224	122.2	1.6614	0.4841	
140	F.	204	122.4	1.5960	0.4204	
141	F.	142	125.1	1.6988	0.4505	
142	F.	109	125.7	1.7096	0.4116	
143	F.	164	126.9	1.6859	0.4941	
144	F.	159	129.0	1.6740	0.4614	
145	F.	175	129.2	1.5553	0.4520	
146	F.	176	134.6	1.5768	0.4459	
147	F.	186	137.9	1.7627	0.4713	
148	F.		139.1	1.6226	0.5210	
149	F.	166	139.9	1.8068	0.4836	
150	F.	109	140.4	1.7858	0.4574	
151	F.	224	142.3	1.7380	0.5181	
152	F.		143.7	1.6750	0.4540	
153	F.	176	145.7	1.6798	0.5059	
154	F.		147.7	1.7738	0.5467	
155	F.	166	148.1	1.6555	0.4950	
156	F.	85	150.4	1.7500	0.5100	
157	F.		151.7	1.8162	0.5546	Over 7 months
158	F.		151.8	1.6350	0.5360	
159	F.		152.3	1.5885	0.5112	
160	F.	80	155.0	1.8150	0.4800	
161	F.	80	155.0	1.8500	0.5250	
162	F.	224	155.3	1.7707	0.5162	
163	F.	166	156.9	1.7954	0.5458	
164	F.		157.5	1.6630	0.4950	
165	F.		158.7	1.7900	0.5200	
166	F.	109	161.9	1.8308	0.4420	
167	F.	100	162.8	1.6800	0.4920	
168	F.	186	163.7	1.9014	0.5113	
169	F.	224	164.2	1.8892	0.5137	
170	F.	90	164.3	1.7170	0.5760	
171	F.	100	165.9	1.6350	0.5320	
172	F.	328	166.1	1.6966	0.5896	
173	F.	211	166.7	1.8387	0.5398	
174	F.	128	167.0	1.6921	0.5642	
175	F.	90	168.0	1.7830	0.5360	
176	F.	305	171.4	1.8010	0.6382	
177	F.		171.7	1.7176	0.5491	
178	F.	176	171.9	1.8342	0.5277	
179	F.	396	172.6	1.8003	0.6001	
180	F.	328	174.0	1.7840	0.6329	
181	F.		176.0	1.9499	0.6677	Over 10 months

GENERAL TABLE—Continued.

Series 1. No.	Sex.	Age in days.	Weight in gms. Body.	Brain.	Cord.	Remarks.
182	F.	224	180.7	1.8190	0.5239	
183	F.		181.0	1.6120	0.4990	Over 10 months
184	F.	380	182.4	1.7803	0.5817	
185	F.	110	183.7	1.6810	0.5390	
186	F.		184.7	1.9290	0.6146	Over 10 months
187	F.		186.5	1.7314	0.5623	
188	F.	211	188.4	1.8627	0.5310	
189	F.		190.7	1.7950	0.6380	
190	F.	345	191.2	1.8007	0.5878	
191	F.	328	191.7	1.8731	0.6302	
192	F.	361	195.0	1.6944	0.5746	
193	F.	110	195.2	1.7090	0.5210	
194	F.	380	196.0	1.8426	0.6177	
195	F.	380	196.4	1.7878	0.6311	
196	F.	361	197.0	1.9005	0.6219	
197	F.		200.9	1.7950	0.6100	Over 10 months
198	F.	165	202.7	1.7460	0.5680	
199	F.	320	203.0	1.8426	0.5989	
200	F.	345	203.0	1.7937	0.5718	
201	F.	221	203.4	1.9975	0.6053	
202	F.	165	205.5	1.7390	0.5990	
203	F.	309	212.0	1.9330	0.6375	
204	F.	309	212.0	1.9510	0.6205	
205	F.	345	219.4	1.8108	0.5945	
206	F.	320	231.0	1.8265	0.6615	
207	F.	380	231.4	1.7120	0.5807	
208	F.	361	232.0	1.7740	0.6047	
209	F.		233.0	1.9350	0.6570	Over 10 months
210	F.		234.4	1.9800	0.6180	Over 10 months
211	F.		242.6	1.8610	0.6600	Over 10 months
212	F.	320	243.0	1.8622	0.6185	
213	F.	309	245.5	1.9704	0.6340	
214	F.	309	247.0	1.9694	0.6370	
215	F.	309	280.0	2.0130	0.7034	

Series 2. No.	Sex.	Age in days.	Weight in gms. Body.	Brain.	Cord.	Remarks.
216	F.		53.8	1.4260	0.3386	Injected with lecithin HATAI, '03.
217	F.		60.4	1.3810	0.3520	" "
218	F.		62.1	1.4500	0.3220	" "

BIBLIOGRAPHY.

BISCHOFF, T. VON.

1880. Das Hirngewicht des Menschen. S. 1–171, *P. Neusser, Bonn.*

BLAKEMAN, J.

1905. A study of the biometric constants of English brain-weights, and their relationships to external physical measurements. *Biometrika*, vol. 4, pp. 124–160.

BOAS, F.

1905. Statistics of growth. *Report of Commissioner of Education U. S. A.*, chapter 2, pp. 25–132.

BOYCOTT, A. E. AND DAMANT, G. C.

1908. A note on the total fat of rats, guinea-pigs and mice. *Journ. Physiol.*, vol. 37, pp. 25–26.

BUNGE, C.

1902. Physiologic and pathologic chemistry. 2d edition (edited by E. A. Starling). *P. Blakiston's Son & Co., Phila.*

DAVENPORT, C. B.

1904. Statistical methods with special reference to biological variation. 2d ed., *John Wiley & Sons, N. Y.*

DONALDSON, H. H.

1895. The growth of the brain: A study of the nervous system in relation to education. 4th ed., *Charles Scribner's Sons, N. Y.*

1901. Chapter on "central nervous system" in vol. ii, p. 279, of Am. Text Book of Physiol., Howell. *W. B. Saunders & Co., Phila.*

1906. A comparison of the white rat with man in respect to the growth of the entire body. Boas Memorial Volume, p. 5–26, *G. E. Stechert & Co., N. Y.*

DUBOIS, E.

1898. Ueber die Abhängigkeit des Hirngewichtes von der Körpergrosse bei den Säugethieren. *Archiv für Anthropologie*, Bd. 25.

GLADSTONE, R. J.

1905. A study of the relations of the brain to the size of the head. *Biometrika*, vol. 4, pp. 105–123.

GREENWOOD, M.

1904. A first study of the weight, variability and correlation of the human viscera, with special reference to the healthy and diseased heart. *Biometrika*, vol. 3, pp. 63–83.

HANDMANN, ERNST.

1906. Ueber das Hirngewicht des Menschen auf Grund von 1414 im pathologischen Institut zu Leipzig vorgenommenen Hirnwägungen. *Archiv f. Anat.* [u. *Physiol.*], S. 1–40.

HATAI, S.

1903. The effect of lecithin on the growth of the white rat. *Am. J. of Physiol.*, vol. 10, pp. 57–66.

1904. The effect of partial starvation on the brain of the white rat. *Am. J. of Physiol.*, vol. 12, pp. 116–127.

1907. On the zoölogical position of the albino rat. *Biol. Bull.*, vol. 12, pp. 266–273.

1907A. Studies on the variation and correlation of skull measurements in both sexes of mature albino rats (Mus norvegicus var. albus). *Am. J. of Anatomy*, vol. 7, pp. 423–441.

1907B. Effect of partial starvation followed by a return to normal diet, on the growth of the body and central nervous system of albino rats. *Am. J. of Physiol.*, vol. 18, pp. 309–320.

1908. Preliminary note on the size and condition of the central nervous system in albino rats experimentally stunted. *J. of Comp. Neurol.*, vol. 18, pp. 151–155.

LAPICQUE, L.

1908. Tableau général des poids somatique et encéphalique dans les espèces animales. Le poids encéphalique en fonction du poids corporel entre individus d'une même espèce. *Memoires de la Société d'Anthropologie de Paris*, p. 249–344.

MIES, J.

1893. Ueber das Gewicht des Rückenmarkes. *Centralbl. f. Nervenheilkunde u. Psychiatrie*, S. 1–4.

PEARL, R.

1905. Biometrical studies on man. I. Variation and correlation in brain-weight. *Biometrika*. vol. 4, pp. 13–104.

PFISTER, H.
1903. Zur Anthropologie des Rückenmarks. *Neurol. Centralb.*, Bd. 22, S. 757 und 819.
RUBNER, M.
1908. Das Wachsthums problem und die Lebensdauer des Menschens und einiger Säugetiere vom Energistischen Standpunkt aus betrachtet. *Archiv für Hygiene. Bd. lxvi.*
1908A. Probleme des Wachstums und der Lebensdauer. *Beiblatt zu den Mitteilungen der Gesellschaft für innere Medizin und Kinderheilkunde in Wien.* VII. Jahrgang, no. 3, S. 58–82.
VIERORDT, H.
1890. Das Massenwachsthum der Körperorgane des Menschen. *Archiv für Anat. [u. Physiol.]*, S. 62–94.
WATSON, J. B.
1905. The effect of the bearing of young upon the body-weight and the weight of the central nervous system of the female white rat. *J. of Comp. Neurol.*, vol. 15, pp. 514–524.
ZIEHEN, T.
1899. "Nervensystem" p. 10 in "Handbuch der Anatomie des Menschen," von Bardeleben. *Gustav Fischer, Jena.*

EXPLANATION OF PLATE II.

CHART 1. The upper continuous line represents the brain weight according to body weight, calculated by formula [1]. The separate entries (. males and × females) show the individual observations so far as they can be entered without confusion. The lower continuous line represents the spinal cord weight according to body weight, calculated by formula [3]. The same scale is used for the two curves.

CHART 2. The continuous line represents the spinal cord weight according to body weight. The separate entries (. males and × females) show the individual observations so far as they can be entered without confusion. The scale of the ordinates is twice that used in chart 1.

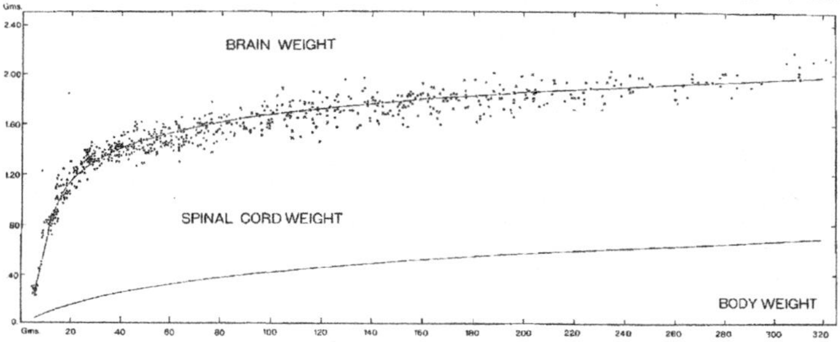

Chart 1.

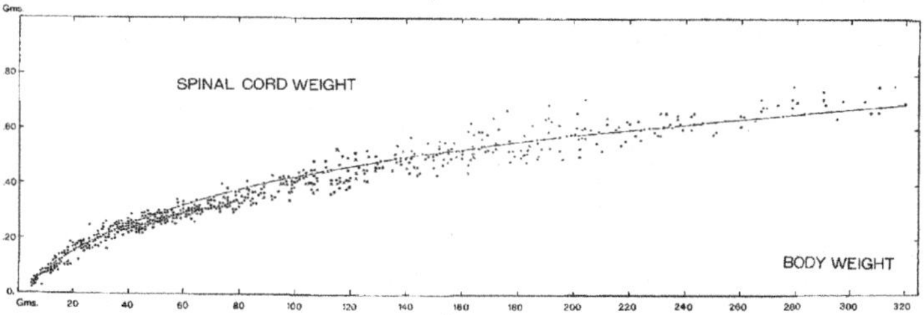

Chart 2.

Explanation of Plate III.

Chart 3. In the upper curves, the continuous line (*C*), shows the weight of the brain according to body weight as calculated by formula [1], the broken line (*M*), the means of the observed brain weights.

In the lower curves, the continuous line (*C*) shows the weight of the spinal cord according to body weight as calculated by formula [3], the broken line (*M*) the means of the observed spinal cord weights. The same scale is used for both curves.

Chart 5. On the base line representing age, one day of rat life is given the same value as 30 days of human life (i.e., 1 day rat = 30 days man). The lower curves represent the weight of the brain according to age (in days) for the rat. The solid line for the male, and the broken line for the female. The upper curves represent the weight of the human brain (at yearly intervals) the actual values of the ordinates being reduced to 1-700th in order to make the curves comparable with those of the rat. Continuous line, male; broken line, female.

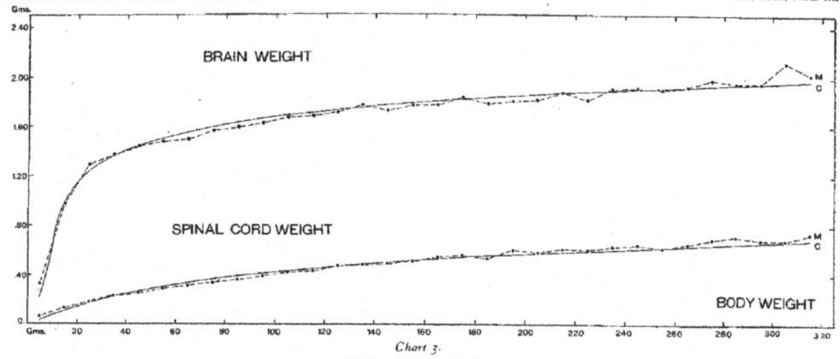

Chart 3.

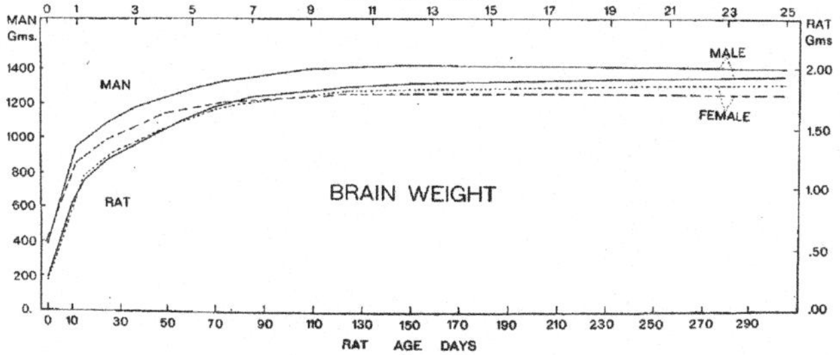

Chart 5.

THE MORPHOLOGICAL SUBDIVISION OF THE BRAIN.

BY

C. JUDSON HERRICK.

(*From the Anatomical Laboratory of the University of Chicago.*)

The great diversity in the internal organization of different parts of the nervous system makes demands on the morphologist for a much more minute regional subdivision than is necessary with most other organs; and at the same time this diversity, together with the intricate interrelations between the various parts, makes such an analysis the more difficult.

The subdivision of the human brain, as made by the first anatomists on the basis of gross external form, has a certain functional as well as morphological basis; but when the attempt was made to study the regions so named comparatively, the morphological imperfections of the scheme became at once apparent. Any further attempt to utilize uncritically in morphology the "regions" as customarily defined in the older human neurology is misdirected energy.

The clear recognition of this fact early led Professor HIS to seek in his embryological studies a safer guide to cerebral morphology, and the result of his labors as finally formulated is incorporated in the BNA nomenclature.

But the cerebral analysis of Professor HIS is based almost wholly on *human* embryology and therefore inevitably shares some of the same defects as a scheme based on the human adult; for the human embryonic brain at all stages is very far indeed from giving a true picture of ancestral phylogenetic conditions. And there can be no sound morphology which does not rest on a phylogenetic basis. And moreover the dynamic element in modern morphology requires throughout a closer correlation between structure and function than any purely embryological or anatomical scheme can effect.

Accordingly, comparative anatomy, comparative embryology, and comparative physiology must be appealed to before we can feel that we are on safe ground in making a morphological subdivision of the nervous system. The correlation of these subjects is attempted in the modern functional analysis of the nervous system as effected by recent students of comparative neurology. All of these subjects are now sufficiently far advanced to justify a reëxamination of the question of the subdivision of the brain from the phyletic standpoint and with especial reference to the application of the BNA terms in comparative studies.

The earliest form of regional differentiation of the nervous system in the phylogeny was probably the concentration of a centralized or integrated system from the primitive diffuse type. This differentiation appears in some form in all but the simplest metazoa. In vertebrates the diffuse system also exhibits a certain degree of integration on its own account and appears as the sympathetic system.

It is not intended to imply here the specific homology of the ganglia of the sympathetic chain or any other part of the vertebrate sympathetic system with the diffuse nervous system of any particular invertebrate type—and much less with any invertebrate central nervous system. It is very probable that the highly complex sympathetic nervous system of vertebrates has been elaborated subsequently to the differentiation of the central nervous system. It is maintained, however, that the diffuse vertebrate nervous system, especially the peripheral sporadic ganglia (and possibly the sympathetic as a whole), is in a general way comparable with the diffuse nervous system of such animals as the Cœlenterata. Even if it should prove, as now seems probable, that the entire sympathetic system is in the ontogeny differentiated from the same embryonic tissue (medullary plate) as the cerebro-spinal system, I think the comparison will still hold, though we must recognize that the ontogenetic development deviates far from a recapitulation of the phylogeny of the sympathetic system. This deviation is in the direction of a concentration of all of the embryonic nervous rudiments, looking forward to the functional intimacy between the cerebro-spinal and the sympathetic systems of the adult, and thus is parallel with the general trend of the evolution of the whole nervous system.

Our primary subdivisions, then, are (1) the sympathetic (autonomic) and (2) the cerebro-spinal nervous systems.

The cerebro-spinal or integrated nervous system in early phylogenesis separated from the diffuse nervous system to serve three great types of bodily reactions, and gave rise to three systems of sense organs (receptors) and their associated centers and return motor paths. These, adopting SHERRINGTON's nomenclature, we may designate as follows:

1. *Exteroceptors*, systems for reaction to stimuli impinging upon the outer surface of the body. The source of the stimulus may not be in contact with the body, in which case the sense organs in question are termed by SHERRINGTON distance receptors. The

modalities of sense primarily represented here are: (1) cutaneous sensation (touch, temperature, etc.); (2) hearing; (3) vision. These are termed the somatic, as distinguished from the visceral, senses, their most essential characteristic being the fact that the reaction evoked is somatic, i. e., a movement of the body as a whole (in lower animals usually locomotor in type) with reference to the source of the stimulus.

In certain cases, as relatively recent phylogenetic adaptations, sense organs of the visceral type have secondarily appeared in the skin and assumed exteroceptive functions, as in the cutaneous taste buds, of certain teleosts. But these exceptions must be recognized as such and need not confuse our primary subdivisions. The organ of smell presents a somewhat similar atypical instance, which I have considered in more detail in another article.[1]

2. *Proprioceptors.*—These systems were evolved parallel with the exteroceptors and subsidiary to them. Their sense organs lie chiefly in the organs of somatic response, i. e., in the muscles, joints, tendons, etc., and are adapted to assist in the correlation of the movements of the body, reporting to the central nervous system the exact degree of contraction of every muscle, the degree of tension on the joints, tendons, etc. They are therefore indispensable to all delicate muscular adjustments involving accurate regulation of movement, strain, etc. Their organs of response are obviously identical with those of the exteroceptors with whose functions they are associated. The most highly specialized member of the group is the mechanism of equilibration in the auditory labyrinth.

3. *Interoceptors.*—These comprise the visceral systems of sense organs chiefly concerned with stimuli received in the digestive tract, and exciting visceral responses thereto. The most typical sense modalities are taste and a group of ill-defined pneumo-gastric sensations (suffocation, nausea, hunger, thirst, etc.). Smell and taste (the chemical senses) are clearly related physiologically and probably have a common origin from a primitive undifferentiated chemical sense.

The purpose of the differentiation in vertebrates of two chemical senses so closely similar physiologically and psychologically and so very dissimilar anatomically has given morphologists a world of trouble. SHERRINGTON suggests the most natural explanation when he calls attention to the fact that, while the organ of taste is a typical interoceptor, the olfactory organ is a distance receptor. In other words, the chemical sense has differentiated along two lines determined simply by whether the source of

[1] On the phylogenetic differentiation of the organs of smell and taste. *Journ. Comp. Neurol. and Psych.*, vol. 18, no. 2. 1908.

the stimulus is in contact with the body (taste) or at a distance (smell)—a practically important distinction from the standpoint of the type of reaction which should follow, though one which may rest on a relatively trivial physico-chemical difference in the stimuli themselves. Accordingly, the cerebral centers for smell, which normally call forth locomotor responses, have differentiated far from those of taste, where typically visceral responses alone are involved. (Cf. the article last cited, On the Phylogenetic Differentiation of the Organs of Smell and Taste.)

Attention should be called in this connection to the fact that in the hypothetical ancestral vertebrate from which the olfactory and gustatory reflex systems were evolved both the effectors and receptors were in an unspecialized condition and in later phylogeny their elaboration occurred simultaneously. Responses to chemical stimuli of an unspecialized or "total" sort would follow in such an animal and the differentiation of the responses into visceral and somatic types occurred *pari passu* with the functional differentiation of the sense organs into interoceptors and distance receptors. And of course the cerebral conduction pathways for smell and taste became clearly defined and separate from the general unspecialized central gray only gradually as the functional need arose. In a similar way all of the well defined reflex paths within the higher brains are shown by comparative studies to have taken form from the more diffuse type of central nervous structure found in the lowest vertebrates. The central gray and formatio reticularis grisea et alba are survivals in higher brains of this primitive unspecialized nervous tissue.

The first two of SHERRINGTON'S physiological groups, then, compromise in general the systems of nerves and end organs designated by recent morphologists the somatic systems; his third group includes the visceral or splanchnic systems.

From the preceding analysis it clearly appears that the basis for the most fundamental divisions of the cerebro-spinal nervous system is found in the contact of the organism with the environment. In other words, the organization of the cerebro-spinal nervous system has been shaped by its peripheral end organs.

Early in the process of the resultant differentiation the vertebrate central nervous system responded to these peripheral influences by developing a series of structurally defined longitudinal zones. In the first place, the dorsal part of the nerve tube and the related nerves were devoted to the afferent or sensory limb of the reflex arc and the ventral part to the efferent or motor limb, separated by the *sulcus limitans* (HIS). The early recognition of this relation in the spinal cord by Sir CHARLES BELL led to the formulation of BELL'S law of the composition of the spinal nerve roots. These have been further subdivided by GASKELL and his followers, so that in passing from the dorsal to the ventral surface of the spinal cord we now recognize successively, in both gray and white matter, the following longitudinal zones:

1. Somatic sensory, including exteroceptive and proprioceptive centers and conducting paths.

2. Visceral sensory, including the interoceptive centers and their pathways. These are feebly developed and still very imperfectly understood in the spinal cord. Afferent impulses enter by

the dorsal roots, probably by way of the rami communicantes of the sympathetic system.

3. Visceral, or splanchnic motor. The region of the intermediate zone and columna lateralis grisea. The efferent fibers are pre-ganglionic sympathetic nerves, which leave the central nervous system mainly at least by the ventral roots and which excite the intero-effectors, including viscero-motor, vaso-motor, excito-glandular fibers, etc.

4. Somatic motor. The columna ventralis grisea and associated fiber paths. The efferent impulses go out by way of the ventral roots to the extero-effectors (somatic or skeletal muscles).

Parallel with this longitudinal differentiation of the neural tube a transverse segmentation took place called forth by the metamerism of the body. In all vertebrates, especially in the spinal cord, this transverse segmentation is much more evident anatomically than the longitudinal divisions. Accordingly, morphologists in the past have devoted their attention almost exclusively to it in the elaboration of metameric schemata of nervous organization. But in reality transverse segmentation is far less important to cerebral morphology save for convenience of anatomical description, as will appear from a consideration of the genesis of the two types of specialization in question.

We cannot hope to elaborate a nomenclature reflecting perfectly the relations in a low type of brain like the lamprey's which will at the same time be adequate for the human neurologist; but we should seek to devise a scheme which is sufficiently elastic to permit of adaptation to both with no change of fundamental plan. A system based on transverse segmentation, while in many respects better adapted for the lowest vertebrates, breaks down completely when applied in the Mammalia.

Metamerism is primarily mesodermal in origin. It arose as an aid to locomotion of the vermiform type in very primitive animals. The segmentation of the skeletal, nervous and vascular systems, etc., is all secondary to that of the body muscles, and where these latter disappear, as in the head of higher vertebrates, the neuromeres lose their individuality also. The longitudinal functional divisions, on the other hand, are primarily nervous. They represent the most fundamental factors in the architecture of the vertebrate body and exert a more potent influence on cerebral structure the higher we go in the evolutionary series. Metamerism is more primitive and dominates the nervous system of lower animals, but becomes relatively less important as we pass up the phyletic series, whose higher types come into progressively more varied relation with the environment. Nevertheless the influence of metamerism is always apparent, even in the human brain; and the fact that two tendencies, independent in origin and often exerting antagonistic influences on the course of the differentiation, have operated in the architectonic of the vertebrate nervous system is a cause of great perplexity and confusion in attempting the analysis of cerebral structure in higher animals. A purely metameric scheme can in the nature of the case be no more satisfactory than one based on adult or embryonic human structure, even though it is based on a correctly interpreted phylogeny.

The matter is still further complicated by the fact that, either primarily in chordate evolution or secondarily, there appeared a difference between the metameres in the rostral part of the body and those in the more caudal part; viz: the presence of gills in the rostral portion. The structural unit here is the branchiomere. The delimitation of the branchial region from the rest of the body gives the most fundamental plane of differentiation which crosses the mid-line of the body.

Our problem, then, is first to analyze the two elementary factors in the phylogenesis, metamerism and longitudinal functional differentiation, and then to endeavor to trace the influence exerted by each and to construct a scheme of cerebral structure which shall hold good both in lower and in higher vertebrates and take due account of both of these directive influences.

In the spinal cord region, where the primitive relations seem to have suffered the least modification in the course of phylogenesis, the two factors referred to above can be quite readily distinguished. The primitive metamerism is highly modified in the peripheral distribution of the spinal nerves, but it is preserved almost unchanged at the surface of the spinal cord, as shown by the serial arrangement of the spinal roots and their ganglia. Again, internally the demand for correlation between the different levels has produced longitudinal arrangements which largely obscure the transverse segmentation. Only in the early embryo is the internal structure of the neural tube evidently segmental. There is a stage when neuromeres are clearly defined in the neural tube as a transient beading of its contour. The internal longitudinal differentiation is however so nearly uniform in the adult throughout the length of the spinal cord that the subdivision of that structure can most conveniently be effected in terms of the transverse segments, these being named after their nerves and the corresponding vertebræ, as defined by the classic nomenclature.

In the rostral end of the body the segmental plan is modified in all lower vertebrates, as we have seen, by the presence of gills. These are visceral structures and their innervation belongs wholly to the splanchnic systems.[2] In lower vertebrates this branchiomeric system may coexist with the typical somatic systems in the same segments, but in higher forms segments possessing gills, or their derivatives in land vertebrates, have usually lost the somatic components or else these latter have suffered so great modification as to be with difficulty recognized as such. The reduction in number of gills took place early in the phylogeny (they never exceed seven in gnathostomes, and usually are less than five) and the surviving members of the series are so closely associated with cranial structures that the whole gill region in gnathostomes may be considered a part of the head.

The branchiomeric type of nerve is preserved with least modification in the region of the medulla oblongata, bounded rostrad by the isthmus; and this structural type is clearly evident, though in a highly modified form, in the human medulla oblongata. Accordingly, the rhombencephalon of HIS is a natural subdivision phyletically as well as embryologically considered.

The cerebellum and its associated pons are derivatives of the somatic sensory column in the cephalic part of the same region, called forth primarily by the vestibular apparatus (and allied sense organs in fishes). This gives a sound genetic basis for the metencephalon of HIS, as distinguished from the remainder of the rhombencephalon (the medulla oblongata).

The metencephalon should be limited to the cerebellum and its immediate dependencies, a structure which has been added to the much older *primary* rhombencephalon, or branchiomeric brain. This usage will necessitate some revision of the limits of the metencephalon as set by the BNA. It can no longer be regarded as a

[2] These structures have suffered considerable secondary modification. For example, the gustatory system has been derived from the unspecialized visceral sensory, and the visceral motor has in part been specialized parallel with the development of striated branchial muscles from the splanchnic mesoderm. The branchio-motor nerves lack the post-ganglionic neurone and structurally resemble the somatic motor nerves in their mode of connection with their end-organs.

transverse segment of the neural tube, but as a dorsal structure which reaches down into the lateral walls and floor of the older brain stem like a girdle. The metencephalon should, therefore, include the cerebellum, pons, corpus restiforme, brachium conjunctivum and some of the nuclei in immediate contact with these structures. It should not include the longitudinal conduction paths in the brain floor above the pons, nor the nuclei of the cranial nerves of the same region.

The term medulla oblongata should be applied to that part of the rhombencephalon lying between the spinal cord and the isthmus, exclusive of the parts here enumerated as belonging to the metencephalon. That portion of the medulla oblongata lying caudad of the VIII nerves and their chief primary nuclei may be called the myelencephalon (which is practically the usage of the BNA), thus limiting this latter term to the region of the typical gill bearing segments in the true fishes. If a distinctive name is required for the preauditory part of the medulla oblongata coördinate with myelencephalon for the postauditory part, the term *pars facialis medullæ* may be suggested. The region so designated includes that portion of the *brain stem* (exclusive of the metencephalon as I here define it) comprised approximately within the metencephalon and isthmus rhombencephali of the BNA, a region which receives the cerebral nerves of the skin and muscles of the face and facial skeleton.

The medulla oblongata, as here defined, extends sufficiently far forward to include the roots and chief nuclei of the V, VI and IV cerebral nerves and the superior secondary gustatory nucleus (nucleus visceralis cerebelli, JOHNSTON). It is bounded rostrad by a constriction, the isthmus, which marks the adult position of the groove between the embryonic second and third cerebral vesicles.

The subdivision of the isthmus region is very difficult. I believe that the use of that term for an encephalic region cannot be justified. The word is, however, a convenient descriptive term for the constriction in question and if retained in our nomenclature at all it should be used only in that sense.

Rostrad of the rhombencephalon the evidence of the primary metamerism has almost entirely disappeared from the adult human brain. In lower brains, even in their embryonic conditions, it is very difficult to decipher the vestiges of metamerism in these regions.

The primary longitudinal functional zones have likewise suffered extreme distortion by reason partly of the development of the organs of special sense but more especially on account of the elaboration of the massive organs of correlation.

The diencephalon and the mesencephalon, from our present point of view, are not natural regions, nor can any other transverse division of the brain be made which will satisfy the conditions, for the primary metamerism has ceased to be an important factor in the problem. The terms mid-brain, thalamus, etc., will in any event of course continue in use as convenient topographic designations. But from the point of view of a broad comparative morphology, I believe that they are confusing rather than helpful.

Functionally and genetically, the retinæ, optic nerves, chiasma and tracts and the optic thalamus (*sensu stricto*) should be associated with the optic tectum of the mid-brain to form an *ophthalmencephalon* whose boundaries cross freely those of the classic encephalic regions.

The principle upon which this term is based is the same as that which led HIS to adopt (in the BNA) the term fasciculus cerebrospinalis in place of the older terms, pyramid and pyramidal tract; viz: the association of neurones belonging to the same functional system. The advantages of this usage over any purely topographical designation are so clearly brought out in the discussion of the terms pyramids, etc., that we refer the reader here to the words of HIS.[3]

The term ophthalmencephalon as proposed here is analogous with *rhinencephalon* as used by the BNA for the whole olfactory apparatus of the forebrain except the olfactory portion of the pallium.[4]

Aside from the visual centers, there are in the midbrain the pedunculus cerebri, colliculus inferior and other correlation centers less clearly defined and still imperfectly known. It would be premature in the present state of our knowledge to attempt a final detailed subdivision of this difficult part of the brain; but the following simple outline may serve as a working basis.

In the mammalia the part of the brain between the tectum opticum and the cerebellum may best be divided into two regions, the

[3] *Archiv f. Anat.* [*u. Physiol.*], Suppl. Bd., 1895, p. 163.

[4] On the rhinencephalon, BNA, see beyond under the subdivision of the forebrain.

pedunculus cerebri lying ventrally and the colliculus inferior dorsally.

The *colliculus inferior* is in mammals chiefly an acoustic center, forming the brain roof (epencephalic area of HIS, Entwickelung des menschlichen Gehirnes, 1904, p. 23) over a portion of the pedunculus cerebri. In the lowest vertebrates this region is feebly differentiated, if at all, from the tectum opticum (colliculus superior). The latter structure in lower fishes contains other types of sensory structures besides the optic. But in higher fishes the differentiation of the optic centers (colliculus superior) from the other systems is practically complete. The tectum opticum of these higher fishes embraces a massive basal structure (the torus semicircularis or "colliculus" of the teleostean anatomists) which receives secondary sensory tracts other than optic—tracts which reach the tectum in the lower fishes. These secondary tracts come chiefly from the tuberculum acusticum. Although the latter structure of fishes is not exactly homologous with the tuberculum acusticum of man, yet the relations are such as to justify us in regarding the so-called colliculus of teleosts as in a general way homologous with the colliculus inferior of the mammals. The latter organ attains its highest development in mammals parallel with the evolution of the cochlea. In the brains of the lowest vertebrates, where the colliculus inferior has not yet differentiated from the colliculus superior, the whole structure may be designated simply colliculus, a region which would include a part of the ophthalmencephalon as well as the rudiments of the colliculus inferior.

In mammals, where the colliculus inferior is well defined as an acoustic center, the question arises, Why not recognize an acustencephalon analogous with the rhinencephalon and ophthalmencephalon? This question has been carefully considered, but found very difficult of practical realization. To carry out the analogy with the two systems last mentioned it would be necessary to include the peripheral auditory nerves and primary centers in the medulla oblongata. Such a region may be defined in terms of neurone systems and as a physiological unit, and has great value as such. But the acoustic (cochlear) pathways and centers are in all cases more or less confused structurally with the vestibular, and even in mammals their physiological independence seems to be less than has sometimes been supposed.'[5] Furthermore, as we pass down to the lower vertebrates, the cochlear and the vestibular systems not only become progressively less clearly separable from each other, but they are also related to other still less highly differentiated types of sensory systems, notably the organs of the lateral line, and still lower in the series to the organs of ordinary tactile sensation. In short, the acustencephalon in the lower members of the phyletic series possesses no individuality, and in the higher members where it is both structurally and functionally better defined the system as a whole is not structurally sufficiently separate from adjacent parts to justify giving it place in a scheme of regional subdivision, however valuable it may be as a physiological unit.

This suggests a consideration of the relative merits of two morphological standards which are not always easily reconciled, viz: the functional system of neurones as contrasted with topographic form relations of the nervous system. The real unit of the nervous system is unquestionably the functional system of neurones and the fruitfulness of this unit in morphology has been well illustrated by the treatment adopted by BARKER in his text-book on the Nervous System (*New York*, 1901) and by the results attained by the students of the so-called functional morphology, especially in America. Any topographic subdivision of the brain will be fully satisfactory as a basis for both morphological and physiological work just in proportion as its divisions express the underlying functional relations. But such a subdivision must take also into account the phylogenetic history of these functional units and of the brain as a whole, primary metamerism, the influence of various mechanical factors which have affected the differentiation of the nervous system and also the practical consideration of anatomical and didactic convenience. These factors are seldom in complete accord. This indeed is what makes the problem of cerebral regional anatomy so difficult. The relative weight to be given to the different factors of the problem is a matter calling for the exercise of the utmost skill and in the nature of the case the question will be answered differently by specialists in different departments. The preceding discussion of the ophthalmencephalon

[5] Cf. the recent results of C. WINKLER. The central course of the nervus octavus and its influence on motility. *Verh. Kon. Akad. van Wettenschappen, Amsterdam* (2 Sec.), Deel 14, no. 1, 202 pp., 24 pl. 1907.

and acustencephalon illustrates my own view of the proper balance to be struck between the claims of the functional systems and of convenience in the recognition of topographic landmarks for didactic and descriptive purposes, in a general scheme of the subdivision of the brain.

The *pedunculus cerebri* is a convenient, but purely artificial, topographic region, including the tegmentum, substantia nigra, basis pedunculi, III nerve, etc.—in short all of the mesencephalon of the BNA after the subtraction of the colliculus superior and colliculus inferior (or the single colliculus of lower vertebrates where there is but one). It will doubtless be found possible to subdivide it or distribute it to neighboring regions, but at present it may be better to use this term as a makeshift than to propose a new subdivision, on the basis of our very imperfect knowledge of the comparative anatomy of this part of the brain.

The nomenclature of the diencephalon, after the separation of the optic centers to form the ophthalmencephalon, requires very little modification of existing usage. The term *medithalamus* may be applied to such parts of the thalamus and metathalamus (BNA) as remain after subtraction of the optic centers (pulvinar, geniculatum laterale, etc.) The term *hypothalamus* may be retained for all of the region so designated in the BNA save the optic chiasma. The epithalamus remains as defined by the BNA. The medithalamus is a derivative of the central gray of the first, second and third neuromeres. The hypothalamus includes the hypophysis and important correlation centers, chiefly of visceral sensation (olfactory and gustatory). The *epithalamus* includes the epiphysis and important olfactory centers in the habenula.

The exact boundaries of these thalamic subdivisions for vertebrates in general cannot be defined in the present state of our knowledge. JOHNSTON has discussed the limits of these regions in lower vertebrates in his text book,[6] to which the reader is referred for a summary review of our present knowledge regarding this difficult subject. Much more extensive comparative study will be necessary before attempting anything but a provisional analysis of the thalamus region and classification of its nuclei and fiber tracts.

The *telencephalon* is well named. It is terminal, not only in position, but also in point of time, having been added relatively late

[6] JOHNSTON, J. B. The nervous system of vertebrates, *Philadelphia*. 1906, chapter 17.

in the phylogeny to the rostral end of the original neural tube. The BNA has done well to omit from it the pars optica hypothalami which was originally tabulated as part of this region by Professor HIS. Originally developed as primary and secondary olfactory centers, it has added successively more and more complexity during the whole course of phylogenetic history.

The most ancient, or stem portion of the telencephalon we may term, following EDINGER,[7] the hyposphærium, and the dorsal correlation centers episphærium. The term rhinencephalon may properly be applied to the whole of the *primary* olfactory brain, including the olfactory bulbs, tracts and basal centers and also the anterior commissure and lamina terminalis. The only important structure of the mammalian hyposphærium not included in the rhinencephalon is the corpus striatum. The episphærium is subdivided after ELLIOT SMITH into archipallium and neopallium, the former including the olfactory cortex (hippocampus, etc.) and the fornix, the latter including the more recently developed cortical representation of the other senses and their association centers and the corpus callosum. Our subdivision of the telencephalon, then, is as follows:

Hyposphærium:

Rhinencephalon (incl. lamina terminalis, ant. com. etc.)
Corpus striatum.

Episphærium:

Archipallium (including fornix).
Neopallium (including corpus callosum).

The confusion in the nomenclature of the forebrain is so great that one hesitates to recommend any of the old terms, for all have been used with diverse significations, and that too very often by authors who supposed they were using them similarly. The BNA terms seem to have suffered especially from this latter form of misuse. Yet one shrinks from adding to the burden by coining new terms which would be free from these confusing implications. The best course is to select the most fit of the current terms and try to give them precision by accurate definition. One of the most valuable contributions in this direction is that of ELLIOT

[7] EDINGER, L. Ueber die Herkunft des Hirnmantels in der Tierreihe. *Berliner klin. Wochenschr.*, 1905, no. 43.

SMITH.[8] He presents in the work cited a very cogent argument for extending the application of the word rhinencephalon to comprise the whole olfactory apparatus, including the olfactory cerebral cortex, or archipallium. This course has much to recommend it, for the olfactory cerebral cortex is more intimately connected with the lower olfactory centers than are the cortical representations of the other sensory systems with their lower centers. In the case of the latter functional systems the cortical centers are widely separated topographically from their respective lower centers, so that the cortex cannot be joined to the lower centers in a regional subdivision of the brain. For the sake of uniformity it seems to me better to treat the entire cerebral pallium in the same way, even though the limits between olfactory pallium and lower olfactory centers cannot always be clearly drawn, as ELLIOT SMITH has shown.

I, therefore, recommend that the rhinencephalon be limited to the basal or stem portion of the olfactory system and that the olfactory pallium be excluded from the rhinencephalon. The olfactory pallium is indeed very distinct structurally as well as physiologically, from the remainder of the cerebral cortex and therefore the separation of the cortex into archipallium and neopallium is worthy of recognition in our nomenclature. The usage here recommended is apparently similar to that of the BNA, but ELLIOT SMITH has shown in the paper cited that neither Professor HIS as reporter for the BNA nor the neurologists who have subsequently adopted his terms are consistent in their use of them.

Confusion of this sort can be avoided only by establishing a definition of pallium which is independent of accidental form relations. The term was first applied to the thin free forebrain roof, as contrasted with the more massive basal ganglia. In lower vertebrates this criterion is of no morphological value. KAPPERS and THEUNISSEN[9] have developed in a fruitful way a histological conception of the distinction between pallium and basal forebrain centers. In the case of the olfactory system the primary and secondary olfactory centers are considered as belonging to the basal or stem portion of the telencephalon, while olfactory centers of

[8] G. ELLIOT SMITH. Notes upon the natural subdivision of the cerebral hemisphere. *Journ. Anat. and Physiol.*, vol. 35, no. 4, pp. 431–454, 1901.

[9] Zur vergleichenden Anatomie des Vorderhirnes der Vertebraten. *Anat. Anz.*, Bd. 30, 1907; and Die Phylogenese des Rhinencephalons, des Corpus striatum und der Vorderhirnkommissuren. *Folia Neuro-biologica*, vol. 1, no. 2, 1908.

the third order are treated as pallium. The application of this principle, however, proves to be very difficult, for there are many cases where olfactory tracts of the third order run to basal regions which have none of the other characteristics of pallium, and on the other hand the archipallium (hippocampus) of all higher vertebrates, including man, is said on good authority to receive olfactory fibers of the second order, as well as of higher orders.

It is clear that in mammals the distinction between hyposphærium and episphærium is capable of tolerably precise anatomical definition and is easily recognized. While accepting this as an important anatomical fact, the writer is of the opinion that our knowledge of both the histology and the phylogeny of the mammalian pallium is still too imperfect to permit of its final morphological interpretation. The phylogeny of the pallium is now under active investigation in a number of important neurological laboratories and until our knowledge of its early phylogenetic stages is more complete it is better to avoid dogmatism and await fuller knowledge before attempting to elaborate in detail the morphology of the telencephalon. The unsettled state of opinion regarding the fundamental character of cerebral localization further emphasizes the need of caution in forebrain morphology.

We may, then, summarize our scheme of subdivision of the vertebrate nervous system as follows:[10]

Systema nervorum sympathicum (BNA).
Systema nervorum cerebro-spinale.
Systema nervorum periphericum (BNA).
Systema nervorum centrale (BNA).
- Medulla spinalis (BNA).
- Encephalon (BNA).
  - Rhombencephalon (BNA).

[10] It should be borne in mind that the terms adopted in this table are not coördinate in morphological value from any standpoint. They are offered simply as a practicable suggestion for a terminology which shall be available for all vertebrates and which deviates as little as possible from the most widely used standard of our time.

The writer would add, moreover, that in recommending this subdivision and nomenclature he by no means pleads for its final adoption as a whole. This scheme is manifestly defective in many places, and it is hoped and expected that further discussion will clarify the more obscure points. These pages are offered primarily as a contribution to method and it is earnestly maintained that the principles here illustrated are vitally important and must be taken into account in all future morphological work in this field. The exact limits of the regions and the names to be applied to them are subsidiary considerations upon which unanimity of opinion can hardly be expected until our anatomical and physiological knowledge is far more complete.

- Medulla oblongata (BNA).
  - Myelencephalon (BNA).
  - Pars facialis medullæ.
- Metencephalon (BNA).

Cerebrum (BNA).

- Pedunculus cerebri (BNA).
- Colliculus inferior (BNA).
- Ophthalmencephalon.
- Medithalamus.
- Hypothalamus.
- Epithalamus (BNA).
- Telencephalon (BNA).
  - Hyposphærium.
    - Rhinencephalon (BNA)
    - Corpus striatum (BNA).
  - Episphærium (= Pallium BNA).
    - Archipallium.
    - Neopallium.

ON THE COMMISSURA INFIMA AND ITS NUCLEI IN THE BRAINS OF FISHES.

BY

C. JUDSON HERRICK.

(From the Anatomical Laboratory of the University of Chicago.)

WITH TWELVE FIGURES.

It has long been known that the brains of fishes possess an extensive mass of fibers which cross the median line above the ventricle in the region of the nuclei of the dorsal funiculi where the medulla oblongata joins the spinal cord. This is the commissura infima of HALLER. The commissure discovered in a similar position in the mammals by CAJAL (1896, p. 43) was at once recognized by fish neurologists as a homologous structure; and since the commissure of CAJAL and its associated nucleus clearly belong to the visceral sensory system (receiving sensory fibers from the visceral roots of the vagus and glossopharyngeus nerves), it has been commonly assumed that in the fishes the commissure is visceral sensory.

Renewed examination shows that this assumption is correct, but it is not the whole truth. The commissura infima includes two morphologically distinct elements: (1) a visceral sensory commissure, comprising a decussation of vagus root fibers and a commissure of secondary elements from the visceral sensory centers; and (2) a somatic sensory commissure, consisting of secondary fibers from the funicular nuclei and adjacent centers of tactile sensation. Each commissure has associated with it a nucleus. The visceral nucleus was discovered by CAJAL and named by him the nucleus commissuralis. The somatic nucleus was first reported in my paper (1906) on the tactile and gustatory centers of fishes, to which the reader is referred for the general topography of this region.

In the paper last mentioned this commissure of Ameiurus, the common horned pout or cat-fish, was analyzed incidental to the study of the adjacent centers, and found to be very complex. The purpose of this paper is the examination of a number of types of fishes in which the various elements of the commissure are differently developed, for the purpose of reaching a more complete understanding of this critical region of the brain. We shall begin with a review of the conditions in Ameiurus, where the visceral and somatic elements are about equally developed, and then examine other species showing various deviations from this type.

The visceral commissura infima of Ameiurus.—The visceral commissural nucleus is a dorsal unpaired structure forming a protuberance in the triangular space behind the vagal lobes and between the funicular nuclei. The cells of the nucleus, which are rather small, are more thickly arranged near the external borders of the lobe (fig. 1). Their dendrites ramify through its whole substance and freely cross the median line. The fibers of the descending vagus root, which enter the nucleus at its lateral borders, terminate chiefly uncrossed by widely branched aborizations, but some of these endings cross to the opposite side before terminating (fig. 3). This commissure also receives short secondary tracts, either unmedullated or with slight medullation, from the vagal lobe, which end partly crossed and partly uncrossed (cf. HERRICK 1906, fig. 7). The cells of its nucleus therefore may receive visceral impulses either directly from the periphery by way of the descending vagus root or indirectly by way of the descending secondary visceral tract from the vagal lobes. Both types of afferent fibers, as well as dendrites of the cells of the nucleus, participate in the formation of the commissure, which is diffuse and chiefly unmedullated. The efferent tract from this nucleus is by tolerably compact bundles of unmedullated fibers which curve downward near the median line around the cephalic end of the median funicular nucleus to effect connection with the adjacent formatio reticularis under the vagal lobe (fig. 1).

The somatic commissura infima of Ameiurus.—This commissure lies ventrally of the preceding and dorsally of the ventricle. The two commissures are for the most part very distinct anatomically, though at the cephalic end there is some mingling of their fibers. The somatic commissure is chiefly, though by no means wholly, a commissure of the funicular nuclei (fig. 2). Its most dorsal part is a strong compact bundle from the dorso-lateral fasciculus and lateral funicular nucleus (fig. 2, *y*; fig. 3). Among these fibers and ventrally of them are dendrites of the cells of the median and lateral funicular nuclei and delicately medullated fibers connecting the two median funicular nuclei. Farther ventrally and extending nearly down to the canalis centralis are fine fibers, chiefly unmedullated, passing between the formatio reticularis of the two sides (fig. 2). Among the latter fibers are numerous cells, some very small and some large (fig. 3; cf. fig. 5 of Catostomus). These cells of the somatic commissural nucleus constitute an extension of the adjacent formatio reticularis grisea which, accordingly, is broadly continuous across the median line above the canalis centralis. The large cells of this nucleus send their neurites ventrally into the ventral cornu and perhaps the ventral funiculi. The smaller cells seem to connect with the adjacent formatio reticularis.

The splanchnic and somatic divisions of the commissura infima are thus seen to be very similar, save in the fact that the somatic division is not known to contain root fibers. Each contains commissural fibers from adjacent primary centers and a median nucleus whose efferent tracts reach the adjacent formatio reticularis. These nuclei seem to have been differentiated from the formatio reticularis grisea, the splanchnic nucleus being the more highly specialized.

I have elsewhere described these commissures in Gadus (1907, p. 75–78) and found the typical relations with neither the somatic nor the visceral elements especially differentiated, the somatic component being considerably larger than the visceral (see figs. 3, 5, 6 and 7 of the paper cited).

In Haploidonotus grunniens I find the same conditions as in Gadus save that the visceral component is considerably larger.

The commissura infima of cyprinoids.—In these fishes the visceral element of this complex is greatly enlarged, without a corresponding modification of the somatic element.

In the carp, Cyprinus, the morphologic relations are the same as in Ameiurus with such modifications only as are caused by the larger vagal lobes. The nucleus intermedius vagi seems to be included in the motor layer of the vagal lobes (HERRICK 1905, p. 433). The visceral commissural nucleus is both larger and more distinct than in Ameiurus. It is continuous headward with the motor layer of the vagal lobes and under the caudal border of the vagal lobes appears as a pair of pyriform swellings whose larger ends are fused over the ventricle (fig. 4). The area of fusion contains the visceral portion of the commissure infima, consisting of sparse medullated fibers and numerous unmedullated tracts. The narrow end of each nucleus is directed laterally and receives the most caudal (descending) sensory vagus root. This root is unusually large and spreads throughout the substance of the lobe. Numerous fascicles of root fibers pierce the lobe to cross in the commissura infima above the ventricle. These lobes receive, as in Ameiurus, large numbers of feebly medullated diffuse tracts from the layer of secondary tracts of the vagal lobes. These end partly uncrossed and partly crossed.

These visceral nuclei pass forward into the nucleus intermedius vagi, as just stated, and ventrally they are bounded by the somatic sensory centers and their commissure. The dorsal part of the somatic commissura infima is blended with the ventral part of the visceral commissure. The visceral commissure is diffuse and chiefly unmedullated. The somatic commissure contains numerous separate fascicles of heavily medullated fibers, some of which come from the dorso-lateral fasciculi, and also more lightly medullated fibers which connect the adjacent funicular nuclei and formatio reticularis.

The somatic sensory centers are nearly as large as in Ameiurus, but do not show as much differentiation. The lateral funicular nucleus is distinct; the median nucleus and nucleus of the spinal V tract are indistinguishably fused. They extend far forward under the visceral nuclei, the latter having been crowded backward by the enlarged vagal lobes. Both funicular nuclei receive descending secondary gustatory fibers from the facial lobe. Most of these fibers end in the median nucleus, while the lateral nucleus receives a much larger proportion of fibers from the dorso-lateral fasciculus. Both nuclei are, therefore, as in Ameiurus, correlation centers for tactile and gustatory sensation from the outer skin (HERRICK 1906). Their secondary connections are as in Ameiurus and in addition there is a large tract of small medullated fibers from the median nucleus and the formatio reticularis ventrally of it, which crosses at once in the ventral part of the somatic commissura infima and then passes headward along the lateral border of the ventricle to end in the nucleus ambiguus. This is evidently a direct path from the tactile-gustatory correlation center to the nucleus of the gill muscles, and may be termed the tractus funiculo-ambiguus (fig. 4).

In Catostomus the general plan is much as in Cyprinus, but more compact. The visceral commissural nucleus (nucleus of CAJAL) is not well developed in the median line, but is a paired structure lying close to the meson at the ends of the commissura infima, whose visceral portion is short, thick and more compact than in the carp. The descending sensory vagus root is very large and most of its root fibers decussate in this commissure. The visceral commissural nucleus is not as clearly separate from the surrounding structures as in the carp. This nucleus and the adjacent parts, both forward and backward, are connected by a strong tract of unmedullated external arcuate fibers with the funiculus ventralis. The nature of this connection I have not been able to make out. It is present in Ameiurus and contains some feebly medullated fibers. The dorsal end of this tract is shown in fig. 6 of my recent paper on Ameiurus (1906), passing down laterally of the tracts marked *f. l.* It seems to be the path from the visceral sensory centers to the motor nuclei of the muscles of the fins and trunk.

The great enlargement of the visceral centers of cyprinoids is confined to the medulla oblongata. The spinal visceral centers are even smaller than in some other fishes, and the commissural nucleus cannot be traced far back into a clearly defined area of the spinal cord.

The somatic portion of the commissura infima is less compact, being but little different from the dorsal commissure of the spinal cord, which is well developed. The somatic commissural nucleus is developed about as in Ameiurus. One of its neurones is shown in fig. 5.

The gold fish, Carassius auratus, resembles the carp very closely, the chief difference being the greater distinctness of the visceral and somatic components of the commissura infima. The tractus funiculo-ambiguus is well developed and there is a conspicuous well medullated descending tract from the visceral commissural nucleus, which passes directly back from the commissure close to the median line. It passes caudad among the transverse fibers of the somatic commissura infima and disappears before reaching the caudal end of the funicular nuclei. This tract is present in all of the fishes which I have examined, but is exceedingly variable in size. Usually it is unmedullated. GOLGI sections of various species have shown sparse fibers passing from it ventro-laterally into the formatio reticularis. It probably is the path of conduction between the visceral sensory centers of the oblongata and the corresponding areas of the spinal cord.

The commissura infima of the Anguillidæ.—The brain of the eel is strikingly selachian in aspect and by reason of its extreme elongation presents some features of the oblongata more clearly than any other type. My material consists of two series of the brain of Conger cut transversely and stained by WEIGERT's method, two of Anguilla cut transversely and stained by DELAFIELD's and MALLORY's hæmatoxylin respectively and two of Anguilla cut horizonally and stained by DELAFIELD's and WEIGERT's hæmatoxylin respectively.

The somatic sensory centers of these brains are very highly developed, the visceral sensory centers moderately. The vagal lobe is greatly elongated; it receives in front the visceral sensory root of the facial nerve, the sensory IX root about midway and the sensory X roots at the caudal end, beyond which it passes directly into the visceral commissural nucleus. There is no specially differentiated facial lobe. The ascending secondary gustatory tract arises from the whole extent of the vagal lobe in the typical manner. From the cephalic or facial end of the lobe a descending tract appears to pass back ventrally of the spinal V tract into the dorso-lateral fasciculus for the funicular nucleus and spinal cord, like the descending secondary tract from the facial lobe of Ameiurus; but the sections do not make the relations perfectly plain.

The vagal lobes at their caudal ends fuse and from the point of fusion the visceral commissural nucleus extends far backwards as a compact mass of dense neuropil containing few medullated fibers, many cells and unmedullated tracts running in various directions, many of which cross the median line. Some unmedullated vagus terminals probably reach this commissure, though the WEIGERT sections do not permit a demonstration of this. Strong unmedullated tracts pass from the whole length of this nucleus into the formatio reticularis of the same side (figs. 6, 7, 8). This visceral nucleus extends far caudad, diminishing in size and sharply marked off structurally from the adjacent somatic centers. The nucleus ambiguus accompanies its lower border for some distance and can be followed back as far as the level of the second spinal nerve (figs. 6 to 10). With the enlargement of the funicular nuclei in this region the visceral area shrinks to a mere vestige which can, however, be followed more than five millimeters back into the spinal cord (fig. 10). This area corresponds to the descending tract from the visceral commissural nucleus described above for the gold fish.

The somatic sensory centers of both the oblongata and spinal cord are greatly enlarged, as well as elongated. The tuberculum acusticum is large and extends far caudad over the vagal lobes. From its lower end a somatic sensory field (fig. 6) extends back laterally of the vagal lobe to connect with the lateral funicular nucleus, which is well developed about as in Ameiurus. The somatic commissura infima begins close behind the vagal lobes, over-arching the visceral commissure and its nucleus (fig. 6). This part of the somatic commissure contains a strong tract of medullated fibers which pass from the caudal end of the tuberculum acusticum back through the lateral sensory field just referred to, crossing as the most cephalic fibers of the somatic commissure and terminating in the lateral funicular nucleus of the opposite side (fig. 6, *com. tub. acust*). I have not observed such a tract in any other fish which I have studied. Even in Gadus, which also has very large tubercula acustica, there is no field of gray matter extending between the tuberculum acusticum and the lateral funicular nucleus and no tract can be distinguished in this region for the commissura infima, though such a tract may pass from the tuberculum acusticum indistinguishably mingled with the others in the dorso-lateral fasciculus. In Gadus the tubercula acustica fuse dorsally over the ventricle and there is a strong medullated acoustic commissure in this area of fusion (see KAPPERS 1906, figs. xcv to xcix). In Conger there is also a broad dorsal fusion of the tubercula acustica but no medullated fibers appear in it; nor is there an acoustic commissure of unmedullated fibers except at the extreme cephalic end. It may therefore be that the descending medullated tract for the commissura infima in Conger is a compensation for the absence of an acoustic commissure in the tubercula acustica themselves. But we should have more precise knowledge of the courses of these fibers before accepting this suggestion.

The somatic commissura infima is larger than in any other fish which I have examined except Prionotus. Its cephalic part is large and very compact and lies dorsally of the visceral nucleus and commissure (figs. 6, 7, 8). Its commissural nucleus is not developed at this level. The first part of the commissure contains, in addition to the acoustic commissural tract above described, more numerous finer fibers which connect the lateral funicular nuclei and the underlying dorso-lateral fasciculi (figs. 7 and 8). Even as far back as the first dorsal spinal root (fig. 8) the somatic centers are not greatly enlarged except the lateral funicular nucleus. The lateral funicular nucleus extends far caudad and behind fuses with the median funicular nucleus. The fibers of the first dorsal spinal root terminate chiefly in this lateral nucleus. A slender fascicle accompanies these root fibers and passes farther ventrally to connect with the nucleus ambiguus (fig. 8, *rx XI?*). These appear to belong to the spinal accessory nerve, though their peripheral destination is unknown.

As we pass caudad from this level both the dorsal somatic commissure between the lateral funicular nuclei and the visceral commissural nucleus disappear, the place of the latter being taken by the somatic commissural nucleus and the commissure between the median funicular nuclei, which have meanwhile rapidly enlarged. The spinal V nucleus is indistinguishable from the funicular nucleus. The somatic commissural nucleus is only moderately developed (figs. 9, 10), and in this region the somatic commissura infima is represented by scattered fascicles of medullated fibers between the funicular nuclei and formatio reticularis. The funicular nucleus lies unusually far caudad and farther back passes over into the dorsal cornu, which continues to be large far back into the spinal cord, where it becomes enveloped by the large dorsal funiculus.

Anguilla, the common eel, is essentially the same as Conger, the visceral system being relatively smaller. These fishes illustrate a very high development of the somatic longitudinal conduction paths of the somatic sensory centers of the medulla oblongata.

The commissura infima of Prionotus I have elsewhere described (1907a). In this species, as in the other gurnards, the visceral sensory system is rather poorly developed, but the somatic sensory systems, especially the spinal tactile centers, are very extensive. The "accessory lobes" of the cephalic end of the spinal cord are enlarged dorsal cornua and the first lobe includes also the spinal V nucleus. The funicular nucleus is very greatly enlarged. The visceral commissura infima and commissural nucleus are rather small but of the typical teleostean type. The somatic commissura infima contains at the cephalic end a large, heavily medullated tract from the funicular nucleus, and farther back extensive more diffuse connections between the first and second accessory lobes. The somatic commissural nucleus is also very large and associated with the tracts last mentioned. The commissural nucleus does not extend back beyond the cephalic end of the second accessory lobe, though there is a well developed medullated dorsal commissure between all of the accessory lobes.

The commissura infima of Amia calva.—In this fish both the visceral and the somatic commissural systems are moderately developed and the whole region is simpler and probably more primitive than in any teleost examined.

The vagal lobes are elongated and their caudal ends rise up to form the lips of the fourth ventricle (tænia ventriculi quarti) in the calamus region, fusing together behind the ventricle to form the visceral commissural nucleus (fig. 11). Sparse unmedullated fibers cross the medial plane within this nucleus, constituting the visceral commissura infima. This nucleus contains but few cells and is broadly connected with the homolateral formatio reticularis ventrally of the funicular nucleus.

The funicular nuclei are elongated, extending from the level of the visceral commissural nucleus caudad far into the spinal cord. The somatic commissural nucleus (fig. 12) contains many densely crowded cells extending from the ventro-mesial borders of the two median funicular nuclei and across the meson ventrally of the most caudal part of the visceral nucleus and close above the canalis spinalis. There are no medullated fibers in the somatic commissural nucleus and very few unmedullated fibers.

KAPPERS (1907, p. 490) has briefly referred to the commissura infima of Amia, but he was not able to demonstrate definite commissural nuclei. He says, referring to the visceral commissural nucleus, "Ich möchte ihn lieber nicht als besonderen Kern betrachten." It is true that the visceral nucleus is not so large and clearly defined in Amia as in the teleosts which I have examined, yet its individuality is perfectly clear, as I have seen in an extensive series of sections of Amia kindly loaned to me by Prof. CHARLES BROOKOVER. These include preparations by the methods of WEIGERT, NISSL, CAJAL, BIELCHOWSKY and the iron hæmatoxylin method of HAIDENHAIN. I have chosen for illustration (figs. 11 and 12) a teries through a young specimen stained by the method last mensioned. The adult relations are similar.

In young *Lepidosteus* about 9 cm. long the relations of the commissura infima region are about as in the adult Amia as far as can be determined by the study of hæmatoxylin preparations. In the adult the disposition of the parts has been considerably altered. The adult vagal lobe does not reach back to the lower end of the fourth ventricle; but from its caudal end a diffuse tract of fibers, both medullated and unmedullated, extends caudad mesially of the funicular nucleus and adjacent to the wall of the fourth ventricle, to the commissura infima. Here this tract rises up in the tænia ventriculi quarti and decussates above the caudal end of the fourth ventricle to end in the visceral commissural nucleus farther caudad and ventrad. These visceral fibers are the only elements in the commissura infima of this region so far as our preparations demonstrate. The visceral commissural nucleus does not lie in the median plane in association with its commissure, but some sections farther caudad and ventrad as a paired structure lying ventromesially of the funicular nucleus of each side near the dorsal fissure. The efferent tract from this nucleus, consisting of both medullated and unmedullated fibers, passes out ventro-laterally into the adjacent formatio reticularis.

The funicular nucleus lies for the most part cephalad of the calamus region instead of caudad of it, as in Amia. Some fibers from the funicular nucleus or somatic sensory field may cross along with the visceral fibers in the commissura infima, though I have not been able to demonstrate them. The size of the somatic commissura infima is reduced; possibly the strongly developed efferent tracts from the funicular nuclei forming internal arcuate fibers below the ventricle may compensate for the reduction of the dorsal commissure.

The combined nucleus of the spinal V tract and funicular nucleus extends, as mentioned above, far cephalad of the commissura infima into a somatic sensory field laterally of the caudal end of the vagal lobe. It is quite sharply separate from the tuberculum acusticum, which occupies the somatic sensory field cephalad of it. The efferent fibers from the funicular nucleus can be easily followed. They leave the nucleus in compact strands of heavily medullated fibers which pass downward into the formatio reticularis, ventral cornu, and in still larger numbers cross the ventral raphé as internal arcuate fibers entering the fasciculus longitudinalis medialis. The funiculus dorsalis is unusually large in

Lepidosteus and it terminates for the most part in the cephalic part of the funicular nucleus.

For my preparations of the brain of Lepidosteus I am again indebted to the kindness of Professor BROOKOVER.

CONCLUSION

In effecting the functional analysis of the somatic and visceral centers at the lower end of the medulla oblongata difficulty has been experienced on account of the diffuse character of these nuclei and their tracts and the intricacy of their interrelations. At the time of the publication of my report on the central gustatory paths in fishes (1905) these relations were very imperfectly understood. Further comparative study showed that different fishes exhibit very unequal degrees of development of the somatic and visceral elements in the funicular nucleus region and that by comparing types with the maximal visceral elements, like the carp, with those showing maximal development of the tactile system, like the gurnards, the obscurity was largely cleared up, and forms with an approximately equal development of both systems could then be understood.

For the convenience of the reader I have presented these results in the reverse order of that adopted in the research, having published the findings in Ameiurus (1906) before those in the more highly specialized species, like Cyprinus and Prionotus. The latter type has proven especially helpful, since, while the *brain* is nowhere very highly specialized, the *spinal* tactile path is greatly enlarged and may therefore be isolated for study simply by comparison with other unmodified types of fishes. Accordingly I have made a more detailed examination of this species (1907a). In the present paper I have brought together the data scattered through my previous articles so far as they bear upon this region, together with some new observations on these and other fishes.

It appears that the commissura infima of fishes (HALLER) and of mammals (CAJAL) and its nucleus, so far from being a purely visceral structure, as believed by recent critics, has an important somatic sensory component, which in some fishes far exceeds in size the visceral element. The visceral commissure and its nucleus are extensions of the deep layer of the vagal lobes (nucleus intermedius vagi) and also of an imperfectly known visceral zone of the spinal cord, which probably sends some fibers into the dorsal commissure of the spinal cord. If the nucleus dorsalis (CLARKE'S column) of the spinal cord represents the visceral sensory center of the spinal cord, as supposed by some recent anatomists, then the spinal root fibers which cross in the dorsal commissure to terminate in the heterolateral nucleus dorsalis are comparable with the sensory root fibers of the vagus which cross in the commissura infima. But our knowledge of the visceral sensory system in the spinal cord is still too imperfect to permit of final decision of any of these questions. The somatic component of the commissura infima is no doubt comparable with the somatic sensory fibers which make up the greater part of the dorsal commissure throughout the spinal cord.

It is probable that in fishes (and in higher vertebrates) visceral centers which primitively were arranged, like the somatic centers, in a metameric way throughout the spinal cord have become concentrated in the medulla oblongata, the intestinal branch of the vagus and various sympathetic connections of the cranial nerves having to a large extent supplanted the original segmental visceral nerves of the spinal cord. If, as appears probable, the primitive vertebrate ancestor had gills extending down the greater part of the length of the trunk region, as in Amphioxus, the explanation for this concentration of the visceral nerves is clearly apparent in the progressive loss of the more caudally placed gills as we ascend the vertebrate series.

In the extensive region of the enlarged visceral sensory area of the medulla oblongata the roof plate is membranous and widely extended laterally. This feature prevents the visceral sensory commissural fibers of the medulla oblongata from crossing at their own level and necessitates their passage caudad to the region behind the calamus scriptorius, where they are concentrated as the visceral commissura infima. Both root fibers of the vagus and secondary visceral tracts from the vagal lobes are involved in this movement, which has carried with it a certain number of cells pertaining to these commissural fibers. These cells make up the visceral commissural nucleus and are probably mainly terminal nucleus cells for the vagus root fibers of the commissure. The secondary connections of this nucleus are substantially the same as those of the visceral sensory nucleus of the vagus in the medulla oblongata.

The somatic sensory centers of the spinal cord have shown no tendency to migrate into the head, like that seen in the visceral sensory centers. Even in Prionotus, where the most cephalic segments of the spinal cord alone are greatly enlarged, there is no especial tendency for them to be concentrated in the medulla oblongata. In the oblongata, however, the somatic sensory centers have always suffered extreme modification. The somatic sensory zone in the cephalic end of the oblongata has given rise to the tuberculum acusticum and cerebellum and because of the crowding due to the presence of these structures, or for some other reason, in teleosts practically all of the tactile, or unspecialized somatic sensory nerves from the skin of the head pass back in the spinal V tract to terminate in the funicular nucleus region. In some fishes there is a separate and well defined spinal V nucleus; in others this is indistinguishable from the dorsal cornu and funicular nucleus. The latter is a derivative of the formatio reticularis, which also gives rise to the somatic commissural nucleus. The somatic commissura infima is a continuation of the commissure of the dorsal cornua and fasciculi proprii of the spinal cord. It receives also large additions from the funicular and spinal V nuclei. The commissure of the tubercula acustica in fishes like Gadus and Haploidonotus, where these fuse above the ventricle, is probably homologous, and also a part of the commissural fibers of the cerebellum.

LITERATURE CITED.

CAJAL, S. R. Y.

1896. Beitrag zum Studium der Medulla oblongata, des Kleinhirns und der Ursprungs der Gehirnnerven. Trans. by BRESLER. *Leipzig.*

HERRICK, C. JUDSON.

1905. The central gustatory paths in the brains of bony fishes. *Journ. Comp. Neur. and Psychol.*, vol. 15.

1906. On the centers for taste and touch in the medulla oblongata of fishes. *Ibid.*, vol. 16

1907. A study of the vagal lobes and funicular nuclei of the brain of the codfish. *Ibid.* vol. 17.

1907a. The tactile centers in the spinal cord and brain of the sea robin (Prionotus carolinus L) *Ibid.*, vol. 17.

KAPPERS, C. U. A.

1906. The structure of the teleostean and selachian brain. *Journ. Comp. Neur. and Psychol.* vol. 16.

1907. Untersuchungen über das Gehrin der Ganoiden Amia calva und Lepidosteus osseus *Abhandl. Senckenbergischen Naturf. Gessellschaft, Frankfurt,* Bd. 30, Heft 3, 1907

FIG. 1. Parasagittal section through the oblongata of a specimen of Ameiurus nebulosus about 2 cm. long, showing the visceral commissural nucleus and the secondary tract from it to the formatio reticularis. GOLGI method. ×42.

FIG. 2. Section through the widest part of the somatic commissura infima and funicular nuclei of the adult Ameiurus nebulosus. Method of WEIGERT-PAL. ×35. From the *Journ. Comp. Neurol. and Psychol.*, vol. 16, p. 425 (HERRICK '06, Fig. 4).

The lateral funicular nucleus appears external to the spinal V tract and its nucleus, receiving many fibers from the fasciculus dorso-lateralis and sending large tracts to the formatio reticularis and the commissura infima. This commissure receives also a large mass of fibers from the median funicular nucleus and probably also from the spinal V nucleus. Fibers are seen passing from the spinal V tract into its nucleus, and some pass through this nucleus to end in the median funicular nucleus.

com. inf., somatic part of commissura; *f.d.l.*, fasciculus dorso-lateralis; *f.l.*, fasciculus lateralis; *f.l.m.*, fasciculus longitudinalis medialis; *f.r.*, formatio reticularis; *f. v.*, fasciculus ventralis; *n.amb.*, caudal end of nucleus ambiguus; *n.fn.l.*, lateral funicular nucleus; *n.fn.m.*, median funicular nucleus; *n.sp.V.*, spinal V nucleus; *s.*, secondary tracts from spnial V nucleus and median funicular nucleus; *sp.V.tr.*, spinal V tract; *tr. sp.b.tect.*, rectus spino- et bulbo-tectalis (lemniscus); *w.* tractus bulbo-spinalis; *x.*, tractus spino-cerebellaris; *y.*, secondary tracts from lateral funicular nucleus and other elements for the commissura infima; cf. Fig. 3.

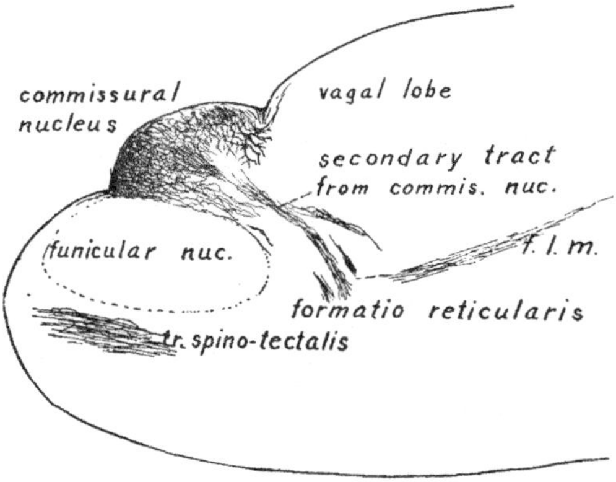
commissural nucleus
vagal lobe
secondary tract from commis. nuc.
funicular nuc.
f. l. m.
formatio reticularis
tr. spino-tectalis

FIG. 1.

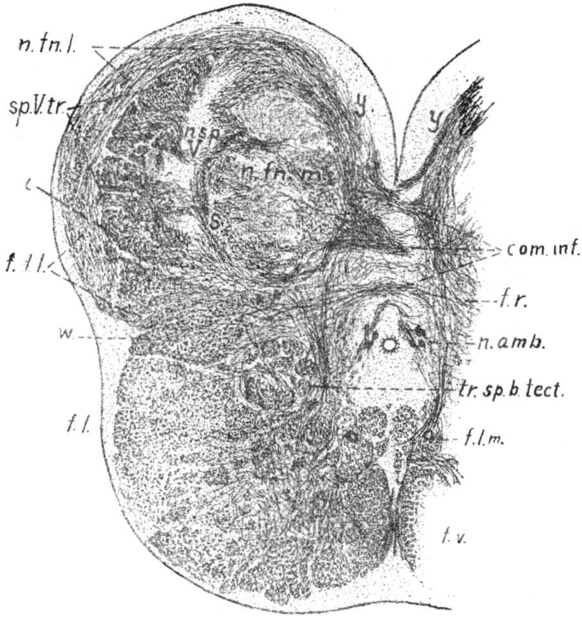
n. fn. l.
sp. V. tr.
n. sp. V
n. fn. m.
y
y
s
com. inf.
f. r.
w.
n. amb.
tr. sp. b. tect.
f. l.
f. l. m.
f. v.

FIG. 2.

FIG. 3. Transverse section through the funicular nuclei and visceral commissural nucleus of CAJAL of Ameiurus nebulosus. GOLGI method. ×75. The drawing is a composite of two adjacent sections, the right side drawn wholly from one and the left from the other. From the *Journ. Comp. Neurol. and Psychol.*, vol. 16, p. 437 (HERRICK '06, Fig. 13).

The endings of the descending sensory vagus root in the commissural nucleus are shown and of the spinal V tract and dorso-lateral fasciculus in their nuclei. One big neurone of the somatic commissural nucleus is impregnated and a few neurones in the lateral funicular nuclei. Dendrites (and probably axones) of the latter accompany fibers from the dorso-lateral fasciculus into the somatic commissura infima in the tract marked *y* in Fig. 2. Cf. HERRICK '06, Fig. 14, for another similar section.

FIG. 4. A transverse section through the middle of the visceral commissural nucleus of the carp, Cyprinus carpio, showing the decussation of fibers of the descending sensory root of the vagus in the commissura infima, also the decussation of fascicles from the dorso-lateral fasciculus (*dec. f.d.l.*) and funicular nuclei in the somatic portion of this commissure. The area designated *funicular nucleus* contains also the spinal V nucleus and receives in this section thick bundles of fibers from the descending secondary gustatory tract from the facial lobes (*desc. sec. VII*). Method of WEIGERT-PAL. ×30.

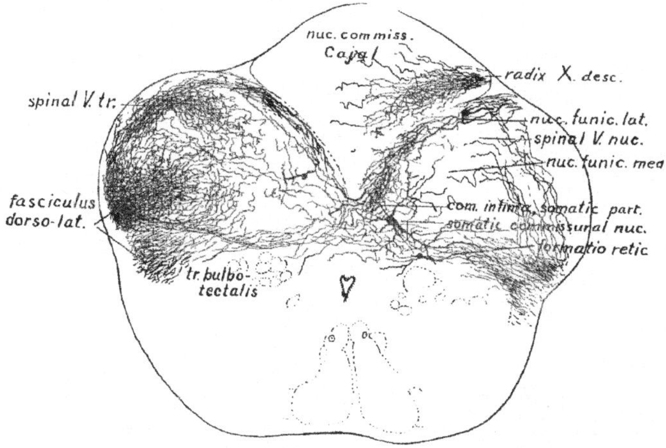
nuc. commiss. Cajal
radix X. desc.
spinal V. tr.
nuc. funic. lat.
spinal V. nuc.
nuc. funic. med
fasciculus dorso-lat.
com. infima, somatic part.
somatic commissural nuc.
formatio retic
tr. bulbo-tectalis

FIG. 3.

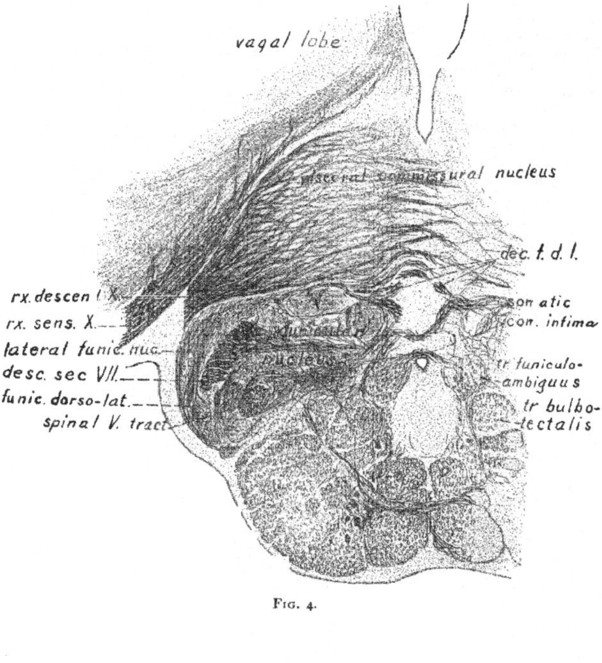
vagal lobe
visceral commissural nucleus
dec. t. d. l.
rx. descend. X
rx. sens. X
lateral funic. nuc.
desc. sec VII
funic. dorso-lat.
spinal V. tract
somatic com. infima
tr. funiculo-ambiguus
tr. bulbo-tectalis

FIG. 4.

FIG. 5. A neurone of the somatic commissural nucleus of Catostomus commersoni. GOLGI method. X 130.

The cell body lies just to the right of the median line, in the same position as the neurone of Ameiurus marked *somatic commissural nucleus* in Fig. 2. Several dendrites extending across the meson soon pass out of the plane of the section. One large dendrite spreads throughout the whole ventral part of the combined funicular nucleus and spinal V nucleus; another spreads throughout the whole of the dorsal part of the formatio reticularis.

FIGS. 6 TO 10. A series of transections through the region of the commissura infima in the adult conger eel (Conger conger), all drawn to the same scale. WIEGERT method. X 26.

FIG. 6. Section through the brain of Conger just caudad of the vagal lobes, passing through the visceral commissural nucleus which is composed of a dense mass of neuropil and a cluster of small cells in the middle. Unmedullated fibers cross the meson, constituting the visceral commissura infima; others pass as uncrossed secondary visceral tracts to the formatio reticularis. Dorsally of this nucleus is the most cephalic part of the somatic commissura infima, consisting of medullated fibers from the caudal ends of the tubercula acustica. The somatic sensory field laterally of the nucleus extends caudad from the tuberculum acusticum to the funicular nuclei. It contains both neuropil and fine medullated fibers diffusely distributed. Internal arcuate tracts pass from this field into the ventral commissure.

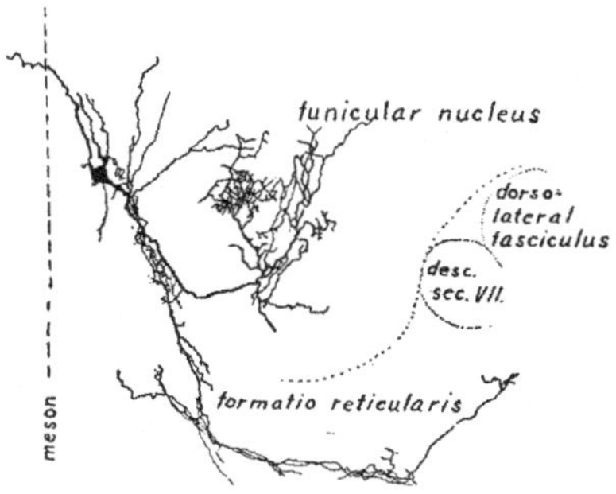

funicular nucleus
dorso-lateral fasciculus
desc. sec. VII.
formatio reticularis
meson

FIG. 5.

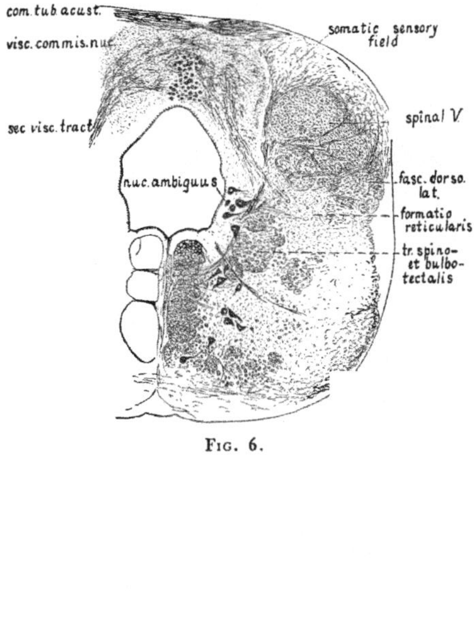

com. tub. acust.
visc. commis. nuc.
sec visc. tract
nuc. ambiguus
somatic sensory field
spinal V
fasc. dorso. lat.
formatio reticularis
tr. spino- et bulbo-tectalis

FIG. 6.

FIG. 7. Section through the brain of Conger .22 mm. farther caudad, showing the visceral commissura infima and nucleus, and the somatic commissura infima passing between the somatic sensory fields. The latter area at this level may be regarded as lateral funicular nucleus. Internal arcuate fibers pass from it to the ventral commissure.

FIG. 8. Section through the brain of Conger .4 mm. farther caudad at the level of the first dorsal root of the first spinal nerve. At this level the visceral commissural nucleus is less sharply separated from the adjacent somatic centers. A root passes from the nucleus ambiguus out into the first dorsal spinal root (*rx. XI?*). Its peripheral course is unknown. The somatic sensory field is desginated *nuc. funiculi lateralis*, and the cephalic end of the median funicular nucleus appears ventrally of the spinal V tract. This nucleus receives termini of the spinal V tract and of the first spinal nerve and is related with the fasciculus dorso-lateralis. Internal arcuate fibers from both the lateral and the median funicular nuclei cross in the ventral commissure and largely enter the ventral funiculi within which they turn caudad for the most part.

FIG. 9. Section of the brain of Conger .3 mm. farther caudad, through the somatic commissural nucleus. The visceral commissura infima and its nucleus have disappeared, vestiges only being apparent dorsally of the somatic commissural nucleus. The portion of the somatic commissura infima which crosses between the somatic sensory fields dorsally of the visceral nucleus (Figs. 6, 7, 8) has also disappeared, but tracts passing forward into that commissure from the lateral funicular nucleus are seen at *a*. A broad diffuse somatic commissura infima appears ventrally of the visceral nucleus associated with the somatic commissural nucleus. The latter is also closely associated with the formatio reticularis and median funicular nucleus.

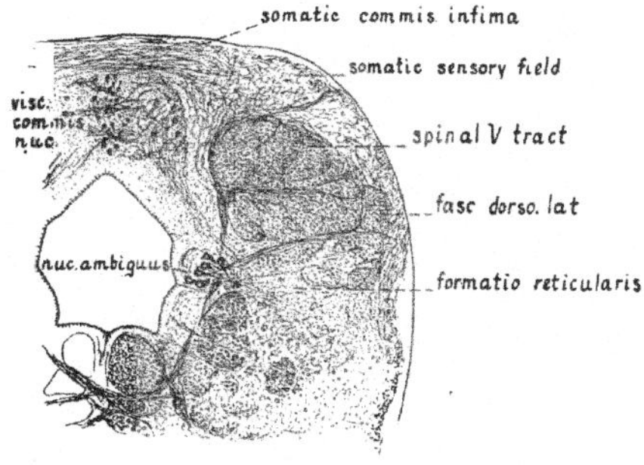

FIG. 7.

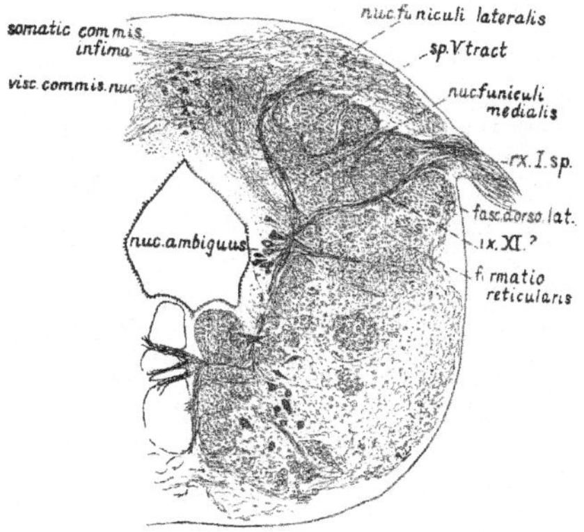

FIG. 8.

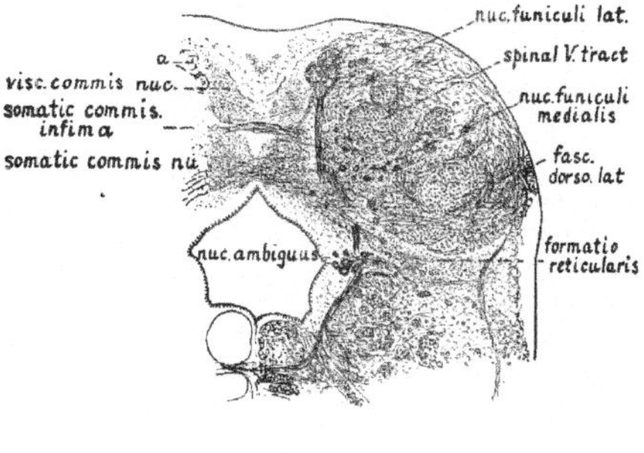

FIG. 9.

FIG. 10. Section through the brain of Conger .6 mm. farther caudad, passing through the median funicular nucleus at its widest part and including the lowest spinal vestiges of the somatic and visceral commissural nuclei and the nucleus ambiguus. The funicular nucleus includes also the spinal V nucleus and passes caudad directly into the dorsal cornu.

FIGS. 11 AND 12. Transverse sections through the commissural nuclei of Amia calva L. Drawn from sections of a young fish about 7 cm. long stained in iron hæmatoxylin, kindly loaned to me by Professor CHARLES BROOKOVER.

FIG. 11. Section through the visceral commissural nucleus at the level of the most caudal rootlet of the vagus nerve. The area marked *somatic sensory field* contains the extreme cephalic part of the funicular nucleus and the *fasc. dorso-lat.* contains the spinal V root in addition to other elements. ×55.

FIG. 12. Section through the somatic commissural nucleus at the level of the first dorsal spinal root. The visceral nucleus has entirely disappeared at this level. ×55.

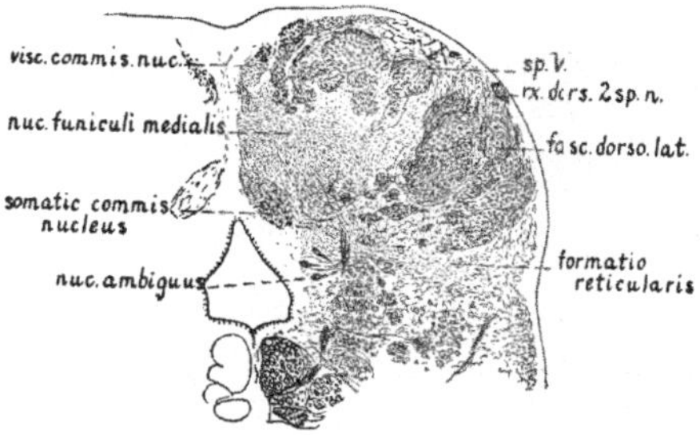

FIG. 10.

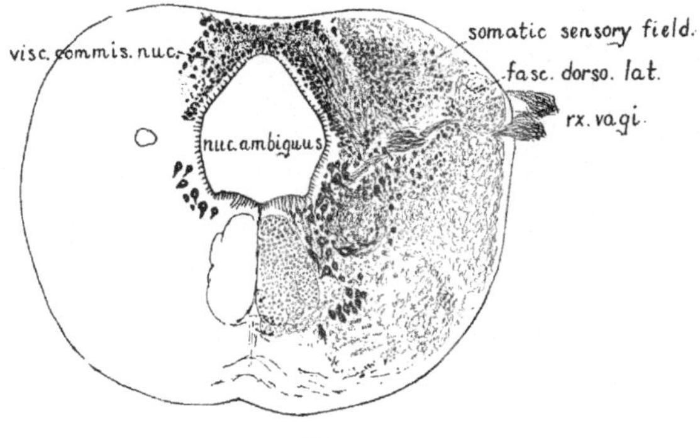

FIG. 11.

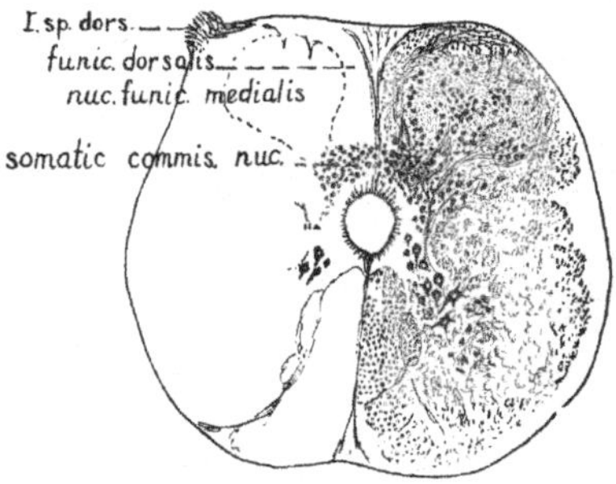

FIG. 12.

EVERSION AND INVERSION OF THE DORSO-LATERAL WALL IN DIFFERENT PARTS OF THE BRAIN.

BY

C. U. ARIËNS KAPPERS, *Amsterdam.*

WITH FIVE FIGURES.

In the *Anatomische Anzeiger*, *Bd.* xxx, 1907, and in a more extensive way in the *Folia Neuro-biologica*, Heft 2, 1908, THEUNISSEN and I have given a comparative description of the different forms of forebrain, as they occur in vertebrates, from which it resulted that in Cyclostomes and in Selachians the upper part of the lateral wall of the forebrain is bent in a medio-dorsal direction, forming a sort of pallium, whereas in Ganoids and Teleosts this same part is bent ventro-laterally, so that the primitive brain-mantle which is inverted in the former is everted in the latter. I further called attention to the fact that this primitive mantle should be called *palæo-pallium*, as it is older than the archipallium, receiving only secondary olfactory fibers, whereas the archipallium receives tertiary olfactory fibers, and further that in those animals where the palæo-pallium is everted it is always greatly reduced in size compared to the inverted palæo-pallium of Cyclostomes and especially of Selachians, and that this reduction of the palæo-pallium gives rise to the formation of the medial epistriatum[1] of Ganoids and Teleosts, which has a vicarious function. For this reason we find either a large and inverted palæo-pallium and a feebly developed epistriatum or an everted palæo-pallium and a large epistriatum.

The ontogenetic development of the prosencephalon in Ganoids (ALLIS, v. KUPFFER) makes it probable to me that these differences find their origin in the form of the skull in embryos, which in a certain stage of development probably pressed on the brain, so that an extensive growth of the dorso-lateral part of the forebrain wall was made impossible. The ventral part of it took then a great deal of its function by means of an enlargement of the striatum

Not visible in Fig. 2, which is drawn after a section anterior to the epistriatum.

(epistriatum) by which again the dorso-lateral wall was more pushed in a ventro-lateral direction.

That really the large epistriatum is an important factor for the eversion of the dorso-lateral forebrain wall is clearly proved by the Teleosts, where the epistriatum on an average is larger than in the Ganoids, and consequently the eversion of the palæo-pallium is also more striking, The same is seen in one of the bony Ganoids, Amia calva, where the epistriatum is larger in the middle of the forebrain, where consequently the eversion of the latero-dorsal brain wall is also more striking[2] than in the frontal

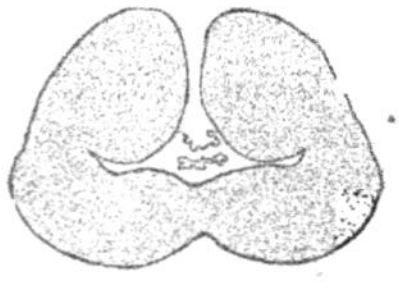

Fig. 1.

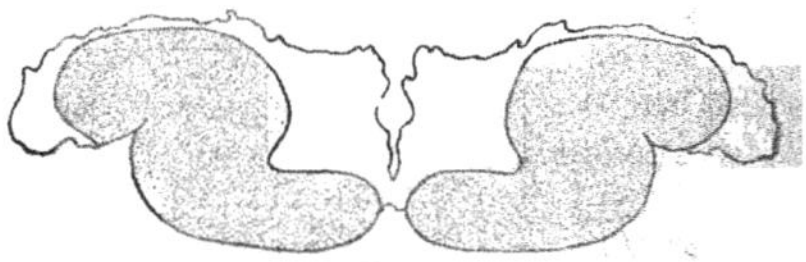

Fig. 2.

Fig. 1. Frontal section through the posterior part of the forebrain of Galeus canis.
Fig. 2. Frontal section through the anterior part of the forebrain of Amia calva.

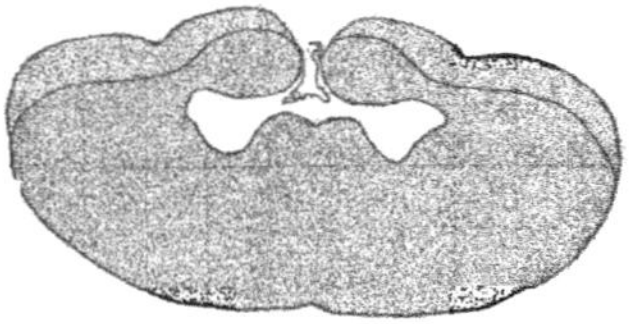

Fig. 3.

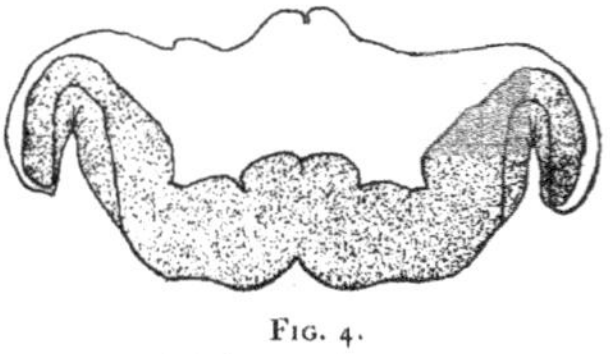

Fig. 4.

Fig. 3. Frontal section through the medulla oblongata of Galeus canis.
Fig. 4. Frontal section through the medulla oblongata of Hexanchus griseus.

or caudal parts of the fore-brain. These differences and homologies were proved by an exact study of the course of the afferent forebrain tracts, which (namely, the tr. tæniæ) proved that this interpretation was right.

Nearly the same ideas about the different forms of forebrain have been published by Studnicka, who however gave only morphological proofs for this conception and whose interpretations were not generally, or rather were generally not, accepted, probably because the conception of a palæo-pallium was never

[2] Chimæra monstrosa has a forebrain of which the morphology shows both forms of development, as in the frontal part of the forebrain the palæo-pallium is large and inverted, whereas in the posterior part of it, it is everted and reduced.

exactly defined; and he with FRIEDRICH MEYER regarded the olfacto-habenular tract, which in Cyclostomes for the greater part originates in the palæo-pallium, as a homologue of the cortico-habenular tract of Reptiles and Mammals, not making an exact distinction between the pallium of fishes and the archipallium of higher vertebrates, which was not right, an archipallium and archicortex being entirely absent in the former. Amongst others Prof. J. B. JOHNSTON criticised this point of STUDNICKA's work and I can only join him in this.[3]

Referring for the description of the forebrain tracts in different animals (I studied the vertebrate series from the Cyclostomes to the Chiroptera) to my work in the *Folia Neuro-biologica*, I here only want to draw attention to the fact that the differences above mentioned for the dorso-lateral forebrain wall also occur in other parts of the brain.

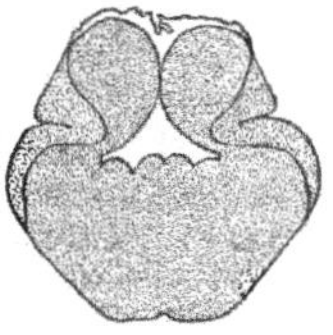

FIG. 5.

FIG. 5. Frontal section through the medulla oblongata of Chimæra monstrosa.

If we compare figs. 1 and 2, which represent two different forebrain types (Galeus canis and Amia calva) with figs. 3 and 4 which are made after sections through the medulla oblongata of different Plagiostomes, we see at once that the same contrast concerning the form of the dorso-lateral wall is also present between the hindbrain of Galeus canis on one side and Hexanchus griseus on the other.

The nucleus of the nervus lateralis anterior, which in most sharks is bent inward, so that it lies under the cerebellum, is turned outward in Hexanchus, and Chimæra monstrosa (fig. 5) exhibits a character which keeps about the middle between these extreme forms, as the same nucleus, though inverted, is not nearly

[3] On the other hand I do not consider Professor JOHNSTON's nomenclature, as far as concerns this question, very happily chosen, as he would better make a difference between the epistriatum of fishes and the primitive mantle-portion, palæo-pallium, which morphologically are very different things.

as far inverted in this animal as it is in Galeus, Mustelus, Scyllium and other sharks.

Contrary, however, to what is seen in the forebrain, that the eversion always accompanies a reduction of the palæo-pallium, the nucleus nervi lateralis anterior in Hexanchus is not smaller than in other sharks, or, better, this nucleus in Galeus and other Selachians is not larger than in Hexanchus, being never very large in transverse sections (in comparison with inverted palæopallium forms) so that reduction of its size and a replacing of its function, for instance, by further development of the end-region of the VIII and posterior lateral nerve seems not at all necessary in this case. It is not easy to say which factor exactly has caused this strange position of the nucleus lateralis anterior in Hexanchus; it might be a traction in the lateral direction by the anterior lateralis root, about which, however, we have no certainty.

LITERATURE CITED.

STUDNICKA.

Beiträge zur Anatomie und Entwickelungsgeschichte des Vorderhirns der Cranioten. *Sitzungsberichte der böhmischen Gesellschaft der Wissenschaften. Mathem.-naturwissenschaftliche Classe.* 1895, 1896.

Nach einige Worte zu meinen Abhandlungen über die Anatomie des Vorderhirns. *Anat. Anzeiger,* Bd. 13. 1898.

JOHNSTON.

The Brain of Acipenser. *Fischer, Jena,* 1901.

The nervous system of vertebrates. *P. Blakiston's Son & Co., Philadelphia.* 1906.

ARIËNS KAPPERS.

The structure of the teleostean and selachian brain. *Journal of Comparative Neurology and Psychology,* vol. 16. 1906.

Untersuchungen über des Gehirn der Ganoiden Amia calva und Lepidosteus osseus. *Abhandlungen der Senckenbergischen Naturforschenden Gesellschaft in Frankfurt a/M.* Bd. 30. 1907.

ARIËNS KAPPERS and THEUNISSEN.

Zur vergleichenden Anatomie der Vorderhirns der Vertebraten. *Anat. Anzeiger.* Bd. 30. 1907.

ARIËNS KAPPERS (unter Mitwirkung von THEUNISSEN).

Die Phylogenese des Rhinencephalons, des Corpus striatum und der Vorderhirn commissuren. *Folia Neuro-biologica.* Heft 2, 1908. (W. Klinkhardt, Leipzig.)

ALLIS.

The cranial muscles and cranial and first spinal nerves in Amia calva. *Journal of Morphology,* vol. 12. 1897.

v. KUPFFER.

Die Morphogenie des Central-nervensystems. (Abdruck aus dem Handbuch der vergleichenden und experimentellen Entwickelungsgeschichte der Wirbelthiere von O. Hertwig. Bd. II).

The Journal of
Comparative Neurology and Psychology

Volume XVIII NOVEMBER, 1908 Number 5

THE RELATIONS OF COMPARATIVE ANATOMY TO COMPARATIVE PSYCHOLOGY.[1]

BY

LUDWIG EDINGER.

(Translated from the German by H. W. Rand.)

With Five Figures.

The relation between animal psychology and human psychology constitutes an old problem. It has interested me since my earliest years of study. However, when I endeavored to learn from the literature more precisely how brain anatomy and psychic phenomena are related to one another in the lower animals, I discovered something very surprising. It is true that I found in all the text-books very promising illustrations of the brains of sharks, frogs, rabbits, and other animals, yet I remember as if it were today the lively undeception which I experienced when I found that in all the books, even in Wundt's great work, the psychological part of the text made no reference to these illustrations. I discovered that psychology had made no further use of comparative anatomy than, so to speak, as a means of illustrating its texts. I gradually discovered the reason for this. In reality anatomy has had nothing to offer to psychology.

The ideal goal of the study of brain anatomy is a very ambitious one. We desire so thoroughly to understand the organ with which psychic processes are associated that we shall be able to predict its functions, so that where observations are impossible—and that is really the case for a large part of the psychology of the lower vertebrates—we may even infer these functions. To be sure, we are still very far from this goal. When we consider what we know about the human brain, its overwhelming complexity seems even simple compared to what we have observed of its activities. But I hope today to be able to point out that, at least in the realm of comparative psychology, anatomy, pursued always in connection with observation of the living animal, can explain much

[1] An address before the Third Congress for Experimental Psychology.

which has hitherto been unknown, and that particularly it is a source of much stimulation and clarification in the realm of sense psychology. You will recognize with me that even today the constitution of the brain in the lower vertebrates enables us to predict most of the activities which we observe in these animals.

I divide the brain into the *palæëncephalon* and the *neëncephalon.* The palæëncephalon appears, with all its characteristic subdivisions, from cyclostomes to man. No part is ever entirely absent; its type remains unchanged whether we have before us the brain of a shark or the brain of an elephant. It is the oldest portion of the entire central nervous system, and many animals possess nothing but it. The neëncephalon, however, develops

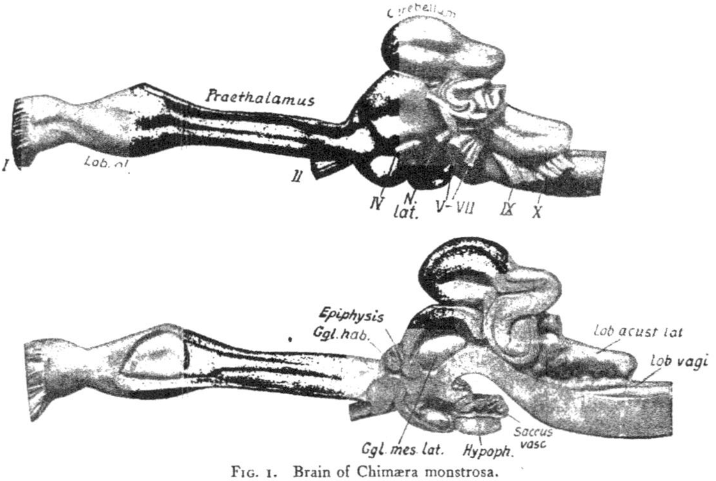

FIG. 1. Brain of Chimæra monstrosa.

above fishes. From very small beginnings in the selachians it increases to that enormous organ, the cerebrum, which in man fills almost the entire skull.

I will illustrate the palæëncephalon by reference to a figure of the brain of Chimæra monstrosa. *This fish in reality possesses nothing but the palæëncephalon.* From the nasal cavities in front, the olfactory nerves lead into the olfactory lobes and terminate there. Behind and above the olfactory lobes lies the corpus striatum covered by a thin plate whence in other animals the neëncephalon is developed. From these parts of the brain fibers extend backward through a long stalk, and in this stalk probably lie also tracts which lead into the forebrain from the peripheral region supplied by the trigeminal nerve. Then ventrally there is a much folded sac, the hypothalamus, upon which lies the hypophysis, while dorsally there is a hollow sphere, the roof of the mid-brain, wherein terminate the optic nerves which may be seen emerging from the chiasma just in front of the hypothalamus. The cerebellum rises in prominent folds over the roof of the mid-brain and behind it one sees a large lobe situated laterally on the oblongata. Here terminate the nerves for the sense of hearing and for the lateral line sense. Below is seen the oblongata which is very well developed, because even in Chimæra the cranial nerves which emerge from it are extraordinarily large.

This apparatus is thoroughly suited by its inner connections, which are now well known, for the reception of sense impressions from the outside world and for conveying them to various places whence groups of motor ganglion cells send out their nerves to the muscles. It also includes a number of special regulatory mechanisms, amongst which the cerebellum is most important. The motor mechanisms are everywhere united into *motor-complexes* in such a way that a sensory impulse brings about the movement, not of a single muscle, but always of a group of muscles adapted for some special action.

Even isolated parts of the palæëncephalon are capable of simple reactions. For example, a ring cut from the neck region of a male frog embraces the female, if the skin of the breast comes in contact with the skin of the female, exactly as the whole animal does (GOLTZ). In fact, the embracing reflex may be induced even if one rubs the skin of the male with the juice of eggs. I need not in this place point out that all the mechanisms for movement—swimming, flying, and the like—are so lodged in the palæëncephalon that the animals are able to execute these movements for some time after the removal of the neëncephalon. This was demonstrated two thousand years ago by the ostriches which ran about Rome's arena with their heads pierced by arrows.

No part of the palæëncephalon can be absent without a corresponding function becoming lost, and all parts develop in size according to the demands which the activities of the animal make upon them. A knowledge of the degree of development is of the highest importance for sense psychology, as may readily be shown by a single example.

That part of the brain which in man and other mammals is undoubtedly concerned with the sense of smell exhibits a constant arrangement and microscopic structure, not only in them but in all vertebrates down to the cyclostomes. We are therefore thoroughly justified in the conclusion that an animal which possesses this part smells, even though from its behavior nothing may safely be inferred. Indeed we may judge of the importance of the sense of smell to the animal according as this organ is large or small in relation to the remainder of the brain.

The olfactory lobes vary greatly amongst the mammals, and the following example which I select from the lizards enables you to see that here, too, very considerable differences in the sense of smell occur in different species. In Chameleon, which obviously seeks its prey by means of the eyes, the roof of the mid-brain, where the optic nerves end, is very large while the olfactory lobes are extremely small. The nearly related lizards have enormous olfactory lobes. It has long been disputed that birds possess a sense of smell. Anatomy however shows us that they possess true, although small, olfactory lobes. This simply and satisfactorily settles the much discussed question, and in fact there are today observations enough which make the presence of an olfactory sense at least very probable. The vulture and the eagle are attracted by concealed prey and ravens find dead animals in a thicket, even when they are deeply buried or covered with snow. ROTHE saw in Lithuania, at a temperature of 24°, that the sea-eagle scented out a dead animal deeply covered with snow, uncovered it, and devoured it. The woodcock finds worms which are deeply buried. It inserts the beak only to withdraw it with a worm. A blackbird pecks industriously at the ground and digs out a grub which lay fully five centimeters below the surface. It must be that our ducks smell under water, for they dive suddenly down into the mud and come up with full bills. The refraction of the water must prevent their seeing anything on the bottom.

The degree of development of the various parts of the palæëncephalon will always give us information as to the possible activities of the animal.

If, however, anatomy solves one problem, it suggests, as the following shows, new problems in sense psychology. Probably in the lizards, and certainly in the birds, a large fiber-tract leading from the nucleus of the trigeminus terminates in a field situated close behind the olfactory apparatus. This field, the lobus parolfactorius, attains an enormous size in birds and the question arises as to what function this structure can serve. The importance of the beak, which is innervated by the trigeminus, the extraordinarily rich trigeminal supply about the mouth and in the tongue, and the further circumstance that stimulation of the lobus parolfactorius produces movements of the beak, lead one to the conclusion that we have here a center for the territory innervated by the trigeminal nerve, that is, a hitherto quite unknown feature of the brain. I am now engaged, together with Dr. KAPPERS, in tracing out this apparatus and we are able even now to declare that *in all vertebrates up to the mammals there must exist an as yet scarcely studied sense which is localized about the mouth and has its center in the lobus parolfactorius.* In the chameleon, with very small olfactory nerves, this lobe is almost as large as in birds, and we should remember that this animal catches its food by extending its tongue. We know how significant in fishes is the investigation of food by means of the barbels and the tip of the snout, how serpents are guided by touching with the tongue, and as we trace these functions, which we may tentatively designate as the *oral sense,* upwards in the scale we find, not without surprise, that even the mammals possess in this same locality a brain structure which is small and atrophied in those whose snout plays no important rôle (man, the apes, and ruminants), but which becomes a giant structure in mammals, of the most diverse orders, so far as they make extensive use of the snout. In the brains of the hedgehog, the mole, the armadillo, also in swine and the elephant, the lobus parolfactorius is strongly developed. In man it has completely disappeared except for a vestige in the atrophied lamina perforata anterior. So much for the oral sense. It is little enough, yet it shows that sense psychology acquires an entirely new problem from anatomical studies.

According to the degree of development of the olfactory lobes in mammals much may be inferred as to the condition of the olfactory sense. Since these matters have been well known from the time of BROCA, I will now only briefly refer to the fact that these lobes in the lower mammals make up more than half of the entire brain, that in the beasts of prey they play an important rôle, that in man and the apes they are reduced to a small structure, while in the aquatic mammals they are completely absent. No one will deny that the various degrees of development of the parts of the palæëncephalic olfactory apparatus furnish most important problems for psychology.

I should like to show you by means of two more examples how the development of the palæëncephalon is dependent upon the demands of the outside world. The roof of the mid-brain, which receives upon the one hand the optic nerves and upon the other hand secondary sensory tracts, is much more strongly developed in birds and fishes than in any other vertebrates, but in blind animals it may be atrophied. In cases where we find such anatomical atrophy, we should be stimulated to investigate the capacity for sight. Then it appears that in animals (Proteus) which are entirely blind certain tracts of other sensory mechanisms are especially strongly developed. Their mode of operation presents still other problems.

The size of the cerebellum is so completely dependent upon life habits that in some sedentary animals it has completely disappeared, while in weak swimmers (eel, flounder) it is very small, but in the strong swimmers and fliers it attains enormous size. In so nearly related animals as the land and water chelonians the former have a cerebellum less than half as large as that of the latter. Much futile work on the physiology of the cerebellum would have been spared us if we had regarded these facts of comparative anatomy.

Finally, let me refer again to the important apparatus of the lateral line sense of the fish. Because of the obviousness of its end organs in the skin, this sense fortunately has found many investigators and now we know through the investigations of FUCHS, and of HUBER that this entire apparatus enables the animal to detect changes of pressure in the water, particularly the resistance which it encounters in swimming. Here anatomy has led to physiological investigation.

The palæëncephalon alone is present in the bony fishes. The activities which depend upon it we will designate as *palæëncephalic activities.* Since in all other vertebrates, with the appearance of the neëncephalon, quite different—neëncephalic—activities make their appearance, it is of the greatest importance to study thoroughly the activities of fishes. The central nervous apparatus of the fish doubtless serves for all the receptions necessary to the animal, for all regulations, and for all the movements which the animal's relation to the outside world demands, that is, for locomotion, for obtaining food, and for the reproductive activities. Not only all the activities which we commonly designate as reflex, but also all instincts, are localized in the palæëncephalon. Flight when surprised, migrations, nest-building, courtship, and many other activities are to be observed in the bony fishes.

On theoretical grounds it has particularly interested me to ascertain if fishes learn. From my own observations, from the literature, and from hundreds of contributions which I have received in response to an inquiry, it is now well established that new kinds of receptions, provided that they affect the inherited motor mechanisms with sufficient intensity or sufficiently often, stimulate them. The animals *learn* in a very moderate degree to modify their activities. One can tame them, and train them not to flee, so that they allow themselves to he held in the hand; or they may be called to food at a certain place or a certain time. They can learn to swim to a particular person who feeds them. These associations become so well established that, for example, my Macropoda, which I never feed myself, swim up as soon as I appear because, five months before, they had always been fed by anyone who approached. A pike which has several times escaped the spear becomes more cautious and learns to avoid it. But fishes always return to the hook so long as the bait presents the same appearance, for it is not the fish which attracts the prey but the prey which attracts the fish. If the bait is unusual in appearance it does not attract. All the experience of anglers goes to show that fish will not go to poorly arranged bait. That, however, does not necessarily indicate intelligence, for if they possessed the genuine intelligence which has occasionally been attributed to fishes, we should expect that sometimes they would be caught by inappropriate bait. As a rule fishes respond to particular sensory stimuli by the execution of certain definite combinations of movements. But their brain is able to relate a new sense impression with a movement combination which formerly had not answered to it. I propose to designate this lowest kind of association by the term *establishing of relations*, but to reserve the term *connecting of associations* for those totally different processes of the brain which we observe after the appearance of the neëncephalon. Such very unlike mechanisms are required for the two processes that this distinction seems well justified.

Since it is certain that the palæëncephalon persists quite unchanged even after a well developed neëncephalon has been added to it, there is no ground for regarding those activities which we recognize as palæëncephalic in one class of animals as anything else or as otherwise localized in higher animals. *Furthermore, we may regard an entire series of acitvities as common to all vertebrates, and we may then seek to ascertain how other activities are added to these when a new structure is added to the palæëncephalon. All sense impressions and movement combinations belong to the palæëncephalon. It is able to establish simple new relations between the two, but it is not able to form associations, to construct memory images out of several components. It is the bearer of all reflexes and instincts.*

Through the separation of palæëncephalic and neëncephalic activities we gain an entirely new point of view and statement of the problem for sense physiology. If the palæëncephalon can not form associations, then those animals which depend upon it entirely, or almost entirely, must remain unaffected by many sense impressions to which, according to our own experiences or according to our knowledge of the sense organs of these animals, we should expect them to give some response by movement. A lizard which listens to the slight rustling of an insect in the grass remains quite at rest, as my own investigations have shown, when one pounds upon a stone just over its head, or when one calls loudly, sings, or makes an uproar. The animal, otherwise so shy that an unexpected shadow or a slight shaking of the ground caused by my step makes it disappear, does not flee. With these new sounds, which biologically it never encounters, it associates nothing, just as a warning placard written in Chinese could never save me from an abyss. The mechanism for conveying new stimulations to the old inherited movement complexes is entirely lacking to it. The reptiles must all appear to us practically deaf, although they do hear. It is said that turtles react to music, but that is yet to be proved. YERKES has demonstrated to us that amphibians do not flee from noises and the sound of a bell. Yet his talented researches have shown us that the acoustic nerve is in some way stimulated by these sounds. It is well known, however, that frogs call loudly at mating time in order to attract the female, and Professor BÖTTCHER has informed me that he was able to attract a tree-frog by imitating its cry with a metal mortar. Clearly, then, these animals hear very well that which concerns them. It has also been shown by PIEPER that in fishes, which according to all the accounts hitherto given, appear so deaf, negative variation in the auditory nerve is caused by the sounding of a tuning fork. How much work has been done entirely in vain because we have not as yet fully appreciated the fact that, in the absence of a mechanism for association, nothing but the biologically adequate stimulus can bring about movement! Why should a fish flee, as we have always hitherto expected, at the sound of a bell or of a tuning fork? Sounds of that kind mean nothing to the animal unless—and I consider this possible—it has been brought into relation with them by training.

Thus we find ourselves compelled to divide sense stimuli into *those which are biologically adequate* and *those which operate only by association.* As one readily sees, here arise new problems for investigation. But now we have reached the limit of the possibilities of the palæëncephalon.

I suspect that thus far in my address you have been under the impression that what I have been presenting is not psychology but physiology. I am entirely in accord with that, if we undertake to draw sharply the line between psychology and physiology upon the ground of our newly acquired anatomical knowledge. No objection can be made if, not for all time, but tentatively, we exclude all the above mentioned activities and also all instincts from purely psychological consideration. As a knowledge of the literature continually reminds me, the instincts hitherto have rendered difficult a consideration of the truly psychological phenomena of animals. In the literature—and one need think only of what has been written about birds—they are continually intruding to pervert our general views. This proposition to regard the simple activities and the instincts of animals as sharply separated from the other psychological processes, a proposition to which I have been able to come only through a comparison of the anatomy with the activities, is not a fundamental one but only methodological. It will call forth your objection. But I hope in the second part of this discussion, which will concern itself with the neëncephalon, to be able to show you that it is not so entirely impracticable.

The neëncephalon, the bearer of the cortex, develops in the roof of the brain, beginning as a rudiment which is evident even in the selachians and becoming more and more conspicuous in amphibians and especially in reptiles. By reference to fig. 2 one may see how the palæëncephalon persists unchanged underneath the very important neëncephalon.

In the neëncephalon of reptiles there appears for the first time, and very definitely, a mechanism which by means of numberless connections within itself provides the possibility for association. In the first rudiment of the cortex these connections are already so numerous that they can scarcely be overlooked. Even in the lizards the number of associations rendered possible by their network is inconceivable.

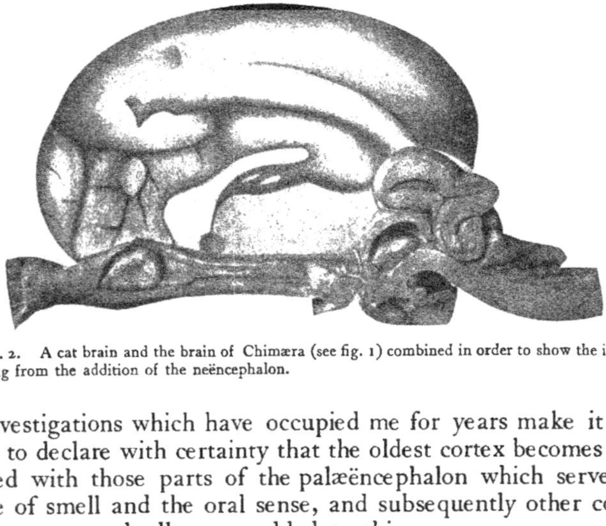

FIG. 2. A cat brain and the brain of Chimæra (see fig. 1) combined in order to show the increase resulting from the addition of the neëncephalon.

Investigations which have occupied me for years make it possible to declare with certainty that the oldest cortex becomes connected with those parts of the palæëncephalon which serve the sense of smell and the oral sense, and subsequently other cortex regions are gradually superadded to this.

With the appearance of the neëncephalon the behavior of the animal becomes completely changed. Let us first consider the obtaining of food, because that is the best of the activities to study —indeed the lower animals present to the observer no other form of activity so often as this. We have recognized as the characteristic of the palæëncephalon that when stimulus and disposition are the same, the same activities always result, so that they may be predicted.

Hungry animals if they possess only the palæëncephalon seize food under all circumstances, provided the stimuli which proceed from it are appropriate, but only then. An animal which is incited to seizing only by a moving body never recognizes the

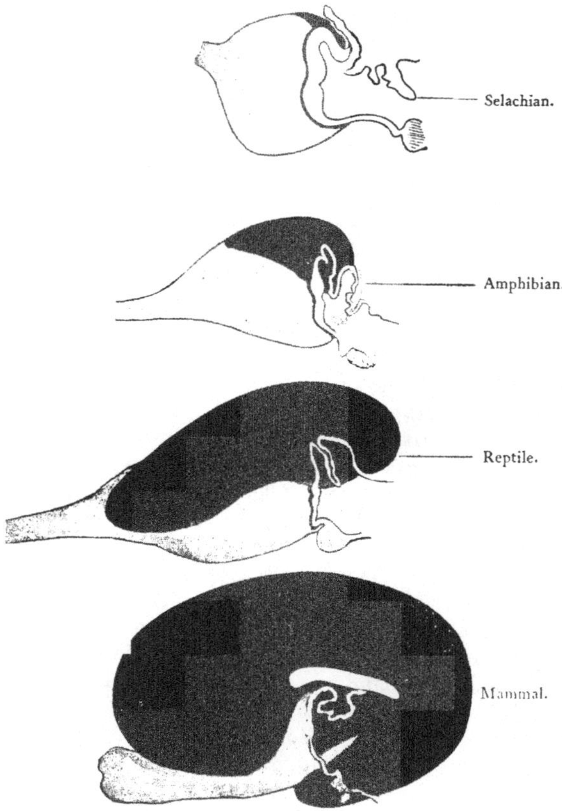

FIG. 3. The evolution of the neëncephalon (black) and the regression of the palæëncephalon (gray).

same body if it is at rest. All of these animals can be caught with bait if one has ascertained the proper stimulus. Fishes which, like the trout, go toward swiftly moving and glittering insects can easily be caught by imitations of such insects constructed of metal and feathers, providing the angler rightly imitates the hopping movement. The entire art of angling, concerning which we possess large volumes, depends upon knowledge of the proper stimulus and upon the excluding of disturbing stimuli, such for example as a thick fish-line. Frogs may be caught by means of heath-berries dangled before them on a string. Even the frog has a rudimentary neëncephalon, but so far as my observations go, it plays no part in the obtaining of food. It still eats palæëncephalically. No matter how hungry a frog may be, it seizes the earthworm only when it crawls, or catches the fly only when it makes some movement. One may lay a worm on the frog's snout or may in any way bring the two in contact, but eating does not result. The worm acts as a stimulus only when crawling, otherwise it is not recognized. One can very clearly observe how the entire operation of eating results from the addition of very definite reflexes. The crawling worm first induces, by way of the optic nerve, possibly also by way of the acoustic nerve, a turning of the head; if it crawls further a new stimulus is added, and then the body is turned, the head sinks and, if the stimulation continues, the seizing results. If, as frequently happens, the animal misses the prey it does not immediately strike at it again. The worm must crawl further and the entire series of reflexes must be repeated. If the worm stops crawling, the series is at that moment interrupted. Upon the other hand, an object which is not eatable gives rise to the seizing act provided that the same stimuli proceed from it as from the worm. Hanau saw a toad follow and repeatedly snap at a blind-worm's tail for hours.

The sense stimuli which lead reptiles to the obtaining of food are not materially different from those just described. Most serpents, all lizards, and the carnivorous turtles appear not to see motionless prey. However, they do not rush recklessly at moving prey as does the frog at the heathberry, but they orient themselves with respect to their food by sniffing or by touching with the tongue. In some cases the stimulus is received through the sense of hearing. But the serpents appear not to use this sense, for a very hungry animal does not change its attitude when a mouse squeaks or a bird calls.

What, then, are the differences in behavior which depend upon the presence of the cortex in serpents? Is it possible from the structure of the cortex to formulate problems for observation, and how far does observation of the behavior of the animals accord with the fact that a new mechanism has been added to the palæëncephalon? The cortex of serpents consists of several layers of cells which are manifoldly connected by means of countless fibers. Judging from our knowledge of mammals, we should naturally take the view that such a mechanism renders possible the holding back of an impression and the associating of one impression with another. The tracts which lead to this association mechanism come from the center for the olfactory and oral senses. Others, such as tracts from the center for the optic nerve, have not yet been found in the animals which have been studied. This is in accord with the fact that the reptiles recognize prey optically usually only as some definite combination, such as the moving mouse, the moving frog. The optic impression of the resting mouse does not by itself suffice to induce seizing. But one readily sees that these animals use the olfactory and oral senses very differently from the amphibians. A serpent, by touching with the tongue, determines whether it has an animal of one kind or of another. It marks where a food animal has been resting and finds it by following it to its lair. By testing the surface of the water with the tongue the ringed-snake determines if there are fish in the water. The oral sense is very much used. Occasionally the animals try to devour pieces of wood upon which prey has rested and left its odor, but after touching they turn away. Zamenis, by touching, selects a pigeon egg from amongst a number of turtle eggs of the same size. The hungry serpent is restless, it makes slight movements, touches the ground, it seeks food—something which the frog is never observed to do.

In another respect reptiles differ from those animals which do not possess a cortex. If the latter fail to obtain the prey which has stimulated them to eat, they remain quiet until a new stimulus appears. Not so with reptiles. Serpents, once stimulated by a jumping frog or a running mouse, *follow their prey* at least for a time and, guided by the olfactory and oral senses, they are able amongst a number of holes to find that particular one into which the prey has crept. Finally, there first appears in reptiles something which indicates that they occasionally foresee what may follow from a certain experience. Many lizards and serpents assume an attitude of defense when danger threatens. They direct the head toward the enemy, raise the forward part of the body, and open the mouth to bite. I have never observed anything of that kind in a palæëncephalic animal.

Probably it is due to the neëncephalon, too, that first in the reptiles we meet with individual differences. Within the same species there are indolent and excitable, dull and lively individuals. Everyone who has kept many mud turtles knows this.

Reptiles learn more easily and quickly than fishes. One can teach turtles to come to be fed at the sound of clapping. They also learn to follow correctly the path which leads to good food and will work all day long against obstacles. SIEGWART's turtles worked themselves repeatedly through successively narrower gratings to an aquarium containing Proteus. They even climbed over fences which were interposed and placed themselves on edge in order to get between the bars. Finally, reptiles which naturally follow only jumping prey learn to recognize resting prey.

Aside from the obtaining of food, the life of reptiles consists merely in resting and sunning themselves. Therefore, in so far, we recognize no very marked differences between reptiles and amphibians. Most important in the psychological behavior of reptiles is the fact that the animals are no longer always dependent upon the sense impression of the moment, but that earlier impressions influence them. Further, they associate certain sense impressions which lie within the realm of the olfactory and oral senses, and turn them to account; they learn more easily than fishes and amphibians; occasionally they foresee; and they exhibit individual differences. There can be no doubt that all of these facts are referable to the appearance of a cortex in the nëencephalon.

So far as our observations go at present, genuinely psychological processes make their appearance at this point in the animal series. It is certainly possible that they may occur even in the selachians and particularly in the amphibians in connection with the beginnings of the cortex, but they are so rudimentary that they will probably be found only when attention is especially directed toward them.

From the brain of the reptiles two different types of brain are derived. One, the type found in the lower mammals, develops by increase of the cortex; the other is the avian type.

In birds the cortex is more highly developed than it is in reptiles. The increase in the bulk of the brain, however, actually results from the enlargement of the palæëncephalon whose various parts here reach a perfection which they attain nowhere else. We know that in birds nearly all parts of this palæëncephalon are connected with the cortex; that particularly the brain-part connected with the oral sense (the lobus parolfactorius) is of enormous size; and finally, that from the optic termini an especially large number of fibers lead to the cortex.

A priori one would infer from this structure that the instinctive actions must be of much greater variety and perfection, and that also the capacity for forming associations must be much greater than in reptiles.

As a matter of fact, the investigation of the psychic behavior of birds—I am speaking now of nest-building, migration, and courtship—has met certain difficulties in the numerous strong instincts whose perfection is so great that it has not always been possible to distinguish them from activities which are dependent upon the cortex. Although we possess many works dealing with the behavior of birds, the observers have only very seldom endeavored to maintain an objective point of view. I regard the works of WURM and GREPPIN as among the best. If one leaves out of account instinctive actions, one is struck with the fact that the new (as compared with reptiles) connections of the palæëncephalic optic termini with the neëncephalic cortex play the all important rôle in the behavior of the animal. Birds see and recognize; a single visual characteristic of the object often enables them to judge of the whole. Their actions are for so long a time influenced by a thing seen that one must infer that they possess and make use of memory images. Ducks soon recognize the hunter's screen and avoid it after several of their number have been killed. EIMER relates that on the first day he caught thirteen sparrows in a newly constructed sparrow trap, but afterwards no more. Two years later the trap was again set up but not a bird went into it. Game birds learn so well to recognize the hunter that they distinguish him from wood-choppers, wagons, horses and the like, just as do wild mammals. Upon this fact are based many of the tricks of hunting, such as stealing up behind a horse or arranging a trap under a screen upon which is painted a cow. When partridges see the falcon they crouch down anxiously. Thus it is often the practice to arrest a scattered covey by means of a painted paper kite and then to kill the birds (WURM). Only birds can be frightened from fields by scarecrows, only the bird of prey recognizes its victim in the far distance, only amongst birds do we find creatures which, like the carrier pigeons, retrace the path once seen.

Everyone who in winter strews crumbs of bread from a window observes how upon all sides the birds watch his action, but approach only after he has closed the window.

Accordingly these animals, which are the first to possess an optic tract leading from the palæëncephalon into the cortex, are likewise the first to so far understand and retain optic impressions that they may long afterward be employed to bring about relatively complicated actions depending upon associations of many kinds.

But when SCHRADER deprived his falcon of its cortex it fell at once into the condition of a palæëncephalic animal. Running mice were readily caught by the injured bird, but mice which had crept under the falcon's wings remained unrecognized and gradually devoured their host, which, as a merely palæëncephalic animal, could no longer recognize them associatively.

Birds hear very well. It is probably only a palæëncephalic hearing when the female follows the call of the male; but magpies, ravens, and parrots learn to come when their names are called and birds of many kinds learn to imitate whistled melodies or even pronounced words. In spite of many anecdotes, there is as yet no conclusive proof that parrots understand language, but there can be no doubt that they employ the same words upon similar occasions.

It is anatomically uncertain if the oral mechanism is connected with the cortex, and the behavior of the animals scarcely indicates that it is. The action of a bird in digging up worms which are six centimeters under ground can just as well be accomplished through the mechanism of the palæëncephalon.

Quite new as compared to the reptiles are certain indications of true intelligence. Of course it is difficult here to avoid being deceived as to the significance of acts. But when a parrot learns always to plunge its hard bread into water before eating, and when animals which have been repeatedly disturbed at one nesting place remove the nest and seek a place inaccessible to the danger first experienced, we can find no other name for this kind of association-formation than intelligence. This intelligence is very clearly seen when an animal assures itself of safety, a matter which GREPPIN has especially studied. Every bird before alighting or before taking food looks about on all sides very knowingly. This inspection is not an inherited habit but, as shown by GREPPIN's observations on young blackbirds, is acquired. Very young birds at every jar, every noise, stretch out the head and open the bill. It is only later that they learn the opposite behavior. Ground birds acquire the habit of assuring themselves much earlier than birds of the air. Visual images and associations surely play an important rôle in this assurance. Crows, which remain quietly at rest as one approaches, fly away as soon as one of them is shot, and thereafter they can be shot only from ambush.

Birds very carefully seek out a place for their nest and often reinforce it purposefully with very remarkable supports. It is not conceivable that all these actions should take place without the participation of the cortex, for they involve numerous memories and associations.

It must also depend upon the presence of the cerebral cortex that birds are particularly easy to tame and that they may be trained to a large number of performances. Thus, they learn to modify the old hereditary behavior; in fact such activity rules in close relation with instinct, as one may see in the feeding of nestlings by the mother, or in the teaching of young storks to fly.

What the anatomy of the bird brain leads one to expect is in excellent accord, as one may see, with the results of studying the behavior. The differences between reptiles and birds are easily referable to anatomical differences in the brain. To be sure, it must be the task of further observation to elaborate what is here set forth; above all things, to determine what activities of the lower vertebrates are palæëncephalic and what are neëncephalic. Accordingly, reptiles and birds must be studied much more thoroughly than they have been hitherto, because we have demonstrated the first appearance in them of activities which depend upon a cortex and these activities occur in relative simplicity. It is also an important question whether neëncephalic reflexes and instincts exist.

We have come to know fishes as strictly palæëncephalic animals. In reptiles and birds a small neëncephalon coöperates. Finally, in the mammals we meet a brain which has so large a neëncephalon that we may well expect a subordination of reflexes and instincts to associative and intelligent actions. That, in fact, is the case with those mammals in which the neëncephalon includes much more than half the bulk of the entire brain. But in many families there is very little advance beyond the condition prevailing in birds, for example in the hedgehogs and the moles. In the mice, the rabbits, in fact in nearly all the rodents the two parts are about evenly balanced. What we know of the intelligence of these animals—and that is little enough—is in very close accord with the condition of the brain. In fig. 4 is represented a hedgehog brain in which one may readily see that the two portions of the brain, somewhat separated from one another by a horizontal furrow, are of approximately equal size.

It would exceed the limits within which I must keep, if I should more than cursorily outline the task of him who undertakes, through anatomy, to be of use to mammalian psychology. The

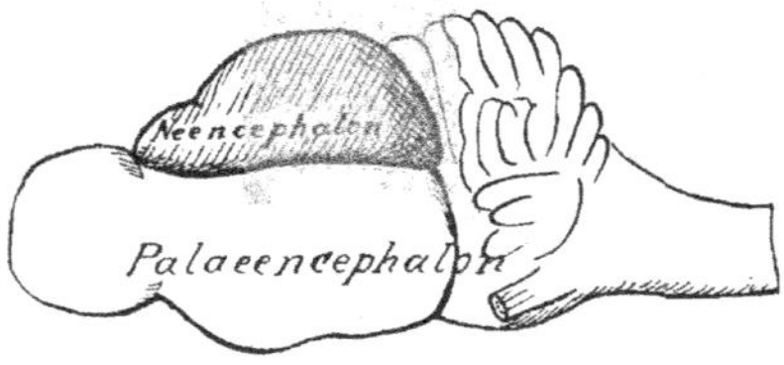

Fig. 4. Brain of a hedgehog.

oldest part of the neëncephalon, namely the olfactory and parolfactory or oral centers which are present in reptiles, persists as the horn of Ammon. As shown in fig. 5, with the evolution of the other portions of the cortex it becomes pushed to the median plane and rolled upon itself. It is very probable that the function of this part of the brain, which because of its age has been designated as the *archipallium* in distinction to the remaining cortex or *neopallium*, remains unchanged, but unfortunately we lack any far-reaching observations on animals which have been subjected to operations. After all, we know that the archipallium is well developed only in those animals which are guided to a large extent by their olfactory sense and that it is smaller but not lacking in animals which have an atrophied olfactory mechanism. The whales have rudimentary olfactory nerves but the archipallium has not entirely disappeared. Since we know that also the central connections of the oral mechanism (around the snout) terminate here, the

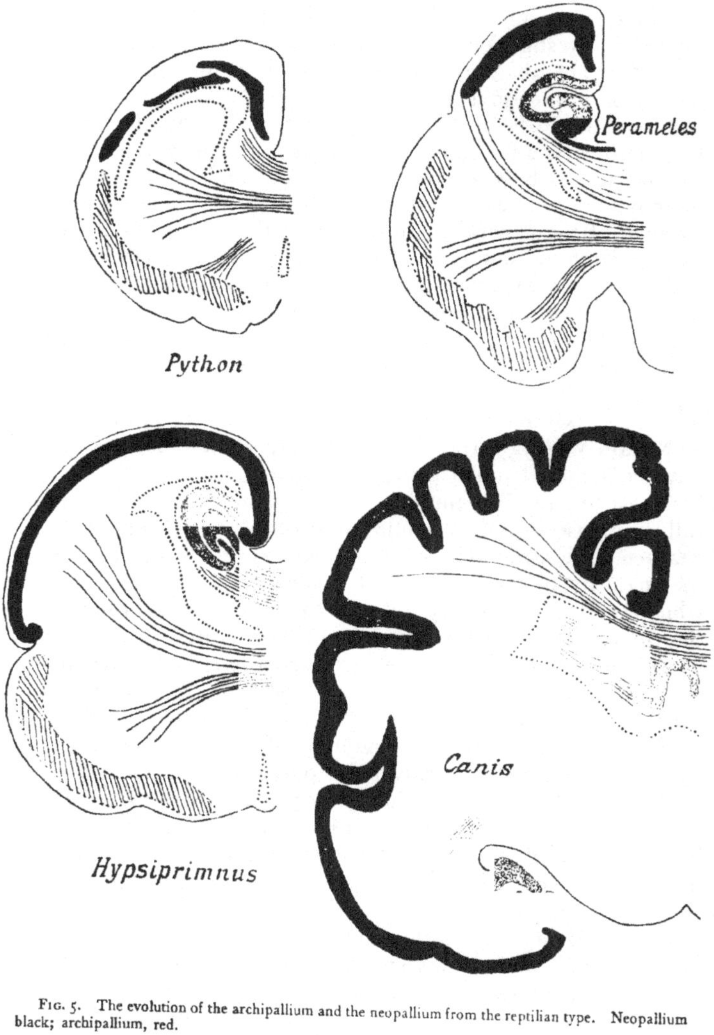

FIG. 5. The evolution of the archipallium and the neopallium from the reptilian type. Neopallium black; archipallium, red.

study of the whale brain in its psychological relations informs me that these animals, which in life are entirely beyond the reach of investigation, do not possess the olfactory sense, but that they are able to behave associatively and intelligently by means of impressions received through the trigeminal nerve.

Now it would be of the greatest importance if we also knew well the anatomy and functions of the parts of the neopallium, because the differences between the human brain and the brain of other mammals depend essentially upon the degree of its development. In spite of the general impression that the mammalian cortex is well understood, not only anatomically, but, through numerous and celebrated experiments, also physiologically and clinically, I must here assert that we know surprisingly little as soon as one inquires how much our knowledge tells us about its functions. It is only in recent years that the researches of FLECHSIG have shown us how extensive is the region of the cortex which we must regard as essentially the association field. Further, the works of S. RAMÓN Y CAJAL, BRODMANN, MOTT, CAMPBELL, and others, have contributed much new information regarding the cortex of the most different mammals. Only recently has it been successfully demonstrated that definite parts of the cortex exhibit definite characteristic structures. We do not yet know the function of the majority of the cortex fields which are today distinguished, which looks rather badly for our knowledge of the functions of the cortex. But there are two definite structures whose rôles are so well understood that wherever these structures are encountered one may expect the same functions. These are the so-called motor cortex and that type of cortex which encloses the visual areas.

Here we stand at the very beginning of important and necessary researches and one can only point out where they may possibly lead. It is, however, absolutely necessary that we undertake them, for only such investigations, carried on at the same time both anatomically and psychologically, can help us where observations of the living animal are impossible. I here call attention to the huge brain of Tursiopsis tursio, a dolphin, of which we know little more than that it swims about ships. Its much folded surface is doubtless much greater than that of the human brain. But of its operations we know absolutely nothing. It is entirely conceivable, however, that further adequate study of the cortex may show us what its possibilities are. The same thing holds true for animals which are better known to us. I well know what the American authors, in particular, have accomplished in the way of exact observation of mammals. But their observations as well as those of popular psychology, which are by no means to be neglected, show clearly how far we are from our goal and how difficult it is to make rapid progress by means of such studies. For what do we know even of the small mammals which are so familiar to us, such as the mice and the rabbits which live about us in our laboratories?

Even now the anatomy of the mammalian brain seems to have afforded one definite result. It is assuredly an error to attribute to man the greatest power of association in all fields. The degree of development of certain parts of the cortex makes it appear highly probable that, as the popular mind has long held, many mammals far excel man in their capacity for observation and association in certain fields.

At this point, where anatomy is still found wanting, where in future it will be called upon to render the highest of service, I will conclude. I hope that I have succeeded in showing how much more rapidly we shall progress if psychological observation and anatomical study are united first in that field where even now some tangible results would be forthcoming from such union—namely in the study of those animals which possess only a very simple neëncephalon. Where the two have hitherto attempted to work together, so unattainable has been the problem set that it was quite impossible to gain any valuable results.

There is much more to be done; and in doing it let us never forget GOETHE's words, "Willst du ins Unendliche schreiten, geh' erst im Endlichen nach allen Seiten."

THE RELATION OF STRENGTH OF STIMULUS TO RAPIDITY OF HABIT-FORMATION

BY

ROBERT M. YERKES AND JOHN D. DODSON.

(From the Harvard Psychological Laboratory)

WITH FIVE FIGURES.

In connection with a study of various aspects of the modifiability of behavior in the dancing mouse a need for definite knowledge concerning the relation of strength of stimulus to rate of learning arose. It was for the purpose of obtaining this knowledge that we planned and executed the experiments which are now to be described. Our work was greatly facilitated by the advice and assistance of Doctor E. G. MARTIN, Professor G. W. PIERCE, and Professor A. E. KENNELLY, and we desire to express here both our indebtedness and our thanks for their generous services.

The habit whose formation we attempted to study quantitatively, with respect to the strength of the stimulus which favored its formation, may be described as the white-black discrimination habit. Of the mice which served as subjects in the investigation it was demanded that they choose and enter one of two boxes or passage-ways. One of the boxes was white; the other black. No matter what their relative positions, the subject was required to choose the white one. Attempts to enter the black box resulted in the receipt of a disagreeable electric shock. It was our task to discover (1) whether the strength of this electric stimulus influences the rapidity with which dancers acquire the habit of avoiding the black passage-way, and if so, (2) what particular strength of stimulus is most favorable to the acquisition of this habit.

As a detailed account of the important features of the white-black visual discrimination habit in the dancer has already been published,[1] a brief description of our method of experimentation

[1] YERKES, ROBERT M. The dancing mouse. New York: The Macmillan Company. See especi ally p. 92, et seq. 1908.

will suffice for the purposes of this paper. A sketch of the experiment box used by us in this investigation appears as fig. 1, and a ground plan of the box with its electric attachments, as fig. 2.

This apparatus consisted of a wooden box 94 cm. long; 30 cm. wide; and 11.5 cm. deep (inside measurements), which was divided

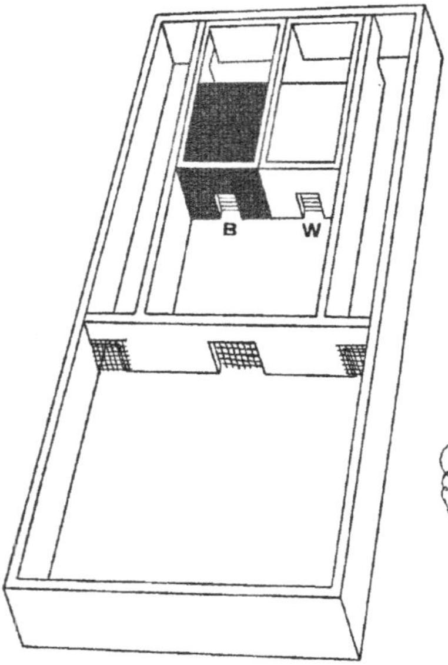

Fig. 1.

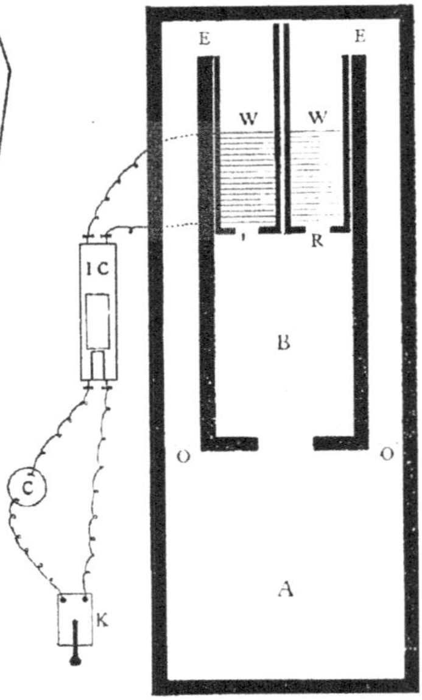

Fig. 2.

Fig. 1. Discrimination box. *W*, electric box with white cardboards; *B*, electric box with black cardboards.

Fig. 2. Ground plan of discrimination box. *A*, nest-box; *B*, entrance chamber; *W W*, electric boxes; *L*, doorway of left electric box; *R*, doorway of right electric box; *E*, exit from electric box to alley; *O*, swinging door between alley and *A*; *IC*, induction apparatus; *C*, electric battery; *K*, key in circuit.

into a nest-box, *A*, (fig. 2) an entrance chamber, *B*, and two electric boxes, *W*, *W*, together with alleys which connected these boxes with the nest-box. The doorways between the electric boxes and the alleys were 5 by 5 cm. On the floor of each electric box, as is shown in the figures, were the wires of an interrupted circuit which could be completed by the experimenter, by closing the key *K*, whenever the feet of a mouse rested upon any two adjacent wires in either of the boxes. In this circuit were an electric battery and a Porter inductorium. One of these electric boxes bore black cards, and the other white cards similarly arranged. Each box bore two cards. One was at the entrance on the out-

TABLE 1.

Positions of white cardboards for two preference series and twenty-five training series.

Tests → / Series ↓	1	2	3	4	5	6	7	8	9	10
A	l	r	l	r	l	r	l	r	l	r
B	r	l	r	l	r	l	r	l	r	l
1	r	l	r	l	r	l	r	l	r	l
2	l	l	r	r	l	r	l	l	r	r
3	r	r	l	r	l	l	r	l	r	l
4	l	l	l	r	r	r	l	r	r	l
5	r	l	r	l	r	l	r	l	r	l
6	l	l	r	l	r	r	l	r	l	r
7	r	l	l	l	r	r	r	l	r	l
8	r	r	l	l	r	l	r	l	r	l
9	r	r	r	l	l	l	r	l	r	l
10	l	l	l	l	r	r	r	r	l	r
11	r	l	r	r	r	l	l	l	r	l
12	r	l	r	l	r	r	l	l	r	l
13	r	l	r	l	l	l	r	r	r	l
14	l	l	l	l	r	r	r	r	l	r
15	r	l	r	r	r	l	l	l	r	l
16	l	r	l	l	l	r	r	r	l	r
17	r	r	r	r	l	l	l	l	r	l
18	l	r	l	r	r	l	l	r	l	r
19	r	l	r	l	r	l	r	l	r	l
20	l	l	l	r	l	r	l	r	r	r
21	r	l	l	r	r	l	l	r	r	l
22	l	l	r	r	l	l	r	r	l	r
23	r	l	l	l	l	r	r	r	r	l
24	l	r	l	l	l	r	r	r	l	r
25	r	r	r	r	l	l	l	l	r	l

side of the box and the other on the inside, as fig. 1 indicates. The latter consisted of three sections of which two constituted linings for the sides of the box and the third a cover for a portion of the open top of the box. In no case did these inside cards extend the entire length of the electric boxes. The white and black cards were readily interchangeable, and they never were left on the same electric box for more than four consecutive tests. The order in which they were shifted during twenty-five series of ten tests each, in addition to the preference series *A* and *B*, is given in table 1. In case a mouse required more than twenty-five series of tests (250 tests), the same set of changes was repeated, beginning with series 1. In the table the letters *r* and *l* refer to the position of the white cards; *r* indicates that they marked the electric box which was on the right of the mouse as it approached the entrances of the electric boxes from the nest-box; *l* indicates that it marked the left electric box.

The way in which this apparatus was used may be indicated by a brief description of our experimental procedure. A dancer was placed in the nest-box by the experimenter, and thence it was permitted to pass into the entrance chamber, *B*. The experimenter then placed a piece of cardboard between it and the doorway between *A* and *B* and gradually narrowed the space in which the animal could move about freely by moving the cardboard toward the electric boxes. This, without in any undesirable way interfering with the dancer's attempts to discriminate and choose correctly, greatly lessened the amount of random activity which preceded choice. When thus brought face to face with the entrances to the boxes the mouse soon attempted to enter one of them. If it happened to select the white box it was permitted to enter, pass through, and return to the nest-box; but if, instead, it started to enter the black box the experimenter by closing the key, upon which his finger constantly rested during the tests, caused it to receive an electric shock which as a rule forced a hasty retreat from the black passage-way and the renewal of attempts to discover by comparison which box should be entered.

Each of the forty mice experimented with was given ten tests every morning until it succeeded in choosing the white box correctly on three consecutive days, that is for thirty tests. A choice was recorded as wrong if the mouse started to enter the black box and received a shock; as right if, either directly or after running from one entrance to the other a number of times, it entered the white box. Whether it entered the white electric box or the black one, it was permitted to return to the nest-box by way of the white box before another test given. Escape to the nest-box by way of the black box was not permitted. A male and a female, which were housed in the same cage between experiments, were placed in the experiment box together and given their tests turn about

Almost all of the mice used were between six and eight weeks old at the beginning of their training. The exact age of each, together with its number, is stated in table 2. This table shows also the general classification of our experiments. They naturally fall into three sets. These are designated by the roman numerals

TABLE 2.

Age in days, at the beginning of training, of each mouse, with a statement of the conditions of training.

Condition of discrimination.	Strength of stimulus.	MALES.		FEMALES.	
		Number.	Age in days.	Number.	Age in days.
Medium	Weak 125± 10	128	50	127	50
		134	50	133	43
Set I	Medium 300± 25	192	47	191	47
		194	47	193	47
Medium	Strong 500± 50	130	36	129	36
		132	44	131	37
	135	268	52	267	52
		274	50	269	52
Great	195	266	50	263	50
		418	48	265	50
Set II	255	260	43	259	43
		262	43	261	43
Easy	375	396	48	189	41
		398	48	195	41
	420	280	40	279	40
		412	74	281	43
Slight	135	290	44	199	53
	195	288	45	223	25
Set III	255	286	42	285	42
Difficult	375	284	42	283	42

I, II, and III in the table, and will throughout the paper be referred to as the experiments of set I, set II and set III. As is suggested by the heading "condition of discrimination," at the top of the first vertical column of table 2, these sets of experiments differ from one another first of all as to condition of visual discrimination or, more explicitly stated, in the amount by which the two electric boxes differed from one another in brightness. For set I this difference was medium, in comparison with later conditions, and discrimination was therefore of medium difficultness. For set II the difference was great, and discrimination was easy. For set III the difference was slight, and discrimination was difficult. It is clear, then, that the series of words, medium, great, slight, in the table refers to the amount by which the electric boxes differed in brightness, and the series medium, easy, difficult, to the demand made upon the visual discriminating ability of the mice.

For the sake of obtaining results in this investigation which should be directly comparable with those of experiments on the modifiability of behavior in the dancer which have been conducted during the past three years, it was necessary for us to use the same general method of controlling the visual conditions of the experiment that had previously been used. This we decided to do, notwithstanding the fact that we had before us methods which were vastly superior to the old one with respect to the describability of conditions and the accuracy and ease of their control. To any experimenter who wishes to repeat this investigation with other animals we should recommend that, before recourse is had to the use of cardboards for the purpose of rendering the boxes distinguishable, thorough tests be made of the ability of the animal to discriminate when the boxes are rendered different in brightness by the use of a screen which excludes a measurable amount of light from one of them. We have discovered that the simplest and best method of arranging the conditions for such experiments with the dancer as are now to be described is to use two electric boxes which are alike in all respects and to control the amount of light which enters one of them from the top. It is easy to obtain satisfactory screens and to measure their transmitting capacity. We regret that the first use which we wished to make of our results in this investigation forced us to employ conditions which are relatively complicated and difficult to describe.

For the sake of the scientific completeness of our paper, however, and not because we wish to encourage anyone to make use of the same conditions, we shall now describe as accurately as we may the conditions of visual discrimination in the several sets of experiments.

The cards at the entrances to the electric boxes were the same in all of the experiments. Each card (the black and the white) was 11.5 cm. in height and 5.4 cm. in width, with a hole 3.5 by 3.5 cm. in the middle of its lower edge as is shown in fig. 1. These entrance cards were held in place by small metal carriers at the edges of the electric boxes. The area of white surface exposed to the view of a mouse as it approached the entrances to the electric boxes was 49.85 sq. cm. and the same amount of black surface was exposed. The white cardboard reflected 10.5 times as much light as the black cardboard.

Special conditions of set I. The inside length of each electric box was 28.5 cm. the width 7 cm. and the depth 11.5 cm. The inside cards extended from the inner edge of the front of each box a distance of 13.5 cm. toward the back of the box. Consequently there was exposed to the view of the mouse a surface 13.5 cm. by 11.5 cm. (the depth of the box and of the cardboard as well) on each side of the box. The section of cardboard at the top measured 13.5 cm. in length by 6.5 cm. in width. The total area of the white (or black) cardboard exposed on the inside of an electric box was therefore $13.5 \times 11.5 \times 2$ (the sides) $+ 13.5 \times 6.5$ (the top) $= 398.25$ sq. cm. If to this we add the area of the entrance card we obtain 448.10 sq. cm. as the amount of surface of cardboard carried by each electric box.

But another condition, in connection with the amount of cardboard present, determined the difference in the brightness of the boxes, namely, the amount of open space between the end of the inner cardboards and the end of the experiment box. The larger this opening the more light entered each box. In the case of the experiments of set I this uncovered portion of each electric box was 15 cm. long by 7 cm. wide; its area, therefore, was 105 sq. cm.

Special conditions of set II. Both the outer and the inner cardboards were precisely the same in form and arrangement as in the case of set I, but in order that discrimination might be rendered easier, and the time required for the acquisition of the habit thus shortened, a hole 8.7 cm. long by 3.9 cm. wide was cut in the middle or top section of the white cardboard. This greatly increased the amount of light in the white electric box. The difference in the brightness of the boxes was still further increased by a reduction of the space between the end of the cardboard and the end of the box from 15 cm. to 2 cm. or, in terms of area, from 105 sq. cm. to 14 sq. cm. This was accomplished by cutting 13 cm. from the rear end of the experiment box. For the experiments of set

II the black box was much darker than it was for those of set I, whereas the white box was not markedly different in appearance.

Special conditions of set III. The experiments of this set were conducted with the visual conditions the same as in set II, except that there was no hole in the white cardboard over the electric box. This rendered the white box much darker than it was in the experiments of set II, consequently the two boxes differed less in brightness than in the case of set II, and discrimination was much more difficult than in the experiments of either of the other sets.

In the second column of table 2 the values of the several strengths of electrical stimuli used in the investigation are stated. To obtain our stimulus we used a storage cell, in connection with gravity batteries, and with the current from this operated a PORTER inductorium. The induced current from the secondary coil o- this apparatus was carried by the wires which constituted an interrupted circuit on the floor of the electric boxes. For the experiments of set I the strengths of the stimuli used were not accurately determined, for we had not at that time discovered a satisfactory means of measuring the induced current. These experiments therefore served as a preliminary investigation whose chief value lay in the suggestions which it furnished for the planning of later experiments. The experiments of sets II and III were made with a PORTER inductorium which we had calibrated, with the help of Dr. E. G. MARTIN of the Harvard Medical School, by a method which he has recently devised and described.[2]

On the basis of the calibration measurements which we made by MARTIN'S method the curve of fig. 3 was plotted. From this curve it is possible to read directly in "units of stimulation" the value of the induced current which is yielded by a primary current of one ampere for any given position of the secondary coil. With the secondary coil at 0, for example, the value of the induced current is 350 units; with the secondary at 5.2 centimeters on the scale of the inductorium, its value is 155 units; and with the secondary at 10, its value is 12 units. The value of the induced current for a primary current greater or less than unity is obtained by multiplying the reading from the calibration curve by the value

[2] MARTIN, E. G. A quantitative study of faradic stimulation. I. The variable factors involved. *Amer. Jour. of Physiol.*, vol. 22, pp. 61–74. 1908. II. The calibration of the inductorium for break shocks. *Ibid.*, pp. 116–132.

of the primary current. The primary current used for the experiments of sets II and III measured 1.2 amperes, hence the value of the stimulating current which was obtained when the secondary coil stood at 0 was 350 × 1.2 = 420 units of stimulation.

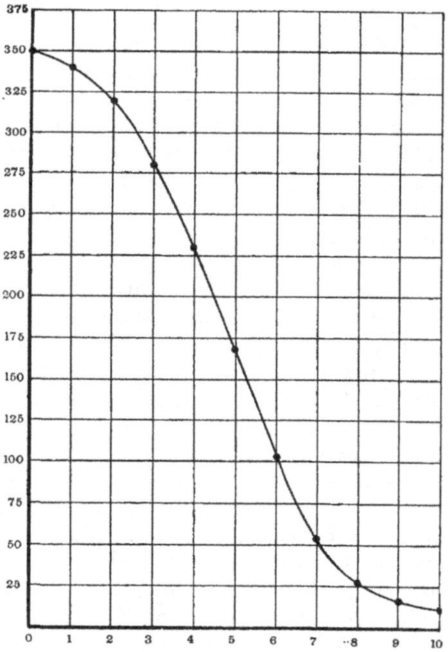

FIG. 3. Calibration curve for PORTER inductorium. The numbers below the base line refer to the position of the secondary coil with reference to the primary. The positions are read, as on the scale of the inductorium, in centimeters. The numbers in the margin represent values of the induced current in terms of MARTIN's unit of stimulation.

As conditions for the experiments of set I, we chose three strengths of stimuli which we designated as weak, medium, and strong. The weak stimulus was slightly above the threshold of stimulation for the dancers. Comparison of the results which it yielded with those obtained by the use of our calibrated inductorium enable us to state with a fair degree of certainty that its value was 125 ± 10 units of stimulation. The strong stimulus was decidedly disagreeable to the experimenters and the mice reacted to it vigorously. Its value was subsequently ascertained to be 500 ±50 units. For the medium stimulus we tried to select a value which should be about midway between these extremes. In this we succeeded better than we could have expected to, for comparison indicated that the value was 300 ± 25 units. Fortunately for the interpretation of this set of results, the exact value of the stimuli is not important.

By the use of our calibrated inductorium and the measurement of our primary current, we were able to determine satisfactorily the stimulating values of the several currents which were used in the experiments of sets II and III. The primary current of 1.2 amperes, which was employed, served to actuate the interrupter of the inductorium as well as to provide the stimulating current. The interruptions occurred at the rate of 65 ± 5 per second. We discovered at the outset of the work that it was not worth while to attempt to train the dancers with a stimulus whose value was much less than 135 units. We therefore selected this as our weakest stimulus. At the other extreme a stimulus of 420 units was as strong as we deemed it safe to employ. Between these two, three intermediate strengths were used in the case of set II, and two in the case of set III. Originally it had been our intention to make use of stimuli which varied from one another in value by 60 units of stimulation, beginning with 135 and increasing by steps of 60 through 195, 255, 315, 375 to as nearly 425 as possible. It proved to be needless to make tests with all of these.

We may now turn to the results of the experiments and the interpretation thereof. Before the beginning of its training each mouse was given two series of tests in which the electric shock was not used and return to the nest-box through either the white or the black box was permitted. These twenty tests (ten in series A and ten in series B) have been termed preference tests, for they served to reveal whatever initial tendency a dancer possessed to choose the white or the black box. On the day following preference series B, the regular daily training series were begun and they were continued without interruption until the dancer had succeeded in choosing correctly in every test on three consecutive days.

Results of the experiments of set I. The tests with the weak stimulus of set I were continued for twenty days, and up to that time only one of the four individuals in training (no. 128) had acquired a perfect habit. On the twentieth day it was evident that the stimulus was too weak to furnish an adequate motive for the avoidance of the black box and the experiments were discontinued.

A few words in explanation of the tables are needed at this point. In all of the tables of detailed results the method of arrangement which is illustrated by table 3 was employed. At the top of the table are the numbers of the mice which were trained under

TABLE 3.

The results of the experiments of set I, stimulus weak (125 ± 10 units).

Series.	Males.			Females.			General Average.
	No. 128	No. 134.	Average.	No. 127.	No. 133.	Average.	
A	6	7	6.5	4	5	4.5	5.50
B	5	5	5.0	6	4	5.0	5.00
1	3	5	4.0	4	4	4.0	4.00
2	6	6	6.0	6	7	6.5	6.25
3	5	4	4.5	2	5	3.5	4.00
4	4	5	4.5	6	4	5.0	4.75
5	3	7	5.0	3	5	4.0	4.50
6	2	5	3.5	4	4	4.0	3.75
7	3	4	3.5	4	7	5.5	4.50
8	2	2	2.0	2	3	2.5	2.25
9	5	5	5.0	3	3	3.0.	4.00
10	1	2	1.5	4	2	3.0	2.25
11	0	3	1.5	3	5	4.0	2.75
12	1	1	1.0	3	2	2.5	1.75
13	1	2	1.5	2	2	2.0	1.75
14	1	1	1.0	0	3	1.5	1.25
15	1	3	2.0	1	3	2.0	2.00
16	0	0	0.	1	0	0.5	0.25
17	0	1	0.5	0	0	0.	0.25
18	0	.0	0.	2	1	1.5	0.75
19		1	0.5	2	1	1.5	1.00
20		3	1.5	2	3	2.5	2.00

the conditions of stimulation named in the heading of the table. The first vertical column gives the series numbers, beginning with the preference series A and B and continuing from 1 to the last series demanded by the experiment. In additional columns appear the number of errors made in each series of ten tests, day by day, by the several subjects of the experiments; the average number of errors made by the males in each series; the average number of errors made by the females; and, finally, the general average for both males and females. In table 3, for example, it appears that male no. 128 chose the black box in preference to the white 6 times in series A, 5 times in series B, 3 times in series 1, 6 times in series 2. After series 15 he made no errors during three consecutive series. His training was completed, therefore, on the eighteenth day, as the result of 180 tests. We may say, however, that only 150 tests were necessary for the establishment of a perfect habit, for the additional thirty tests, given after the fifteenth series, served merely to reveal the fact that he already possessed a perfect habit. In view of this consideration, *we shall*

TABLE 4.

The results of the experiments of set I, stimulus medium (300 ± 25 units).

Series.	Males.			Females.			General Average.
	No. 192.	No. 194.	Average.	No. 191.	No. 193.	Average.	
A	4	8	6.0	3	7	5.0	5.50
B	6	6	6.0	4	6	5.0	5.50
1	4	4	4.0	4	5	4.5	4.25
2	3	3	3.0	4	2	3.0	3.00
3	4	5	4.5	5	6	5.5	5.00
4	3	4	3.5	6	3	4.5	4.00
5	2	4	3.0	5	7	6.0	4.50
6	2	0	1.0	2	2	2.0	1.50
7	2	2	2.0	0	3	1.5	1.75
8	1	0	0.5	1	0	0.5	0.50
9	0	2	1.0	0	0	0.	0.50
10	0	0	0.	0	0	0.	0.
11	0	0	0.	0		0.	0.
12		0	0.				0.

take as a measure of the rapidity of learning in these experiments the number of tests received by a mouse up to the point at which errors ceased for at least three consecutive series.

Precisely as the individuals of table 3 had been trained by the use of a weak stimulus, four other dancers were trained with a medium stimulus. The results appear in table 4. All of the subjects acquired a habit quickly. Comparison of these results with those obtained with the weak stimulus clearly indicates that the medium stimulus was much more favorable to the acquirement of the white-black visual discrimination habit.

In its results the strong stimulus proved to be similar to the weak stimulus. All of the mice in this case learned more slowly than did those which were trained with the medium strength of stimulus.

The general result of this preliminary set of experiments with three roughly measured strengths of stimulation was to indicate that neither a weak nor a strong electrical stimulus is as favorable to the acquisition of the white-black habit as is a medium stimulus.

TABLE 5.

The results of the experiments of set I, stimulus strong (500 ± 50 units).

Series.	Males.			Females.			General Average.
	No. 130.	No. 132.	Average.	No. 129.	No. 131.	Average.	
A	7	6	6.5	5	1	3.0	4.75
B	6	4	5.0	4	4	4.0	4.50
1	3	5	4.0	5	5	5.0	4.50
2	3	1	2.0	3	3	3.0	2.50
3	5	3	4.0	3	3	3.0	3.50
4	3	2	2.5	2	3	2.5	2.50
5	2	2	2.0	2	4	3.0	2.50
6	3	1	2.0	2	2	2.0	2.00
7	3	0	1.5	2	4	3.0	2.25
8	4	0	2.0	1	2	1.5	1.75
9	3	2	2.5	2	1	1.5	2.00
10	2	3	2.5	1	1	1.0	1.75
11	1	1	1.0	2	0	1.0	1.00
12	1	2	1.5	0	0	0.	0.75
13	1	1	1.0	2	2	2.0	1.50
14	0	0	0.	2	2	2.0	1.00
15	2	0	1.0	0	1	0.5	0.75
16	0	0	0.	0	2	1.0	0.50
17	0		0.	0	1	0.5	0.25
18	0		0.		2	1.0	0.50
19					1	0.5	0.25
20					1	0.5	0.25
21					0	0.	0.
22					0	0.	0.
23					0	0.	0.

Contrary to our expectations, this set of experiments did not prove that the rate of habit-formation increases with increase in the strength of the electric stimulus up to the point at which the shock becomes positively injurious. Instead an intermediate range of intensity of stimulation proved to be most favorable to the acquisition of a habit *under the conditions of visual discrimination of this set of experiments.*

In the light of these preliminary results we were able to plan a more exact and thoroughgoing examination of the relation of strength of stimulus to rapidity of learning. Inasmuch as the training under the conditions of set I required a great deal of time, we decided to shorten the necessary period of training by making the two electric boxes very different in brightness, and the discrimination correspondingly easy. This we did, as has already been explained, by decreasing the amount of light which entered the black box, while leaving the white box about the same. The influence of this change on the time of learning was very marked indeed.

With each of the five strengths of stimuli which were used in set II two pairs of mice were trained, as in the case of set I. The detailed results of these five groups of experiments are presented in tables 6 to 10. Casual examination of these tables reveals the fact that in general the rapidity of learning in this set of experiments increased as the strength of the stimulus increased. The weakest stimulus (135 units) gave the slowest rate of learning; the strongest stimulus (420 units), the most rapid.

TABLE 6.

The results of the experiments of set II, stimulus 135 units.

Series.	Males.			Females.			General Average.
	No. 268.	No. 274.	Average.	No. 267.	No. 269.	Average.	
A	9	7	8.0	8	7	7.5	7.75
B	8	6	7.0	4	6	5.0	6.00
1	6	4	5.0	6	4	5.0	5.00
2	2	3	2.5	2	4	3.0	2.75
3	2	4	3.0	4	6	5.0	4.00
4	1	4	2.5	0	1	0.5	1.50
5	0	3	1.5	2	2	2.0	1.75
6	0	2	1.0	0	0	0.	0.50
7	0	1	0.5	1	1	1.0	0.75
8		0	0.	0	0	0.	0.
9		0	0.	0	0	0.	0.
10		0	0.	2	0	1.0	0.50
11				1		0.5	0.25
12				1		0.5	0.25
13				0		0.	0.
14				0		0.	0.
15				1		0.5	0.25
16				0		0.	0.
17				0		0.	0.
18				0		0.	0.

TABLE 7.

The results of the experiments of set II, stimulus 195 units.

Series.	MALES.			FEMALES.			General Average.
	No. 266.	No. 418.	Average.	No. 263.	No. 265.	Average.	
A	6	6	6.0	6	4	5.0	5.50
B	6	7	6.5	8	3	5.5	6.00
1	6	7	6.5	5	7	6.0	6.25
2	5	1	3.0	1	1	1.0	2.00
3	3	5	4.0	1	4	2.5	3.25
4	2	2	2.0	2	1	1.5	1.75
5	1	1	1.0	0	2	1.0	1.00
6	2	1	1.5	1	0	0.5	1.00
7	1	1	1.0	0	0	0.	0.50
8	1	0	0.5	0	0	0.	0.25
9	0	0	0.	0		0.	0.
10	0	0	0.				0.
11	0		0.				0.

TABLE 8.

The results of the experiments of set II, stimulus 255 units.

Series.	MALES.			FEMALES.			General Average.
	No. 260.	No. 262.	Average.	No. 259.	No. 261.	Average.	
A	5	5	5.0	5	6	5.5	5.25
B	7	6	6.5	5	5	5.0	5.75
1	6	7	6.5	9	3	6.0	6.25
2	4	7	5.5	4	3	3.5	4.50
3	1	4	2.5	3	1	2.0	2.25
4	0	2	1.0	4	0	2.0	1.75
5	0	2	1.0	0	2	1.0	1.00
6	0	0	0.	0	1	0.5	0.25
7		0	0.	0	1	0.5	0.25
8		0	0.		1	0.5	0.25
9					0	0.	0.
10					0	0.	0.
11					0	0.	0.

TABLE 9.

The results of the experiments of set II, stimulus 375 units.

Series.	Males.			Females.			General. Average.
	No. 396.	No. 398.	Average.	No. 189.	No. 195.	Average.	
A	6	6	6.0	6	7	6.5	6.25
B	5	3	4.0	5	6	5.5	4.75
1	6	6	6.0	4	5	4.5	5.25
2	5	1	3.0	5	3	4.0	3.50
3	5	3	4.0	8	2	5.0	4.50
4	0	4	2.0	3	1	2.0	2.00
5	0	3	1.5	1	4	2.5	2.00
6	0	0	0.	0	0	0.	0.
7		1	0.5	0	0	0.	.25
8		0	0.	0	0	0.	0.
9		1	0.5				.25
10		0	0.				0.
11		0	0.				0.
12		0	0.				0.

TABLE 10.

The results of the experiments of set II, stimulus 420 units.

Series.	Males.			Females.			General Average.
	No. 280.	No. 412.	Average.	No. 279.	No. 281.	Average.	
A	5	5	5.0	4	6	5.0	5.00
B	6	6	6.0	6	4	5.0	5.50
1	5	5	5.0	5	5	5.0	5.00
2	4	5	4.5	1	0	0.5	2.50
3	2	5	3.5	2	4	3.0	3.25
4	1	3	2.0	0	2	1.0	1.50
5	0	3	1.5	0	1	0.5	2.00
6	0	0	0.	0	0	0.	0.
7	0	0	0.		0	0.	0.
8		0	0.		0	0.	0.

The results of the second set of experiments contradict those of the first set. What does this mean? It occurred to us that the apparent contradiction might be due to the fact that discrimination was much easier in the experiments of set II than in those of set I. To test this matter we planned to use in our third set of experiments a condition of visual discrimination which should be extremely difficult for the mice. The reader will bear in mind that for set

II the difference in brightness of the electric boxes was great; that for set III it was slight; and for set I, intermediate or medium.

For the experiments of set III only one pair of dancers was trained with any given strength of stimulus. The results, however, are not less conclusive than those of the other sets of experiments because of the smaller number of individuals used. The data of tables 11 to 14 prove conclusively that our supposition was correct. The varying results of the three sets of experiments are explicable in terms of the conditions of visual discrimination. In

TABLE 11.

The results of the experiments of set III, stimulus 135 units.

Series.	MALE.	FEMALE.	Average.
	No. 290.	No. 199.	
A	6	4	5.0
B	4	7	5.5
1	4	6	5.0
2	5	2	3.5
3	3	6	4.5
4	4	2	3.0
5	7	4	5.5
6	4	4	4.0
7	7	7	7.0
8	7	5	6.0
9	4	4	4.0
10	4	2	3.0
11	4	1	2.5
12	5	3	4.0
13	3	2	2.5
14	2	4	3.0
15	4	3	3.5
16	3	0	1.5
17	2	2	2.0
18	0	2	1.0
19	1	1	1.0
20	3	3	3.0
21	1	1	1.0
22	1	0	0.5
23	2	0	1.0
24	1	0	0.5
25	3		1.5
26	1		0.5
27	1		0.5
28	0		0.
29	0		0.
30	2		1.0

TABLE 12.

The results of the experiments of set III, stimulus 195 units.

Series.	MALE.	FEMALE.	Average.
	No. 288.	No. 223.	
A	4	4	4.0
B	7	8	7.5
1	5	7	6.0
2	3	6	4.5
3	5	6	5.5
4	6	3	4.5
5	6	7	6.5
6	4	4	4.0
7	5	3	4.0
8	2	2	2.0
9	0	0	0.
10	3	1	2.0
11	2	1	1.5
12	1	0	0.5
13	1	0	0.5
14	0	0	0.
15	0		0.
16	0		0.

TABLE 13.

The results of the experiments of set III, stimulus 255 units.

Series.	Male. No. 286.	Female. No. 285.	Average.
A	4	7	5.5
B	4	5	4.5
1	5	6	5.5
2	3	3	3.0
3	2	3	2.5
4	5	5	5.0
5	2	4	3.0
6	2	3	2.5
7	3	2	2.5
8	1	1	1.0
9	1	2	1.5
10	2	1	1.5
11	2	3	2.5
12	3	0	1.5
13	2	0	1.0
14	0	1	0.5
15	3	1	2.0
16	1	0	0.5
17	0	0	0.
18	0	0	0.
19	0		0.

TABLE 14.

The results of the experiments of set III, stimulus 375 units.

Series.	Male. No. 284.	Female. No. 283.	Average.
A	4	5	4.5
B	3	4	3.5
1	6	6	6.0
2	3	2	2.5
3	4	3	3.5
4	4	2	3.0
5	2	5	3.5
6	3	2	2.5
7	6	5	5.5
8	4	2	3.0
9	1	1	1.0
10	1	2	1.5
11	1	2	1.5
12	3	1	2.0
13	1	1	1.0
14	1	1	1.0
15	1	0	0.5
16	1	1	1.0
17	0	1	0.5
18	0	1	0.5
19	0	1	0.5
20		0	0.
21		2	1.0
22		0	0.
23		2	1.0
24		0	0.
25		0	0.
26		0	0.

set III both the weak and the strong stimuli were less favorable to the acquirement of the habit than the intermediate stimulus of 195 units. It should be noted that our three sets of experiments indicate that the greater the brightness difference of the electric boxes the stronger the stimulus which is most favorable to habit-formation (within limits which have not been determined). Further discussion of the results and attempts to interpret them may be postponed until certain interesting general features of the work have been mentioned.

The behavior of the dancers varied with the strength of the stimulus to which they were subjected. They chose no less quickly in the case of the strong stimuli than in the case of the weak, but they were less careful in the former case and chose with less delib-

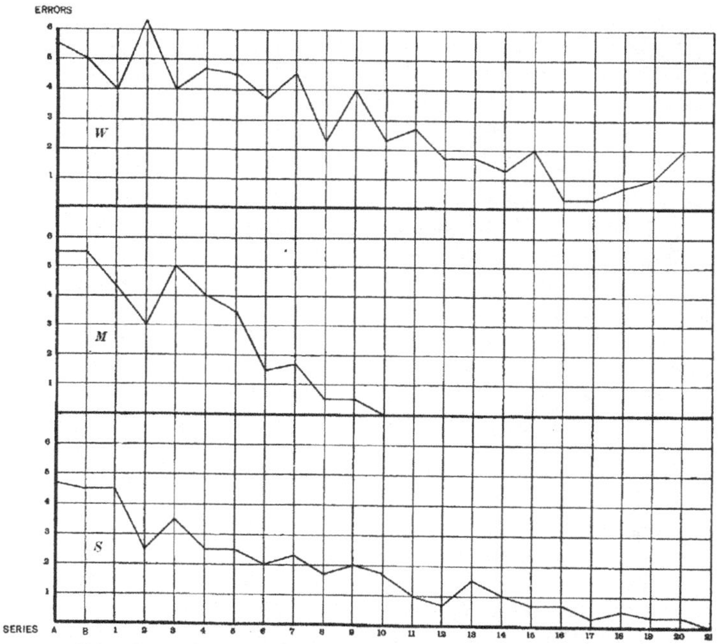

FIG. 4. Curves of learning. Ordinates represent series of ten tests each, and abscissæ represent the average number of errors for four mice in each series. *W*, designates the error curve for the individuals which were trained under the condition of *weak* electrical stimulation; *M*, designates the corresponding curve for the *medium* strength of stimulation; and *S*, that for the *strong* stimulus.

eration and certainty. Fig. 4 exhibits the characteristic differences in the curves of learning yielded by weak, medium, and strong stimuli. These three curves were plotted on the basis of the average number of errors for the mice which were trained in the experiments of set I. Curve *W* is based upon the data of the last column of table 3, curve *M*, upon the data in the last column of table 4; and curve *S* upon the data of the last column of table 5. In addition to exhibiting the fact that the medium stimulus yielded a perfect habit much more quickly than did either of the other stimuli, fig. 4 shows a noteworthy difference in the forms of the curves for the weak and the strong stimuli. Curve *W* (weak stimulus) is higher throughout its course than is curve *S* (strong stimulus). This means that fewer errors are made from the start under the condition of strong stimulation than under the condition of weak stimulation.

Although by actual measurement we have demonstrated marked difference in sensitiveness to the electric shock among our mice, we are convinced that these differences do not invalidate the conclusions which we are about to formulate in the light of the results that have been presented. Determination of the threshold electric stimulus for twenty male and twenty female dancers proved that the males respond to a stimulus which is about 10 per cent less than the smallest stimulus to which the females respond.

Table 15 contains the condensed results of our experiments. It gives, for each visual condition and strength of stimulus, the number of tests required by the various individuals for the acquisition of a perfect habit; the average number of tests required by the males, for any given visual and electrical conditions; the same for the females; and the general averages. Although the numbers of the mice are not inserted in the table they may readily be learned if anyone wishes to identify a particular individual, by referring to the tables of detailed results. Under set I, weak stimulus, for example, table 15 gives as the records of the two males used 150 and 200 + tests. By referring to table 3, we discover that male no. 128 acquired his habit as a result of 150 tests, whereas male no. 134 was imperfect at the end of 200 tests. To indicate the latter fact the plus sign is added in table 15. Of primary importance for the solution of the problem which we set out to study are the general averages in the last column of the table. From this series of averages we have constructed the curves of fig. 5. This figure

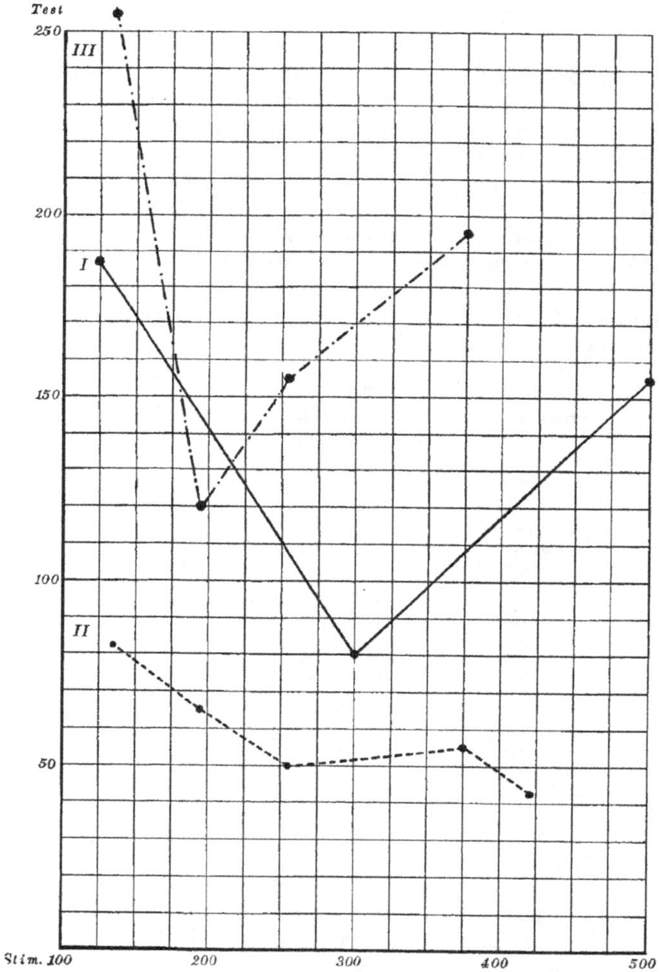

Fig. 5. A graphic representation of the relation of strength of electrical stimulus to condition of visual discrimination and rapidity of learning. Ordinates represent value of electric stimulus in units of stimulation; abscissæ represent the number of tests given. Curve *I* represents the results of the experiments of Set I. Each dot indicates a value of stimulus which was used in the experiments. For example, the first dot to the left in curve *I* signifies that the stimulus whose value was 125 units gave a perfect habit, in the case of the four individuals trained, with 187 tests; the second dot, that for the stimulus value of 300 units 80 tests were necessary; and the third that for the stimulus value of 500, 155 tests. Curves II and III similarly represent the results of the experiments of sets II and III, respectively.

very clearly and briefly presents the chiefly significant results of our investigation of the relation of strength of electrical stimulus to rate of habit-formation, and it offers perfectly definite answers to the questions which were proposed for solution.

In this figure the ordinates represent stimulus values, and the abscissæ number of tests. The roman numerals *I, II, III*, designate, respectively, the curves for the results of set I, set II, and set III. Dots on the curves indicate the strengths of stimuli which were employed. Curve I for example, shows that a strength of stimulus of 300 units under the visual conditions of set I, yielded a perfect habit with 80 tests.

TABLE 15.

The number of tests required by the mice for the acquisition of a perfect habit of discrimination.

Set.	Stimulus.	Males.		Average.	Females.		Average.	Gen. Av.
I	Weak	150	200+	175+	200+	200+	200+	187+
	Medium	80	90	85	80	70	75	80
	Strong	150	130	140	140	200	170	155
II	135	40	70	55	150	70	110	82.5
	195	80	70	75	60	50	55	65
	255	30	50	40	40	80	60	50
	375	30	90	60	50	50	50	55
	420	40	50	45	30	50	40	42.5
III	135	300+			210			255
	195	130			110			120
	255	160			150			155
	375	160			230			195

From the data of the various tables we draw the following conclusions:

1. In the case of the particular habit which we have studied, the rapidity of learning increases as the amount of difference in the brightness of the electric boxes between which the mouse is required to discriminate is increased. The limits within which this statement holds have not been determined. The higher the curves of fig. 5 stand from the base line, the larger the number of tests represented by them. Curve II is lowest, curve I comes next, and curve III is highest. It is to be noted that this is the order of increasing difficultness of discrimination in the three sets of experiments.

2. The relation of the strength of electrical stimulus to rapidity of learning or habit-formation depends upon the difficultness of the habit, or, in the case of our experiments, upon the conditions of visual discrimination.

3. When the boxes which are to be discriminated between differ very greatly in brightness, and discrimination is easy, the rapidity of learning increases as the strength of the electrical stimulus is increased from the threshold of stimulation to the point of harmful intensity. This is indicated by curve II. Our results do not represent, in this instance, the point at which the rapidity of learning begins to decrease, for we did not care to subject our animals to injurious stimulation. We therefore present this conclusion tentatively, subject to correction in the light of future research. Of its correctness we feel confident because of the results which the other sets of experiments gave. The irregularity of curve II, in that it rises slightly for the strength 375, is due, doubtless, to the small numbers of animals used in the experiments. Had we trained ten mice with each strength of stimulus instead of four the curve probably would have fallen regularly.

4. When the boxes differ only slightly in brightness and discrimination is extremely difficult the rapidity of learning at first rapidly increases as the strength of the stimulus is increased from the threshold, but, beyond an intensity of stimulation which is soon reached, it begins to decrease. Both weak stimuli and strong stimuli result in slow habit-formation. A stimulus whose strength is nearer to the threshold than to the point of harmful stimulation is most favorable to the acquisition of a habit. Curve III verifies these statements. It shows that when discrimination was extremely difficult a stimulus of 195 units was more favorable than the weaker or the stronger stimuli which were used in this set of experiments.

5. As the difficultness of discrimination is increased the strength of that stimulus which is most favorable to habit-formation approaches the threshold. Curve II, curve I, curve III is the order of increasing difficultness of discrimination for our results, for it will be remembered that the experiments of set III were given under difficult conditions of discrimination; those of set I under medium conditions; and those of set II under easy conditions. As thus arranged the most favorable stimuli, so far as we may judge from our results, are 420, 300, and 195. This leads us to infer that an easily acquired habit, that is one which does not

demand difficult sense discriminations or complex associations, may readily be formed under strong stimulation, whereas a difficult habit may be acquired readily only under relatively weak stimulation. That this fact is of great importance to students of animal behavior and animal psychology is obvious.

Attention should be called to the fact that since only three strengths of stimulus were used for the experiments of set I, it is possible that the most favorable strength of stimulation was not discovered. We freely admit this possibility, and we furthermore wish to emphasize the fact that our fifth conclusion is weakened slightly by this uncertainty. But it is only fair to add that previous experience with many conditions of discrimination and of stimulation, in connection with which more than two hundred dancers were trained, together with the results of comparison of this set of experiments with the other two sets, convinces us that the dancers would not be likely to learn much more rapidly under any other condition of stimulation than they did with a strength of 300 ± 25 units of stimulation.

Naturally we do not propose to rest the conclusions which have just been formulated upon our study of the mouse alone. We shall now repeat our experiments, in the light of the experience which has been gained, with other animals.

SOME REACTIONS OF DROSOPHILA, WITH SPECIAL REFERENCE TO CONVULSIVE REFLEXES.

BY

FREDERIC W. CARPENTER

(*Zoölogical Laboratory, University of Illinois.*)

WITH ONE FIGURE.

The behavior of the pomace fly (Drosophila ampelophila) in respect to several kinds of stimulation has already been the subject of investigation. BARROWS ('07) has recently shown that the insect is positively chemotropic to certain strengths of odorous substances occurring in fermenting fruit, such as alcohol, acetic and lactic acids, and acetic ether. To light varying in intensity from 5 to 250 candlepowers Drosophila is positively phototropic; and under the influence of gravity it is negatively geotropic (CARPENTER '05).

In the present study of the reactions of Drosophila to stimuli other than those just mentioned, attention was first directed to the behavior of the insects when they pass from a region of optimum temperature into regions relatively warm or cold. The flies were confined in a flat glass box, 38 cm. long, 23 cm. wide, and 8 mm. deep. The edges of the box along the two sides and one end were sealed with aquarium cement and enamel, and thus made water-tight. At the unsealed end of the box a small opening was left through which the flies could pass into the interior. All the movements of the insects could readily be observed through the glass, and the short distance that separated the roof and floor of the box permitted the use of a hand lens when desired.

In the temperature experiments the box was partially immersed in water as shown in the accompanying sectional view of the apparatus.

The water could be heated by means of an alcohol lamp placed beneath the vessel containing it, or it could be cooled by placing in the vessel small pieces of ice. By arranging the apparatus so that the immersed end of the box was directed toward a window the positively phototropic flies, introduced into the elevated end, could be made to creep toward the hot or cold region, since this region lay in the direction of the source of light.

Reaction to increased temperature. The water surrounding the immersed end of the box was raised to 45° C., a temperature that is soon fatal to Drosophila. Flies introduced into the opposite end of the box, which was practically at room temperature, crept more or less steadily toward the light. This movement brought them gradually to a region of increasing temperature. Upon arriving near the lower water-line (*D*) the creeping flies turned about, describing curved paths, and headed back toward the cooler end of the box, their positive phototropism apparently being overcome by the repelling effect of the heat. In no instances did creeping flies, whether on the floor or the roof of the box, pass

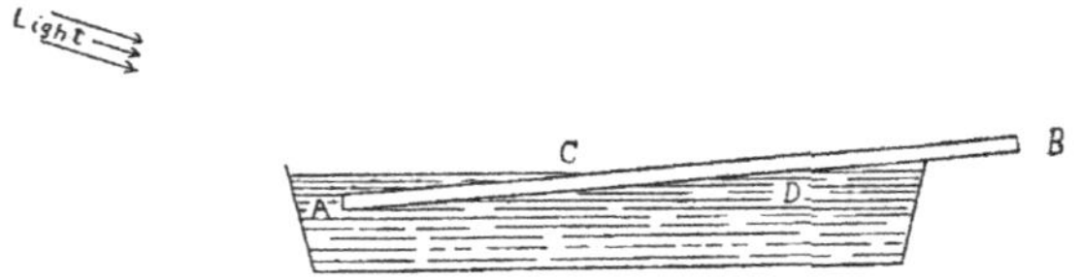

FIG. 1. Sectional plan of apparatus. *A B*, glass box, 38 cm. long, 23 cm. wide, and 8 mm. deep, partially immersed in water; *C*, upper water-line; *D*, lower water-line.

beyond this lower water-line. Occasionally one would creep along the line in a zigzag manner, as though alternating between the two antagonistic directive influences, but finally it would yield to the negative stimulus, and creep back to the region of lower temperature. The majority of the return excursions were made as the result of a uniform and continuous deflection from the heated area. Since there was no satisfactory evidence of random movements involving a "trial and error" method of reaction, this behavior may conveniently be spoken of as a "tropism." This term is here used merely in a descriptive sense for an orderly turning away from a stimulating region. It does not carry with it an implied theoretical explanation of the precise effect of the stimulus on the organism.

Reaction to decreased temperature. When the water in which one end of the glass box was immersed was cooled with ice, the other conditions of the experiment remaining the same, a similar tropic reaction followed. The flies were consistently negative to a temperature of from 5° to 6° C. Occasionally the creeping excursions toward the light were prolonged beyond the lower water-line, but only in very infrequent cases did the flies reach the upper water-line.

As a control for both this and the previous experiment the water vessel of the apparatus was filled with water at nearly room temperature. The flies then, following the light, crept to the end of the immersed portion of the box.

Reaction to unilateral light stimulation. The regular curved paths described by Drosophila in its response to the repelling effect of heat and cold suggested the possibility that this reaction might be explained by the "local action theory of tropisms." It was conceivable that the unsymmetrical stimulation might act locally on the organs of locomotion, presumably through the nervous system. Those locomotor organs on the side subjected to the greater stimulus might move more rapidly than those on the opposite side until the insect should be turned so as to head directly away from the region of stimulation. The two sides of the body would then be equally affected by the heat or cold, and the organs of locomotion would, therefore, move with equal rapidity, and carry the fly away in a straight line.

The adequacy of this simple explanation of the temperature reactions might have been tested if the stimulus could have been confined to one side of the body only. The tropism theory would call for circus movements by the fly as long as the unilateral stimulation was maintained. No satisfactory method for applying this test to the temperature reactions occurred to me. The light reaction, however, presented fewer difficulties, and furnished quite as critical a reflex for the purpose. The light stimulus is effective through the paired eyes of the insect. If one eye is covered circus movements are to be expected under the theory of tropisms, with the uncovered eye toward the centers of the circles, since Drosophila is positively phototropic. Such reactions in insects with one eye blinded have been recorded by HOLMES ('01) for the bluebottle fly, two species of bees, a robber-fly, a horse-fly, and a species of syrphid. PARKER ('03) has observed similar behavior in the mourning-cloak butterfly (Vanessa antiope).

In preparation for this test experiment, one eye of each of several insects was covered by an opaque cap of lampblack and mucilage. To reduce the flies to a temporary state of insensibility and quiet, during which the covering to the eyes could be applied, recourse was made to what might be called a natural anesthetic. If Drosophila is exposed for a short time to a temperature of 0° C. its movements cease, and it becomes apparently insensible. If then brought into the ordinary temperature of the laboratory several minutes will elapse before it recovers, and during this time the operation of covering the eye may be performed under a dissecting microscope. When the recovery takes place it appears to be a complete one, the fly responding in a normal way to all stimuli. Since a low temperature must often be present in the insect's natural environment, it may be supposed that the action of cold will be attended with less risk of altering the nervous system than that of the chemical anesthetics usually employed.[1]

Repeated trials made with flies thus deprived of the sight of one eye showed that under such conditions of unsymmetrical stimulation they, nevertheless, crept in a fairly direct path toward the light, although a tendency to deviate toward the side of the normal eye regularly occurred. The insects generally moved in a peculiar, jerky manner. The tendency to diverge from the direct path toward the side of the uncovered eye was overcome by a series of short, quick turns in the opposite direction, which kept them headed toward the light. Normal flies, used as a control, pursued straight courses, and usually reached the end of the container before the experimental flies. Now and then one of the partially blinded flies performed circus movements; but this conduct was exceptional, and was never persisted in except in the case of a single insect, which had long been active, and showed signs of fatigue.

It is clear that the tropism theory, with its assumption of a local action of the stimulus on the side exposed to its effect, does not furnish a complete explanation of these reactions. Though the persistent tendency to turn toward the side of the functional eye gives some evidence of a purely mechanical reaction to a local stimulation, such a reaction is evidently inhibited and dominated

[1] The resistance of Drosophila to cold is rather remarkable. I have buried a glass vessel containing thirty-three flies in a snow-bank over night, exposing them thus to a temperature of about 0° C. for seventeen hours. Of these thirty-three, all except five recovered when brought again into the ordinary temperature of the laboratory.

by another and more complicated one. The latter belongs in that category of reactions included under the somewhat vague designation of "pleasure-pain" behavior. (For a discussion, see JENNINGS '04, pp. 248–249, and '06, pp. 332, 340.) Of reactions of this kind JENNINGS and others have pointed out abundant examples among the lower organisms. HOLMES ('05) has described in Ranatra, after one eye had been blinded, conduct nearly similar to that observed in Drosophila. He concludes that the "phototaxis" of Ranatra seems in many ways to be "intermediate between purely reflex conduct on the one hand, and conduct of the pleasure-pain type on the other." This conclusion applies equally well to Drosophila.

The production of convulsive reflexes.—In the experiments with an increased temperature, as above described, it sometimes happened that an insect, instead of creeping along the floor or roof of the box, would fly or hop toward the light. If it were under considerable headway it might, in spite of its negative thermotropism, be carried by its momentum into the immersed portion of the box. This rapid form of locomotion could often be induced by tapping on the exposed end of the box. A fly thus carried into a temperature of 45° C. soon became violently active. The high temperature evidently acted as a powerful kinetic stimulus, so that nervous impulses overflowed, as it were, from the sensory nerves concerned with temperature into the entire motor nervous system, producing a convulsive reflex. The wings vibrated with great rapidity, and the legs, abdomen, head and mouth-parts were affected by spasmodic contractions of their muscles. These activities often resulted in a peculiar spinning motion, and carried the fly rapidly about from place to place. Owing to the inclination of the box the insects usually tended toward the lower or immersed end, where they shortly succumbed to the heat, often dying in a characteristic attitude, with wings rigidly extended. Occasionally, however, the energetic, haphazard motor reflexes carried a fly out of the heated area into the cool, elevated region of the box, where its convulsive movements ceased, and it shortly began to creep about, restored, apparently, to its former condition.

When the experiments with reduced temperature were being performed, flies were introduced into the cold portion of the box. Many of these also gave the convulsive reflex before settling down into the quiet, benumbed condition eventually brought on by the low temperature. The reflex, though generally of shorter duration than when produced by heat, was, nevertheless, unmistakable. The characteristic spinning motion, and the final rigid extension of the wings could often be observed. These flies soon became motionless, but they did not die. They could be revived by raising the temperature.

Since extremes of both heat and cold produced, through the nerves concerned with temperature, such marked motor reflexes, I decided to try the effect of intense light acting through the optic nerves. The positive phototropism of Drosophila to light of various intensities has been described in a former paper (CARPENTER '05). In using the highest intensity at that time available, an arc light of 250 candlepower, it was noted that after continued exposure at a distance of 40 cm. an insect became extremely active, flying and hopping about irregularly, and giving little or no evidence of a directive control. It seemed probable that this excessive activity might have led to a true convulsive reflex had the intensity of the light been still further increased. For the purpose of testing this Prof. C. W. HOTTES, of the Botanical Department of the University of Illinois, kindly placed at my disposal an arc light of 480 candlepower, conveniently suspended in a dark room. A small glass box was constructed, consisting of two compartments separated by a vertical glass plate. In one of these compartments the flies were placed; the other was filled with water to serve as a heat screen. A thermometer was placed just behind the heat screen in contact with the vertical glass plate separating the two chambers.

When flies contained in this apparatus were brought to a distance of from 2 to 3 cm. from the arc light their movements at first, while rapid and irregular, were not true convulsive reflexes. The front wall of their compartment had an initial temperature of 25° C.; after about a minute this temperature rose, in spite of the heat screen, to 30° C. The flies then gave the convulsive reflex, tumbling and whirling about on the floor of the compartment, and showing no signs of orientation. Up to this time there had been no evidence of a reversal of their phototropism from positive to negative. Removed from the influences of the light and heat they resumed their ordinary activities.

To determine whether or not the intense light was a factor in inducing this convulsive reflex the flies were subjected to the same temperature in ordinary daylight. Ten insects were placed in a water-tight glass vessel with thin walls, and the latter was immersed in water which had previously been heated to 30° C. The convulsive reflexes did not appear. The temperature was then gradually raised. The flies showed the first convulsive reflexes between 36° and 38°, and these became general between 38° and 40°. The experiment was twice repeated with the same results. It follows, then, that the light of a 480 candlepower electric arc, at a distance of from 2 to 3 cm., calls forth the convulsive reflex at a temperature at which, in ordinary daylight, this reaction does not occur.

In view of the above-described effects of excessive temperature and light stimulation, it seemed probable that certain volatile substances, acting through the end-organs and nerves concerned with chemical sense, might also prove sufficiently stimulating for the production of convulsive reflexes. The following experiment was, therefore, made. The floor of the glass box used in the temperature experiments was moistened near one end along a line drawn from side to side, first with aqua ammonia, and afterward with glacial acetic acid. This end of the box was in each instance turned toward a window, and the flies, at first assembled at the other end, crept in the direction of the light. There was little or no evidence of a negative tropic reaction. But when the insects came close to either fluid the irritating vapors produced violent convulsive reflexes. Some of the insects were whirled into the fluid where they perished; others were carried to a distance and soon recovered. One fly spun about for fifteen seconds during a convulsive reflex induced by acetic acid, and finally reached a position outside the stimulating area. Its excessive activity then ceased, and it was soon creeping about in the usual way.

The student of animal behavior will ask himself how this convulsive reflex is related to other reactions of animals. It deserves this consideration since it is, in Drosophila, a normal reaction. The conditions necessary to call it forth are, it is true, extreme, and usually cause the death of the fly if allowed to continue. The reflex, however, is not a death struggle due to pathological changes in the body. If it removes the insect from the stimulating region the excessive activity gives way to ordinary movements, and the insect appears to be none the worse for its experience.

Mainly through the recent writings of JENNINGS attention has been directed to the relative importance and widespread occurrence of "trial and error" behavior among the lower animals. This kind of behavior is characterized by a repetition of "random movements," certain of which, under ordinary circumstances, are selected and followed up to the advantage of the organism. The convulsive reflex of Drosophila appears to be an instance of behavior of this character, in which, under excessive stimulation, random movements are made with extraordinary vigor and rapidity. There is, however, little evidence of the selection and repetition of those movements which carry the insect in favorable directions. Escape from the stimulating region seems to depend on chance alone.

The haphazard "trials" that are made during the convulsive reflex are the result of a complex reaction which seemingly involves all the movements of which the animal is capable. MAST ('03) saw in planarians subjected to a high temperature nearly all the reactions the worms have at their command appearing one after another. In Drosophila, with its more highly organized and specialized nervous and muscular organs, the reactions are simultaneous, each movable part performing its special function to the limit of its capacity.

Summary. 1. Drosophila is negatively thermotropic to high and low temperatures.

2. When one eye is covered so that the light stimulus is unilateral, Drosophila moves toward the source of light in a fairly direct path, but tends to deviate toward the side of the functional eye. A "pleasure-pain" reaction appears to inhibit and dominate a "tropic" reaction.

3. A violent, uncoördinated motor reaction or convulsive reflex may be induced in Drosophila by stimulating the insect either by a high temperature, or by a low temperature, or by intense light, or by the vapors of such irritating chemical substances as ammonia and acetic acid.

4. The convulsive reflex thus obtained in Drosophila may be regarded as an instance of trial and error behavior, characterized by a complex of vigorous random movements, involving, apparently, all the movable parts of the insect's body. The escape of the insect from the region of stimulation appears to depend on chance

LITERATURE CITED.

ARROWS, W. M.

1907. The reactions of the pomace fly, Drosophila ampelophila Loew, to odorous substances. *Journ. Exper. Zoöl.*, vol. 4, no. 4, pp. 515–537.

ARPENTER, F. W.

1905. The reactions of the pomace fly (Drosophila ampelophila Loew) to light, gravity, and mechanical stimulation. *Amer. Nat.*, vol. 39, no. 459, pp. 157–171.

HOLMES, S. J.

1901. Phototaxis in the Amphipoda. *Amer. Jour Physiol.*, vol. 5, no. 4, pp. 211–234.

1905. The reactions of Ranatra to light. *Journ. Comp. Neurol. and Psychol.*, vol. 15, no. 4, pp. 305-349.

JENNINGS, H. S.

1904. Contributions to the study of the behavior of lower organisms. *Carnegie Institution of Washington.* Publication no. 16, 256 pp.

1906. Behavior of the lower organisms. *New York*, 366 pp. (Columbia University Biological Series, vol. 10.)

MAST, S. O.

1903. Reactions to temperature changes in Spirillum, Hydra, and fresh-water Planarians. *Amer. Jour. Physiol.*, vol. 10, no. 4, pp. 165–190.

PARKER, G. H.

1903. The phototropism of the mourning-cloak butterfly, Vanessa antiope Linn. *Mark Anniversary Volume*, article no. 23, pp. 453–469.

PHOTOTAXIS IN FIDDLER CRABS AND ITS RELATION TO THEORIES OF ORIENTATION.

BY

S. J. HOLMES

(From the Zoölogical Laboratory of the University of Wisconsin.)

In the phototactic movements of animals which orient themselves definitely to the light it is almost always the longitudinal axis of the body which is kept parallel to the rays. It is therefore not without interest in relation to the much discussed "theory of tropisms" to find an animal which orients itself sidewise instead of in the usual manner. Such an animal is the common fiddler crab of the Atlantic coast, Uca pugnax (Smith), which shows a very decided positive phototaxis, especially in strong light. Like many other crabs the members of this species run sidewise, and in fact they are so constructed that they would find great difficulty in any other form of locomotion. When recently brought into the laboratory they seem remarkably attracted by a bright light. They gather on the side of their enclosure nearest the light and often struggle for a long time to get nearer the source of attraction. By changing the position of the light the crabs may be made to follow it about in any desired direction, changing their course promptly whenever the light is moved. There is lateral instead of longitudinal orientation to the direction of the rays.

When the crabs reach the side of the dish they usually do not maintain their lateral orientation. They frequently face the light, holding their eyes erect and moving from side to side in the endeavor to get as near the light as they can. Very often they settle down facing the light, remaining there for a long time as if spellbound. There seems to be no tendency to get into a position of lateral orientation once the animal attains to a position of proximity to the light. This orientation is necessary if a certain object is to be followed, but, so far as I can determine by watching its behavior, the crab does not tend to assume it except during locomotion. When it has gone as far toward the light as it can it settles down preferably in a position of longitudinal orientation.

When a crab that is facing the light is subjected to very strong illumination it will often raise its body so as to stand supported on the tips of its claws. If the light is withdrawn a short distance the body comes to rest again upon the bottom of the dish. The crab may be made to repeat this act many times by causing the light to approach and recede. The behavior of the animal is apparently the involuntary result of the increase in the tension of the leg muscles brought about by strong illumination.

Similar increase in muscular tone by light is shown in the behavior of the eye-stalks. When a fiddler is seized the eye-stalks are drawn back into the orbits and tightly held there. If, however, the crab is brought very close to a strong light the eye-stalks are erected. If the animal is removed a little further from the light the eye-stalks are pulled back into their orbits again. When brought into strong light the crab is apparently no longer able to hold the eye-stalks down and they come up in spite of the instinct to hold them in a protected situation. If the cornea of one eye is blackened over usually only the unblackened eye rises upon exposure to strong illumination, but the blackened one sometimes does so to a greater or less extent, owing perhaps to the fact that the two eyes usually make associated movements.

The experiment of crossing the eye-stalks was tried in order to see what effect would be produced on the animal's reactions to light. The eye-stalks of the fiddler crab are long and they may readily be crossed like the parts of a letter X and tied in the middle. Crabs treated in this way show great confusion in their reactions to visual stimuli. When approached they give signs of alarm and frequently run directly towards the source of danger. When a light is moved suddenly they often run towards it instead of away as they usually do from all moving objects. Normal phototaxis, however, is mainly destroyed. The crabs neither go directly toward or away from the light with any regularity; in fact, they seem to pay little attention to the light, owing perhaps to the discomfort of their unusual predicament. There is no definite reversal of phototaxis, which I thought might occur under the circumstances, although there is often a reversal in the responses to moving objects. The crabs' actions, however, are hesitating and uncertain; often they move a distance one way and then in another as if the result of their movements were something unexpected. In a crab with crossed eyes the behavior of objects in its visual field consequent on its movements is different from what it is accustomed to. Under normal conditions fiddler crabs probably possess binocular vision so far as may be judged by the behavior of the eye-stalks and the arrangements of the facets of the eyes, but when the eyes are crossed the animals have two visual fields which give quite different impressions, thus adding further to their confusion. The visual world of such crabs doubtless seems a hopelessly mixed up affair, and it is not surprising that the animals often sulk as if discouraged with their efforts. Crabs kept for several days in order to ascertain if they would improve in the appropriateness of their responses gave only negative results.

The phototactic reactions of fiddler crabs are very easily checked or overcome by fear. When a light is brought near the animals they often scurry away in great haste, and at first one might be led to interpret this behavior as a manifestation of negative phototaxis, but it is a very different phenomenon. Even when showing a strong positive reaction the fiddlers often flee in apparent alarm upon a slight movement of the light. If the movement is a sudden one they are more apt to beat a retreat, while they follow slower movements without fear. When exposed to strong illumination for some time they become more insensible of their surroundings and are dominated almost entirely by the stimulus from the light without regard to movements made near them which at first would send them scurrying off in great haste. They get, as it were, "warmed up" to the work, becoming not only less responsive to other stimuli, but more vigorous in their phototactic activity. I have described similar phenomena in the water-scorpion, Ranatra, and other observations have shown that it is a not uncommon trait in tropic responses.

I do not wish to add further to the confusion that exists in the use of the term tropism, but I believe that the retreat of the fiddlers from a moving light cannot properly be described as a tropic reaction. It may be related to tropic reactions as vision is according to RADL, but it is not so much a response to the light *per se* as to a sudden movement or appearance of the light. If we described as negative phototaxis all movements however caused which were directed away from the light we should have to include the flight of the fiddlers under this designation, and say that a sudden movement of the light caused a reversal in the sense of the response. But will anyone maintain that when a wild animal runs away in alarm from a sudden blaze it manifests a negative phototaxis? It is with such an action as this, rather than with the reactions commonly included under the head of negative phototaxis that the retreat of the fiddlers is more closely related. The flight of the crabs from light has all the characteristics of their flight from moving objects in general. Whenever one enters a room where the crabs are kept or makes a small movement in their vicinity they promptly scuttle away; and they often detect one's presence on a beach for a distance of several rods and make for their holes. These reactions we commonly attribute to fear, whatever physiological explanation we may offer for them, and anyone who has observed the fiddlers scurrying away from a moving light can hardly fail to ascribe their behavior to the same cause.

The point of principal interest in the phototaxis of the fiddler crabs is the relation of their lateral orientation to the theories of tropisms. Can we regard orientation as a direct response in which the animal is involuntarily forced into line, or is it rather to be considered as coming under the pleasure-pain type of behavior, and as therefore related to the voluntary seeking of a certain end which is exhibited in the behavior of higher forms? In order to explain the orientation of a highly organized form like an insect or crustacean in which, in most cases, response to light takes place through the eyes, we may assume that light falling more strongly on one eye sets up impulses which are transmitted more or less directly to the leg musculature. We may assume that the extensors of the opposite side are stimulated, or the flexors on the same side, or both, and that in consequence of this distribution of impulses the animal moves until its body is in line with the rays. In such a case the movements involved in orientation are the same as those employed in ordinary locomotion only the activity of the legs on one or the other side is accentuated according to the position of the body in relation to the direction of the rays.

In the fiddler crab, however, the case is different, and we cannot explain the phenomenon in this way. The legs of the fiddler move in a plane approximately at right angles to the sagittal plane of the body, but they are capable of a certain amount of forward and backward motion which may be employed to change the direction of locomotion. The movements involved in orientation are different from those employed in ordinary running. They are special movements employed to check deviations from a certain course, a circumstance which would greatly complicate any attempt to explain orientation as a comparatively direct response. The results of observations on fiddler crabs tend to confirm the conclusion reached in studies made on the phototaxis of Ranatra,[1] namely, that light is followed much as an animal pursues any other object of interest such as prey, or its mate, and until we can give a physiological explanation of these phenomena we are not, I believe, in a position to give a satisfactory explanation of orientation to the direction of the rays of light.

[1] The reactions of Ranatra to light. *Jour. Comp. Neur. and Psych.*, vol. 15, 1905.

THE LIMITS OF EDUCABILITY IN PARAMŒCIUM.

BY

STEVENSON SMITH

WITH FOUR FIGURES.

The theory of the phylogenetic development of adaptive behavior in animals has given interest to the question of the development of consciousness in animals, and to the possibility of the two developments having been coexistent throughout. A synthetic study of animal behavior, beginning with the lowest form, points out the possibility of an adaptive phylogenetic development, becoming more and more complex as we ascend in the scale of life, and all this without the coexistence of consciousness.

If we start with the lowest possible form and determine that all its behavor may be described solely by mechanical laws, it may be possible to interpret the more complex behavior of the higher forms as the action of a more complex mechanism. As we have empirical evidence of such facts as inheritance and individual variation we may assume these as factors in evolution and still exclude consciousness as in no way explaining them.

Some writers have looked upon natural selection as involving consciousness in that some organisms, possessing better memory than others, would be educated more easily and so adapt themselves sooner and better to their environment than the rest. But these writers fail to see that memory has another than a psychic side. This is memory as we observe it in others; when a stimulus has acted upon an organism one or more times the next stimulation, of a like kind, produces a different reaction from that which the previous stimulus occasioned. This form of memory is observed in inorganic manifolds as well as in organic. For instance, through continued use an old lock acts differently from the ame lock when it was new. The action of any mechanism is subject to such a change. This change may be to the advantage or to the disadvantage of the organism, but it is still memory.

Thus, more complex and more advantageous memory, though accounted for by evolution, may in no way involve consciousness.

Evolution depends upon those organisms being selected which have that certain kind of memory which enables them to cope with the conditions of their surroundings. Whether this memory is evolved through the inheriting of acquired aptitudes or through the selection of the best adapted variants, or mutations, does not bear on the question. With such factors given, and they are universally accepted laws in biology, the evolution of an adaptive behavior follows of necessity. It is probable that an explanation in these terms seems inadequate only because of the scarcity of empirical data. Observation of the physiological economy and gross behavior of animals is the basis for such a theory of the genesis of adaptive behavior, but the only starting point for a study of consciousness is the consciousness of the observer.

The observer, however, finds that other organisms of his own species have solved problems relating to, and have systematized, this very matter of consciousness, and other matters, which he, though possessing consciousness, might hardly have done. This at first seems one of his strongest grounds for assuming consciousness in his fellows. But by such an assumption he implies that consciousness is an aid to this solving and systematizing and that he without it, could not solve and systematize. This assumption he can make only by a denial of parallelism, for if consciousness in him has an invariable physiological accompaniment it may then be only the indicator of his ability and not an aid to this ability at all.

But if the individual wishes, for convenience or any other reason, to assume consciousness in his own species, he may next consider what grounds he has for attributing it to other species. It is obvious that the probability of the other species possessing consciousness is directly proportional to the similarity of their behavior to that of the species assumed to possess it. In so far as investigation follows this line it is valid.

As the behavior of any species most nearly approximates that of the species next to it in the scale of genesis any speculation as to the origin of consciousness may best be made by studying the species in inverse order of their development. The question then presents itself, does consciousness belong to every order of life? If not, what is the lowest order that possesses it?

Many criteria of consciousness have been suggested by the various writers in this field, and their motives and the value of their assumptions must be considered and agreed upon by animal psychology before we can have an exact classification in that science. LLOYD MORGAN[1] says that consciousness is present wherever we find profit by experience. He feels justified in assuming that pleasurable consciousness is associated with those modes of behavior which through repetition become more vigorous, and disagreeable consciousness with those that are checked.

In man most education takes place in the highest level of the brain, the functions of which are synchronous with consciousness. Hence, in man, profit by experience usually implies consciousness. In many lower animals no such level exists and education is a modification of the lower centers or even of the somatic tissue. If our inference of animal consciousness is based on analogy, is it analogy with the gross behavior of man or with the nervous mechanism of man? If profit by experience is our criterion it would seem that the analogy was to man's gross behavior. But all profit by experience in man is not necessarily accompanied by consciousness. Many adjustments are brought about unconsciously and many more are arrived at without involving our choice, and this is what MORGAN really has in mind. We certainly do not think that the profitable adaptation of our muscle tissue to new conditions renders the muscle "a conscious mechanism." But much adaptation in animals may be of this kind, and in animals possessing no differentiated nervous system any adaptation, or *profit by wear and tear*, must be a modification of body tissue.

How then shall we justify ourselves in saying that all profit by experience is evidence of consciousness in an organism? This special case of inference by analogy is usually described in such words as these: "I infer consciousness in others because I observe their behavior to be similar to that behavior in myself which is accompanied by consciousness." But is the similarity of gross behavior our only ground of inference?

If a race of beings should appear whose gross behavior was similar to our own but whose central nervous system was entirely different would the analogy of gross behavior alone justify the inference of their consciounsess? Though it might do so for the unreflective man, it would not for the ontologist. Why is the doll more than a puppet to the child? Because of the analogy of

[1] Animal Behavior p. 45 ff

shape, of facial expression, and of general human appearance. The closer the analogy the stronger is the child's illusion that the doll possesses a consciousness like her own. Hence the desire for movable limbs and closing eyes. For the child the analogy of shape is enough. For some zoölogists the analogy of gross behavior suffices. If analogy is to be used at all should not the analogy be complete? If the analogy is made complete the criterion of consciousness will involve the possession of a cerebral cortex.

One motive for assuming subjectivity in creatures like our own bodies is the ease of description resulting from the use of subjective terms. The hypothesis is further fixed by our social instincts, of pity for an injured fellow man, of gratification in his welfare, of envy of his good fortune, etc., all of which states in him recall to us the mental accompaniment which we would have were we to experience them. The usefulness of the concept disappears when we extend the theory to apply to the lower forms of animal life.

JENNINGS[2] calls attention to MÜNSTERBERG's suggested criterion of consciousness, namely, that consciousness exists where it is useful to assume it in order to help us to understand and anticipate the behavior of others.

We do not speak of choice in inorganic manifolds for this would be animistic. Water does not choose to run down hill rather than up and an acid does not choose to combine with a metal. The water and the metal follow certain physical and chemical laws and to apply consciousness to them is improper. Yet when the protoplasm of a living organism contracts in one way, on being affected by a physical stimulus or a chemical reagent, rather than in another, some scientists at least[3] are willing to claim that these selective contractions and resulting changes of space relations to the stimuli are evidence of mind in the organism.

What is the meaning of the word selective as used by MORGAN? Certainly it means selective of the most favorable conditions for the life processes of the individual and its community. Some permanency of organization is a feature of living things. Dynamic biology has only begun to solve the question how this structure is maintained. The choosing of conditions is a factor in such maintaining of organization. Bad conditions drive the creature away while good conditions retain or attract it. It is adjusted so

[2] Behavior of lower organisms, p. 335 ff.

[3] MORGAN and ROMANES, e.g.

as to react in these ways. The conditions are physical and the thing's movements are physical. Is the adjustment somethingelse, something half psychical? If it can be shown that a little mind would help in making the movements, then we may busy ourselves in describing the behavior of mind as it lends a helping hand to clumsy protoplasm. But what seems unphysical is not the contraction but the purpose which the contraction serves.

The purpose that some contractions serve is regulation or readjustment and *we may call behavior regulatory when a process having proceeded too far is the cause of its own remedy.* Such readjustment is not without parallels in the world of inorganic manifolds. For instance, I put my coffee pot on an open camp fire, the fire becomes too hot, and the water boils over. But the boiling over of the water regulates the fire so that a fire of nearly constant heat is kept up as long as there is a certain amount of water in the pot. So a pond is kept from freezing to the bottom in winter by a regulation based on the densities of water at different temperatures. As it changes from a temperature of 3.9° C. it becomes, up to a certain point, less dense. Therefore, when the whole pond is at a temperature of 3.9° and the top is cooled, the warmer water from the bottom does not rise. So many machines have been made involving a regulatory principle, such as the burglar alarm or the differential valve, that no mystery surrounds it.

There is a difference most valuable for classification between living organisms, which through metabolism return after some time to, or nearly to, the original state which existed previous to an experience, and those inorganic manifolds which do not return to a normal state but which remain indefinitey changed after they are acted upon. It is this reversion to a normal state in all but the highest nerve centers of living organisms which makes possible their adaptation to often recurring stimuli of the same kind, and the stability of those vital economies which we call adjustment and adaptation.

For instance, let an organism at birth be capable of giving N reactions (*a*, *b*, *c*,—N) to a definite stimulus S and let only one of these reactions be appropriate. If only one reaction can be given at a time and if the one given is determined by the state of the organism at the time S is recevied, there is one chance in N that it is the appropriate reaction. When the appropriate reaction is finally given the other reactions are not called into play, S may cease to act, but until the appropriate reaction is given let the organism be such that it runs through the gamut of the others until the appropriate reaction is brought about. As there are N possible reactions the chances are that the appropriate reaction will be given before all N are performed. At the next appearance of the stimulus, which we may call S_2, those reactions which were in the last case performed, are, through habit, more likely to be again brought about than those which were not performed. Let u stand for the unperformed reactions. Then we have $N-u$ probable reactions to S_2. Habit rendering the previously most performed reactions the most probable throughout we should expect to find the appropriate reaction in response to:

S_1 contained in N.
S_2 contained in $N-u_1$.
S_3 contained in $N-u_1-u_2$.
. . . .
S_n contained in $N-nu$, which approaches *one* as a limit.

Thus the appropriate reaction would be fixed through the laws of chance and habit. This law of habit is that when any action is performed a number of times under certain conditions it becomes under those conditions more and more easily performed.

There are two main roads leading to the hypothesis of animal consciousness. One is traveled by the psychologist in his effort to extend the limits of the introspective science, and the other is followed by the biologist who would find in the conscious fiat some explanation of selective movements or of regulation in behavior. Hence the criteria of consciousness applied by the two investigators are different.

Loeb[4] holds that a psychology of the lower forms must be a science of tropisms, and that even in the higher forms consciousness is no explanation of behavior but merely a function of the mechanism of associative memory.[5] Whether or not the behavior of any form of life has a definite significance to the psychologist, the complexity of its functions must decide its rank in a science of behavior, and for this classification the limits of such complexity

[4] Loeb: Dynamics of living matter, p. 158.
[5] *Ibid.*, p. 6.

must be discovered. With this in view the present study has sought to define the limits of educability of Paramœcium.

EXPERIMENTS IN EDUCABILITY.[6]

The purpose of the following experiments was to determine what kind of modifiability is shown by Paramœcium due to recurring experiences of the same kind. Less interest attaches to that modification of behavior due to fatigue, which is usually a retardation of movement, than to that modification supposedly due to a rearrangement of structure more suitable to perform the movement, which may be called adaptation through practice, and which is usually characterized by more rapid or exact movement.

Aside from the results bearing on modifiability, there are noted below certain movements made by Paramœcium under the conditions of the experiments which, to the best of my knowledge, are not spoken of by other observers.

The experiments fall into three groups: (1) Those in which the animal was stimulated by touch (the meniscus of a capillary tube) the conditions being such that it could react in but two ways in order to escape; and (2) those in which the animal was stimulated by change in temperature; and (3) those in which the animal was frequently made to experience two conditions—say A and B which at first occurred simultaneously, and later made to experience condition A alone, any difference being noted between the reaction to condition A before it had been combined with condition B and the reaction to it after it had been combined with and again separated from condition B.

Reactions to touch. The difficulty of observing Paramœcium, or any such free-swimming organism, when it is allowed to swim unrestrained about a slide is known to all who have attempted it. The uncertainty of any results obtained is proportional to this difficulty. It is not important in these experiments to imitate the conditions of real life, so it was decided to make as fixed as possible the conditions of the experiment.

For this purpose a capillary tube was selected of a bore smaller than the length of the Paramœcium and larger than his width. The animal was caught by the upward suction of the tube and the

[6] These experiments were carried out mainly in the Laboratory of Psychology of the University of Pennsylvania.

tube was then placed on a movable carriage, so the animal could always be kept in the field of the microscope no matter what part of the tube it might be swimming through.

Once in the tube the Paramœcium swims to the forward end and upon reaching the meniscus jerks backward for several times its own length, then approaches again in a wider spiral than before. This backing and approaching takes place at least a dozen times and later the Paramœcium settles down to a pecking movement, revolving anti-screwwise about the meniscus and attacking about five places in its circumference.

In the original approaching and retreating both movements may be either screwwise or anti-screwwise. In approaching, both the screwwise and anti-screwwise movements give about the same width of spiral, namely, a very slight one. If the retreat is made anti-screwwise a relatively straight course is followed, the spiral being hardly noticeable. If the retreat is screwwise a very wide spiral results.

FIG. 1. Paramœcium turning in capillary tube.

In most cases the animal after a varying time bends its anterior end around toward the aboral side (fig. 1), forming a "U" with its body, and after a number of jerks succeeds in reversing the position of its body in the tube. In all cases it turns toward the aboral side, thus using the long creeping cilia near the buccal groove to obtain a hold on the side of the tube.

Due to these movements being of no fixed type but varying greatly from time to time under the same conditions, a satisfactory explanation cannot be made in terms of tropism. The "trial and error" explanation, although the principle is no doubt involved, as it is in the gross movements of all animals, does not seem to satisfy, because the movement of reversing in the tube requires a great deal of effort and perseverance on the part of the Paramœcium and a relatively long time to accomplish. The law of trial and error describes the organism as avoiding any great difficulty and turning to a more easily accomplished movement. The summation of stimuli of many failures probably becomes adequate to cause this unusual reaction.

As this facing about in the tube is repeated, the time taken for each turn may be longer than for the last, the animal finally dying of apparent fatigue, or, if the tube is not so small that too violent an effort is required of the animal, the time may gradually be shortened and a most surprising aptitude of turning be developed. Paramœcia from a vigorous culture give better results than poorly nourished ones. Under optimum conditions I have found a reduction of turning time, after the animals have been in the tube for twelve hours or more, from four or five minutes to a second or two, which is the minimum time in which the turn can be made.

Often the Paramœcium will rest for a long period at one meniscus, slowly circling around with its buccal groove resting against the air surface. When, however, the effort is made to reverse, the shortening of time in the practiced individuals is very apparent.

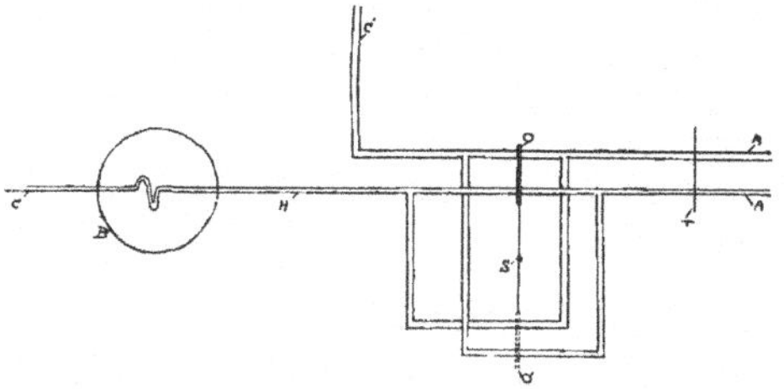

FIG. 2. *A* and *A'*, tubes through which water of alternating temperatures passes and on which the capillary tube rests. *B*, beaker of boiling water in which pipe coil is immersed. *C* and *C'*, cold water supply tubes. *H*, hot water supply tube. *O* and *O'*, interrupters. *S*, switch which alternates temperatures in *A* and *A'*. *T*, capillary tube containing Paramœcium.

At *O* and *O'* are two interrupters so arranged that a single lever raises one and lowers the other. If the interrupter at *O* presses on the pipes the water flows through the loops, the hot water (from *H*) flowing through *A'* and the cold water (from *C'*) flowing through *A*. If the interrupter at *O'* presses upon the pipes the hot water flows through *A* and the cold water through *A'*. The temperature and flow of the water supply are kept constant so that the alternating temperatures at *A* and *A'* do not vary.

Reactions to temperature. In these experiments a capillary tube was selected large enough to allow the Paramœcia to reverse their direction without touching its sides, and in which two Paramœcia could pass each other without difficulty. A number of individuals were taken up in this tube and the tube was placed on a carriage having two large glass supporting tubes through which water of different temperatures was passing and on which the capillary tube rested (fig. 2).

While hot water was flowing through one support and cold water through the other the temperatures could be reversed, by a single movement of the key, so that the cold water would flow through the one and the hot water through the other. Thus the distribution of temperature in the capillary tube could be reversed at will. By cold water is meant water at normal temperature to which the animal gives no reaction.

As soon as the capillary tube is heated at one end by contact with the hot support the Paramœcia at that end dart about at random until they are headed toward the cool end of the tube and even then do not swim to the cool end at first but often turn back to the hot end several times before finally swimming over to the cool water. Once arrived at the cool end the Paramœcia do not stay there but turn and start back to the hot water. In this way a Paramœcium may traverse the length of the tube a dozen times or more before coming to rest at the cool end of the tube. If

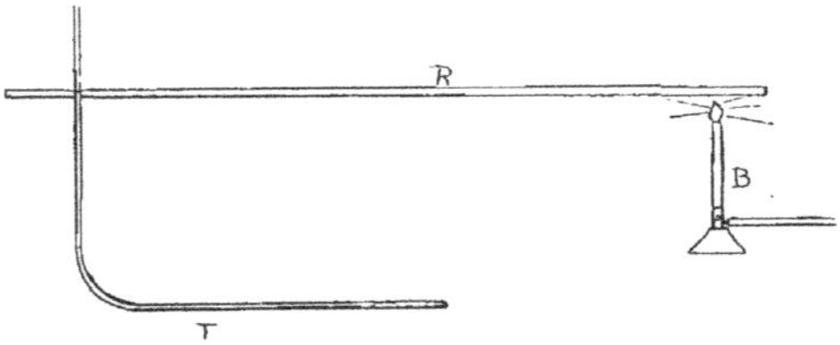

FIG. 3. *T*, L-shaped tube containing Paramœcia. *R*, Brass rod heated and in contact with *T*. *B*, Bunsen burner used to heat *R*.

now the switch (fig. 2) is turned so that the cool end becomes hot, the animal will dart about much as it did when first stimulated and only after many trials will it reach the cool water. As the animals leave their resting place at the warm meniscus in obedience to the repeated reversing of temperatures in the tube, their movements become slower and more regulated and they seldom turn more than once toward the cold water before swimming in that direction. It is not significant to express this modification of behavior in terms of time for, although the time involved in getting away from the heated end of the tube is somewhat reduced as the stimulus recurs, it is the suitability of the movement to accomplish the result which characterizes the later reactions. In these the actual locomotion is slower but the random movements give place to more determined ones.

The influence of an associated past experience upon the reaction to a given stimulus. Although in the following experiments the observations gave nothing but negative results, these results serve to fix the limits of educability in Paramœcium. Although Paramœcium profits by experience, as seen in the above sections, it does not show associative memory such as LOEB would demand as the criterion of consciousness.

The conditions of the first experiment were these. Paramœcia were placed in a trough having an extremely thin glass bottom and this trough was immersed in a partitioned box containing hot and normally cool water on the two sides, so that the bottom of the trough was kept cool on one half and warm on the other (fig. 4). There was a distinct line, not corresponding exactly to the partition of the under box, at which the Paramœcia approaching from

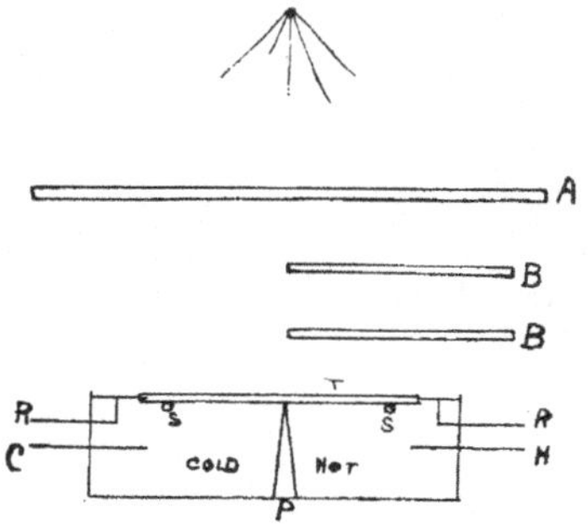

FIG. 4. *A*, alum bath heat screen. *B B*, light screens. *C*, cold water supply pipe. *H*, hot water supply pipe. *L*, electric light. *P*, Knife edge partition. *R R*, overflow return pipes. *S S*, supports for trough. *T*, trough.

the cool side would turn back. A light was fixed above the trough and a screen interposed so that a shadow fell covering the warm area and a minute part of the cool area beyond the reaction line. The white Paramœcium gives no reaction to light or darkness and it was hoped that by allowing the animals to experience darkness whenever they experienced heat they might, when the heat was removed, react negatively to darkness. This they did not do, however, though one group of Paramœcia were allowed to experience the two conditions together for fifteen hours, one for twenty-four hours, and one for forty hours.

Another experiment of a somewhat similar kind was performed in which it was tried to bring about the association of heat and gravity. The conditions were these (fig. 3). A small tube was bent into an L shape and, after some Paramœcia had been drawn up into it, was placed so that one leg was horizontal and the other, rising from this, was vertical. At the top of the vertical leg was placed a hot metal rod in contact with the glass and kept at a constant temperature. If Paramœcia show geotropism,[7] this irritability to gravity should be more easily associated with heat than could light, which, although it must make some impression on the organism, does not cause normally an avoiding reaction.

Whenever the Paramœcia swam up the vertical leg of the tube they received a heat stimulus which caused them at first to jerk backwards and after many random trials to swim downward to the cool water. Although these conditions were kept unchanged for as long as three days the Parmaœcia never learned to avoid the vertical leg of the tube. In the end they did not react as violently to the heat and did not, as at first, swim occasionally past the hot metal rod. Also they seem later to develop greater sensitivity, reacting to the heat before getting as close to the metal rod.

If chemicals were not so diffusible and the conditions so hard to govern, an association might be produced between temperature and some chemical stimulus.

CONCLUSION.

Paramœcium is educable in that its behavior may be modified to show the results of practice, both in a reduction of the time involved in performing a movement and in the increase in suitability of the movement to accomplish the appropriate result.

In so far as the tests here apply, there is no evidence of associative memory in Paramœcium.

The reversing movement above described is in the nature of a *positive reaction.*

Hampden-Sidney College,
Virginia.

[7] Moore: *Am. Jour. Phys.*, vol. 9, pp. 238 ff.

FRENCH WORK IN COMPARATIVE PSYCHOLOGY FOR THE PAST TWO YEARS.

The zeal with which investigations in comparative psychology are being pursued in France is testified to by the contents of the *Bullétin de l'Institut général psychologique*, whereof the reports of the *Groupe d'étude de psychologie zoologique* form something like four-fifths. It is further witnessed by the long bibliography attached to this paper, the extent of which, however, is partly due to the fact that many of the titles refer to short communications in the *Comptes rendus* of the *Société de Biologie* and the *Académie des Sciences*. In some cases the contents of these are repeated in the longer articles.

The work of GEORGES BOHN is, as usual of late years, the most voluminous and the most important French contribution to the science. This writer's earlier papers may be found summarized by Professor YERKES in this *Journal*, vol. 16, p. 231. During the two years since the appearance of that summary, BOHN has devoted himself especially to experiments and observations on actinians and starfish. He has gained what he considers to be new confirmation of one of the cardinal facts upon which he has long insisted: the influence upon the present reactions of an animal of the conditions which have acted upon it in the past, *les causes passées*. Further, he has definitely rejected JENNINGS' conception of "trial and error" in favor of the position of LOEB, and maintains that the oscillations and variations in tropisms which have given rise to the former idea are the effect partly of the influence of past conditions, partly of that "sensibility to difference," or susceptibility to changes in the intensity of a stimulus, which LOEB assumes in addition to the tropism, and which, in BOHN's opinion, LOEB's critics have too much ignored. A special case of the influence of past causes is to be found in the preservation under laboratory conditions of certain oscillations in tropisms which coincide with tidal rhythms and with the alternation of day and night. The study of these oscillations, to which he and others had previously called attention, has been continued by BOHN, and observations bearing on the matter have been made by PIÉRON and by DRZEWINA. BOHN's statements regarding the preservation of the tidal rhythms in the laboratory have been received with some scepticism by L. LAPICQUE.

Of the other contributors whose names are to be found in the bibliography, PIÉRON, in addition to his work on actinians, has been making studies of the sensory factors which play a part in the life of ants, and has also been investigating the phenomenon of "autotomy," or the amputation by an animal of one of its own members. A discussion has arisen in connection with this point between him and DRZEWINA, the nature of which will be later explained. FAURÉ-FRÉMIET has been observing the behavior of certain protozoa. LÉCAILLON has been continuing his studies of the instincts of spiders. HACHET-SOUPLET, pursuing his conception of animal education, for which he has, with the approval of two members of the Academy, adopted the word "*zoopédie*," has succeeded in obtaining the vote of the *Institut général psychologique* for the establishment of a *Section d'étude de dressage scientifique*, the actual formation of which, however, appears to have been postponed.

We shall now take up in order the titles in our bibliography, referring to them by number, and try to indicate the important points brought out in each paper.

1. Of this article by BOHN, the substance is as follows: Those animals which show most rapid transmission of nervous impulse from one segment to others would seem to have most psychic individuality, but this is a vague and obscure concept. The degree of coördination depends on the development of the receptive sensory apparatus.

2 and 3. Acanthia lectularia is a negatively phototropic insect. BOHN shows that when it undergoes a change of illumination, its tendency is to rotate through 180° in a direction which is constant during the day, and changes towards evening. This reaction is but poorly adaptive, since under certain conditions it may lead the animal to turn toward the light. For instance, if the insect is moving along a shadow directly away from the light, and comes to the end of the shadow, it will turn through 180° and thus be brought to face the light. Again, it may either approach or move away from a dark screen, as a consequence of its turning always in the same direction at a given time of the day. The mollusc Littorina, on the other hand, is always, at a given hour, attracted by a dark screen. The superior adaptiveness of Littorina's behavior may be due to the fact that in nature it is accustomed to seek shade, while Acanthia does not come out on bright surfaces at all during the day.

4 and 5. The results stated in these papers were published later in 27.

6. The results of this paper are stated in paper number 15.

7. As between the "teleological" interpretations of JENNINGS and the mechanical explanations of LOEB, BOHN would insist on taking account of the past experience of the animal in determining tropisms.

8. This paper must be taken in connection with that by PIÉRON numbered 59 in the bibliography, and with the paper by BOHN and PIÉRON numbered 28. Actinia equina closes when the sea withdraws, and opens when the sea returns. PIÉRON found that the specimens he kept in the laboratory opened under the following conditions: when sea-water was made to run over them, when the water was agitated, when it was reoxygenated, when food substances were brought near. They closed when they had been some little time dry, when the water was deoxygenated, when they received mechanical shocks, and after grave lesion by toxic substances. In the pools from which they were taken, they expand at the mechanical agitation caused by the first wave of the rising tide that reaches the pool. They close, at falling tide, before the pool is stagnant. They thus show "anticipation," in closing before there is actual need of it. PIÉRON did not find, however, that rhythmic opening and closing in accord with the tides persisted in the laboratory (59). BOHN and PIÉRON, in their joint paper (28), explain the difference between their results in regard to this last point,—BOHN having found a persistence of the tidal rhythm in the laboratory with Actinia equina as with Convoluta roscoffensis,—by the difference in habitat between their specimens. BOHN's were taken from high on the side of a vertical wall, where the contrast between the conditions at high tide and those at low tide was very marked. PIÉRON's were taken from pools not wholly dry even at low tide. The "anticipation" noted by PIÉRON is a step towards the development of such a rhythm as that

observed by BOHN. The actinian first responds to the stimulus of loss of oxygen or drying; then, by anticipation, to a stimulus which regularly precedes this (diminished agitation of the water), and finally to an internal, periodically occurring stimulus (28). In paper number 8 of the bibliography, BOHN gives a further account of the persistence of the tidal rhythm in Actinia. The specimens observed by him spontaneously opened and closed in the laboratory for two or three days. That the rhythm still existed after this time was revealed by the following facts: if an actinian was placed in a current of water, after expanding, it closed, but it closed much more readily at the time of descending tide; when kept in a current for a long time, the actinians remained closed, but opened irregularly and temporarily at the next high tide, and quite generally and persistently at the one after that; mechanical shocks had a tendency to make them open at rising tide and close at falling tide.

L. LAPICQUE, in paper number 43, expresses doubt as to the reality of the periodicity thus observed. He raises the point that the rhythm impressed upon the actinians should be that of the tides on the last day of their sojourn under actual tidal influence, and that this would in the course of the next week bring them quite out of accord with the contemporary tidal periodicity. BOHN replies in paper number 9 by showing that the oscillations correspond in a general way to the contemporary tidal rhythm; and in paper number 38, with FAUVEL, demonstrates that certain diatoms (Pleurosigma aestuarii) exhibit a tidal rhythm in emergence and disappearance which continues in the aquarium. He admits that mathematical exactness in plotting the curves of such periodicities is impossible (12) (13) (18); LAPICQUE, unreconciled, suggests that BOHN had better give up the attempt at it, if this be the case (44) (45).

10. BOHN's study of certain seaside butterflies concerns the relations between phototropism and anemotropism. Satyrus janira orients itself when at rest with head to the wind; its flight is determined by the position of the sun. The more the posterior portion of the eyes is illuminated, the more the wings are spread apart when the insect is at rest, and the more energetically they beat in flight; hence the insect flies away from the sun, especially when it is low. Vanessa cardui shows the same relation between the illumination of the eyes and the beating of the wings, only more strikingly; V. io also shows it, but the relation is the reverse in the case of V. urticæ.

11. This article is a discussion, more or less historical, of the relations between tropisms, instincts, and intelligent acts, in which the conclusion is reached that "JENNINGS is wholly right in considering supposed tropisms as phenomena in general very complex. But these tropisms depend not only on the connections between organs, but on the state of the matter composing the organs. Living matter has a whole history which is responsible for the fact that its reactions are made in accordance with determinate rules. Selection, as JENNINGS conceives it, has but little upon which to exercise itself. The ideas of JENNINGS, far from invalidating those of LOEB, merely supplement them."

14, 15, 16, 17. These papers on the reactions of actinians may be considered together. Paper 15 includes the results of the other three. Actinia equina, as we have seen, was found to preserve a tidal rhythm in the laboratory. A day and night rhythm, usually that of opening at night and closing by day, also showed itself. The influence of *habitat* was indicated by the fact that the day and night rhythm was observable in actinians taken from pools where the tidal changes were less important. It was masked by the tidal rhythm in specimens from high rocks, but might be observed in these when the tidal rhythm had disappeared. The effect of habitat was further shown by the fact that actinians collected from sunny places showed a reverse day and night periodicity, opening by day and closing at night, while those taken from dark places seemed to suffer under the influence of light. The *effect of light* is exhibited by the fact that actinians display more activity, the more light they have been subjected to in the past. *Impurity of the water* increases the effect of light. *Mechanical agitation* seems to destroy a state of inertia in the animals, and may reveal a tidal rhythm. Anthea cereus and Actinoloba dianthus show a more marked response to light than Actinia equina, the former probably because it contains chlorophyll-bearing algæ. A. cereus converges its tentacles towards the light; A. dianthus orients its column in the same direction, but the orientation is reached only after a series of oscillations which suggest trials and errors, but in BOHN's opinion are the absolutely determined effect of past causes, combined with "sensibility to difference" in LOEB's sense. There is a general tendency for actinians to expand under a *thin layer of water;* this is doubtless connected with the fact that the food supply is best near the surface. Tealia crassicornis shows the same general features of behavior as the other actinians mentioned.

19. The most important point in this paper is as follows. The leech observed is positively phototropic; at the outset of its movements toward the light, its orientation is precise, but the further it advances, the more it waves its body from side to side. If these deviations were "trials," they ought, BOHN contends, to diminish rather than increase in number. They are, rather, the effect of the progressive weakening of the light's attraction.

20 and 21. These communications may be summarized in the author's own words. "An animal which has just been immersed in still water or undergone mechanical excitation, if it moves in a restricted area of a constant luminous field, shows in general a progressive and more or less rapid weakening of the effects of phototropism and of sensibility to difference. This weakening is connected with the progressive return of the animal to a state of rest. We should see in this return, not the consequence of" fatigue or an alteration in external circumstances, but "the progressive exhaustion of the nervous effects of the initial mechanical excitation, which momentarily overcame the inertia of the animal." "In a luminous field, phototropic animals follow fatally, in a given direction, certain lines. But during a change of illumination, the animals tend to turn about upon the lines and follow them momentarily in the opposite direction, hence they may deviate for a time. Many of the supposed "trials" of JENNINGS might be explained by applying this law."

22. This is a summary of recent work in America and elsewhere.

23, 24, 25, 27. The echinoderms studied by BOHN were the following: Asterias rubens, Asteriscus verruculatus, Astropecten irregularis, Ophiolepsis ciliata, O. albida, Ophiothrix fragilis, Ophiocnida brachiata. The salient results may be grouped under three heads. (*a*) Sensibility to difference is shown by the tendency of sudden changes of illumination or of slope to change the sign of phototropism and geotropism, producing oscillations which are not properly "trials," but are as fatally determined as the tropisms themselves. Further evidence that these oscillations are not "trials" is to be found, BOHN thinks, in the fact that young starfish show them to a much less extent than older ones; trials should diminish with age. (*b*) The eyes are essential in phototropism; a starfish with one or more eyes sectioned acts as if a black screen were brought near, that is, it moves toward the wound, a tendency which conflicts in an interesting way with the general tendency to move away from a wounded point. (*c*) The formation of habits in the starfish is shown by the following observations. A starfish on a sunny bottom far from any shade converges its arms towards the light in order to protect itself. Those which live normally in sunny regions do this more readily than those which are accustomed to be near shade which they can seek. Further, the starfish is capable of changing the direction of its movement in two ways: by changing the leading arm, and by rotating on itself so as to give its arms a new direction. An individual may be taught to use the latter method by cutting off one or more arms, or by repeatedly stimulating an arm.

26. This paper contains notes on the reproduction of actinians by fission.

29. This article is a more or less popular lecture by BONNIER on the habits of the honey-bee, in which he maintains the singular thesis that individual bees have no intelligence whatever; that intelligence is for the bee a function of the social state, and that it is displayed to a marvelous degree by a "secret committee" which regulates the affairs of the hive.

30 and 31. These articles by Mlle. DRZEWINA will be discussed in connection with numbers 62–67, as they are concerned with PIÉRON's ideas on autotomy.

32. Here DRZEWINA shows that the fortnightly tidal fluctuations are represented by changes of phototropism in the laboratory on the part of the crab Clibanarius misanthropus.

33. Carcinus mœnas put down anywhere on the beach will turn and make for the water, even with eyes blackened, and with the wind from any quarter. This, DRZEWINA thinks, is a case of attraction by humidity, and the influence of past causes is shown by the fact that crabs from high levels are specially sensitive.

34, 35, 36, 37. FAURÉ-FRÉMIET in these articles first surveys the differentiations of structure and of sensory and motor apparatus to be found in the Protozoa. He then classes the reactions of this group under four heads: local and direct response to stimulation, as the withdrawal of a pseudopod; more extended response, involving a considerable portion of the body; general response, involving movement of the entire body; and local but indirect response, such as the retraction of the stem in Vorticella when another part is stimulated. In papers number 36 and 37 the reactions of Colpoda cucullus and Urostyla grandis are described, and the attempt is made to show that they are the resultants of the various ciliary beats.

39. FOREL thinks the following fact shows that bees have memory for time. Some bees learned to visit an out-of-doors dining-table at certain hours of the morning and afternoon during which there were sweets on the table. They continued for several days to come at these hours, although the sweets were no longer placed on the table; then gradually desisted.

40. HACHET-SOUPLET suggests that we may be sure of the purely instinctive, i.e., non-intelligent, character of an act when an animal persists in trying to perform it though one of the essential conditions for its performance is lacking, as when, for example, a hermit crab tries to introduce itself into a smooth glass ball without an opening.

41. In presenting his request for the formation of a special section for the study of animal education, HACHET-SOUPLET makes some observations on the method which he considers best adapted to bring out the highest mental powers of animals: that of *persuasion*, consisting in explaining to the animal, by voice, gesture, or arrangement of surroundings, what it is expected to do.

42. The same writer discusses the method by which dogs are taught to rescue drowning persons.

46, 47, 48. The most important results of the last two papers are included in the first. The instincts treated are the uses of the web, the care of the young, and the courting processes. LÉCAILLON finds that the spiders observed by him show little discrimination in regard to the cocoon, but will accept cocoons of other species and different form from their own; that they are not disturbed if, while they are carrying a cocoon, its wall is cut, allowing the eggs to fall out and decidedly altering the weight of their burden; that they can distinguish at some little distance a strange female occupying their nest.

49. MARAGE places himself on the negative side of the discussion regarding the hearing of fishes. He tested Gobio fluviatilis, Anguilla vulgaris, Esox lucius, Tinca vulgaris, Cyprinus carpio, and Leuciscus rutilus in the aquarium, and, in free water, Alburnus lucidus. The sounds used were the vowels *ou*, *o*, *a*, *é* sung successively on notes from C2 to G6, with energy varying from 0.00045 kgm. to 0.05 kgm., communicated through rubber tubes, the fish not being able to see the experimenter. No response whatever was obtained, though a diver 80 m. away could hear and distinguish the sounds.

50. The chief contribution made by MARTIN to the study of the tidal rhythm in Convoluta is the fact that various influences, such as repeated mechanical shocks, prolonged darkness, colored light, chemicals, etc., may inhibit the rhythm, causing "amnesia," and that "non-amnesic C mingled with a greater number of amnesic C lose their memory, while amnesic C mingled with a greater number of non-amnesic C imitate the oscillatory movements of the latter."

51. This is an unimportant because inexact observation of the attraction of ants to food at a distance.

52. This paper contains definitions of morphological terms and a statement of unsolved problems with regard to the instincts and mental powers of birds.

53 and 55. PIÉRON's study of Actinia equina begins with the question as to what stimuli provoke reaction: he finds, unlike BOHN, that light has no effect, nor has auditory stimulation. There is some response to food held very close to but not in contact with a tentacle. Contact with food produces the feeding reaction; some individuals will attach themselves to any mechanical stimulus, while others give withdrawing movements to any but a food stimulus. A portion of another actinian will not be swallowed. As regards the localization of sensibility, the tentacles are sensitive to both mechanical and chemical stimuli, as are the peristome and mouth; the foot is very sensitive to mechanical stimuli, and the column insensitive to both mechanical and chemical excitants. Varieties, individuals, and ages differ in sensibility. Foul water and drying affect the response to stimulation, as do certain internal factors, such as digestion, regurgitation, and parturition. Reaction ceases when a mechanical stimulus is repeated. Paper 55 is a study of the movements of A. equina and of their synergy.

54. This paper discusses, without reaching a positive conclusion, the problem

as to whether the crab or the actinian started the fashion of the latter's taking up its abode on the former's back.

56, 57, 58, 61. The first of the reports by PIÉRON to the Society of Biology regarding his studies on ants states that he has confirmed with eighteen hitherto untested combinations of species, BETHE's experiment in which an ant was received into a foreign nest when dipped in the juices of ants from that nest. In the second paper (57), he notes various circumstances which modify the reaction to strangers. Certain species are inclined to be tolerant, such as Aphænogaster barbara nigra and Formica cinerea with regard to other nests of the same species, and Myrmecina latreillei with regard to other species. Sometimes an ant of the same nest is attacked "erroneously." Attacks are more frequent near the nest than at a distance from it. A solitary ant tends to run away rather than to attack, save in the case of a very small one meeting an ant of a larger species, when the former clings to the legs of the latter. Males do not distinguish strangers from nestmates, and a female after the nuptial flight is received in a foreign nest. There are also individual differences in reaction. Most of these modifying circumstances have an adaptive significance (58), for instance the tolerance of Formica cinerea is doubtless connected with the fact that its nests are ordinarily built close together, and that of Myrmecina may be due to its hard chitinous armor. As regards the problem of nest finding, PIÉRON would distinguish three types of ants: visual (Formica fusca, *e.g.*, which cannot find the nest when blinded), olfactory (Lasius fuliginosus, *e.g.*), and muscular (Aphænogaster barbara, which if carried out of its path will continue, when set down, until it reaches a point where the opening of its nest would have been found if the ant had not been moved).

59, 60, 63. We have already noted under (8) the contents of paper 59 and PIÉRON's distinction between "anticipation," or reaction which is made ahead of time because it has become associated with an an external stimulus occurring before the original stimulus to the reaction, and rhythmic reaction, where the response is made to an internal state of the organism, which has come to be periodically produced. In papers 60 and 63 this distinction is amplified and the general significance of physiological rhythms considered.

62, 64, 65, 66, 67. The chief point of importance brought out in the discussion of autotomy or self-amputation is PIÉRON's differentiation of a form of the phenomenon which he calls "psychic autotomy," unlike reflex autotomy in the facts that it is made in response to slight stimulation, such as merely holding the member fast, and that it does not occur if the commissures connecting the cerebral ganglia with the ventral ganglia are cut. DRZEWINA (30, 31) sees no reason to distinguish this phenomenon from ordinary reflex autotomy, and has found it occurring after section of the commissures.

68. RETTERER states that actinians in northern seas where the effect of the tides is less marked do not show a tidal rhythm in the laboratory.

69. This paper is a study of the manner in which certain seaside Diptera are adapted to their surroundings. The bodies of most of them are impervious to water; those which are by their manner of life exposed to the wind have a marked tendency to orient to it, or to hide behind shells when it is very strong; other forms resist it by taking very short flights or by bracing themselves with their legs.

70. The following three points are brought out in this study of Actinia equina. (*a*) The actinian tends to resume the position it had in nature when placed in

the reverse position in the laboratory. (*b*) An actinian would not swallow a bit of mollusc attached to a morsel of cork until the cork was removed. A bit of mollusc and a piece of another actinian, of a different species, being placed on different parts of the disk, the former was swallowed and the latter rejected. (*c*) The foot "prefers" to attach itself to a rough surface if offered the "choice" between this and a smooth one.

BIBLIOGRAPHY.

BOHN, G.

1. L'individualité psychique chez les Vers, les Échinodermes, et les Insectes. *Bull. Inst. gén. psych.*, vol. 6, p. 115. 1906.
2. Sur le photrotopisme de l'Acanthia lectularia Fabr. *C.r. Soc. Biol.*, vol. 60, p. 520. 1906.
3. Sur l'adaptation des réactions phototropiques. *C.r. Soc. Biol.*, vol. 60, p. 584. 1906.
4. Sur les courbures dues à la lumière. *C. r. Soc. Biol.*, vol. 61, p. 420. 1906.
5. Sur des mouvements de roulement influencés par la lumière. *C. r. Soc. Biol.*, vol. 61, p. 468. 1906.
6. Mouvements en relation avec l'assimilation pigmentaire chez les animaux. *C. r. Soc. Biol.*, vol. 61, p. 527. 1906.
7. La finalité dans l'étude des mouvements. *C. r. Soc. Biol.*, vol. 61, p. 570. 1906.
8. La persistance du rythme des marées chez l'Actinia equina. *C. r. Soc. Biol.*, vol. 61, p. 661. 1906.
9. Le rythme des marées et la matière vivante. Réponse a M. Lapicque. *C. r. Soc. Biol.*, vol. 61, p. 708. 1906.
10. Observations sur les papillons du rivage de la mer. *Bull. Inst. gén. psych.*, vol. 6, p. 285. 1906.
11. Les tropismes, les réflexes, et l'intelligence. *Ann. psych.*, vol. 12, p. 137. 1906.
12. Quelques chiffres rélatifs au rythme vital des Convoluta. *C.r. Soc Biol.*, vol. 62, p. 51. 1907.
13. Sur l'impossibilité d'étudier avec une précision mathématique les oscillations de l'état physiologique chez les animaux littoraux. *C. r. Soc. Biol.*, vol. 62, p. 211. 1907.
14. L'influence de l'éclairement passé sur la matière vivante. *C. r. Soc. Biol.*, vol. 62, p. 292. 1907.
15. Introduction à la psychologie des animaux à symétrie rayonnée. I. Les etats physiologiques des Actinies. *Bull. Inst. gén. psych.*, vol. 7, pp. 81, 135. 1907.
16. L'influence de l'agitation de l'eau sur les Actinies. *C. r. Soc. Biol.*, vol. 62, p. 395. 1907.
17. Le rythme nycthéméral chez les Actinies. *C. r. Soc. Biol.*, vol. 62, p. 473. 1907.
18. À propos du procès-verbal. Le ralentissement et l'accélération des oscillations des Convoluta. *C. r. Soc. Biol.*, vol. 62, p. 564. 1907.
19. Les tropismes, la sensibilité differentielle, et les associations chez le Branchellion de la Torpille. *C. r. Soc. Biol.*, vol. 63, p. 545. 1907.
20. À propos des lois de l'excitabilité par la lumière. I. Le rétour progressif à l'état d'immobilité après une stimulation mécanique. *C. r. Soc. Biol.*, vol. 63, p. 655. 1907.
21. À propos des lois, etc. II. Du changement du signe du phototropisme en tant que manifestation de la sensibilité differentielle. *C. r. Soc. Biol.*, vol. 63, p. 756. 1907.
22. L'acquisition des habitudes chez les animaux. *Ann. psych.*, vol. 13, p. 170. 1907.
23. Sur la rôle et la protection des organes des sens chez les échinodermes. *C. r. Soc. Biol.*, *Biol.*, vol. 64, p. 277. 1908.
24. Sur les mouvements rotatoires des Étoiles de mer et les Ophiures. *C. r. Soc. Biol.*, vol. 64, p. 532. 1908.
25. De l'acquisition des habitudes chez les Étoiles de mer. *C. r. Soc. Biol.*, vol. 64, p. 633. 1908.
26. Scissiparité et autotomie chez les Actinies. *C. r. Soc. Biol.*, vol. 64, p. 936. 1908.
27. Introduction à la psychologie des animaux à symetrie rayonnée. II. Les essais et erreurs chez les Étoiles de mer et les Ophiures. *Bull. Inst. gén. psych.*, vol. 8, p. 21. 1908.

BOHN, G. and PIÉRON, H.

28. Le rythme des marées et le phénomène de l'anticipation réflexe. *C. r. Soc. Biol.*, vol. 61, p. 660. 1906.

BONNIER, G.
29. Le socialisme chez les abeilles. *Bull. Inst. gén. psych.*, vol. 7, p. 397. 1907.
DRZEWINA, A.
30. Sur la prétendue autotomie psychique. *C. r. Soc. Biol.*, vol. 63, p. 459. 1907.
31. Y a-t-il une différence effective entre la prétendue autotomie psychique et l'autotomie réflexe? Réponse a M. Piéron. *C. r. Soc. Biol.*, vol. 63, p. 493. 1907.
32. Les variations périodiques du signe du phototropisme chez les Pagures misanthropes. *C. r. Acad. Sci., Paris*, Dec. 9, 1907.
33. De l'hydrotropisme chez les Crabes. *C. r. Soc. Biol.*, vol. 64, p. 1009. 1908.
FAURÉ-FRÉMIET, E.
34. Organisation, fonctionnement, et réactions individuelles chez les Cytozoaires. *Bull. Inst. gén. psych.*, vol. 6, p. 305. 1906.
35. L'organisation, le fonctionnement, et les réactions individuelles chez les Cytozoaires. *Bull. Inst. gén. psych.*, vol. 7, p. 75. 1907.
36. Réactions de deux Infusoires ciliés. *Bull. Inst. gén. psych.*, vol. 7, p. 308. 1907.
37. Les conditions organiques de comportement chez les Cytozoaires: l'Urostyla grandis. *Bull. Inst. gén. psych.*, vol. 7, p. 441. 1907.
FAUVEL, P. and BOHN, G.
38. Le rythme des marées chez les Diatomées littorales. *C. r. Soc. Biol.*, vol. 62, p. 121. 1907.
FOREL, A.
39. La mémoire du temps et l'association des souvenirs chez les Abeilles. *Bull. Inst. gén. psych.*, vol. 6, p. 258. 1906.
HACHET-SOUPLET.
40. Le critérium de l'instinct. *Bull. Inst. gén. psych.*, vol. 7, p. 72. 1907.
41. Sur un projet de création d'une Section d'étude expérimentale de dressage scientifique à l'Institut général psychologique. *Bull. Inst. gén. psych.*, vol. 7, p. 187. 1907.
42. Les chiens sauveteurs et la Zoopédie. *Bull. Inst. gén. psych.*, vol. 7, p. 451. 1907.
LAPICQUE, L.
43. Sur les fonctions rythmiques des animaux soumis à l'alternance des marées. Observavation sur la note de M. Bohn. *C. r. Soc. Biol.*, vol. 61, p. 707. 1906.
44. Sur la précision dans la question du rythme des marées. *C. r. Soc. Biol.*, vol. 62, p. 302. 1907.
45. Réponse a M. Bohn, *C. r. Soc. Biol.*, vol. 62, p. 475. 1907.
LÉCAILLON, A.
46. Les instincts et le psychisme des Araignées. *Bull. Inst. gén. psych.*, vol. 6, 127. 1906.
47. Notes complémentaires sur les moeurs des Araignées. I. Influence de la nutrition sur la réproduction d'Ageléna labyrinthica Cl. *C. r. Soc. Biol.*, vol. 62, p. 334. 1907.
48. Notes complémentaires, etc. II. Nature et importance des soins que certaines femelles donnent à leur progeniture. *C. r. Soc. Biol.*, vol. 63, p. 668. 1907.
MARAGE.
49. Contribution a l'étude de l'audition des poissons. *C. r. Acad. Sci. Paris*, vol. 143, p. 852. 1906.
MARTIN, L.
50. La mémoire chez Convoluta roscoffensis. *C. r. Acad. Sci., Paris*, vol. 145, p. 555.
MÉNÉGAUX, A.
51. Une observation sur le sens olfactif à distance chez les Fourmis. *Bull. Inst. gén. psych.*, vol. 6. p. 302. 1906.
52. Questions de morphologie et de psychologie chez les oiseaux. *Bull. Inst. gén. psych.*, vol. 7, p. 427. 1907.
PIÉRON, H.
53. Contributions à la psychologie des Actinies. *Bull. Inst. gén. psych.*, vol. 6, p. 40. 1906.
54. Contribution à l'étude des rapports éthologiques des Crabes et des Actinies. *Bull. Inst. gén. psych.*, vol. 6. p. 98. 1906.
55. Contribution à la psychophysiologie des Actinies. Les réactions de l'Actinia equina. *Bull. Inst. gén. psych.* vol. 6, p. 146. 1906.
56. Généralité du processus olfactif de reconnaissance chez les Fourmis. *C. r. Soc. Biol.*, vol. 61, p. 385. 1906.
57. Exceptions et variations dans le processus olfactif de reconnaissance chez les Fourmis. *C. r. Soc. Biol.*, vol. 61, p. 433. 1906.

PIÉRON, H.
58. Le mécanisme de la reconnaissance chez les Fourmis. Rôle des données olfactives. *C. r. Soc. Biol.*, vol. 61, p. 471, 1906.
59. La réaction aux marées par anticipation réflexe chez Actinia equina. *C. r. Soc. Biol.*, vol. 61, p. 658. 1906.
60. La question des rythmes spontanés et des phenomènes d'anticipation en biologie. *C. r. Biol.*, vol. 62, p. 51. 1907.
61. L'adaptation à la récherche du nid chez les Fourmis. *C. r. Soc. Biol.*, vol. 62, p. 216. 1907.
62. Autotomie et 'autospasie'. *C. r. Soc. Biol.*, vol. 63, p. 425. 1907.
63. Des phenomènes d'adaptation biologique par anticipation rythmique. *C. r. Acad. Sci., Paris*, vol. 144, p. 138. 1907.
64. Sur une prétendue réfutation de l'autotomie psychique. Résponse à Mlle. Drzewina. *C. r. Soc. Biol.*, vol. 63, p. 461. 1907.
65. L'autotomie protectrice réflexe chez les Orthoptères. *C. r. Soc. Biol.*, vol. 63, p. 463. 1907.
66. L'autotomie volontaire des décapodes. Quelques idées et quelques faits. *C. r. Soc. Biol.*, vol. 63, p. 517. 1907.
67. L'autotomie évasive chez les Orthoptères. *C. r. Soc. Biol.*, vol. 63, p. 186. 1907.

RETTERER, E.
68. À propos du rythme des marées et de la matière vivante. *C. r. Soc. Biol.*, vol. 62, p. 186. 1907.

ROUBAUD.
69. Instincts, adaptation, résistance au milieu chez les Mouches des rivages maritimes. *Bull. Inst. gén. psych.*, vol. 7. p. 60. 1907.

VAN DER GHINST, I.
70. Quelques observations sur les Actinies. *Bull. Inst. gén. psych.*, vol. 6, p. 267. 1906.

MARGARET FLOY WASHBURN.

LITERARY NOTICES.

Pricer, Jno. L. The Life History of the Carpenter Ant. *Biol. Bull.*, vol. 14, pp. 177–218, 1908.

This paper. which is interesting to both the biologist and the psychologist, embodies the results of a year's study of the carpenter ant (Camponotus pennsylvanicus and C. ferrugineus). The contents of numerous nests collected during the winter were examined and counted, and colonies were raised from individual females captured immediately after their nuptial flight. The subject is treated under the following heads: "Life history of the colony," "polymorphism," "division of labor," "food," "relation to light and color," "guests and parasites," "instinct and intelligence."

The author confirms the statement of Dr. WHEELER[1] and others that the fertilized female fasts from the time it enters its brood chamber until its first offspring have reached maturity.

He also agrees with Dr. WHEELER[2] that the polymorphism of this ant is ontogenetic. At the time of laying, all eggs of fertilized females are essentially alike. The size of the resulting offspring depends upon the quantity of food fed the larvæ.

Mr. PRICER's observations seem to support the view that sexually mature individuals are not produced until the colony is more than two years old, and that the brood of females produced one summer remains in the nest until the following spring.

The food of the ants consists chiefly of honey dew obtained from aphids. But this is supplemented by insect food and plant juices. The aphids are never domesticated, nor are their eggs stored in the nest over winter. Insects are never captured alive and the head is the only part of insects fed upon that is ever carried into the nest.

In his experiments on light the author employed the much used devise of a central corridor, on each side of which are rooms illuminated by light passing through glasses of different colors He used deep red, green, deep blue, indigo-blue and a cell of carbon disulfid (this latter to exclude the ultra-violet rays). At the beginning of the experiment these glasses were arranged in reverse order in the two sets of rooms. During the experiments the glasses were manipulated in various ways. As a control he used ants the eyes of which had been rendered opaque. The ants preferred to collect under the red glass, and they avoided the blue and light of shorter wave length. In this respect his results harmonize with those of LUBBOCK and Miss FIELDE; but PRICER's experiments seem to show that the ants perceive the red, etc., whereas Miss FIELDE[3] claims that ants are blind to all rays of greater wave length than the violet.

[1] WHEELER, WM. M. On the founding of colonies by queen ants, etc., *Bull. Am. Mus. of Nat. Hist.*, vol. 22, p. 39, 1906.

[2] WHEELER, WM. M. The polymorphism of ants. *Ibid.*, vol. 23, pp. 66–75, 1907.

[3] FIELDE, A. M. Notes on an ant. *Proc. Acad. Sci. Phil.*, vol. 54, p. 615, 1902; Effects of light-rays on an ant. *Biol. Bull.*, vol. 6, p. 309, 1904.

He demonstrated that ants enclosed in a wire gauze cylinder could be exposed to an arc light, at a temperature of 40° C., for an hour without being injured. From this he concludes that the nocturnal habits of this ant are the result not of necessity, but of preference. This is a broader conclusion than the experiment seems to warrant.

McCook and others have held that this ant has an architecture of its own, and some have claimed that it damages trees and timber. The observations of Mr. Pricer seem to show conclusively that it injures neither trees nor timber and that it has no architecture of its own, but lives in the abandoned burrows of wood borers and, unless the wood is rotten, it never alters the shape of these burrows.

Five experiments upon instinct and intelligence led to the following conclusions:

1. Ants are capable of tracking themselves and others of the colony; but they are incapable of distinguishing the direction in which the trail was first laid down.

2. Ants do not depend entirely upon following trails in finding their way about, but are guided often by memory of the location of things and probably depend, as a last resort, on a sense of direction.

3. Ants ordinarily pay very little attention to trails when traveling from the nest.

In the main the above conclusions harmonize well with the results stated in my paper on the Homing of Ants.[4]

He also claims that there is no evidence of anything akin to reason.

From the standpoint of comparative psychology, probably the most interesting portion of the paper is the record of the experiments on the power of communication. The author placed a number of larvæ upon a small island which was connected, by means of bridges, with the food chamber of the nest. An ant discovered these and returned, empty handed, to the nest. There it butted against several workers and then retraced its steps to the larvæ. The ants thus saluted, and no others, followed it. This was repeated several times with practically the same result. Once, however, two out of eight ants saluted reached the larvæ before the ant that discovered them had retraced its steps that far. This most interesting experiment led to the conclusion that ants can communicate. One cannot help wishing that the author had devised an experiment which precluded the possibility of the response being due to an odor conveyed by the ant from the discovered larvæ to the ants saluted.

The following epitome of an unpublished experiment of mine upon an allied species (C. herculeano-ligniperdus) will emphasize the importance of this precaution. The colony, which was housed in a Janet nest, usually kept a guard in the entrance. One day some strange ants (Formica fusca var. subsericea) forced their way past the guard to some food which I had placed just inside the nest. The guard, after fighting them for a while, retreated into the inner chamber, rushed about among the ants and then returned to the fray, followed by several others. This looked like communication. To test the matter the following experiment was devised. I heated dissecting needles and glass stirring rods red-hot, to destroy any odor, and, as soon as they were cool, fought the guard with them. Soon it retreated into the inner chambers, rushed about among the ants and then returned, alone, to the outer chamber. Then I dipped the needle or the stirring rod into oil of cloves and again fought the guard. It again retreated to the inner chamber,

[4] Turner, C. H. *Jour. of Comp. Neur. and Psy.*, vol. 17, p. 423, 1907.

rushed around among the ants, and returned to the outer chamber. In this case, however, it was followed by several of its companions. This was repeated several times with similar results. So interested was I in this experiment that I called in Professor MEAD, of the University of Chicago, and performed it before him. Evidently, in this case, the following reaction was a response to an odor. Whether ants do or do not communicate in any other way is a subject upon which I have no opinion that I am prepared to publish. I mention this experiment merely to show that this question is too complex to be solved by any experiment which is not so planned as to preclude the possibility of the reaction being a response to an odor.

The author is to be complimented for the originality displayed in devising apparatus.

C. H. TURNER.

Jennings, H. S. Behavior of the Starfish Asterias forerri de Loriol. *Univ. of Calif. Publ. in Zoölogy*, vol. 4, no. 2, pp. 53–185, 19 text figures. 1907.

The present investigation on the Pacific Coast starfish, Asterias forreri, is another thorough, analytical contribution to the subject of animal behavior, in which field Professor JENNINGS has already done such masterly work. His general plan of investigation here is the same as that which first gave him an insight to the behavior of Paramecium and the Protozoa in general, and which led to the conception of the "motor reaction" as a stereotyped, almost universal mode of reaction to stimuli among these lower oragnisms—namely, a preliminary careful, minute, descriptive study of the behavior of the animal.

The list of the headings which cover the descriptive portion of his paper will give an idea of the range of the investigation; these are: "Respiration and its protection by the pedicellariæ," "Detailed behavior of the pedicellariæ," "Capture of food," "Behavior of the starfish in selecting conditions of existence," "Reaction to light," "Positive reactions," "The righting reaction," and "Formation of habits in the starfish." Of these most attention is directed to the righting reaction, while the reactions to light are least fully worked out. There is such a wealth of detailed observational results that only the barest selection can be made in a review of this character. They form, however, most interesting reading, in spite of their detail and of the fact that the author has in places relapsed somewhat from his usual literary care, as, for example, in the description of the capture of food, where the tense changes with confusing rapidity. Frequent reference back and forth to the interrelated phenomena assists, even at the expense of some repetition, to keep in the reader's mind the relationship of numerous factors which go to make up the complicated behavior of the starfish as a whole. For the author comes very decidedly to the conclusion—and the reader can hardly disagree with him—that the behavior of the starfish cannot be attributed to simple direct responses to obvious stimuli. On the contrary, besides the external stimuli, internal factors, depending upon past actions, etc., may determine the method of behavior. The author himself states that perhaps the most important thing developed in his paper is "the demonstration of the variability, modifiability, unity and adaptiveness in the main features of the behavior of the starfish."

The unity of the parts of the starfish in performing its various actions is an important point, upon which much emphasis is placed. Thus the behavior of the pedicellariæ remind one very strongly of the behavior of individuals in a colony, such as of bees or ants, all of which work together to accomplish a definite end. When a starfish is placed in a new situation, where there is a problem to solve, the movements are at first varied, but soon a definite impulse to act in a certain way appears to be formed, after which all the parts work together with a unity to bring about the results on this line. JENNINGS calls this the *unified impulse*. When this is once established the movements tend to continue along this course, even if the conditions be changed and new stimuli introduced. This latter tendency the author believes is "evidently akin to the formation of a habit." Moreover, he has actually been able to demonstrate habit formation in the starfish by training individuals to use different rays in righting themselves from those which they naturally employ. The effect of this laborious training, however, soon disappears.

Considerable space is devoted to a discussion of these results in relation to DRIESCH's postulation of a "Psychoid" or "Entelechy," a sort of vitalistic principle, which that author believes the only way of explaining these adaptive reactions. While JENNINGS does not claim to have analyzed all the factors which enter in, he nevertheless maintains—and very properly, it seems to the reviewer—that DRIESCH's would-be explanation is "merely a way of collecting all the difficulties together and giving the bundle a name." "The Entelechy would be a problem not a solution," while "to accept the Entelechy unanalyzed and unexplained is merely to give up the problem as insoluble." The author's alternative answer to the question is pregnant with suggestion and may well be quoted:

"The only other answer that can be given is that the precise way each part shall act under the influence of the stimulus must be determined by the past history of that part; by the stimuli that have acted upon it, by the reactions which it has given, by the results which these reactions have produced (as well as by the present relations of this part to other parts, and by the immediate effects of its present action). In other words, this complex harmonious working of the parts together is only intelligible on the view that there is a history behind it; that it is a result of development. We can not look upon it as a final thing ('etwas Letztes, Naturgegebenes'), because there *is* a history behind it, and we know as solidly as we know anything in physiology that the history of an organ does modify it and its actions—in ways not yet thoroughly understood, doubtless, yet none the less real. The starfish that we have before us has an actual history of untold ages, in which it has existed as germ plasm or otherwise, and there can be no greater mistake in physiology than to leave this out of account. The modifications induced in organisms by their experiences, either while existing as germ plasms or as individuals, are as clearly a part of physiology as is the study of digestion, and their existence is not less doubtful."

LEON J. COLE.

Buttel-Reepen, H. v. Are Bees Reflex Machines? An experimental contribution to the natural history of the honey-bee. Translated from the German by MARY H. GEISLER. Pp. 48, $0.50. *The A. I. Root Company, Medina, Ohio.* 1907.

Students of animal behavior and comparative psychology will welcome this translation of VON BUTTEL-REEPEN's noteworthy discussion of the behavior and psychology of the bee. The monograph may now be used to advantage in connection with introductory courses in Animal Psychology. Apparently the translator has done her work with commendable care, but despite this fact many minor errors appear in the text.

It is to be noted in this connection that von Buttel-Reepen has recently published,[1] in the form of a monograph, the results of several years of work on the biology of the honey bee.

Cole, Leon J. An experimental Study of the Image-Forming Powers of Various Types of Eyes. *Proc. of the Amer. Acad. of Arts and Sciences*, vol. 42, pp. 335-417. 1907.

The image-forming power is here studied not directly, but indirectly by means of the responses of the animals in question to light-stimuli of equal intensity but unequal areas. The apparatus was so arranged that the individual under investigation was free to move either toward a small source of light or (in the opposite direction) toward a luminous field of about 10,000 times the area but of the same total intensity. This intensity varied in the course of the experiments from 5 to 1.25 candle meters. Thus an animal which can form no optical image of a source of light and which responds therefore to intensity alone, would respond to either of the stimuli here offered, indifferently. And this was found to be the case in an eyeless form, the earthworm.

Animals having "direction eyes" (Bipalium kewense, Periplaneta americana, larva of Tenebrio molitor, and larva of the wood-borer) were found to respond almost wholly to *intensity* of light. The few doubtful cases are explicable, since certain forms and arrangements of direction eyes "may be considered as the beginning of a crude image-forming apparatus" (p. 362). Animals with image-forming eyes (Vanessa antiopa, Ranatra fusca, Acris gryllus, and Rana clamata) were found to be mostly positive; they move toward the stimulus of larger area. Whether positively or negatively phototropic, their preference for the smaller or the larger simulus was well marked.

The author studies and discusses many more than the above-mentioned species, both experimentally and from the available literature. In pp. 402–412 are presented several interesting theoretical considerations, particularly those relating to three types of response, to *intensity* of light, to luminous *area*, and to "*definite objects*." The investigation is ingenious, careful, and suggestive.

E. B. H

[1]Apistica. Beiträge zur systematik Biologie, sowie zur geschichtlichen und geographischen Verbreitung der Honigbiene (*Apis mellifica, L.*), ihrer Varietäten und der übrigen Apis-Arten *Mitth. d. kgl. Zool. Museums, Berlin.* 1906.

The Journal of
Comparative Neurology and Psychology

VOLUME XVIII DECEMBER, 1908 NUMBER 6

THE CRANIAL NERVES OF AMPHIUMA MEANS.

BY

H. W. NORRIS.

WITH PLATES IV, V, VI, VII AND VIII.

I. INTRODUCTORY.

The only descriptions of the cranial nerves of Amphiuma, with the exception of more or less incidental mention, are to be found in the papers of FISCHER (1864), WILDER (1892), KINGSLEY (1902a), and DRÜNER (1904), but none of these attempts a systematic account from the standpoint of nerve components.

The present paper is an outline of the more salient features of the origin and peripheral distribution of the cranial nerves of Amphiuma means with reference to their components. Inasmuch as the nervous systems of but two of the urodele amphibians have been analyzed into their components (see BOWERS 1900, and COGHILL 1902), it seems to the writer that the time is not ripe for making detailed comparisons. For that reason the present account deals chiefly with facts of description.

The material studied consisted of individuals varying from 55 to 300 millimeters in length, sectioned through the head transversely and sagittally. Fixation was for the most part in VOM RATH's picric-acetic-osmic-platinic mixture. Plate IV is from plottings of a series of cross-sections, ten micra in thickness, prepared by the paraffin method, supplemented and checked up by a number of series sectioned in celloidin. Plates V to VII are chiefly from material sectioned in celloidin, sections fifteen and twenty micra in thickness. Counter-stains, when used, were either gold-orange or acid fuchsin. As individuals small enough to make sectioning of the entire body feasible are seldom obtained, there is still lacking detailed information regarding some of the cranial nerve branches that pass back into the trunk of the body. The nervous system of Amphiuma is favorable for study because of the differential staining that results from fixation in VOM RATH's fluid. The lateralis components become intensely black; motor fibers are a dark gray; general cutaneous fibers are brown-black; and the communis system with slight development of myelin is but lightly colored. As the methods employed do not differentiate the sympathetic fibers, no description of the sympathetic system is attempted here.

There is evidently a need for a thorough revision of the nomenclature of the sub-divisions of the cranial nerves of Amphibia, but more facts are needed upon which to base comparisons. In this paper the older names are employed as far as consistent with the existing state of information upon the subject.

2. THE OLFACTORY NERVE.

Little need be added to the account given by KINGSLEY. A little anterior to the level of the posterior border of the eyeball fibers begin to arise from the olfactory glomeruli on the dorso-lateral border of the brain (figs. 13 and 20). From this point to the anterior end of the brain the emerging fibers constitute a dorsal part of the olfactory trunk. A ventral smaller part is formed by fibers that leave the more ventral glomeruli. Anteriorly, before the nerve leaves the brain, the two portions have become indistinguishably united; in fact there appears to occur an interlacing of fibers. Other smaller roots join these two main portions. As the nerve passes into the nasal capsule (fig. 3) it is seen to consist of about eight branches, arranged in two or three bundles. The ventral branch in the more ventral of the three bundles passes to JACOBSON's organ (figs. 2 and 20) giving off also some branches to the ventral olfactory epithelium. The nerve to JACOBSON's organ passes to the anterior lateral ventral border of that structure, thence dorsally and posteriorly along its dorsal wall. The other branches in the ventral group innervate the ventral olfactory epithelium; the fibers of the middle bundle supply the median wall of the olfactory epithelium; and the dorsal bundle goes to the dorsal wall, except that immediately on entering the nasal capsule one of the dorsal branches sends a large twig to the ventral wall (fig. 20). It may be true that the branch innervating JACOBSON's organ comes from the ventral portion (or root) of the nerve, but it is certainly too small to contain all the fibers that arise by that root. It will thus be seen that the condition in Amphiuma contributes little in answer to the query whether or not there are two morphologically distinct elements in the olfactory nerve.

3. THE OPTIC NERVE.

The account given by KINGSLEY is confirmed throughout. Myelinic sheaths are not in evidence. Sagittal sections show that as the nerve leaves the brain a small ventricular diverticulum reaches out to the point of emergence of the nerve, but apparently the cavity has no peripheral extension.

4. THE EYE-MUSCLE NERVES.

Of these KINGSLEY was able to find but one, the oculomotorius. I find that the oculomotorius, the trochlearis and the abducens nerves arise in the typical manner, but are much reduced in size. The oculomotor nerve leaves the lateral wall of the anterior part of the medulla oblongata, although its deep origin is ventral (fig. 13). Soon after its emergence it comes into intimate relation with the gasserian ganglion and more anteriorly with the ramus ophthalmicus profundus V, for some distance being embedded in the median border of these structures (figs. 13 and 6). Passing through its own foramen in the skull (fig. 5) it runs a short distance to the vicinity of the origins of the eye-muscles and there divides into two branches, the dorsal of which supplies the superior rectus muscle, and the ventral the inferior and internal rectus and the inferior oblique muscles (fig. 22). It will be seen (fig. 13) that the dorsal branch passes dorsal to the main trunk of the r. ophthalmicus profundus while the ventral branch runs ventral to this nerve. The trochlear nerve, consisting of two or three fibers, arises at the extreme posterior border of the dorsal part of the mid-brain (figs. 13, 7 and 9) and passes anteriorly to its foramen of exit closely pressed against the inner wall of the skull (figs. 5 and 6). It ends in the superior oblique muscle (fig. 22). The abducens is extremely attenuated. It takes its exit from the ventral surface of the brain a little posterior to the level of the origin of the seventh nerve as a few fibers, usually two in number (figs. 7 and 13). These I have been able to follow but a short distance. But anteriorly at the point where the r. mandibularis of the fifth nerve leaves the gasserian ganglion (figs. 6 and 13) there may be found leaving the ganglion or the r. mandibularis a nerve of two fibers that passes out of the skull along with the r. ophthalmicus profundus and ends in the external rectus muscle (fig. 22). Before reaching the muscle the abducens nerve is sometimes in very intimate relation with a small nerve that innervates two small muscles which have their insertions upon the antorbital cartilage. But there seems to be no exchange of fibers between the two nerves. The incomplete development of the eye-muscle nerves precludes any more extended account of their relationships. Of ciliary nerves or ganglia I have found no traces.

5. THE TRIGEMINAL NERVE.

a. The roots of the trigeminal nerve.—The fifth nerve derives its fibers from four sources: (1) From the spinal V tract whose fibers may be traced as far posteriorly as the level of the second spinal nerve. (2) From fibers just dorsal to the spinal V tract.

It is doubtful whether this should be considered as a source distinct from the preceding. (3) According to KINGSBURY (1895a) and to OSBORN (1888) there occurs in Necturus a tract of fibers from the so-called trigeminal nidus in the roof of the mid-brain that passes in part into the spinal V tract near the exit of the fifth nerve. I find in Amphiuma a similar tract of large fibers from the mid-brain passing into close proximity to the spinal V tract and apparently giving off fibers to the latter near the motor root of the fifth nerve, but the fibers so given off are few in number, the greater number passing apparently posteriorly mesal to the spinal V tract, as many do in Necturus. (4) From motor fibers in one or two rootlets that come from a nidus of cells lying in the floor of the medulla.

KINGSLEY speaks of the fact that the fifth nerve leaves the brain as three roots: dorsal and ventral small roots and a median large one. The small dorsal root is made up of fibers that compose (2) above. In this region the spinal V tract is reinforced by numerous fibers from the adjoining cinerea. This small dorsal root appears to be merely some of these fibers that delay the union until after emergence from the brain. The small ventral root is the motor component. As it leaves the brain the fifth nerve contains only motor and general cutaneous fibers.

b. The ramus mandibularis V.—The fibers of the fifth nerve leave the gasserian ganglion in three groups. There is first given off the r. mandibularis, composed of motor and general cutaneous fibers, innervating the temporal, masseter, pterygoid, intermandibular (mylohyoid anterior), retractor bulbi and levator bulbi muscles, and supplying the skin of the lower jaw and the side of the head (in part). The r. mandibularis passes out of the cranium through a foramen common to it and the "dorsal VII." It emerges at the posterior dorsal border of the pterygoid muscle, at first enters the masseter muscle, then passes anteriorly, ventrally and laterally between the pterygoid and masseter muscles, finally out through the masseter. On emerging from the skull it gives off a number of small twigs to the pterygoid and masseter muscles. As the main trunk of the nerve is passing through the foramen there is given off a large branch which rising rapidly between the pterygoid muscle and the internal portion of the masseter and giving off branches to the anterior part of the pterygoid muscle passes around to the dorsal side of the cranium and runs posteriorly to supply the temporal muscle on the top of the head. While this branch to the temporal muscle is passing anteriorly there is given off from it a short distance from its origin from the main trunk a small branch which running mesally into the pterygoid muscle, gives off twigs to the latter and descending applies itself so closely to the outer border of the r. ophthalmicus profundus as to be with difficulty distinguished from it. Leaving the r. oph. prof. nerve at the ventral border of the latter, it runs anteriorly to innervate the two small muscles, previously mentioned, which have their insertion upon the antorbital cartilage (figs. 5, 13 and 22, *mb.*).

These two muscles appear to have escaped the notice of previous writers. One of these (figs. 4 and 22, *rtb.*) has its origin on the posterior part of the maxillary bone and the anterior part of the pterygoid cartilage. As seen in the illustrations, its action is to depress the antorbital cartilage. From the position of the tip of the latter ventral to the eyeball the contraction of the muscle will bring about a retraction of the eye, or it acts as a retractor bulbi muscle. Its position and origin make such an homologizing not improbable. The other muscle (*lvb.*) has its origin on the orbitosphenoid and parietal bones. Its contraction opposes that of the retractor muscle, or elevates the antorbital cartilage. It is here designated as levator bulbi muscle. As previously noted, the nerve supplying these two muscles sometimes comes into intimate relations with the abducens nerve. But that there is no fundamental anastomosis between them is shown by the fact that this small branch to the retractor and levator bulbi muscles sometimes arises directly from the gasserian ganglion and passes out of the cranium through the foramen of the oph. prof. nerve and lateral to the latter, thence anteriorly through the pterygoid muscle dorsal and mesal to the oph. prof. without coming in contact with the latter, and nowhere approaching closely to the abducens nerve (fig. 22).

As the r. mandibularis passes out through the masseter muscle it gives off twigs to the latter. One large branch, composed of motor and general cutaneous fibers, runs for some distance nearly parallel with the main nerve. Its motor fibers finally pass to the anterior part of the masseter muscle; its general cutaneous fibers run as a large branch nearly to the level of the mandible, and a little anterior to the angle of the jaw break up into twigs that run in all directions to supply the skin on the side of the head. The main mandibular nerve on reaching the mandible enters a groove on the dorsal side of the latter (fig. 5) and soon divides into two branches. The ventral of these passes directly down through the jaw between the dentary bone and Meckel's cartilage and on emerging between the angulo-splenial and the dentary bones divides anteriorly and posteriorly into branches that supply the mm. intermandibularis (mylohyoideus anterior) anterior and posterior and the skin of the ventral surface covering these muscles. The dorsal division runs along in the groove above Meckel's cartilage between the dentary and the angulo-splenial bones, and soon divides into two branches. The larger of these, of darker staining fibers (*md* (*3a*)), occupies a canal in the dentary bone (fig. 4) and in turn divides into two divisions. Small branches to the skin pass out from the canal, but the two chief divisions (represented as one nerve in fig. 1) do not emerge from the dentary until well near the tip of the jaw where they supply the skin. The smaller lighter stained division of the ramus in the mandibular groove (*md* (*3b*)) shifts mesally and ventrally from the larger division and runs along dorsal to Meckel's cartilage in a groove between the dentary and angulo-splenial bones (fig. 4). There unites with it a branch of the r. alveolaris VII (*alv.* (4)), the combined nerve passing anteriorly just ventral to the teeth and apparently supplying the latter and possibly the lateral floor of the mouth. It may be traced as far as the extreme anterior teeth and always in close relation to the latter. An anastomosis between that part of the mandibularis supplying the intermandibular muscles and the portion of the r. jugularis VII innervating the interhyoideus muscle, such as COGHILL describes in Amblystoma, I do not find in Amphiuma, but it is certain that the two nerves in question approach very close to each other. KINGSLEY was inclined to believe that branches of the ramus mandibularis in Amphiuma supply lateral line sense-organs. I can state with certainty that no such relationship exists.

c. The ramus ophthalmicus profundus V.—The ramus ophthalmicus profundus leaves the extreme anterior portion of the gasserian ganglion and after passing anteriorly and somewhat dorsally into the region of the eye divides into a number of branches, of which there may be said to be five that are fairly constant in their occurrence and relationships. Of these the first given off, the nasalis internus (*op.* (*1*)), arises as a group of nerves, or as a single nerve, that soon divides into branches. The larger of these branches (*op.* (*1a*)), that goes up through the edge of the cranium in a passage-way between the frontal and the prefrontal bones and then runs along in a canal in the edge of the frontal as far as the nasal capsule, was called by KINGSLEY "ethmoideus caudalis." Anteriorly this can be traced along the upper surface of the frontal bone to a point halfway between the eye and the tip of the snout. As it passes along its canal and on the surface of the frontal it gives off numerous twigs to the overlying skin. Arising from the posterior part of the nasalis internus, or directly from the trunk of the oph. prof., are one or two branches (*op.* (*1b*)) that pass to the skin of the dorsum dorsal and a little posterior to the eye. The main portion of the nasalis internus passes anteriorly and enters the dorsal portion of the nasal capsule near its mesal border. A little before its entrance to the nasal capsule it gives off a branch (not figured) that passes forward in the capsule and emerging dorsally from the skull is distributed to the skin near the tip of the snout. After entering the nasal capsule the nasalis internus anastomoses with the r. ophthalmicus superficialis VII, then passing nearly to the ventral side of the nasal capsule divides, one branch ascending and uniting with the r. ophthalmicus superficialis in a second anastomosis, and the other passing out of the anterior end of the capsule to be distributed, like the dorsal division anastomosing with the oph. spf., to the skin of the tip of the snout. It will thus be seen that the distribution of the nasalis internus and its branches is to the skin of the dorsal side of the head from the extreme anterior end to a point some distance posterior to the eye. In its distribution it seems to answer approximately to the ophthalmicus superficialis V of fishes. It probably gives off fibers to structures in the nasal capsule, but I have detected none such. It evidently does not supply the nasal epithelium.

A second branch of the ophthalmicus profundus (*op.* (*2*)) is one that arises usually in part from the nasalis internus and in part from the main trunk. It was designated by WILDER as r. glandularis II, on the supposition that it innervates the lateral nasal gland. It and its branches, of which there are commonly two divisions, anastomose with each other and with the nasalis internus, enter the lateral dorsal portion of the nasal capsule, run anteriorly and emerging from the capsule are distributed to the skin of the side of the snout from near the anterior end of the latter posteriorly about halfway to the eye. The fibers, if any, given off to the lateral gland are certainly few in number, for I have been unable to detect them. The ramus is characteristically cutaneous and the name, r. glandularis II, is a misnomer.

The next branch given off from the oph. prof. ((*op. 3*)) is the one that anastomoses with the r. palatinus VII. The union occurs at the posterior border of the nasal capsule. The palatine rapidly ascends from its ventral position and the trigeminal ramus descends to meet it. As the two nerves approach each other each divides into two parts, and the union occurs between the branches in pairs, so that there result two nerves, each containing general cutaneous and communis fibers (fig. 26). The dorsal of the two nerves thus formed ascends slightly and divides into three branches. The dorsal one of these three branches (*op.-pal.d.*) from its mode of formation and from the appearance of its fibers consists of general cutaneous fibers only. It may divide and its two divisions on entering the nasal capsule run along the lateral wall of the nasal epithelium on the dorsal border of JACOBSON's organ to the extreme anterior end of the latter. The posterior wall of the nasal epithelium is supplied by branches (*op.-pal.pn.*) derived from this same source, of general cutaneous fibers only. The ventral posterior nasal epithelium is supplied by a branch from a division with mixed components (*op.-pal.mn.*), and its exact composition has not been determined. The other two branches of the dorsal of the two nerves resulting from the palatine-trigeminal anastomosis (*op.-pal.l.*) pass to the lateral portions of the roof of the mouth innervating the lateral teeth, etc. The ventral branch of the anastomosis (*op.-pal.m.*) passes ventrally into the inner ventral angle of the nasal capsule and running forwards supplies the median series of teeth, etc. From it branches are also given off to the posterior ventral nasal epithelium (*op.-pal.mn.*). According to the figure given by WILDER this ramus of the trigeminus that anastomoses with the palatine comes from the nasalis internus. The origin of the two from the main trunk of the ophthalmicus profundus is such that it is not improbable that in some cases they may arise by a common branch. The mode of anastomosis, it will be seen, is like that described by COGHILL in Amblystoma.

The remaining portion of the ophthalmicus profundus divides

into two large branches. Of these the larger lateral one (*op.* (*4*)) fuses with one (*buc.* (*1*)) of the two main divisions of the buccalis VII. Sometimes the fusion occurs between the two undivided trunks, but more often each divides into three or four branches which then fuse in pairs or approximately so (fig. 25). The resulting mixed nerves (r. nasalis externus of WILDER) supply the neuromasts of the infraorbital series and the skin of the side of the snout. The other profundus branch (*cp.* (*5*)) comes into close relation with the second division of the buccalis VII (*buc.* (*2*)), but I can find little evidence of actual anastomosis. The combined nerves (r. glandularis I of WILDER) supply the skin and the infraorbital series of neuromasts at the side and tip of the snout.

KINGSLEY refers to this union of branches of the ophthalmicus profundus with the buccalis (maxillaris according to him) as a condition reported only in Amphiuma. A casual examination of the figures which WILDER gives of Cryptobranchus (Menopoma) and of Siren will show that similar (if not identical) anastomoses occur in these forms. The prediction may be safely made that careful study of these forms will reveal the fact that as in Amphiuma it is a union between lateral line (buccalis) and general cutaneous (ophthalmicus profundus) components.

d. Trigeminal fibers entering the dorsal VII.—A third group of fibers leaving the gasserian ganglion is made up of general cutaneous fibers that at once associate themselves with rr. ophthalmicus superficialis VII and buccalis VII. Their subsequent course will be considered in connection with the facial nerve.

6. THE FACIAL AND AUDITORY NERVES.

a. Roots of the facial and auditory nerves.—The fibers of this complex arise by two groups of rootlets. The more dorsal group comprises the lateral line fibers of the seventh nerve and is formed by three rootlets. Of these the dorsal rootlet (*VIIb* (*1*)) enters that portion of the medulla oblongata which in Necturus is designated by KINGSBURY as the "dorsal island," a mass of alba occupying the extreme dorsal part of the medulla (figs. 7–10, *ll.*) This "dorsal island" suggests an homology to the lateral line lobe (lobus lineæ lateralis) of cyclostomes, selachians and ganoids, although JOHNSTON (1906) asserts that the lateral line lobe and the dorsal root of the "dorsal VII" are absent in aquatic amphibians. The other two rootlets (*VIIb (2)* and *VIIb (3)*) enter the sensory column ventral to the dorsal island.

In the second and more ventral group of rootlets we find true facialis and acusticus fibers. There may be recognized five rootlets, three belonging to the auditory and two to the facial nerve (figs. 1, 7 and 8). Of the auditory fibers there are four groups: (1) medium and large fibers that pass posteriorly into the spinal VIII tract (*VIII (1)*); (2) medium fibers that pass anteriorly in the so-called (incorrectly) "descending VIII" tract (*VIII (2)*); (3) medium and small fibers that pass into "tract b" (Necturus, KINGSBURY) (*VIII (3)*); (4) large fibers forming a tract at first distinct from (1) but posteriorly passing into the spinal VIII tract or into very close proximity to it (*VIII (4)*). (1) forms an anterior rootlet; (2) and (3) form a larger posterior rootlet; (4) forms a rootlet dorsal and intermediate to the other two (figs. 1 and 23a). Of these rootlets (1), (2) and (3) supply the sacculus, lagena, macula neglecta and the posterior canal; (4) supplies the utriculus and the anterior and horizontal canals. The peripheral divisions of the auditory branches will not be considered here.

The two facial rootlets consist of a dorsal one of communis fibers (*VIIaa*) entering the brain at the anterior dorsal border of (4) above mentioned, and passing into the fasciculus communis, and a ventral motor rootlet (*VIIab*). The relative positions of these rootlets of the facial and auditory nerves are subject to some variation.

b. Lateral line components of the facial nerve.—From the points of entrance of its fibers into the brain the root of the lateral line portion of the seventh nerve passes anteriorly as a flattened band closely compressed between the brain and the ear capsule. Most if not all of the fibers of its dorsal rootlet, together with part of the fibers of the ventral rootlet, pass ventrally into the acustico-facial ganglion and thence out in the main trunk of the ventral portion of the seventh nerve. The fibers of the middle rootlet and part of those of the ventral rootlet pass anteriorly as the "dorsal VII" into the lateral line ganglion lying dorsal to and confluent with the gasserian ganglion. Passing anteriorly from this dorsal lateralis ganglion the lateral line fibers are joined by general cutaneous fibers from the gasserian ganglion (the third group of fibers mentioned under the head of the trigeminal nerve). These general cutaneous fibers are in two distinct bands, one of which is applied to the ventral and the other to the mesal surface of the lateralis trunk. As the combined nerves pass anteriorly the general cutaneous components shift their positions, the median one becoming dorsal and the ventral one shifting to a lateral position. The main combined trunk soon divides into a dorsal and a ventral division, each consisting of lateralis and general cutaneous fibers. Each division then divides into two rami.

c. Ramus buccalis VII, and ramus maxillaris V.—The ventral, or infra-orbital division, forms a ventrally situated ramus consisting solely of lateralis fibers, the buccalis VII, and a more dorsal one of general cutaneous fibers with a few lateralis fibers, the maxillaris V. The few lateral line fibers mixed with the maxillaris are soon given off to certain neuromasts of the infra-orbital series posterior to the eye. The r. buccalis innervates the neuromasts of the infra-orbital series anterior to the eye. Its two main divisions that anastomose with the r. ophthalmicus profundus V have been described. The maxillaris of Amphiuma seems to be distributed chiefly to the skin of the side of the head posterior and a little anterior to the eye, but it does not run as far anteriorly as in most Urodela that have been figured, in the more anterior parts of the head being replaced by branches of the opththalmicus profundus.

The term, ramus maxillaris V, is here used with the limitations that COGHILL gave it. STRONG (1895) called attention to the striking parallelism between trigeminal and lateral line branches in the tadpole of the frog. This is especially noticeable between certain so-called accessory lateral line branches of the ophthalmicus superficialis and the buccalis and so-called accessory trigeminal twigs. Of these the most striking is the close relationship between the buccalis and the larger of the accessory trigeminal branches. COGHILL's contention that the so-called maxillaris in Urodela is the homologue of this larger accessory trigeminal twig of the tadpole seems to me valid. The distribution of the maxillaris in Amphiuma agrees very closely with the distribution of this accessory twig in the tadpole. The intimate union of maxillaris and buccalis in Amphiuma suggests the parallelism in the tadpole between the buccalis and the accessory trigeminal branch. The distribution of the terminal branches of the ophthalmicus profundus V, palatinus VII and maxillaris V in Amphiuma agrees almost to details with that in Amblystoma. COGHILL's conclusions that "there is no distinct maxillaris branch of the trigeminus in Amblystoma" can be as truly applied to Amphiuma. The smaller accessory general cutaneous and lateralis twigs given off from the combined lateral line and gasserian ganglia in Rana seem to represent possibly a ramus oticus. A determination of their exact terminations would be necessary before coming to any conclusions in the matter. But it must be noticed that in both Amphiuma and Amblystoma there occur some minute rudimentary nerves given off from the gasserian ganglion. These possibly correspond to the smaller accessory twigs in Rana.

d. Ramus ophthalmicus superficialis VII and ramus oticus.— The supra-orbital division of the "dorsal VII" forms a ventral lateralis ramus and a dorsal portion consisting of both lateralis and general cutaneous fibers. The ventral ramus is the r. opthalmicus superficialis VII, that supplies all the neuromasts of the supra-orbital series except some at the posterior end of the series. Anteriorly in the nasal region it anastomoses with the nasalis internus, as already noted. The dorsal portion of the supra-orbital division, consisting of general cutaneous and lateralis fibers, divides into a number of small branches supplying the skin of the dorsal part of the head posterior to the eye, and the neuromasts that form the posterior end of the supra-orbital series and four or five neuromasts at the posterior dorsal end of the infra-orbital series (fig. 24). In its distribution to neuromasts this dorsal division seems to represent the ramus oticus of fishes. In the latter it supplies the neuromasts of the posterior end of the supra-orbital series and the first few neuromasts of the infra-orbital series. In the r. oticus of fishes are found general cutaneous fibers from the gasserian ganglion associated with the lateralis fibers. In Amphiuma these general cutaneous fibers are distributed to a region which in some other Urodela, as Amblystoma, is innervated by branches of the ophthalmicus profundus. In some specimens of Amphiuma there occurs an anastomosis between the maxillaris and this dorsal general cutaneous division just before the latter breaks up into its smaller branches. We have here the suggestion that these dorsal general cutaneous fibers represent in part the posterior portion of the ophthalmicus superficialis trigemini. Both WILDER and STRONG are of the opinion that the ophthalmicus profundus in Amphibia "represents the united ophthalmicus profundus and ophthalmicus superficialis trigemini" of fishes. It is certainly true that the distribution of these dorsal general cutaneous branches and of the nasalis internus collectively corresponds to that of the piscine ramus ophthalmicus superficialis V.

e. The ventral trunk of the facial nerve.—As the ventral division of the facial nerve leaves the acustico-facialis ganglion complex it consists of lateralis, communis and motor fibers. As above noted, the lateral line fibers are derived from the dorsal and ventral rootlets of the dorsal division of the facial nerve. In this acustico-facial ganglion are three distinct groups of cells: (1) the large lateralis ganglion cells of the lateral line component. These are situated mostly on the anterior and dorsal borders of the complex. (2) Small cells of the geniculate ganglion, situated on the anterior mesal border of the complex. (3) Medium sized acustic cells, situated at the posterior and along the dorsal border of the ganglionic mass. Of these the cells of the vestibular branch of the VIII nerve are larger and situated anterior to the others.

f. Ramus palatinus VII.—There is first given off from the ventral facial trunk the r. palatinus of communis fibers from the geniculate ganglion. The palatine runs anteriorly and shortly after its emergence from the cranium receives an anastomosing branch from the r. pretrematicus IX. This anastomosing branch is JACOBSON's commissure, of communis fibers. The anastomosis of the r. palatinus with the r. ophthalmicus profundus V has been already noticed. In Amblystoma according to COGHILL there occurs a ganglion on the r. palatinus at the junction of its lateral division with the lateral division of the trigeminal branch. In Amphiuma there seems to be a ganglion on the palatine nerve shortly before the anastomosis with the trigeminus is reached. From this region (fig. 26) are given off a number of small nerves some of which pass to the roof of the mouth and the median series of teeth. Some of these small nerves contain general cutaneous as well as communis fibers, but the exact composition of all of them was not determined.

g. The ramus alveolaris VII.—The second branch given off from the ventral facial trunk is the r. alveolaris of communis fibers. From the r. pretrematicus IX it receives two anastomosing branches (*alv.* (*1*) and (*2*)). Near the angle of the jaw it gives off a small branch (*alv.* (*3*)) that passes mesally, ventral to the quadrate cartilage, to the roof of the mouth. There seem to be no peculiarities about the distribution of the alveolaris in Amphiuma. It divides into a number of terminal branches, one of which anastomoses with the mandibularis V as already noted. The other branches supply the ventral lateral wall of the anterior floor of the mouth. In this account the alveolaris is assumed to be a pretrematic nerve.

h. The rami mentalis externus and internus VII.—From the hyomandibular trunk of the facial nerve there are given off two large lateral line rami, of which the anterior dorsal one supplies the angular and the oral series of neuromasts, and the posterior ventral one the gular series of neuromasts. KINGSLEY designated the first as r. mandibularis externus, and the second as r. hyomandibularis accessorius. Each contains lateral line fibers only, and the two evidently correspond to the two nerves which in other Urodela hitherto figured arise by a common trunk from the hyomandibularis, and which in Amblystoma are termed by COGHILL r. mentalis externus and r. mentalis internus. FISCHER seems to have overlooked the internal division in Amphiuma. At a point about opposite the articulation of the lower jaw the r. mentalis internus divides into two branches that pass anteriorly parallel to each other. I find no anastomoses between the r. mentalis internus and the r. mandibularis V, such as KINGSLEY reports, much less can I substantiate the statements of DRÜNER that numerous anastomoses occur between these nerves.

At the point where the r. mentalis externus leaves the hyomandibular trunk there are given off a few small nerves (figured as two in number in the illustrations), partly from the main nerve and partly from the mentalis externus, which supply the post-orbital and jugular series of neuromasts. These small lateralis branches have a very wide distribution for their size. Some of their smaller divisions may be traced through hundreds of sections, yet consist of but two or three fibers each. In their origin and distribution these small lateralis nerves seem to represent opercular branches of fishes. DRÜNER reports a small cutaneous branch given off from the mentalis externus and passing to the skin overlying the depressor mandibulæ muscle. I fail to find such a nerve. Moreover the occurrence of general cutaneous fibers in a branch of the seventh nerve before the anastomosis X ad VII is received is improbable. I do find a small branch of lateralis fibers leaving the r. mentalis internus and supplying the posterior neuromasts of the gular series overlying the depressor mandibulæ muscle.

i. The ramus jugularis VII.—At about the level of the origin of the r. mentalis internus there is given off from the hyomandibular trunk a small motor branch to the anterior division of the depressor mandibulæ muscle. Another motor branch to the same division of the muscle runs along and in the anastomosis X ad VII. At about the place where the hyomandibularis finally breaks up into the larger divisions of the r. jugularis it is joined by the anastomosis from the glossopharyngeal-vagal complex carrying general cutaneous fibers. The origin and composition of this commissure will be considered under the subject of the glossopharyngeus and vagus nerves. The r. jugularis is distributed to the interhyoideus (mylohyoideus posterior), depressor mandibulæ (digastricus) and sphincter colli (levator maxillæ inferioris ascendens of FISCHER, quadrato-pectoralis of DRÜNER) muscles, and in addition carries general cutaneous fibers to the skin overlying these muscles. Lateral line fibers also occur in the r. jugularis.

k. The ramus lateralis VII.—There is one branch of the r. jugularis, if indeed it may be considered as a branch of the latter, that requires especial mention. It was first described by FISCHER as a structure peculiar to Amphiuma, and said to be traced to the hyotrachealis (interbranchialis 4) muscle. KINGSLEY states that it supplies the dorso-trachealis muscle. The writer (1904) gave a brief description of this nerve, showing that it does not end in the dorso-trachealis muscle, but passes posteriorly into the trunk region as far as the pelvis. There was suggested a possible relation to the neuromasts of the trunk, and the nerve was provisionally designated as r. lateralis VII. In the same year (1904) appeared the paper of DRÜNER in which he described the nerve, calling it nervus lateralis VII and asserting that it supplied in part the median series of neuromasts of the trunk, that is, he considered it a lateral line nerve. In the following year the writer in a second paper withheld the name, r. lateralis VII, believing that the evidence of the presence in the nerve of lateral line fibers was not convincing. The statements of DRÜNER as to the composition of the nerve may now be confirmed. It is composed largely, if not entirely, of lateralis fibers; I have not, however, as yet detected any connection between it and the neuromasts of the trunk. Such connection doubtless exists. Most of its fibers come from those branches of the r. jugularis that supply the posterior division (cerato-mandibularis) of the depressor mandibulæ muscle. In addition it is reinforced by fibers from the jugularis branch that supplies the sphincter colli muscle. If the nerve contains any general cutaneous fibers they must come from the latter source. The exact mode of origin of this nerve is subject to a great deal of individual variation, and as it leaves the posterior border of the depressor mandibulæ muscle it may consist of two or more parallel twigs that run for some distance before uniting (as in fig. 1). In passing the thymus gland the nerve is always in two or more divisions. Sometimes a small branch runs back into the trunk along with the main nerve (fig. 1). In some individuals a small branch is seen to leave the main nerve in the neighborhood of the thymus gland, and to pass antero-dorsally anastomosing with the r. supratemporalis X. In one instance the anastomosis was also with one of the twigs of the r. lateralis medius X. The r. lateralis VII seems to be peculiar to Amphiuma among the amphibians. Its origin and distribution suggest a possible homology with the r. lateralis recurrens VII of fishes, but in these latter, as shown by HERRICK (1899, 1900, 1901) and CLAPP (1898) the ramus recurrens, or lateralis accessorius, is primarily a communis nerve, with which a few lateralis or general cutaneous fibers may be associated. As shown above, there may occur an anastomosis between the lateralis VII and the r. supratemporalis X, suggesting a similar relation in fishes. In the trunk region there occur some peculiar relations between the lateralis VII and the spinal nerves, but further study of this region will be necessary before it will be safe to attempt exact comparisons. KINGSLEY (1902b) in comparing the Cæcilians and Amphiuma gives as one of the supposed points of resemblance: "The occurrence in both of a ramus lateralis recurrens branch of the facial nerve," but he gives no further explanation.

All of the branches of the r. jugularis, except those supplying the anterior portion of the depressor mandibulæ muscle, may contain lateralis fibers. From the branches of the jugularis that pass back into the posterior division of the depressor mandibulæ muscle there are given off two or three small lateralis twigs which after emerging superficially from the muscle run posteriorly just beneath the skin to supply a few of the more posterior of the jugular series of neuromasts. These commonly anastomose more or less with the twigs of the lateralis branches that supply most of the neuromasts of this series. The fore-going small lateralis twigs are possibly the "rr. cutanei jugulares" mentioned by DRÜNER. From the descriptions of FISCHER, KINGSLEY and DRÜNER, we are led to infer that the two jugularis branches that pass back through the depressor mandibulæ muscle are distinct from each other, but in fact they anastomose. In the dorsal one, the r. lateralis VII which may be double, there are motor fibers associated with the lateralis fibers. The ventral one, primarily motor, may be exclusively motor or it may contain lateral line fibers also. I have not been able to demonstrate general cutaneous fibers in these; in fact in the r. lateralis VII I have been able to demonstrate beyond question that in some individuals no general cutaneous fibers can possibly be present until it receives the branch from the nerve that supplies the sphincter colli muscle. As noted above, it is not impossible that the r. lateralis VII after it has emerged from the muscle may contain general cutaneous fibers. Cross-sections of the main nerve show scattered among the intensely black-stained lateralis fibers that compose the bulk of the nerve some lighter colored ones. It may also contain communis fibers, but I have detected no evidence of this.

7. THE GLOSSOPHARYNGEAL AND VAGUS NERVES.

a. The roots of the IX-X complex.—The IX-X complex in Amphibia is generally described as arising by five roots, but careful comparison by determination of components shows that the roots described in one species by one writer do not always correspond to the roots of a second species described by another writer. Hence a statement that in general the IX-X complex in Amphibia arises by five roots requires some qualifications. In Amphiuma the IX-X nerves arise from the brain by five roots, or rather groups of rootlets (fig. 1); but these roots do not correspond in detail to those described in Necturus by KINGSBURY, nor to those in Amblystoma as described by COGHILL. The five roots are those mentioned by KINGSLEY. The first group of rootlets (figs. 1, 10 and 23) is composed of lateralis (*X* (*1*)), communis (*IX* (*1*)) and motor (*IX* (*2*)) fibers. The communis and motor fibers after passing through the ganglion form the glossopharyngeal or first branchial nerve, with the exception of the general cutaneous component in the latter. The lateralis component supplies all the lateral line fibers of the vagus group. This lateralis component enters the brain by two rootlets which seem to correspond in origin to the median and ventral rootlets of the lateral line component of the facial nerve. The second group of rootlets consists of general cutaneous (*X* (*2a*)), communis (*X* (*2b*)) and motor (*X*(*2c*)) rootlets. The communis rootlet carries all the communis fibers of the second (*X.1.*) and third (*X.2.*) branchial nerves and of the r. intestino-accessorius. The general cutaneous component supplies fibers to the glossopharyngeus nerve, to the ramus communicans cum faciali, to the r. auricularis X, and to the second and third branchial nerves. The motor rootlets supply the second and third branchial nerves and contribute fibers to the r. intestino-accessorius. The third (*X* (*3*)), fourth (*X* (*4*)), and fifth (*X* (*5*)), groups of rootlets are exclusively motor. They may be considered as constituting one root. With some fibers from the second group of rootlets they form the motor component of the r. intestino-accessorius. In some individuals the fifth root is lacking or extremely attenuated on one side (fig. 23). In one specimen there was found a posterior vagal rootlet emerging with the first spinal nerve. A comparison of the IX-X roots in Amphiuma with those in Necturus and Amblystoma is shown in the following table:

AMPHIUMA.	NECTURUS. (KINGSBURY).	AMBLYSTOMA (COGHILL).
IX (1 + 2)	IX^{3+4}	IX
X (1a + 1b)	IX^{1+2}	X.1
X (2a + 2b + 2c)	$X^{1+2+3+4}$ (approximately)	X.2 + 3
X (3), X (4), X (5)	X^{5+6+7}	X.4

The IX-X ganglion is elongate in shape, somewhat oval posteriorly, and anteriorly flattened wedge-shaped where it passes under the ear capsule. The nerves leaving the ganglion are eight in number.

b. The ramus communicans rum faciali.—Of these the ramus communicans cum faciali, or anastomosis X ad VII, will be considered first. It would appear from the different accounts given that this trunk in origin and composition is subject to considerable variation in the amphibians. It is usually described as leaving the ganglion in a common trunk with the glossopharyngeus proper. KINGSLEY and DRÜNER both so describe it in Amphiuma, but I have yet to find an instance in Amphiuma where the two do not emerge from the ganglion separate from each other. In Amblystoma (COGHILL) the r. communicans is composed of general cutaneous and communis fibers; in Spelerpes (BOWERS) of general cutaneous fibers only, apparently. DRÜNER assumes that in the Urodela in general the anastomosis contains motor fibers. For example, in Amblystoma and Triton he describes (1901 and 1904) a motor component in the anastomosis X ad VII, but COGHILL (1906) can find no evidence of motor fibers in the anastomosis in these forms. In Amphiuma DRÜNER believes the anastomosis to consist solely of motor fibers, but I can find nothing to support such an opinion. Beyond question DRÜNER is correct in figuring the r. communicans as giving off fibers to branches of the jugularis VII, but it is not necessary to draw his conclusion that these are motor fibers. In Amphiuma the r. communicans consists chiefly, if not wholly, of general cutaneous fibers. In serial cross-sections these can be followed with precision from the second IX-X root into the ganglion, thence out into the anastomosis. I have not been able to demonstrate with certainty that no communis fibers pass from the first IX-X root into the r. communicans. Such fibers pass very close to the beginning of the latter, and possibly some enter. That none of the coarse deeply staining motor fibers of the first IX-X root enter the anastomosis seems certain. As the anastomsis approaches the VII nerve it is seen to divide into a dorsal and a ventral branch (figs. 9, 9a, 14–16, 18 and 21). These two branches unite with branches of the r. jugularis that innervate the sphincter colli and interhyoideus muscles. The variability in the mode of union is shown in figs. 14–18. In some cases (fig. 17) the r. communicans does not divide on approaching the r. jugularis. FISCHER noticed the double nature of the anterior end of the r. communicans and designated the dorsal portion as "Kopftheil des Sympathicus." DRÜNER also describes a dorsal smaller portion that passes ventrally into the r. jugularis. In Amblystoma (Siredon) he recognizes a sensory component in the r. communicans, and suggests the possibility of some of the fibers being sympathetic.

In Amphiuma motor branches from the facial nerve commonly run posteriorly into the anterior portion of the depressor mandibulæ muscle closely joined with the dorsal branch of the anastomosis. When these branches are finally distributed to the muscle there is an appearance of fibers being given off from the r. communicans but careful search shows that all of the fibers so given off come from the VII nerve. I can find no indication of motor fibers associated with the posterior part of the anastomosis. DRÜNER figures the anastomosis as giving off fibers to the anterior portion of the depressor mandibulæ muscle the direction of the fibers being such that the only interpretation that can be given to the figure is that these fibers come from the IX-X complex. I have sought carefully in all my series of sections, the material being so sharply differentially stained that nerves such as figured by DRÜNER could not escape detection, and I find not a suggestion of the presence of such fibers. In one series of sections, sagittally cut, with most excellent differentiation of components, the motor fibers from the VII nerve that usually run posteriorly along the smaller dorsal portion of the r. communicans are everywhere distinct from it, and not a single twig is given off from the r. communicans until the r. jugularis is reached (figs. 15 and 21).

In a preliminary communication (1908) I stated that the r. communicans contains communis fibers. While I am not yet ready completely to retract that statement, I now regard the presence in the r. communicans of such fibers as highly improbable.

We may summarize the foregoing statements regarding the ramus communicans in Amphiuma as follows: (1) DRÜNER's contention that the r. communicans is exclusively motor will not stand. A large general cutaneous component enters the ramus from the second IX-X root. The general cutaneous fibers in the VII nerve can come from no other source. (2) That the r. communicans contributes fibers to the r. jugularis does not indicate that these fibers are motor. The branches that receive fibers from the r. communicans are those that contain general cutaneous fibers; they must receive them from that source. (3) DRÜNER's statement and figure showing branches given off from the r. communicans to the anterior division of the depressor mandibulæ muscle are incorrect. It has been seen in some instances that this cannot possibly be true; in other cases the fibers so given off do not originate from the IX-X ganglion, but come from the r. hyomandibularis, that is, are not to be considered a part of the r. communicans. (4) That motor fibers enter the r. communicans from the IX-X ganglion has not been demonstrated, and is highly improbable. (5) The r. communicans is composed of general cutaneous fibers from the second IX-X root; it may possibly contain other sensory fibers, communis or sympathetic.

c. The rami supratemporalis and auricularis X.—Passing out from the ganglion with the r. communicans is a small nerve that evidently answers to the r. supratemporalis as described by various writers. Its course out through the cranial wall is correctly described by KINGSLEY. It is exclusively lateralis, and, as KINGSLEY suggests, supplies neuromasts in the occipital region. Anastomosing with the terminal divisions of the r. supratemporalis are branches of a second nerve springing dorsally from the IX-X ganglion. It is composed of lateralis and general cutaneous fibers. The general cutaneous component evidently corresponds to the r. auricularis vagi of the tadpole (STRONG 1895), and to the general cutaneous component of the nerve in Amblystoma termed r. auricularis vagi by COGHILL.

d. The glossopharyngeal nerve.—General cutaneous, communis and motor fibers are contained in the glossopharyngeus, or first branchial nerve, which divides just before or soon after it leaves the ganglion into a r. pretrematicus and a r. posttrematicus. According to DRÜNER the r. pretrematicus is larger than the r. posttrematicus and the r. communicans larger than either. Cross sections of the three as they leave the ganglion show that the r. posttrematicus is the largest of the three, and the r. communicans the smallest. The manner of branching of the r. pretrematicus is so variable that it is difficult to make statements regarding it at all accurate. Shortly after leaving the ganglion the r. pretrematicus gives off its principal branches. One small branch goes ventrally to supply the dorsal wall of the pharynx ventral and posterior to the ear capsule. Another branch which may be called the pharyngeal proper (*ph. IX.*) is given off dorsally. From it there passes a slender nerve anastomosing with the r. palatinus VII, forming JACOBSON'S commissure. The fibers of this anastomosis on entering the r. palatinus are seen to pass some centrally and some peripherally. From the anterior portion of JACOBSON'S commissure near where the latter joins the r. palatinus there is given off a branch to the roof of the mouth. The pharyngeal branch sends an anastomosis to the r. alveolaris VII. The main portion or ramus pretrematicus proper, passes antero-ventrally and mesally until the hyoid arch is reached. Thence, after dividing sooner or later into two branches, it passes anteriorly a little dorsal to the hyoid arch as far as the tip of the latter, supplying the floor of the mouth at the sides. From the pretrematic proper there passes an anastomosis with the r. alveolaris VII. The fibers of the anastomosis on entering the alveolaris pass some centrally and some peripherally. Between the pretrematic proper and the pharyngeal branch there may occur anastomoses (fig. 21). Sometimes the pharyngeal anastomosis with the r. alveolaris comes from the main pretrematicus (fig. 1). From the pharyngeal branch there may pass a branch into the main hyomandibular trunk (fig. 21). Thus there may be formed in the hyoid and mandibular region a plexus between the IX and VII nerves, consisting of communis and possibly sympathetic fibers, all the fibers of which are destined presumably to supply blood vessels and the mucous membrane of the mouth and pharynx. Some of the smaller twigs of this plexus can be traced into the close vicinity of blood vessels of this region. Of isolated ganglia or ganglion cells in this region I find none.

It is the pretrematic division of the IX nerve that KINGSLEY considered the glossopharyngeus proper. He was thereby led into the error of supposing that the IX nerve contains no motor fibers. He failed to discover the actual anastomoses with the seventh nerve. The branch that anastomoses with the palatine he correctly termed the pharyngeal. The main pretrematic branch he designated provisionally the hyoid branch. According to him the r. supratemporalis X is a branch of the glossopharyngeus. I find the former more intimately associated with the r. communicans at its exit from the ganglion.

As the posttrematic division of the glossopharyngeus ascends to the level of the dorsal border of the branchial arches it gives off from its ventral border several small pharyngeal twigs, as described by DRÜNER. As the trunk turns to pass laterally between the dorsal ends of the hyoid and first branchial arches it gives off a small motor branch (*lab. 1*) to m. levator arcus branchialis 1. A little more laterally a small general cutaneous branch runs obliquely postero-laterally over the first branchial arch to the skin of that region. This general cutaneous component of the glossopharyngeus is so small that I have as yet detected it in but one individual. Mention of it was omitted in my preliminary paper. DRÜNER says that the first branchial nerve sends two branches to the m. lev. arc. br. 1. As he does not mention this general cutaneous branch he possibly mistook it for a motor branch. KINGSLEY describes and figures a branch of the first branchial nerve as innervating the posterior portion of the depressor mandibulæ muscle. Such a relation would certainly be anomalous. I do not find such a nerve, but in one specimen I find that the ramus posttrematicus just as it begins to descend along the outer anterior border of the first branchial arch divides into a posterior portion that runs nearly in the usual course, and an anterior portion that becomes almost lost among the fibers of the depressor mandibulæ muscle, rejoining the other division more ventrally. This anterior portion, or possibly a still more aberrant but corresponding branch, may be the one mistaken by KINGSLEY for a division to the muscle. Ventrally the posttrematicus divides into motor branches to the ceratohyoideus internus muscle and communis branches to the lateral pharyngeal wall between the hyoid and first branchial arches, the extreme anterior portion being the ramus lingualis of communis fibers to the tongue.

e. The second and third branchial nerves.—The second and third branchial nerves (*X.1.* and *X.2.*) sometimes leave the ganglion as a common trunk that soon divides; sometimes they originate separately, but close together. They have the same general arrangement of parts. They send branches to mm. levatores arcuum branchialium 2 et 3 respectively. Each divides into a ramus posttrematicus and a ramus pharyngeus. From the latter ramus in each there runs anteriorly a small branch just dorsal to the extreme dorso-lateral angle of the pharynx as far as the preceding branchial arch and thence along the inner border of the latter, supplying the ventro-lateral wall of the pharynx. This answers to a pretrematic ramus in its distribution. DRÜNER's failure to find a distinct pretrematic branch on either of these two nerves is probably due to the fact that in adults, such as he examined, it has become much attenuated. The main portion of the r. pharyngeus is distributed to the dorsal wall of the pharynx. The r. posttrematicus after giving off a branch or branches to its corresponding m. levator arcus branchialis turns ventrally and sending off one or two small general cutaneous branches to the skin, passes antero-ventrally along the outer border of its respective branchial arch.

The second branchial posttrematic sends branches to mm. subarcuales recti 1, 2 et 3 (m. constrictor arcuum branchiarum inferior), and to m. subarcualis obliquus (m. constrictor arcuum branchiarum superior), and continues anteriorly to innervate the m. ceratohyoideus internus. It also sends communis branches to the ventral wall of the pharynx. The third branchial posttrematic according to DRÜNER has no motor fibers in its ventral portion, although in describing the mm. subarcuales he says they receive branches from the second and third branchial nerves. I find that the greater part of the ventral portion of the third branchial posttrematic is of communis fibers distributed, as DRÜNER says, to the ventral pharynx wall at the sides of the larynx; but I also find a small motor branch given off to m. subarcualis rectus 3, and in one instance I have traced fibers to the m. subarcualis obliquus. Between the posttrematic divisions of the second and third branchial nerves and the ramus intestinalis recurrens X there occur anastomoses such that it is difficult to distinguish the source of some of the fibers innervating the subarcual muscles.

f. The rami laterales X.—From the posterior end of the IX-X ganglion there pass out two large nerve trunks. The dorsal of these is composed solely of lateralis fibers, and soon divides forming a smaller dorsal r. lateralis dorsalis supplying the neuromasts of the dorsal series of the trunk of the body, and a larger ventral r. lateralis medius supplying the median series of neuromasts of the trunk. The remaining lateralis component of the vagal group will be described in the following section.

g. The ramus intestino-accessorius X.—The second great trunk passing posteriorly from the IX-X ganglion is the r. intestino-accessorius, composed of lateralis, communis and motor fibers. The communis and motor fibers have a very diverse distribution. A short distance posterior to the ganglion there leaves the main trunk a small nerve of communis and motor fibers, that in part represents fourth and fifth branchial nerves. It sends motor fibers to mm. levator arcus branchialis 4, trapezius and dorso-laryngeus. The branches to the trapezius and dorso-laryngeus muscles may arise separately from the main int.-acc. trunk. After giving off fibers to the m. lev. arc. br. 4, the nerve, now of communis fibers only, passes near the dorsal ends of the third and fourth branchial arches and there divides, one branch running along the anterior median border of the third branchial arch, and evidently forming a fourth ramus pretrematicus (*prt. X. 3*), and a second branch running similarly along the fourth branchial arch and constituting a fifth ramus pretrematicus (*prt. X. 4*). The main intestino-accessorius trunk finally divides (figs. 1 and 12) into a r. lateralis ventralis supplying the ventral series of neuromasts of the trunk; three rr. intestinales that pass posteriorly, one dorsal to the œsophagus and the other two latero-ventral to the same; and a r. intestinalis recurrens that turns anteriorly to supply m. interbranchialis 4 (m. hyotrachealis) and mm. subarcuales. From one of the intestinal branches there turns anteriorly a r. laryngeus recurrens that innervates muscles of the larynx, mm. dorso-laryngeus, constrictor laryngei, etc., and also supplies communis fibers to the pharyngeal wall in the same region. Communis fibers are also given off to the pharyngeal wall from the r. intestinalis recurrens in the laryngeal region. The rr. intestinales were traced posteriorly as far as the heart only and nothing can be stated precisely of their destination.

8. THE FIRST AND SECOND SPINAL NERVES.

The first spinal nerve in its early stages, as described by KINGSLEY, arises by four roots, two dorsal and two ventral, and is thus in origin clearly double (fig. 19). The common trunk formed by these roots passes out through a foramen in the first vertebra. In individuals of 120 mm. length I have found the first spinal nerve to possess two very rudimentary dorsal roots, two large ventral roots, and a small ganglion (fig. 19). In individuals of 140 mm. length the dorsal roots have disappeared, but the ganglion remains. In individuals of 175 mm. length the ganglion has disappeared. In older individuals the nerve appears to arise by four ventral roots in two groups. According to DRÜNER the hypobranchialis (hypoglossus) nerve is derived from the first and second spinal nerves. Careful search through my preparations fails to show any anastomosing between these two nerves. I have found no instance where they come into contact even. The nearest approach to contact is between a general cutaneous division of the second spinal nerve and the main hypoglossal trunk of the first spinal nerve. The hypoglossus nerve is formed solely from the first spinal nerve, and contains only motor fibers. The main ventral portion of the first spinal nerve soon after it emerges from the spinal canal passes posteriorly and after giving off a few small branches runs back to the region where the r. int.-acc. divides into its several branches (figs. 1 and 12). Here the hypoglossus comes into intimate relations with the r. lateralis ventralis and the r. intestinalis recurrens X, but there is no fusing between them such as KINGSLEY describes and figures. In some cases there may be a temporary mingling of fibers, but in other instances the distinction between hypoglossal and other nerves is absolutely clear throughout, so that we may confidently deny the occurrence of any anastomosing between the hypoglossal and vagal nerves. From the point of the branching of the r. int. acc. trunk the hypoglossal nerve runs antero-ventrally, giving off no branches until it reaches the anterior section of the sternohyoid muscle. This it innervates and then runs along in the geniohyoid muscle supplying it, to end anteriorly in the genioglossal muscle. The main ventral branch of the second spinal nerve passes posteriorly in a direction nearly parallel with that of the hypoglossal trunk. A short distance posterior to the place of branching of the r. int.-acc. it turns sharply about and running antero-ventrally comes into close relations with the r. intestinalis recurrens X. It receives a general cutaneous branch from the third spinal nerve and then divides into a general cutaneous and a motor division. The general cutaneous branch supplies the latero-ventral skin in the anterior post-branchial region; the motor branch divides into a nerve that runs anteriorly to innervate the anterior segment of the sternohyoid muscle, and a second branch that runs ventrally to supply other sections of the same muscle, after anastomosing with a motor branch of the third spinal nerve. The brachial plexus is formed from branches of the third and fourth spinal nerves. The ramus lateralis ventralis X becomes very intimately associated with portions of the brachial plexus, but it is very clearly seen that no anastomosing occurs, such as BOWERS describes (doubtless incorrectly) in Spelerpes.

9. CONCLUDING STATEMENTS.

It is evident that the arrangement of the cranial nerves of Amphiuma gives support to the view that this species represents in many respects a primitive amphibian type. The group of nerves here designated as ramus oticus; the nasalis internus V; the clear differentiation of pretrematic, posttrematic and pharyngeal rami in the branchial nerves; possibly also the ramus lateralis VII; the lateral line lobe; all these have distinctly fish-like characteristics. Although the nerves connected with the eyes are degenerate, this cannot be said of the other cranial nerves. As compared with other Urodela, Amphiuma shows in the arrangement of its cranial nerves a tendency towards great diffuseness and individual variability. A nerve trunk in one individual may break up into a number of divisions later to become consolidated into a main trunk again. In another specimen the same nerve may show no such tendency to diffuse subdivsion. General conclusions based upon one or two specimens, in Amphiuma at least, are likely to be much in error.

The resemblances between the cranial nerves of Amphiuma and those of Amblystoma are very striking. When the components of the cranial nerves of Cryptobranchus, Necturus, Siren, and one or two more of the Salamandridæ have been carefully worked out, we shall be in a position to define the urodele type of cranial nerves, and in the opinion of the writer, not until then.

The distribution of the lateral line organs of Amphiuma has been carefully mapped out and described by KINGSBURY (1895b). The innervation of these organs on the head has been worked out by the writer with precision, corroborating the description of KINGSBURY, except in some details. A detailed account of the innervation of these organs in Amphiuma is withheld from this paper because of the uncertainty as to the exact distribution of the ramus lateralis VII on the trunk of the body.

Iowa College,
June 20, 1908.

LITERATURE CITED.

BOWERS, MARY A.
1900. The peripheral distribution of the cranial nerves of Spelerpes bilineatus. *Proc. Amer. Acad. Arts and Sci.*, vol. 36.

CLAPP, CORNELIA M.
1898. The lateral line system of Batrachus tau. *Jour. Morph.*, vol. 15.

COGHILL, G. E.
1902. The cranial nerves of Amblystoma tigrinum. *Jour. Comp. Neurol.*, vol. 12.
1906. The cranial nerves of Triton tæniatus. *Jour. Comp. Neurol. and Psych.*, vol. 16.

DRÜNER, L.
1901. Studien zur Anatomie der Zungenbein-, Kiemenbogen-, und Kehlkopfmuskeln der Urodelen. I Theil. *Zool. Jahrb.*, Bd. 15.
1904. Studien zur Anatomie der Zungenbein-, Kiemenbogen-, und Kehlkopfmuskeln der. Urodelen. II Theil. *Zool. Jahrb.*, Bd. 19.

FISCHER, J. G.
1864. Anatomische Abhandlungen über die Perennibranchiaten und Derotremen. *Hamburg*

HERRICK, C. JUDSON.
1899. The cranial and first spinal nerves of Menidia. *Jour. Comp. Neurol.*, vol. 9.
1900. A contribution upon the cranial nerves of the codfish. *Jour. Comp. Neurol.*, vol. 10.
1901. The cranial nerves and cutaneous sense-organs of the North American siluroid fishes. *Jour. Comp. Neurol.*, vol. 11.
JOHNSTON, J. B.
1906. The nervous system of vertebrates. *Philadelphia.*
KINGSBURY, B. F.
1895a. On the brain of Necturus maculatus. *Jour. Comp. Neurol.*, vol. 5.
1895b. The lateral system of sense organs in some American Amphibia, and comparisons with Dipnoans. *Proc. Amer. Micr. Soc.*, vol. 17.
KINGSLEY, J. S.
1902a. The cranial nerves of Amphiuma. *Tufts College Studies*, no. 7.
1902b. The systematic position of the Cæcilians. *Tufts College Studies*, no. 7.
NORRIS, H. W.
1904. The so-called dorsotrachealis branch of the seventh cranial nerve in Amphiuma. *Proc. Iowa Acad. Sci.*, vol. 10.
1905. The so-called dorsotrachealis branch of the seventh cranial nerve in Amphiuma. *Anat. Anz.* Bd. 27.
1908. The cranial nerve components in Amphiuma. *Science*, n. s., vol. 27, no. 702.
OSBORN, H. F.
1888. A contribution to the internal structure of the amphibian brain. *Jour. Morph.*, vol. 2.
STRONG, O. S.
1895. The cranial nerves of Amphibia, *Jour. Morph.*, vol. 10.
WILDER, H. H.
1892. Die Nasengegend von Menopoma alleghaniense und Amphiuma tridactylum. *Zool. Jahrb.*, Bd. 5.

EXPLANATION OF THE PLATES.

REFERENCE LETTERS.

acc. branch of the X nerve supplying the anterior part of the trapezius muscle.
ags. angulo-splenial bone.
alv. r. alveolaris VII.
alv.(1). first anastomosing branch between r. alveolaris VII and r. pretrematicus IX.
alv.(2). second anastomosing branch between r. alveolaris VII and r. pretrematicus IX.
alv.(3). branch of r. alveolaris VII that supplies roof of mouth.
alv.(4). branch of r. alveolaris VII that anastomoses with r. mandibularis V
am. muscles of the arm.
ao. antorbital cartilage.
auc. cartilage of the ear capsule.
auo. ossifications of the ear capsule.
aur. r. auricularis X.
bhy. basihyal cartilage.
bplx. branches of the brachial plexus.
1br.,2br.,3br. first, second and third branchial arches.
buc. r. buccalis VII.
buc.(1),buc.(2). branches of the r. buccalis VII that anastomose with the r. oph. prof. V.
ch. cerebral hemisphere.
chi. m. ceratohyoideus internus and branches of IX nerve innervating it.
hi.X.1. branch of the second branchial nerve innervating m. ceratohyoideus internus.
chy. ceratohyal.
cor. coracoid cartilage.
dent. dentary bone.
dien. diencephalon.
dl. m. dorsolaryngeus.
dln. branches of the X nerve innervating the dorsolaryngeus muscle.
dm. m. depressor mandibulæ.
dma. anterior division of m. depressor mandibulæ and the nerve innervating it.
dmp. posterior division of m. depressor mandibulæ and nerve innervating it.
fc. fasciculus communis.
flm. fasciculus longitudinalis medius.
fr. frontal bone.
gac. ganglion acusticum.
gen. ganglion geniculatum.
gg. ganglion Gasseri.
gh. m. geniohyoideus.
g.1.sp. first spinal ganglion.
g.VIIba. ganglion on "dorsal VII."
g.VIIbb. lateral line ganglion fused with ganglion acusticum.
g.VII-VIII. acustico-facial ganglion.
g.IX-X. ganglion common to IX and X nerves.
h. humerus.
hgl. n. hypoglossus.
hhy. hypohyal cartilage.
hm. tr. hyomandibularis VII.
hy. hyoid arch.
ib.4. m. interbranchialis 4 = m. hyotrachealis.
ih. m. interhyoideus and branches of r. jugularis VII innervating it = m. mylohyoideus posterior.
im. m. intermandibularis = m. mylohyoideus anterior.
int. rr. intestinales X.
int.-acc. r. intestino-accessorius X.
int.rec. r. intestinalis recurrens X.
io. m. obliquus inferior.
jc. JACOBSON's commissure.
jc.(a). branch of JACOBSON's commissure innervating the roof of the mouth.

jgl. r. jugularis VII.
jo. JACOBSON's organ.
lab.1,2,3, and 4. mm. levatores arcuum branchialium and the nerves innervating them.
lag. lagena and branch of the VIII nerve innervating it.
lar. rec. r. laryngeus recurrens X.
lat.d. r. lateralis dorsalis X.
lat.m. r. lateralis medius X.
lat.v. r. lateralis ventralis X.
lat.VII. r. lateralis VII.
lg. communis branch of r. posttrematicus IX to the tongue.
ll. lobus lineæ lateralis.
lvb. m. levator bulbi.
mas. m. masseter.
max. r. maxillaris V.
mb. branch of r. mandibularis V innervating the retractor and levator bulbi muscles
mck. MECKEL's cartilage.
md. r. mandibularis V.
md.(2). branch of r. mand. V innervating skin of side of head.
md.(3). branch of r. mand. V running just above and in the mandible.
md.(3a). branch of preceding innervating skin near tip of lower jaw at the side.
md.(3b). mandibular branch anastomosing with r. alveolaris VII.
md.(4).,md.(5). branches of r. mand. V innervating m. intermandibularis and overlying skin.
mes. mesencephalon.
mplx. metaplexus.
mtl.ext. r. mentalis externus VII.
mtl.int. r. mentalis internus VII.
mx. maxilla.
na. neuromasts of the angular series.
nas. nasal bone.
nas.int. rm. nasalis internus, first terminal division of r. profundus V.
nc. nasal cartilage.
ngm. median nasal gland = JACOBSON's gland (?).
nio. neuromasts of the infra-orbital series.
nio.(a). neuromasts of the infra-orbital series innervated by the r. oticus.
nso. neuromasts of the supra-orbital series.
nso.(a). neuromasts of the supra-orbital series innervated by the r. oticus.
op. r. ophthalmicus profundus V.
op.(1). first terminal division of the r. oph. prof. V. = rm. nasalis internus.
op.(1a). branch of the preceding that passes into a canal in the edge of the frontal bone = ethmoideus caudalis of KINGSLEY.
op.(1b). branch of op. (1) innervating the skin of dorsum posterior and dorsal to eyeball.
op.(2). second terminal division of r. oph. prof. V = r. glandularis II of WILDER.
op.(3). third terminal division of r. oph. prof. V, anastomosing with r. palatinus VII.
op.(4) and (5). Fourth and fifth terminal divisions of r. oph. prof. V, anastomosing with the two main divisions of r. buccalis VII.
op-pal. anastomosis of the third terminal division of r. oph. prof. V with r. palatinus VII.
op.-pal.d. branch or branches from the oph-pal. anastomosis passing antero-dorsally innervating the dorsal lateral nasal epithelium and JACOBSON's organ.
op.-pal.l. branches from the oph.-pal. anastomosis supplying the lateral (maxillary), series of teeth and roof of mouth.
op.-pal.m. branch from the oph.-pal. anastomosis supplying the median (vomero-palatine) series of teeth and the roof of the mouth.
op.-pal.mn. branch from the oph.-pal. anastomosis supplying the median nasal epithelium.
op.-pal.ph. small branches from the oph.-pal. anastomosis passing chiefly to the roof of the mouth at the posterior border of the nasal capsule.
op.-pal.pn. branch from oph.-pal. anastomosis supplying the posterior nasal epithelium.
os. r. ophthalmicus superficialis VII.
osph. orbitosphenoid bone and cartilage.
ot. r. oticus.
pa. parietal bone.

pal. r. palatinus VII.
pf. prefrontal bone.
ph.IX. r. pharyngeus IX.
ph.X.1. r. pharyngeus of the second branchial nerve
ph.X.2. r. pharyngeus of the third branchial nerve.
pmx. premaxilla.
pn. postnares (edge of wall).
pplx. prosoplexus.
prt.IX. r. pretrematicus IX.
prt.X.1. r. pretrematicus of the second branchial nerve.
prt.X.2. r. pretrematicus of the third branchial nerve.
prt.X.3. branch of the r. intestino-accessorius X, representing a pretrematic ramus of a fourth branchial nerve.
prt.X.4. a nerve associated with the preceding and representing a pretrematic ramus of a fifth branchial nerve.
psph. parasphenoid bone.
pst.IX. r. post-trematicus IX.
pst.X.1. r. post-trematicus of the second branchial nerve.
pst.X.2. r. post-trematicus of the third branchial nerve.
pt. m. pterygoideus.
ptc. pterygoid cartilage.
ptn. nerve innervating m. pterygoideus.
pto. pterygoid bone.
rext. m. rectus externus.
rinf. m. rectus inferior.
rint. m. rectus internus.
rs. m. rectus superior.
rtb. m. retractor bulbi.
ar.1,2,3. mm. subarcuales recti 1, 2 and 3, = constrictor arcuum branchiarum inferior.
sc. scapula.
so. m. obliquus superior.
se. saccus endolymphaticus.
1sp. first spinal nerve.
1sp.d. dorsal branch of first spinal nerve.
1sp.rd. dorsal roots of first spinal nerve.
1sp.rv. ventral roots of first spinal nerve.
1sp.v. ventral branch of first spinal nerve.
2sp. second spinal nerve.
3sp. third spinal nerve.
sp.V. tractus spinalis trigemini.
p.VIII. tractus spinalis acusticus.
sphc. m. sphincter colli and the nerve innervating it.
spt. r. supratemporalis X.
sq. squamosal bone.
ssc. suprascapula.
sthy. m. sternohyoideus.
stp. stapes.
th. thymus gland.
thr. thyroid gland.
tm. m. temporalis.
tma. anterior division of m. temporalis.
tmn. branch of the r. mandibularis V innervating m. temporalis.
tmp. posterior division of m. temporalis.
tmpt. tendon of the posterior division of m. temporalis.
tr. trachea.
tra. "tractus a" of Kingsbury.
trap. m. trapezius, posterior division.
trapa. m. trapezius, anterior division.
trb. "tractus b" of Kingsbury.
trc. tracheal cartilages.

trm.	muscles of the trunk.
vp.	vomero-palatine bone.
I.	n. olfactorius.
Ijo.	branch of n. olfactorius innervating JACOBSON's organ.
Irrt.	roots of n. olfactorius.
II.	n. opticus.
III.	n. oculomotorius.
IV.	n. trochlearis.
V(2).	motor root of the V nerve.
VI.	n. abducens.
VIIa.	the ventral roots of the facial nerve, communis and motor components.
VIIaa.	the communis portion of the preceding.
VIIab.	the motor root of the facial nerve.
VIIa+VIIbb.	the ventral or main trunk of the facial nerve.
VIIb.	the lateral line portion of the facial nerve.
VIIb(1+2+3)	the three rootlets of the preceding.
VIIba.	the "dorsal VII."
VIIbb.	that portion of the lateral line component of the facial nerve that passes ventrally and joins VIIa.
VIIba+V.	the trunk formed by the union of general cutaneous fibers from the gasserian ganglion with the "dorsal VII."
VIII.	n. acusticus.
VIIIap.	branch of VIII nerve to the posterior ampulla.
VIIIlag.	branch of VIII nerve to the lagena.
VIIImn.	branch of the VIII nerve to the macula neglecta.
VIIIpb.	branch of the VIII nerve to the pars basilaris.
VIIIs.	branches of the VIII nerve to the sacculus.
VIIIv.	vestibular branch of the VIII nerve.
VIII(1).	fibers of the VIII nerve that pass into the tractus spinalis VIII.
VIII(2).	fibers of the VIII nerve that pass anteriorly into the acusticum.
VIII(3).	fibers of the VIII nerve that enter "tract b."
VIII(4).	coarse fibers that enter the sp. VIII tract distinct from VIII (1).
IX(1).	communis root of the IX nerve.
IX(2).	motor root of the IX nerve.
X.1.	second branchial nerve.
X.2.	third branchial nerve.
X.3+4.	branch of r. int.-acc. representing the fourth and fifth branchial nerves.
X(1).	first or lateral line root of the X nerve.
X(2)	second root (group of rootlets) of the X nerve.
X(3),X(4),X(5).	third, fourth and fifth roots of the X nerve.
X.adVII.	r. communicans between the X and VII nerves.
X.adVII.d.	dorsal division of r. communicans.
X.adVII.v.	ventral division of r. communicans.

EXPLANATION OF PLATE IV.

FIG. 1. A projection upon the sagittal plane of the V, VII, VIII, IX and X cranial nerves, together with portions of the first and second spinal nerves, of Amphiuma means. The roots and ganglia are slightly schematic for the sake of clearness. In only a few cases have the positions of nerve trunks been slightly changed. The scale above the figure indicates the serial numbers of the transverse sections employed in the reconstruction, the sections being 10 micra thick × 32.

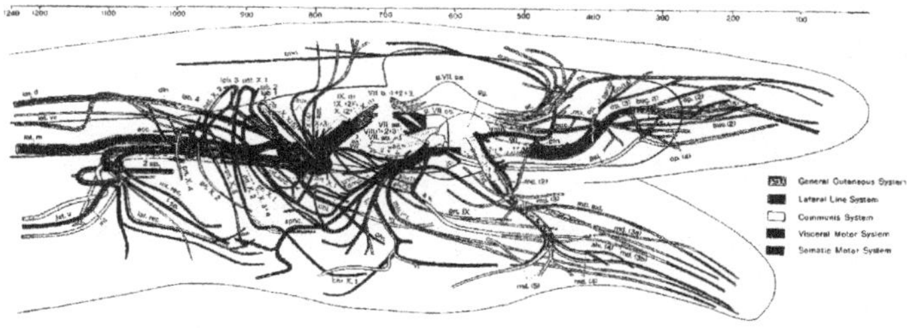
1240
1200
1100
1000
900
800
700
600
500
400
300
200
100
General Cutaneous System
Lateral Line System
Communis System
Visceral Motor System
Somatic Motor System

EXPLANATION OF PLATE V.

FIGS. 2 to 10. Cross-sections of the left half of the head at different levels. The outlines were drawn with a camera lucida; the details are schematic. These were made from a different series of sections than that from which fig. 1 was plotted. After the description of each figure is given the number of the section in fig. 1 which corresponds approximately to the section described.

FIG. 2. Cross-section through the nasal capsule at the level where the branch of the olfactory nerve that innervates JACOBSON's organ is passing around to the dorsal side of the latter structure. Section 200. × 30.

FIG. 3. Cross-section through the nasal capsule at the point where the olfactory nerve trunk breaks up into its terminal divisions and enters the capsule. Section 260. × 30.

FIG. 4. Cross-section cutting through the posterior portion of the eyeball. To show the retractor and levator bulbi muscles and their insertion on the antorbital cartilage. Section 350. × 20.

FIG. 5. Cross-section at the level where the r. mandibularis V enters the lower jaw. Section 460. × 20.

FIG. 6. Cross-section through the origin of the r. mandibularis V. The abducens nerve is seen separating from the gasserian ganglion. Section 560. × 30.

FIG. 7. Composite cross-section through the roots of the VII and VIII nerves. Sections 675-680. × 30.

FIG. 8. Cross-section through the roots of the lateral line component of the VII nerve. Section 685. × 30.

FIG. 9. Cross-section slightly posterior to that of fig. 8. × 20.

FIG. 9*a*. An enlargement of a part of the preceding, showing the r. communicans.

FIG. 10. Cross-section at the level where the roots of the lateral line components of the X nerve enter the brain. Section 710. × 30.

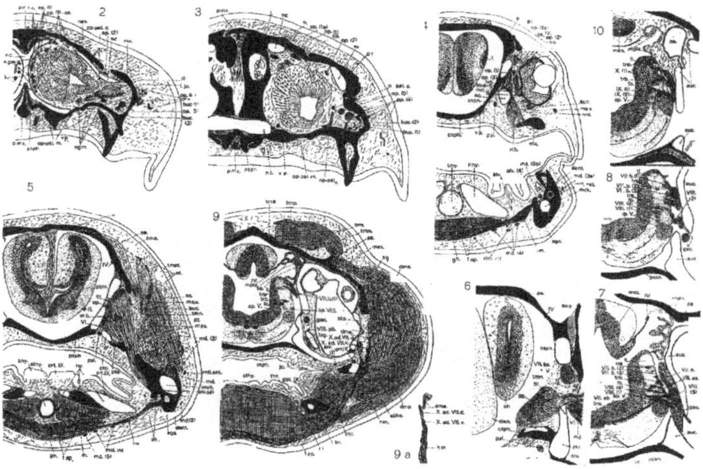

EXPLANATION OF PLATE VI.

FIG. 11. Cross-section through the gill-cleft. Section 960. × 19.

FIG. 12. Cross-section at the level where the r. intestino-accessorius breaks up into its branches. Section 1090. × 19.

FIG. 13. A projection upon the sagittal plane to show chiefly the olfactory, optic and eye-muscle nerves. The outlines of the acustico-facial, gasserian and dorsal lateral line ganglia are drawn in their correct relations.

FIGS. 14 to 18. Projections upon the sagittal plane of the r. hyomandibularis VII in the region where it divides into its chief branches. These are from four different individuals. Figs. 14 and 15 are from the same individual. All are represented as seen from the right side. These sections are to show primarily the variable mode of fusion between the anastomosis X ad VII and the r. jugularis VII.

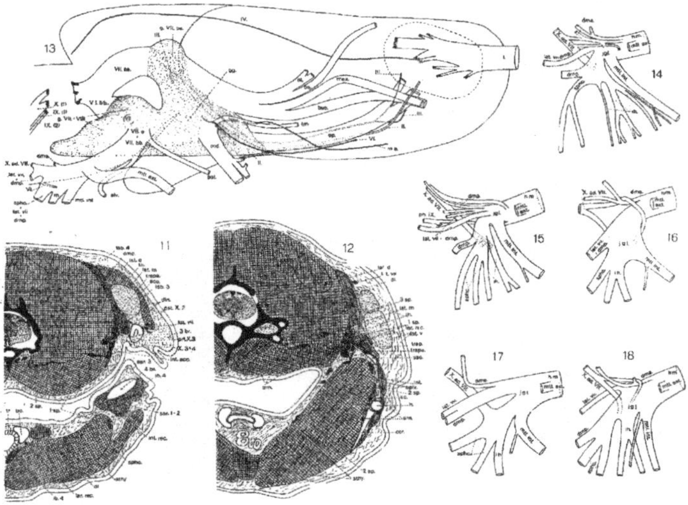
13
14
15
16
11
12
17
18

EXPLANATION OF PLATE VII.

FIG. 19. A projection upon the sagittal plane of the origin of the first spinal nerve, in an individual of 120 mm. length, seen from the right side.

FIG. 20. A projection upon the sagittal plane of the olfactory nerve, its roots and principal branches.

FIG. 21. A projection upon the sagittal plane of the IX-X, and V-VII-VIII ganglionic complexes, and the plexus developed between the IX -X, and VII nerves. The roots of the ganglia are omitted. In the individual from which the plotting was made JACOBSON's commissure is double; from the r. pharyngeus IX there passes an anastomosis to the tr. hyomandibularis VII. The V-VII-VIII ganglionic mass is much more compact than in younger individuals, such as shown in figs. 1, 13 and 24.

FIG. 22. Composite sagittal section through the insertion of the retractor and levator bulbi muscles upon the antorbital cartilage, as seen from the left side. The eyeball and the greater portion of the eye-muscles are situated lateral to the section and are represented in dotted outlines. For the sake of clearness the branches of the r. ophthalmicus profundus V and of r. ophthalmicus superficialis VII are omitted. The outlines of the section were drawn with a camera lucida.

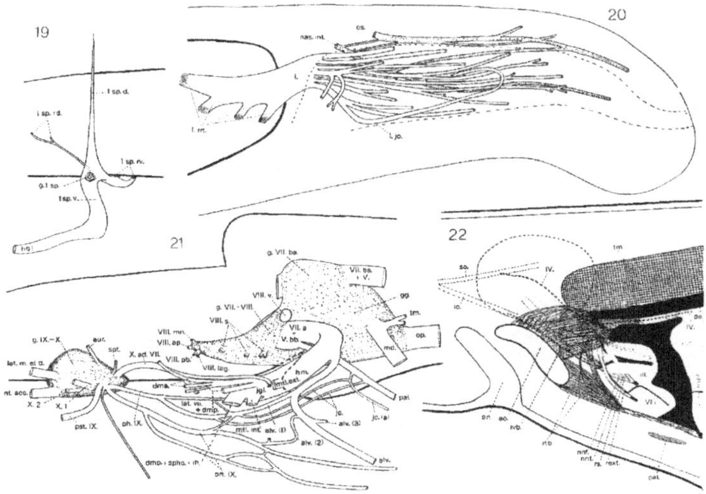
19
20
21
22
nas. int.
os.
g. VII. ba.
VII. ba. + V.
g. IX.-X
aur.
spt.
X. ad. VII.
lat. m. et d.
nt. aco.
pst. IX.
ph. IX.
dmp. + spho. + ih.
pal.
tm.
op.
md.
gg.
IV.
VI.

EXPLANATION OF PLATE VIII

FIG. 23. A projection upon the horizontal plane of the roots, ganglia and principal trunks of the V, VII, VIII, IX and X nerves.

FIG. 23a. The roots of the VIII nerve.

FIG. 24. A projection upon the sagittal plane of the ramus oticus and its branches, together with the neuromasts which it innervates.

FIG. 25. A projection upon the sagittal plane of the anastomoses between the r. oph. prof. V and the r. buc. VII. The condition here is more complicated than in fig. 1.

FIG. 26. A projection upon the sagittal plane of the anastomosis between the r. oph. prof. V and the r. pal. VII. The exact composition of a few of the smaller branches is uncertain; hence they are only partly colored.

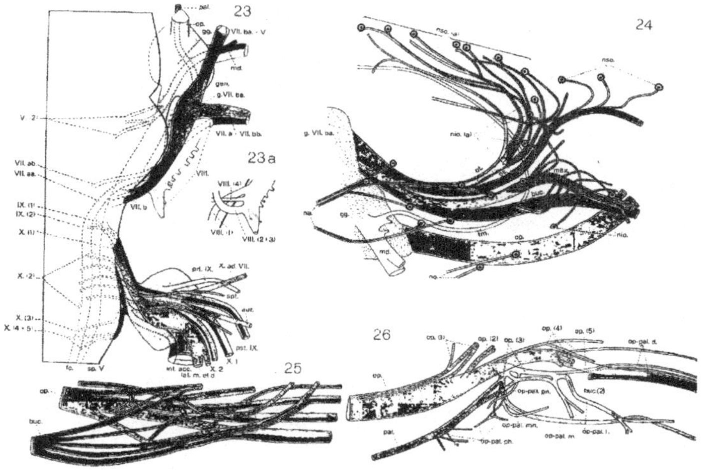
23
23a
24
25
26

ADDITIONAL NOTES ON THE CRANIAL NERVES OF PETROMYZONTS.[1]

BY

J. B. JOHNSTON.

University of Minnesota.

WITH THIRTY-ONE FIGURES.

In the year 1897 I prepared by the GOLGI method a number of series of sections of the head of Lampetra wilderi. These were used for the study of the brain (1902) and were reviewed in connection with the study of the components of the cranial nerves in Petromyzon dorsatus (1905). Owing to the fact that certain fibers are impregnated and others not, the GOLGI preparations are not in themselves suitable for a complete study of cranial nerve components and such differences were found between the two species that it was thought best not to incorporate any of the facts from the Lampetra series in the description of the nerves of P. dorsatus. The GOLGI preparations were therefore laid aside with the hope that they could be supplemented by new preparations and the cranial nerves thoroughly gone over by this method. The time for making these new preparations now seems more remote than ever and I have decided to publish certain results which are entirely clear from the preparations in hand. The animals used were adults. All were taken on their nests just after spawning.

The general relations of the cranial nerves may first be reviewed by means of figs. 1 to 11. These are camera sketches from a series of horizontal sections. The anterior part of the right half of the head is shown, including the first two gill sacs. The figures are not schematized at all except that detail had to be omitted, and in figs. 3 and 8 a little is added from the sections adjacent to those drawn and fig. 11 is a reconstruction from several sections. The left half of this figure represents sections farther ventrad than those drawn on the right. The entire series consists of 109 sections and the section drawn is indicated beneath each figure. In all

[1] Neurological Studies, University of Minnesota, no. 2.

the figures cartilage is shaded with oblique lines and the larger muscles are indicated by light lines running in the direction of the muscle fibers. The following letters are used in addition to

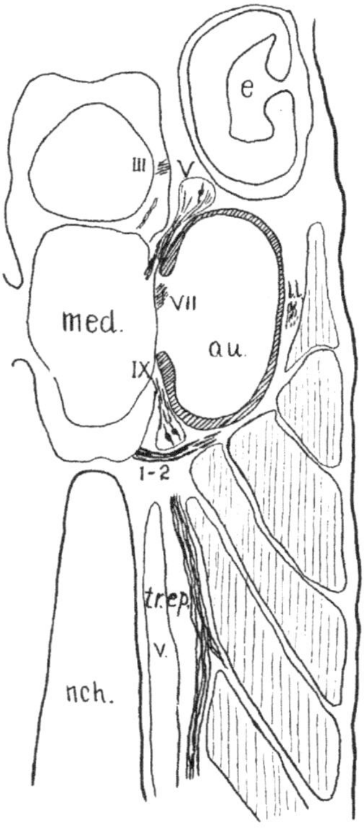

FIG. 1. Section 39.

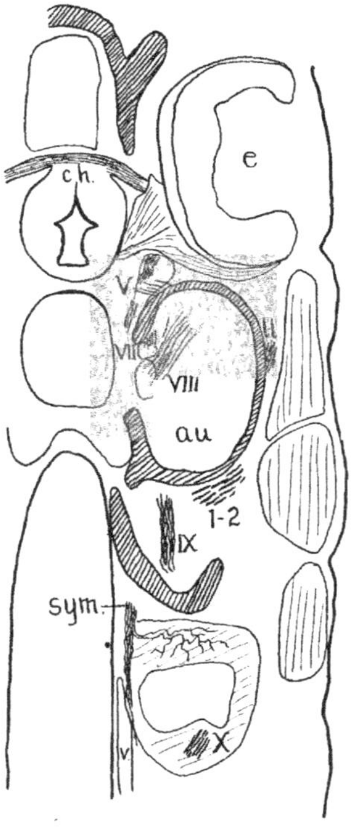

FIG. 2. Section 36.

FIGS. 1 to 11. A series of horizontal sections of the right half of the head to show the general arrangement of the nerves. Explanation in the text. Magnification, 8.5 diameters.

those indicating the nerves, which will be explained below: *a.o.*, aortic arches: *au*, auditory capsule; *m.c.*, mouth cavity; *e*, eye-ball or orbit; *nch.*, notochord; *o*, œsophagus; *r.t.*, respiratory tube; *v.*, blood vessels; *th.*, thyroid gland.

In fig. 1 appear the ganglia and roots of the V and IX nerves, the root of the VII nerve within the auditory capsule, the VII-X connective of lateral line fibers (*ll*) on the outer surface of the auditory capsule, the combined roots of the 1 and 2 ventral spinal nerves (1–2) just behind the glossopharyngeus, and a segment of the epibranchial trunk (*tr.ep.*) with the stump of the first bran-

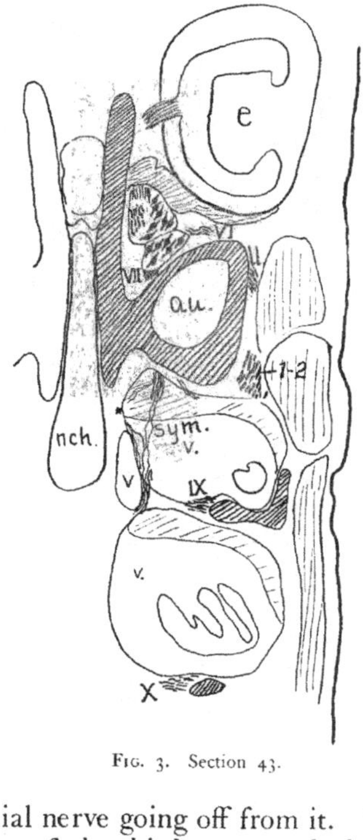

FIG. 3. Section 43.

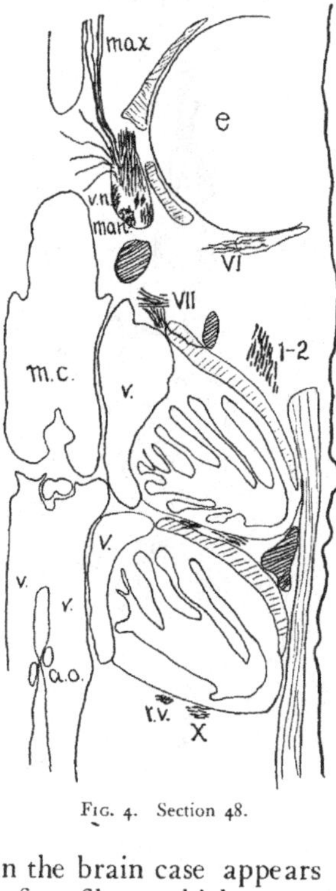

FIG. 4. Section 48.

chial nerve going off from it. Within the brain case appears the root of the third nerve and also two fine fibers which are apparently related to the meninges.

The section from which fig. 2 was drawn passes through the optic chiasma (*ch.*). A part of the root of the trigeminus is still present and in its ganglion appear the cells and fibers of the velar nerve to be described below. The fibers which cross the root and ganglion at right angles are the fibers coming from the lateral line VII ganglion and going to join the ophthalmicus profundus V (1905, p. 152). Within the auditory capsule are the roots of the VII and VIII nerves and behind the capsule the nerves spinal ventral 1 and 2, IX and X whose relations will be readily understood. Along the dorso-mesal surface of the second gill sac runs the trunk of the sympathetic (*sym*).

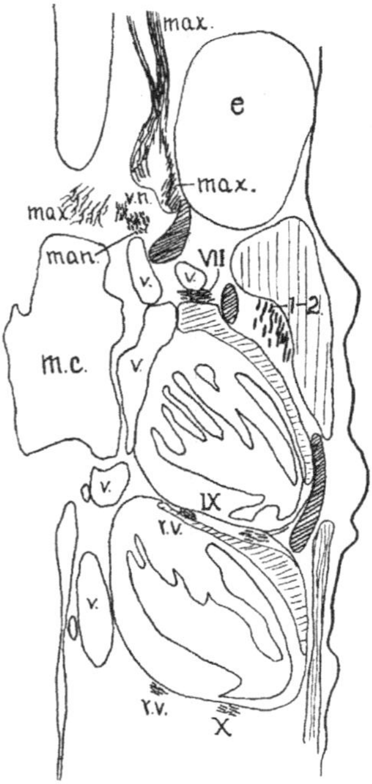

Fig. 5. Section 51.

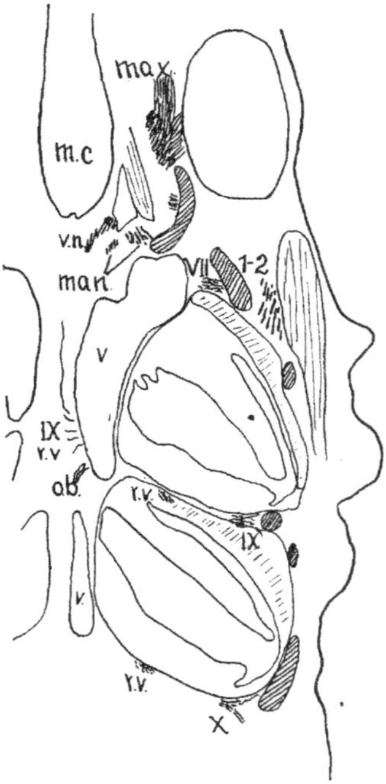

Fig. 6. Section 54.

In fig. 3 the optic nerve enters the eyeball. The trigeminus appears in three parts, at the anterior border the velar nerve, in the mesal portion the coarse fibered motor root, and in the lateral portion the ganglion of the maxillo-mandibular division. Between this and the auditory capsule lies the ganglion of the neuromast division of the VII nerve. Lateral to this is the so-called VI nerve which in the previous section comes from the trigeminal ganglion. Behind the neuromast ganglion is the root of the VII nerve proper which has emerged from the auditory capsule. Latero-caudal to the capsule are the spinal ventral nerves 1 and 2

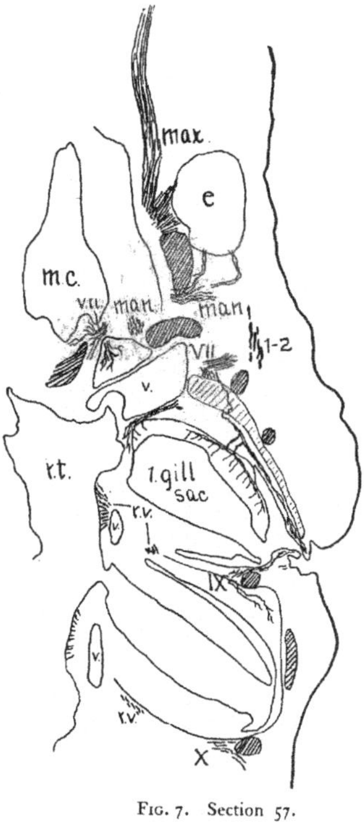

FIG. 7. Section 57.

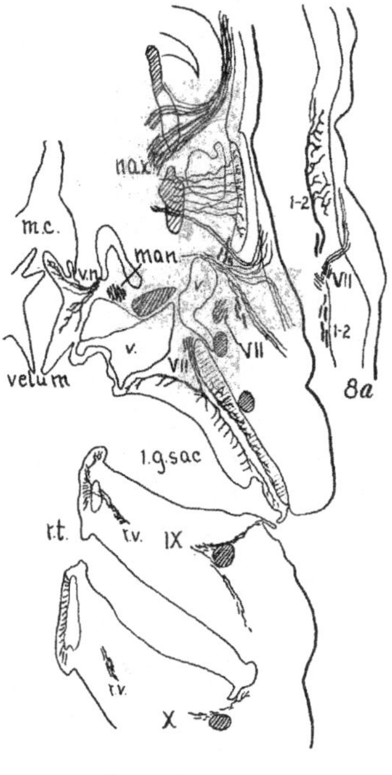

FIG. 8. Section 60.

Fig. 8a. Sections 66, 67, 68; this includes the skin and parietal muscle beneath the orbit and corresponds in position to that part of Fig. 8 opposite to which it stands.

which are passing cephalad with the ventral divisions of the first and second postotic myotomes (NEAL 1897). The trunk of the sympathetic here continues forward over the first gill sac which comes into view for the first time. The IX and first division of the X nerve are passing down mesal to the branchial cartilages.

In fig. 4 the maxillo-mandibular ganglion sends its first branches (*max.*) cephalad and mesad to supply muscles and the lining of the roof of the mouth cavity. Motor and sensory fibers seem to be intermingled. The root of the velar nerve (*v.n.*) is turning ventro-caudad and lies on the mesal surface of the ganglion. Behind it is the pure motor root of coarse fibers (*man.*). The so-called VI nerve enters the posterior rectus muscle in this section.

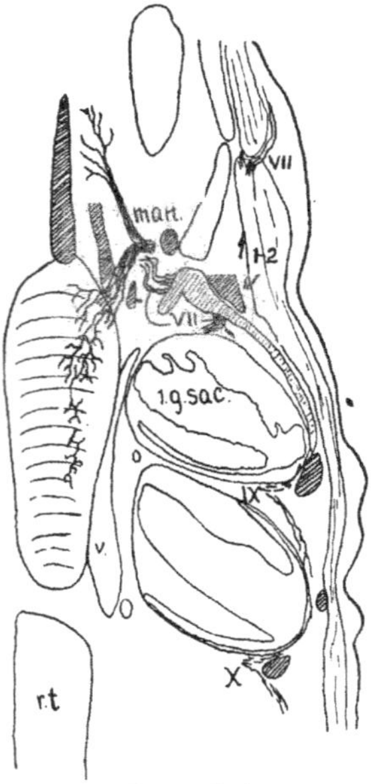

FIG. 9. Section 71.

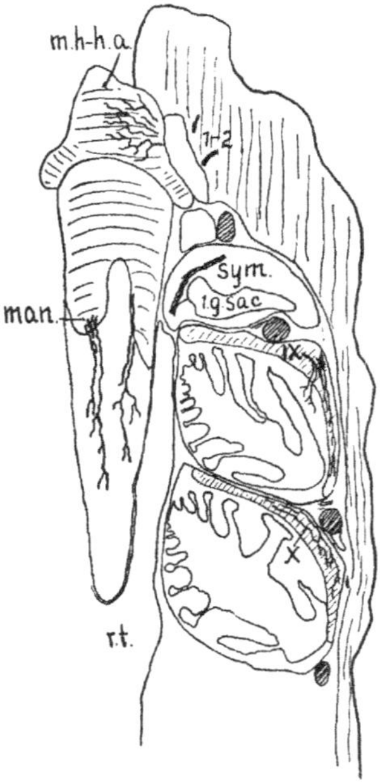

FIG. 10. Section 82.

The VII nerve gives off a large branch caudad which is the root of the sympathetic trunk. The IX and the first division of the X nerve each send mesad a visceral ramus (*r.v.*), the main trunk of the nerve continuing down as before.

Fig. 5 shows two large maxillary rami going forward. The mesal one is motor, the lateral goes directly forward to the dorsal and anterior surface of the buccal funnel and is chiefly or wholly cutaneous. The mesal fibers of fig. 4 now break up in the lining of the buccal cavity. The velar nerve (*v.n.*) and the pure motor root (*man.*) are now separate from the maxillo-mandibular ganglion. The other nerves in this drawing need no further comment.

In fig. 6 the maxillo-mandibular complex is in three parts. The large cephalic ramus runs parallel with that mentioned in fig. 5

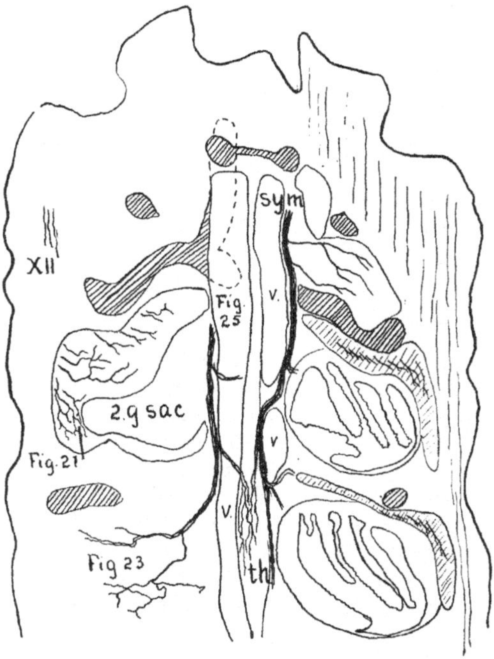

FIG. 11. Combination. On the face of this figure are indicated the areas from which Figs. 21, 23 and 25 are drawn.

and the two together doubtless innervate the skin of both surfaces of the buccal funnel, ventral to the area supplied by the ophthalmic nerve, and especially the tentacles around its border. A large lateral ramus starts to go down beneath the orbit, and a smaller ramus goes caudally and ventrally behind the orbit. The pure motor ramus of V (*man.*) has now given off a mesal branch which in fig. 7 is seen entering the m. velo-hyomandibularis internus of P. FÜRBRINGER. The dorsal branch of the visceral ramus of the IX nerve now appears in the dorsal wall of the water tube. A small bundle destined to the œsophagus is marked *o.b.* in the figure.

In figs. 7 and 8 the final branches of the maxillo-mandibular complex are shown. Large cutaneous branches go forward in addition to those mentioned above. A large number of fibers pass laterad beneath the orbit and form a rich plexus in the connective tissue covering the infra-orbital prolongation of the parietal muscle and forming the ventro-lateral wall of the orbit. From this plexus arise the fibers to the cornea and small nerves into the skin before and behind the eye. Several bundles, one of which was noted in fig. 6 as the posterior lateral branch, pass laterad behind the orbit, cross the spinal ventral nerves 1 and 2, pierce the parietal muscles and diverge to the skin beneath and behind the orbit. The velar nerve reaches its destination in these sections, entering the velum and innervating the epithelial covering of the velum and its tentacles. The VII nerve divides in fig. 7 into cutaneous and visceral rami. The visceral ramus gives off branches which supply the whole lining of the anterior half of the first gill sac with its gill lamellæ. Other branches supply the muscle sheath of this gill sac. The cutaneous ramus goes downward and forward, passes beneath the cutaneous rami of the trigeminus just described, and passes out to the skin below the orbit in several bundles (figs. 8a and 9). In figs. 7 and 8 the visceral ramus of the IX nerve reaches the wall of the water tube and joins with its dorsal branch mentioned above. It has also a ventral branch, not shown in this series of figures, which supplies the floor of the water tube (fig. 18). The visceral rami of the several divisions of the X nerve behave in a similar manner.

In fig. 9 the motor ramus of V divides into a smaller anterior branch to the protractor muscles of the "tongue" and a larger posterior branch to the circular and retractor (fig. 10) muscles of the same organ. The continuation of the visceral ramus of the VII nerve has meantime divided into two branches. The anterior of these consists of relatively coarse fibers seen in fig. 9 just lateral to the tongue muscle nerve. This supplies the m. hyo-hyoidens anterior (see below). The posterior branch of the visceral ramus gives off fibers to the lining and muscle of the first gill sac, then (figs. 10 and 11) descends beneath this sac, runs meso-caudad and then caudad and gives off branches to the ventral wall of each gill sac and to the thyroid gland and blood vessels. In figs. 8 to 11 the IX and X nerves are seen dividing into anterior and posterior branches which supply the half-gills adjacent to the arches in which the nerves run. Some fibers of these nerves also reach the skin of the ventral surface.

THE CUTANEOUS COMPONENTS.

The general cutaneous components have been traced to the skin in the Golgi preparations and the results confirm the findings in P. dorsatus. These components are found in the V, VII, IX and X nerves. Those in the facial nerve are shown in figs. 8a and 9, passing to the skin of the ventro-lateral surface below and behind the orbit. Dorsal to this area the trigeminus sends fibers much farther caudad, so that the trigeminus seems to encroach upon the innervation territory of the facialis at first dorsally, and in higher vertebrates has completely taken over the cutaneous innervation of the facial or hyoid segment. Doubtless the appearance of an operculum covering in the hyoid segment has led to the disappearance of the cutaneous component in the facial nerve.

The velar nerve should be specially described. As already indicated, its ganglion forms the antero-mesal portion of the gasserian ganglion. The ganglion cells are much smaller than those of the maxillary nerve and the fibers are uniformly medium fine. Beyond the ganglion the nerve bundle is compact and distinct from other portions of the maxillo-mandibular ganglion and roots. It turns ventro-caudad on the mesal surface of the gasserian ganglion. It runs down to the velum, remaining independent of the motor nerve of the "tongue" on the one hand and of motor and sensory branches of the maxillaris for the roof of the mouth on the other. The distribution of the nerve to the roof of the velar orifice and to the tentacles is shown in fig. 12 and the endings of its fibers in the border of the velum are drawn in fig. 13. I have looked carefully for fibers from the VII nerve to supply at least the caudal surface of the velum, but if such fibers exist they are not impregnated. Along this surface of the velum lies the m. velo-hyomandibularis externus which is inserted in the border of the velum. It is supplied by a small branch from the motor

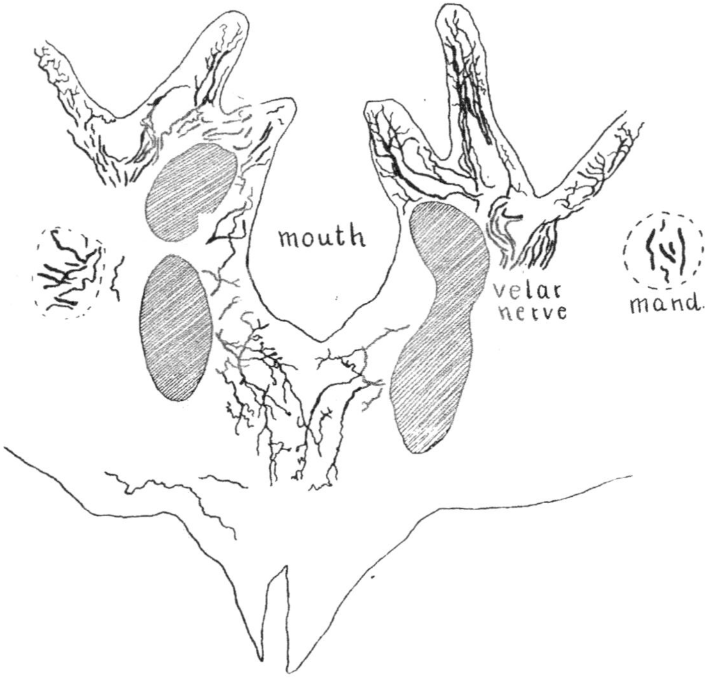

FIG. 12. A horizontal section through the velum just dorsal to its orifice. The fine branches of the velar nerve in the tentacles are not drawn. *mand.*, mandibular nerve. Magnification, 60 diameters.

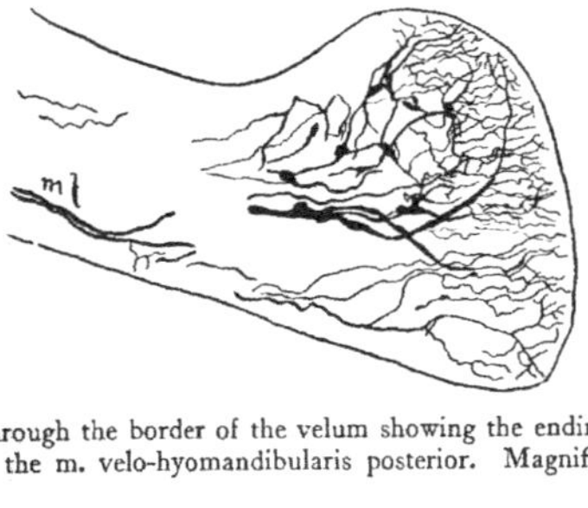

FIG. 13. A section through the border of the velum showing the endings of the velar nerve and two fibers, *mf*, supplying the m. velo-hyomandibularis posterior. Magnification, 60 diameters.

nerve of the tongue and these are the only fibers entering the velum besides those of the velar nerve.

The fibers of the lateral line system have been traced only in part and nothing is to be added to my former descriptions (1902 and 1905).

THE SOMATIC MOTOR COMPONENTS.

The results to be reported here concern the distribution and mode of ending of the motor fibers in the spinal nerves of the branchial region. The postotic myotomes persist throughout life, going to form the great parietal muscle. The ventral spinal nerves innervating this show certain features of peculiar interest.

As has been shown by the embryological studies of NEAL and KOLTZOFF, the first two postotic myotomes reach forward dorsal and ventral to the orbit so that they lie far forward from the roots of the first spinal nerves. In the number and segmental relations of the ventral spinal nerves Lampetra seems to agree closely with P. dorsatus. The first ventral spinal nerve appears to belong to the second postotic myotome and the first three myotomes are innervated by the first two spinal nerves, which are much larger than the following ones. In my previous study I was unable, with the method used, to trace these nerves very far forward with the anterior prolongation of the first myotomes. In the Golgi sections, however, several fibers of the combined trunk of the first and second nerves are impregnated and their general course is indicated in the figures above described. The endings of the fibers in the first and second myotomes are beautifully impregnated and one or two fibers are traced to the extreme anterior end of the first myotome.

The ventral nerve roots consist of a small number of very coarse fibers. Both these features have strongly impressed me since my first study of the lamprey. It is not easy to make a count of the number of fibers in a ventral nerve from sections in any of the conventional planes because, owing to the direction taken by the root, the section is never transverse to the root. In the ammocœtes of P. dorsatus a reasonable estimate of the number of fibers in the ventral roots in the branchial region seems to me to be about twenty. In Lampetra the conditions are a little more favorable for counting, because the fibers arising from the motor cells run caudad just within the ventral surface of the spinal cord and finally emerge as a compact root. On this account a transverse section of the head passing just cephalad from the root gives a transverse section of nearly or quite all of its fibers. The first two nerves are larger than the others, the second containing about 70 fibers. In six following roots counted, part on the right and part on the left side, the count varied from 31 to 40. Even a cursory examination shows that the nerves are much larger in Lampetra than in the ammocœtes of P. dorsatus, but whether this is a difference in species or a difference between the ammocœtes and the adult is an interesting question which I have no means of deciding. If we accept 40 as a reasonable estimate of the number of fibers in the ventral root, it will be seen at once that an interesting relation exists between this and the number of fibers to be innervated in the myotome.

The parietal muscle compared with other organs of the body is large in the lamprey, as in the fishes generally. Each myotome is made up of bands or plates of muscle fibers. Within each muscle band are distinguished parietal and central fibers. The parietal fibers are smaller and are provided with sarcolemma. The central fibers are without sarcolemma (?) and have the muscle fibrillæ more closely packed than in the parietal fibers. The central fibers also vary greatly in size. The number of fibers in each band varies from 7 to 8 in the small bands at the lower border of the myotome to 55 to 60 in the large bands. As the majority of the bands are large, having above 35 to 40 fibers, a fair average for the muscle bands would be about 40 muscle fibers. The total number of muscle bands in a myotome and the corresponding segment of the ventral muscle in the branchial region varied in the counts made between 107 and 120. From these figures the round number 4000 might be taken as fairly indicating the number of muscle fibers to be innervated by each ventral spinal nerve in the branchial region. In other words, each nerve fiber of the ventral spinal nerve must innervate about 100 muscle fibers. These figures are to be regarded as only a rough approximation but they express a general relation between the nerve and its muscle of which there can be no doubt. The mode of distribution and ending of the motor fibers is evidently influenced by this relation.

As the axones leave the motor cells in the spinal cord they are fibers of moderate thickness and increase somewhat in diameter before they enter the ventral roots. When the ventral roots are traced in sections stained in hæmatoxylin, they pass to the inner surface of the myotomes, and divide into dorsal and ventral rami which spread over this surface of the myotome so that the individual fibers are soon lost. It is readily seen, however, that the fibers increase still more in thickness before they reach the myotome. The great thickness of the motor fibers before they enter the myotome is well brought out in Golgi preparations. In all cases, before the fibers begin to give off branches to their endings in relation with the muscle fibers, they have acquired an enormous thickness which can best be appreciated by examining figs. 14 and 15. In fig. 15, which represents a horizontal section of the right half of the body, the size of these fibers relative to that of the myotomes and of the whole body can be seen. The fibers are a little longer in proportion to their thickness than appears here, because six sections of 75μ are projected on one plane. Some of these fibers have a thickness of 24μ, which is about one-fourth of the thickness of the largest muscle bands in the myotomes. How much the fibers increase in thickness in their course between the spinal cord and the myotome can be seen from fig. 14, in which at *A* is drawn the right half of a transverse section of the spinal cord at the same magnification as the rest of the figure. Near the ventral surface of the cord are seen the fibers of the ventral root in cross section. The colossal size of these fibers can be indicated again by calling attention to the fact that although naked fibers are measured here, they equal in diameter the coarsest medullated fibers in man and mammals. The explanation of the thickness of the fibers is found in the fact already pointed out that each motor fiber must supply a large number of muscle fibers.

Figs. 14 and 15 show two forms of nerve endings in the myotomes. One of these is so different from typical motor endings that it was at first thought that it might serve the muscle sense. Both kinds of fibers, however, are traced with certainty to the ventral roots and furthermore the larger part of the myotome is free from the typical endings and must be innervated by the simple endings to be described below.

Typical motor endings.—Those fibers which spread over the inner surface of the myotomes present endings which, although simple in form, resemble typical motor end plates. In figs. 14 and 15 are shown the spreading of such fibers and their ending in short branches or knobs. It is seen that the fibers do not penetrate far into the myotome but spread around the ends of the myotome in the intermuscular septa. Here the end plates are found and it may be said that most of these endings stand in relation with the ends of the muscle fibers. To understand the

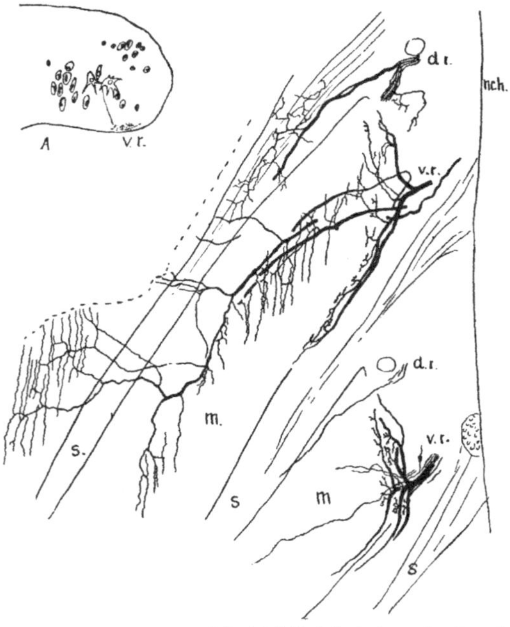

FIG. 14. A horizontal section of part of the left half of the body passing through the notochord to show the relation of the motor fibers to the myotomes. *d.r.*, dorsal roots; *v.r.*, ventral roots, *m.*, myotomes, *s.*, intermuscular septa; *nch.*, notochord. In the intermuscular septum above, a musculo-sensory ending (?). At *A* is shown the cross section of the right half of the spinal cord at the same magnification as the rest of the figure. The motor cells and the Müllerian fibers may be compared with the peripheral motor fibers. Magnification, 40 diameters.

disposition of the fiber drawn in fig. 15 it must be remembered that the inner surface of the myotomes slant obliquely outward and downward from the vertebral region, so that the endings shown actually lie on the ventro-mesal surface of the myotomes, and are drawn as if seen through the mesal part of the myotomes. I have not drawn any of the end plates of these fibers at a high magnification, but they are essentially similar to the end plates in the specialized muscle (see fig. 20). The endings in the specialized muscles are more highly developed than these in the myotomes.

Simple end-branching in the myotomes.—In figs. 14 and 15 are shown other fibers which plunge directly into the myotomes. These sometimes divide into two large branches. The coarse fibers then run through the myotome toward the external surface

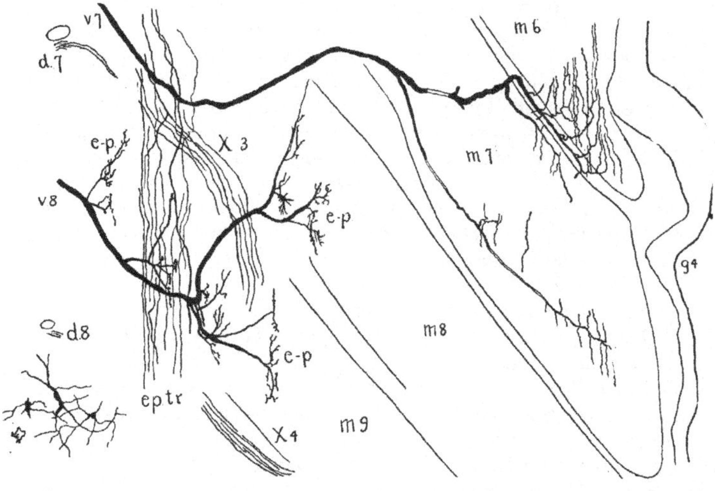

FIG. 15. A section similar to the last, taken from the right side of the body. At the right is the outer surface, at the left the section does not quite reach the notochord. The proximal portion of the motor fibers, the epibranchial trunk and the dorsal nerves were drawn in from sections dorsal to that from which the main part of the figure was taken. *d.7, d.8,* dorsal nerves; *v.7, v.8,* ventral nerves; *e-p.,* motor end-plates; *ep.tr.,* epibranchial trunk (vagus-lateralis-hypoglossus complex); X_3, X_4, third and fourth branchial nerves of the vagus complex coming off from the epibranchial trunk; *m.6, m.7,* postotic myotomes; *g.4,* point dorsal to the fourth gill opening. At the lower left hand corner four motor cells from the spinal cord drawn at the same magnification. Magnification, 40 diameters.

of the body and give off numerous branches to either side which run parallel with the muscle fibers. These final branches are moderately slender, slightly varicose fibers which show no sign of any special end organ. I presume that they penetrate the connective tissue separating the muscle bands and lie upon the muscle fibers. In fig. 14 part of two fibers of this kind are drawn. One of them divides into two branches. One branch is at first relatively slender and thickens greatly as it enters the myotome. Only a short part of this branch is drawn. It continues in the next adjacent section parallel with the other fibers and sends off a great number of branches which, if inserted here, would have rendered the drawing very obscure.

These two forms of endings and their distribution doubtless indicate the beginning of specialization in the myotomes. As the mesal portion of the myotome is the first part to develop muscle fibrillæ in the ontogeny and as this part becomes the muscle plate in the embryos of higher vertebrates, so here the mesal part of the myotome is supplied with special motor endings, while the lateral and greater part has only the simple end branches. In the phylogeny, with the development of hard parts in the vertebral column, the adjacent mesal parts of the myotomes come to be especially efficient in body movements and so become specialized skeletal muscles. The beginning of this is at least foreshadowed in the lamprey.

Segmental relations of the ventral nerves.—In figs. 14 and 15 it is clearly shown that the motor fibers of both types may innervate two myotomes. These nerves are not impregnated in a sufficiently large number of segments or of specimens to enable me to say whether the distribution of one fiber to two myotomes is common, but the fact that the majority of the fibers which are impregnated send branches to two myotomes is sufficient indication that it is at least a normal arrangement. The significance of this seems to the writer to lie not so much in its bearing on questions of segmentation as upon the question of the factors determining the distribution of nerve fibers and directing them to their endings. There seems to be no definite or constant arrangement of these motor fibers. They pass in a haphazard fashion to one or two myotomes, branch once, twice or three times, etc. In studying the peripheral nerves of Amphioxus with methylene blue I gained the general impression that the nerves in that animal showed still less regard for segmental relations. The sensory nerves run more or less straight ventrad over several obliquely placed myotomes and present endless variations in the way in which they reach the same general area of distribution. The obstacle in the way of HENSEN and his followers accepting HIS's theory of the outgrowth of nerve fibers has been that they can not see how peripheral nerve fibers can find their way out from the central nervous system to the organs which they should innervate. To remove this obstacle it is only necessary to recognize that peripheral fibers do not, "unerringly" as HENSEN said, find their way where they "should" go. Recent studies of regeneration give evidences of this and the facts here presented have, I believe, the same significance. In these lowly organized vertebrates the nerve fibers push out much as the pseudopodia of a protozoön are thrust out, and the word haphazard can be applied to the nerve fibers with whatever truth can be claimed for it when applied to pseudopodia. As the pseudopodia of a motor neuroblast grow out they are directed by the forming organs of the body and perhaps by chemical influences. It appears that they do not always take the same course or reach the same end. They go sometimes to one myotome, sometimes to two; sometimes the greater part of a fiber remains in the first myotome which it enters; sometimes it gives off smaller branches to this myotome and runs through it to enter a second. It should be noticed that in the myotome stage of body musculature about the only thing necessary in the way of definite and constant innervation is a general bilateral symmetry. All that is secured is an alternating contraction of the muscles of the two sides passing along the body in waves, giving the fish the undulating movement by which it swims. As the muscles become specialized in the phylogenetic series—and the same is true of other organs—the influences directing the course of nerve fibers as they grow out increase in definiteness to keep pace with the evolution of the organism. Indeed, this is one factor upon which survival would depend. At no stage of evolution, however, so far as the writer can see, is it necessary to suppose that nerve fibers should be unerringly directed to their proper destinations. Why should not some nerve fibers go astray like sheep and be lost? Why should nerve fibers be exempt from the otherwise universal law, the law of occasional failure? Does not the method of trial and error hold here?

ENDINGS OF MUSCLE SENSE.

Occasionally there are found in the intermuscular septa free endings of fibers of the dorsal spinal nerves. One is shown in fig. 14. These fibers seem to be in the proper position to serve the muscle sense. Might not these fibers be stimulated by the contraction of the adjacent myotomes and might not the impulses sent in by them serve to bring about inhibition on the one side and the contraction of the corresponding myotomes of the opposite side ?

THE VISCERAL SENSORY COMPONENTS.

These components can be treated best by describing the sensory portions of the IX and VII nerves. The X nerve need not be described, since its branchial divisions after they leave the epibranchial trunk (see 1905 paper) are like the IX in every way. The last two divisions of the vagus have not been studied in these preparations.

The roots of the IX nerve emerge from the cranium just behind the auditory capsule and enter the ganglion. Beyond the ganglion the sensory and motor fibers cannot be distinguished until they reach their destinations. The trunk descends over the first gill sac and enters the first branchial arch. Here it descends mesal to the branchial cartilage (figs. 1 to 3) and gives off a visceral sensory ramus mesad (figs. 4 to 6). The visceral ramus reaches the wall of the water tube somewhat above its middle and divides into dorsal and ventral branches. The dorsal branch runs up in the wall of the water tube to its dorsal surface where it forms a rich ramification (fig. 16) extending forward in the roof of the water tube to the velum. In fig. 16 it is seen that the water tube is much wider opposite the gill bars than between the gill pouches and this accounts for the arrangement of the fibers. Some of the fibers of this branch do not stop in the roof of the water tube but pass on up to the wall of the œsophagus. A special bundle of such fibers is marked *o.b.* in fig. 16 and its continuation is shown in fig. 17, where it ramifies in the wall of the œsophagus. Fig. 16 is taken from the same section as fig. 6, where this same bundle is indicated (*o.b.*). These endings in the wall of the water tube and of the œsophagus are slender simple fibers or bunches of fine fibers branching out from a small knot or varicosity. Many of the fibers are so excessively fine that they could not be drawn at this magnification.

The ventral branch of the ramus visceralis descends in the wall of the water tube and ramifies in its floor (fig. 18). Here again very many of the finest fibers are omitted from the drawing. Taste organs are present as in P. dorsatus but are not well impregnated.

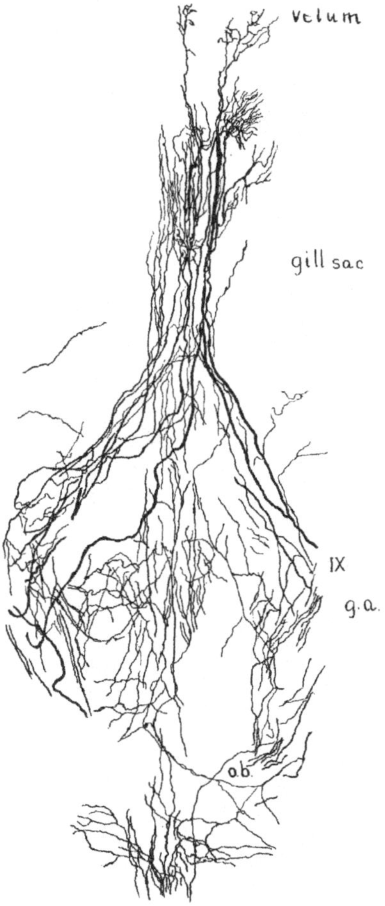

FIG. 16.

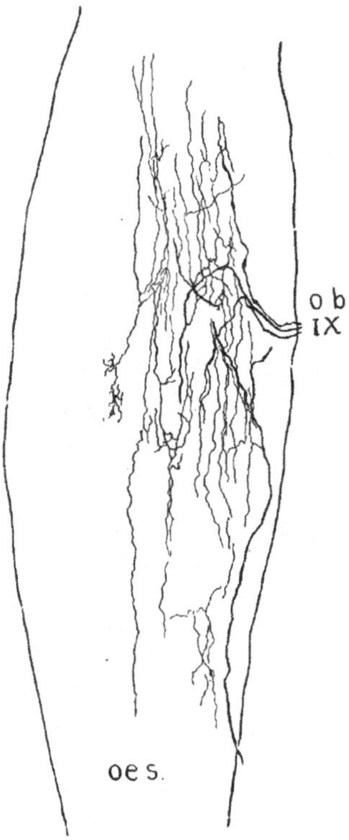

FIG. 17.

FIG. 16. Ramification of the dorsal branch of the ramus visceralis of the IX nerve in the dorsal wall of the respiratory tube. *g.a.*, gill arch; *o.b.*, œsophageal bundle continued in fig. 17. Magnification, 50 diameters.

FIG. 17. Ramification of a branch, *o.b.IX*, from the ramus visceralis IX in the œsophagus (*oes.*). Magnification, 80 diameters.

The trunk of the IX nerve continues down in the branchial arch (figs. 7 to 11) and gives off twigs and single fibers but no large branches. The fibers supply the gill lamellæ and whole lining of the posterior half of the first gill sac, forming rich net-

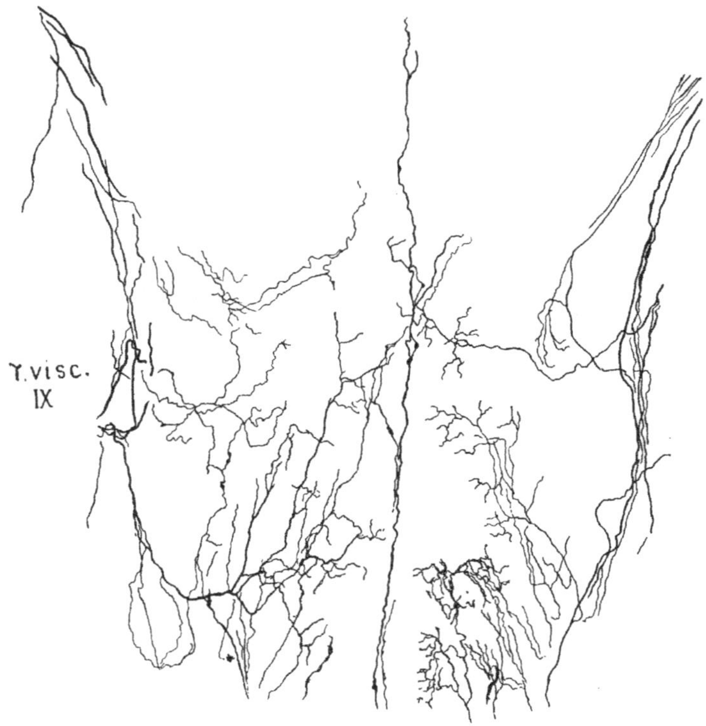

Fig. 18. The ventral branch of ramus visceralis IX ramifying in the floor of the respiratory tube. Magnification, 60 diameters. The finer fibers could not be drawn.

works around both internal and external gill openings, meeting and interlacing at these places with the fibers of the VII nerve. The innervation about the external opening of the first gill sac is shown in fig. 19. This is a horizontal section and the fibers approaching from below are those of the IX nerve, those from above are fibers of the VII nerve. In Fig. 19 *A* is shown the interlacing of the fibers of the two nerves as it appears in a section directly dorsal to the gill opening. Fibers going forward from the IX nerve supply also the thin sheet of muscle covering the posterior wall of the first gill sac.

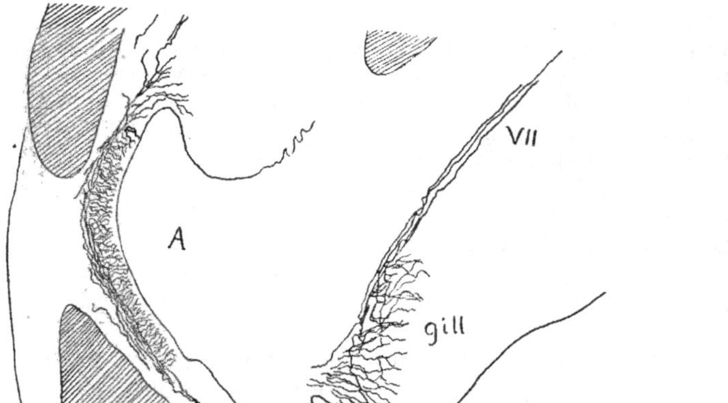

FIG. 19. A horizontal section through the external opening (*ex.o.*) of the first gill sac on the left side, showing the branches of the VII and IX nerves related to it. *b.v.*, blood vessel. At *A* is shown a section passing just dorsal to the same gill opening to show the intermingling of the VII and IX fibers in the epithelium. Magnification, 60 diameters.

After the above fibers are given off the nerve passes down behind the branchial cartilage and supplies the muscular sheath and the lining of the anterior half of the second gill sac. In studying the cranial nerves of P. dorsatus I did not appreciate the importance of the caudal branch of this nerve "which disappears in the muscles" (1905, p. 171). I have reviewed those sections and find this to be a large branch which enters the muscular sheath of the second gill sac in which it dwindles away. The two species agree, then, in that the IX nerve innervates the muscular wall of the halves of the first and second gill sacs adjacent to the first branchial arch in which the nerve runs. Whether this same branch sends fibers into the lining of the second gill sac in P. dorsatus I can not determine. If such fibers are present and are disposed in small bundles or singly as in Lampetra, they could not be traced in transverse sections. It is altogether probable that the arrangement is the same in both species, especially as no other provision is seen for the innervation of the anterior half of the second gill sac.

In P. dorsatus the terminal portion of the IX nerve which goes to the skin is larger than the posterior branch just referred to. In Lamptera a few fibers to the skin, probably general cutaneous, are impregnated in these preparations but the lateral line fibers which presumably run throughout this portion of the nerve (see 1902, p. 47 and 1905, p. 172) are not impregnated.

The VII nerve has its communis or visceral sensory ganglion within the auditory capsule. The trunk, including visceral, cutaneous and motor components, emerges from the capsule and descends on the caudal surface of the lateral line ganglion, bends caudad beneath the capsule and divides into two chief rami. One of these running straight caudad will be described below as the sympathetic trunk. The other is the main trunk of the VII nerve. It descends in front of the first gill sac and at about the level of the dorsal border of the velum (fig. 7) divides into anterior and posterior branches. The anterior branch has been described above as the cutaneous branch. The posterior branch bends inward and backward and gains the inner border of the muscle sheath of the first gill sac (figs. 7 and 8). At the same time it gives off branches which, together with others given off lower down, supply the lining of the whole anterior half of the gill sac, interlacing with the terminal fibers of the IX nerve about both the internal and external gill openings as above described. Other fibers innervate the muscle of the anterior wall of the gill sac. The nerve then gives off the branch to the m. hyo-hyoideus anterior, mentioned above and moves onto the postero-internal surface of the gill muscle farther from the middle plane than before (fig. 9). Here it continues to give off branches to the lining of the gill sac and to the muscle until it reaches the lower border of the gill sac. Here it turns ventro-mesad and runs caudad on the surface of the inferior jugular vein (figs. 10 and 11). The course of this terminal portion of the VII nerve is beautifully impregnated in one transverse and one horizontal series of sections. The transverse sections include only the first two gill sacs and the horizontal sections the first three sacs, and the nerve continues to the end of the series in each case. As it goes backward it gives off branches laterad to ramify on the ventral wall of each gill sac and branches mesad to the walls of the blood vessels and to the thyroid gland.

From this account it is seen that the visceral sensory portion of the VII nerve supplies the lining of the anterior half of the first gill sac. Since the hyomandibular sac, which lies in the embryo in front of the arch in which the VII nerve runs, is aborted, the branchial portion of the nerve corresponds to the posterior branchial branches of the IX nerve. The continuation of the nerve caudad beneath the gill sacs will be considered farther under the head of the sympathetic system.

The supply of fibers from all the branchial nerves to the lining of the gill sacs, lamellæ and filaments is very rich indeed. Great numbers of medium and fine fibers interlace beneath the gill lamellæ and send up fine fibers along the filaments to end between the epithelial cells. It is very difficult in a drawing on one plane to give an adequate idea of the richness of the nerve supply to the gills. The lining of the sac also is everywhere very richly supplied with fibers.

THE VISCERAL MOTOR COMPONENTS.

The motor fibers in the X, IX and VII nerves supplying the gill sacs have been described in a general way in the above paragraphs. As the nerve descends in its branchial arch it gives motor fibers to the half of each adjacent gill sac. The muscle in the anterior wall of the gill sac is much thicker than that in the posterior wall and extends over the dorsal and ventral walls; consequently the posterior branches of each branchial nerve carry many more motor fibers than the anterior branches. In the case of the VII nerve there is no gill sac in front of it to be innervated, but it sends one motor branch forward which goes ventro-mesad in a thick muscle (fig. 10, *m.h-h.a.*) which with its fellow forms a sling depending from the cornual cartilages, in which the front end of the circular muscle of the "tongue" rests. This is the m. hyo-hyoideus anterior of P. FÜRBRINGER which he states is innervated by the internal ramus of the maxillaris. This muscle appears to the writer to be much less closely associated with the large circular muscle of the tongue (hyo-hyoideus posterior) than FÜRBRINGER'S account implies. It would seem to have a special function to raise the tongue during the rasping and sucking movements. The fact that it receives its innervation from the VII nerve is stronger evidence against its being considered a part of the circular muscle, which is innervated by the trigeminus.

It was pointed out in a previous paper (1905) and was already clear from the descriptions by P. FÜRBINGER that the trigeminus of cyclostomes is peculiar in that the maxillary ramus contains motor fibers. The maxillaris indeed innervates the majority of the muscles connected with the whole buccal apparatus and P. FÜRBRINGER failed to recognize a ramus mandibularis, assigning all branches below the ophthalmic ramus to the maxillaris. This was an error. I have shown in the general description above that a pure motor component of the trigeminus can be distinguished in the root and ganglion by its compact rounded form and the coarseness of its fibers. This bundle supplies the internal and external velo-hyomandibular muscles and the protractors, retractors and circular muscle of the tongue. Now if the lingual apparatus corresponds to the lower jaw of higher forms, it is proper to consider this nerve homologous with the motor part of the mandibular ramus of true gnathostomes.

Although my preparations leave some uncertainties, all the other muscles about the buccal and mouth cavities seem to be innervated by branches of the maxillaris.

My preparations are not suitable for the reconstruction of the muscles and I will not attempt any further description of the muscles and their innervation or discussion of their action. I am convinced, however, that in many respects FÜRBRINGER'S descriptions will not apply to Lampetra. It is much to be desired that some one will undertake a thorough study of the muscles and skeleton of our American species of petromyzonts, both before and after the metamorphosis, by the method of serial sections and reconstructions.

Motor endings of the trigeminus.—The fibers of the mandibular ramus are coarse in the ganglion, while the motor fibers of the maxillaris can scarcely be distinguished from the sensory until they reach their muscles. The motor fibers in both maxillaris and mandibularis, however, increase in thickness greatly before they enter their muscles. Many of them are nearly as thick as those of the spinal nerves described above. It is possible in only a few cases to trace out anything like all the branches of one fiber in the muscle, but it is evident in all regions that relatively few fibers enter any given muscle and that each fiber must supply a large number of muscle fibers. In many places enough can be seen of the early divisions of a motor fiber to show that it has a wide distribution and in a few cases some idea can be obtained of the number of end plates supplied by one fiber. In one horizontal section through the tongue muscle one fiber of the mandibular nerve has thirty-nine end plates and in the adjacent section are forty-one more end plates which certainly belong to the same fiber.

The motor end plates are more highly developed in the buccal and lingual muscles than elsewhere. The muscle fibers in the specialized muscles are smaller and more uniform in diameter than those of the myotomes, and the motor end plate commonly is as long or wide as the width of the muscle fiber. The most common form of end plate is something like that of a horseshoe, although the appearance of a closed ring or network is sometimes given by the free ends of the horseshoe overlapping. Several end plates are drawn at a high magnification in fig. 20, which shows another peculiarity, namely, the arrangement of the end plates in chains. This is by no means uncommon but is perhaps not true of the majority of the endings. There are hundreds of such end plates impregnated in these preparations and in all the specialized muscles they are of the same general form. Sometimes they are reduced to a single knob or two and seldom are they any more complex than the two separate ones shown in fig. 20.

In the branchial muscles the endings are simpler but are always characteristic enough, I think, to enable one to distinguish between motor and sensory fibers. The motor fibers are not so thick as are those of the spinal nerves or those of the trigeminus. Even the fibers of the VII nerve which supply the m. hyo-hyoideus anterior, a thick and active muscle, become much thicker than

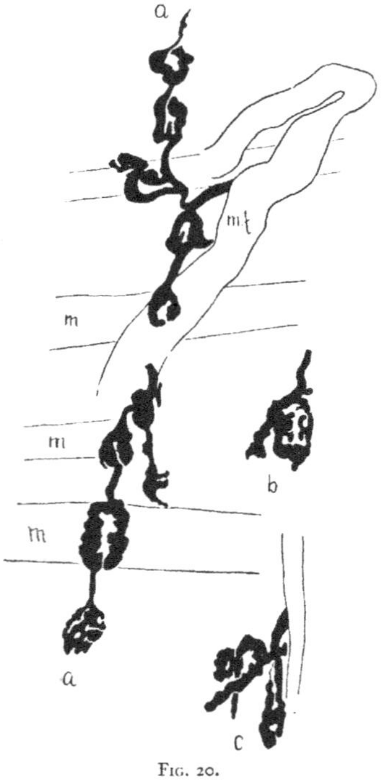

FIG. 20.

FIG. 21.

FIG. 20. Motor end plates from specialized muscles. *a.a,* two chains of end plates connected with one motor fiber, *m.f.* *b,c,* somewhat more complex end plates; *m,* muscle fibers. Magnification, 300 diameters.

FIG. 21. Motor endings in the sheath-muscle of a gill sac. The source of this figure is shown in Fig. 11. Magnification, 150 diameters.

those which supply the branchial muscle. Still the motor fibers are much thicker than the sensory fibers to the lining of the gill sacs. The fibers ramify freely and the functional endings consist of varicosities in the course of the terminal branches and simple knobs on the ends of small branches. This is illustrated in fig. 21 taken from the ventro-lateral wall of the third gill sac.

THE SYMPATHETIC SYSTEM.

The sympathetic trunk.—This has been mentioned above as a bundle of fibers going directly caudad from the trunk of the VII nerve just beneath the auditory capsule. In position this corresponds closely with the sympathetic trunk in P. dorsatus. As the nerve runs caudad it maintains a position a little farther laterad than in P. dorsatus, never approaching so close to the aorta as in that species. Further, the trunk in adult Lampetra is very much larger than in the ammocœtes of P. dorsatus. In addition to this trunk the facialis at its ventral end sends caudad another nerve which extends through at least three branchial segments. This whole system of fibers seems to be related chiefly or exclusively to arteries, veins and blood and lymph sinuses, and it is for this reason that it is spoken of as a sympathetic system. Morphologically the dorsal trunk corresponds to the sympathetic trunk in higher forms and in at least one place ganglion cells are impregnated in it, which confirms the description given for P. dorsatus. The ventral prolongation of the VII nerve caudad, on the other hand, has no parallel known to the writer.

The general course of these nerves is shown in figs. 2 to 11. In fig. 22 is shown the sympathetic trunk as it lies over the first gill sac, from a section between those drawn in figs. 3 and 4. The trunk has just left the VII nerve and is running over the muscle in the anterior wall of the first gill sac. The dotted lines indicate the outline of the sinus at the base of the gill. The full line to the right indicates the border of the muscle whose other border is the sinus. At first sight the fibers given off were taken to be motor fibers, but it was found that the fibers and endings differ from any that are known to be in muscle, that the fibers pierce the muscle to the wall of the sinus beneath, and that in other sections the endings of similar fibers in the walls of blood vessels is perfectly clear. In this figure the fibers which run down toward the left run in a fold or constriction of this sinus and are lost without coming near any other organ whatever. After careful study in all parts of the sections it may be stated with confidence that these fibers are in no way related to muscles. Wherever the muscle upon which they lie is cut transversely the fibers are seen to pierce the muscle and lie directly on the wall of a vessel or sinus. Transverse sections through the very region shown in fig. 22 show this very clearly. It is not so easy to prove whether these fibers may or may not be related to the lining of the gill sac or the epithelium of the gill filaments. The fibers which supply the greater part if not the whole of the lining of the gill sacs, enter the sacs on their

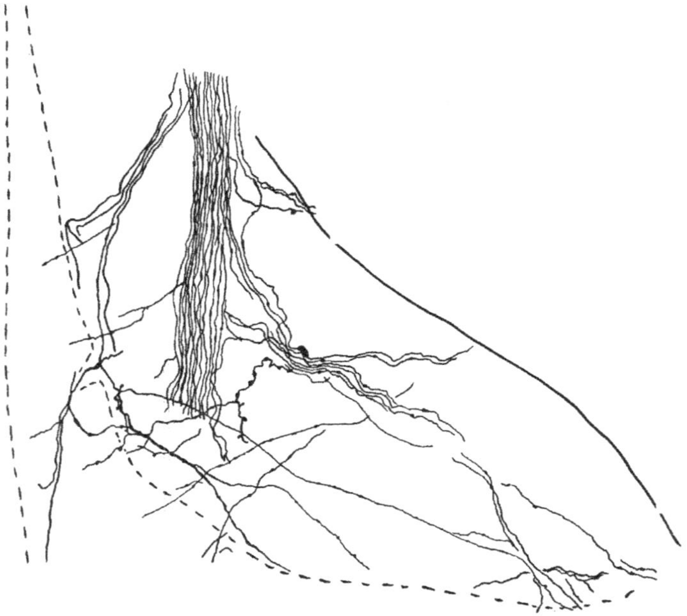

Fig. 22. The proximal portion of the sympathetic trunk of the right side, just behind the auditory capsule and over the first gill sac, in a horizontal section. Magnification, 90 diameters.

anterior and posterior surfaces as above described and spread over the whole wall. The fibers of the sympathetic trunks arrive upon the walls of the gill sacs dorsally and ventrally, but here in many cases at least they mingle with the network of interlacing fibers from the former source and it is no longer possible to distinguish them with certainty. By direct observation the possibility can not be excluded that fibers from what are here called sympathetic trunks supply visceral surfaces merely. However, the facts that these fibers do supply blood vessels and that the fibers which are known to be visceral sensory enter the gill sacs by another route lead me to conclude that when fibers from the sympathetic trunks enter the gills they are still destined to the supply

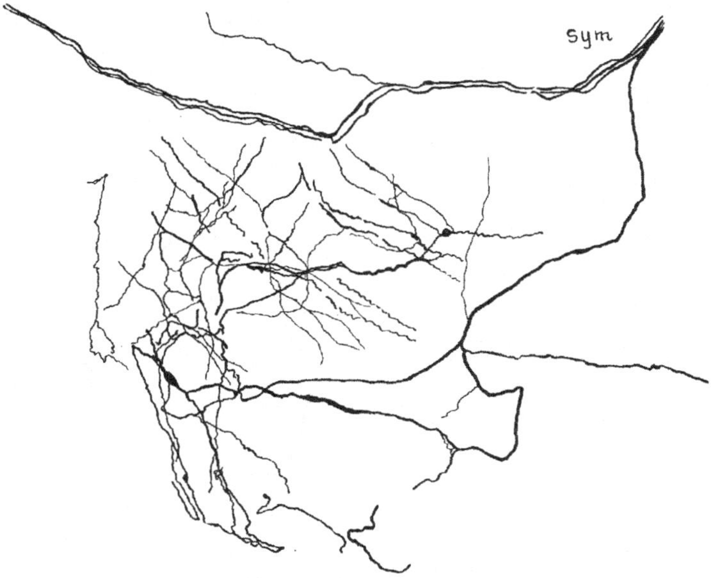

FIG. 23. A branch of the ventral sympathetic trunk of the left side ramifying in the ventral wall of the third gill sac. Compare fig. 11. Magnification, 75 diameters. Many fine fibers are not drawn.

of blood vessels found there. An illustration of the ramification of these fibers on the wall of the sinus at the base of the gills is given in fig. 23, taken from one of the sections drawn in fig. 11 where the position of this bundle will be seen. It lies beneath the third gill sac on the left side and is a branch from the ventral trunk.

Fig. 24.

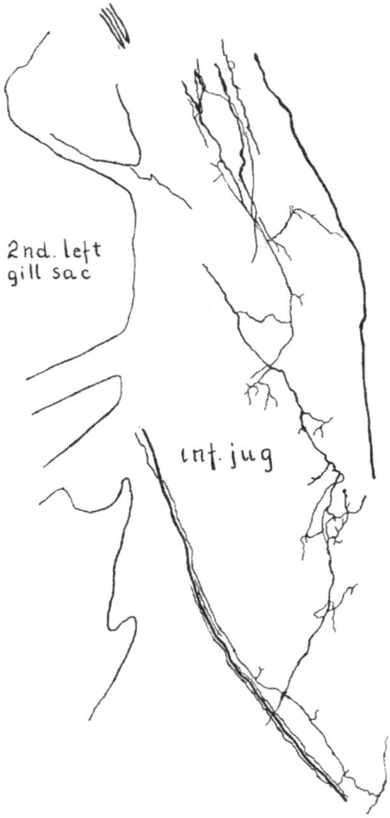

Fig. 25.

Fig. 24. A longitudinal section of the wall of the aorta in a horizontal section of the whole head, showing two fiber endings. Magnification, 250 diameters.

Fig. 25. Endings from the ventral sympathetic trunk of the left side in the inferior jugular vein, *inf. jug.* Magnification, 75 diameters.

For illustration of the endings of the fibers in the walls of blood vessels I have taken one from the wall of the aorta (fig. 24) and one from the wall of the inferior jugular vein (fig. 25). The ending in the aorta is seen in longitudinal section and many such endings are to be found. The fibers which lead to them come from the direction of the sympathetic trunk. The endings in the jugular vein are seen in surface view and all the fibers in the drawing belong to the ventral trunk of the sympathetic. There is no possibility of doubt as to these endings being in the wall of the vein, for the next section above passes through the empty space dorsal to the vein and the next section downward passes through the lumen of the vein containing blood corpuscles. From these facts it may be concluded that in the adult lamprey a large system of fibers makes its exit from the brain with the VII nerve which functions in connection with the control of the circulation at least in the branchial region. Whether fibers join this from other nerve roots still remains questionable, and also to what extent the system is developed in the post-branchial region.

Peripheral ganglia.—Ganglion cells are found impregnated in various places about the head in several series of sections. One is shown in close connection with a ventral spinal nerve root in the lower part of fig. 14. This is the only one I have noticed in this position. A few cells are seen in connection with the sympathetic trunks above described, but not so many are impregnated in connection with these trunks as one would expect. Many cells are found immediately beneath the parietal muscle ventral to the orbit. In one series of sections more than one hundred sharply impregnated cells were counted in fifteen sections through the region ventral to the orbit and lateral and ventral to the subocular cartilage. In fig. 26 are shown some cells in this region. The vertical striations bounded by a dotted line indicate the parietal muscle, the oblique striations the muscle of the first gill sac. The fibers lying on the gill sac are motor, except the finest, which are sympathetic. At the border of the parietal muscle is seen a bundle of fibers which is probably the termination of the hypoglossus. There is one nerve cell among these fibers and farther forward are three cells among fine fibers most of which are of sympathetic nature. These latter cells and fibers are drawn to a higher scale in fig. 27. There are some indications that the fine fibers among which such cells lie follow blood vessels. In fig.

28*a* are shown several cells from the connective tissue beneath the subocular cartilage of the opposite side of the same specimen. The coarse fibers are those of the first ventral spinal nerve nearing their ending in the most anterior portion of the parietal muscle. The cell in the lower right hand corner is more highly magnified

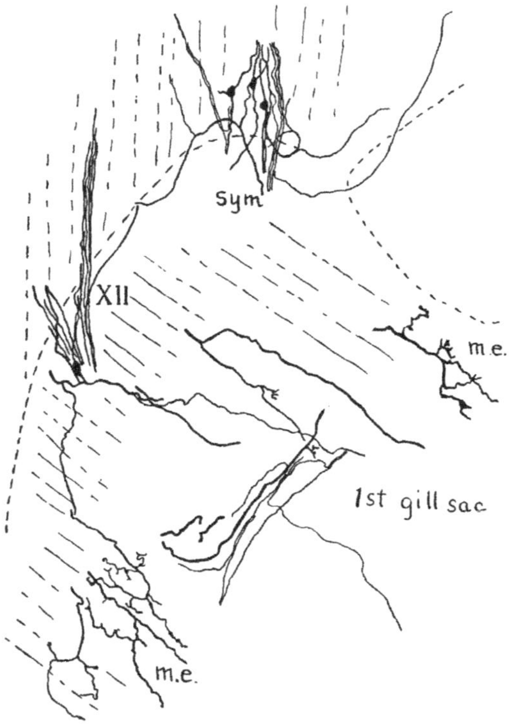

FIG. 26. Motor endings in the ventral wall of the first gill sac of the left side and peripheral ganglion cells. *m.e.*, motor endings; *sym.*, sympathetic fibers and ganglion cells. Magnification, 75 diameters.

at *b*. Many cells are found about the gill sacs, some in the base of the gill filaments (fig. 28*e* and fig. 29) and some on the surface of the muscle (fig. 28*c*). A considerable number of such cells are found in the roof of the mouth cavity beneath the semilunar cartilage, in the connective tissue around the retina especially among the trigeminal fibers in the floor of the orbit, in the subcutaneous tissue dorsal and anterior to the olfactory capsule, etc.

The writer has heretofore been very skeptical regarding the presence of ganglion cells in such places as those mentioned. When these were first seen they were taken to be large varicosities, but thousands of varicosities have been compared with these and all are very much smaller in proportion to the thickness of the fibers on which they occurred. Then it was found that while the majority of these cells are bipolar, a considerable part of them have three, four or five processes. Finally there are in many

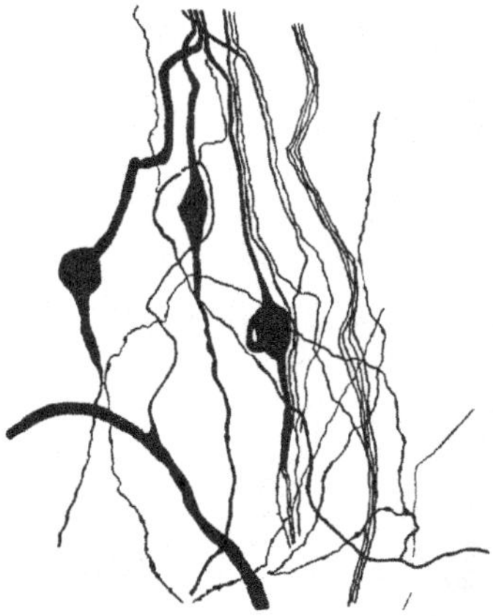

FIG. 27. The ganglion cells of fig. 26 under a higher power. Magnification, 300 diameters.

cases, not in all, the same differences between the processes as we find between the dendrites and neurites of neurones in the central nervous system.

These cells have not yet been exhaustively studied and I cannot state with any confidence either their anatomical connections or their probable functions. They are found both beneath the skin and beneath the mucosa, in the gills and on the surface of both branchial and parietal muscles, in or near the trunks or branches of the V, VII, IX, X, ventral spinal and hypoglossal (?) nerves. In a few cases there are indications that these cells are especially related to blood vessels and this is the only supposition that seems in harmony with their wide distribution.

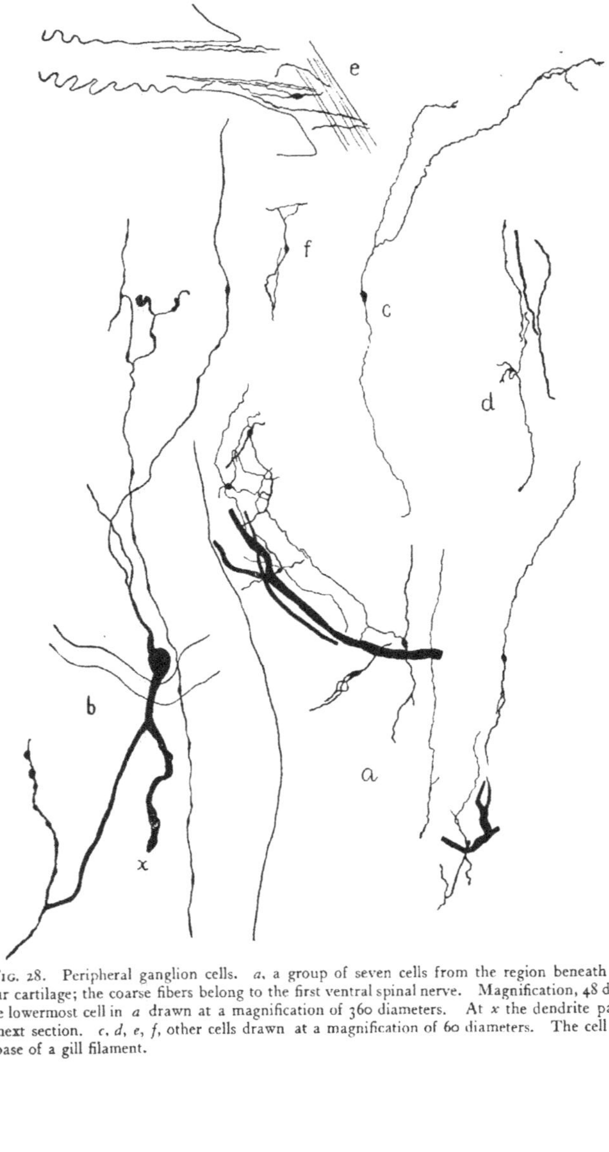

FIG. 28. Peripheral ganglion cells. *a*, a group of seven cells from the region beneath the subocular cartilage; the coarse fibers belong to the first ventral spinal nerve. Magnification, 48 diameters. *b*, the lowermost cell in *a* drawn at a magnification of 360 diameters. At *x* the dendrite passes into the next section. *c*, *d*, *e*, *f*, other cells drawn at a magnification of 60 diameters. The cell at *e* is in the base of a gill filament.

CALIBER OF FIBERS IN THE LAMPREY.

A word should be said regarding the relative thickness of fibers in the peripheral nerves. For this the different figures should be compared and notice taken of the magnification in each. In drawing there is a constant tendency to exaggerate the thickness of the fine fibers and in the case of the finest it becomes physically impossible to represent them except at high magnifications. The thickness of the fibers of the ventral spinal roots has been dwelt upon. The dorsal roots (fig. 31) contain medium (5.2μ), fine (1.5μ) and very fine (0.3μ) fibers. The coarse motor fibers in the muscles are approximately seventy-five times as thick as the finest fibers in the dorsal roots. In the base of the gills and in the walls of the water tube and œsophagus and blood vessels the

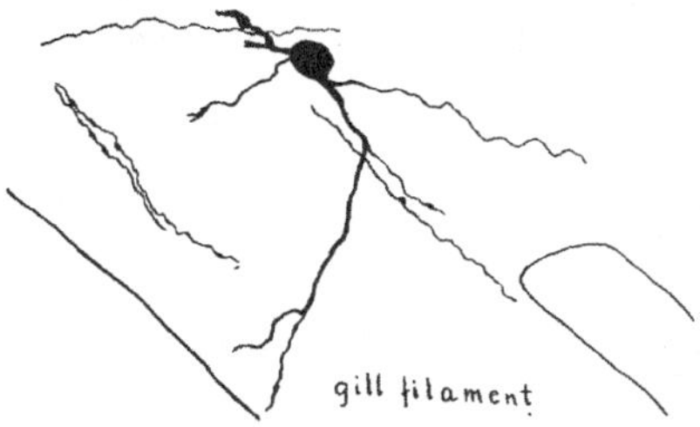

FIG. 29. A ganglion cell at the base of a gill filament. Magnification, 250 diameters.

terminal fibers are just visible under the LEITZ ocular 4 and objective 7. The coarser fibers of the sympathetic trunk are between two and three microns in diameter.

SUMMARY.

The chief results of this study are as follows:

1. The great thickness of the motor fibers and their great increase in thickness before entering the muscles; the existence of two distinct types of motor endings and the increasing complexity of the end plates in the myotomes, branchial muscles, and the muscles of the buccal funnel and the tongue; and the great number of muscle fibers innervated by each motor nerve fiber. FUSARI (1901) has pointed out that the end plates in the myotomes are simpler than those in the other muscles and mentions that after

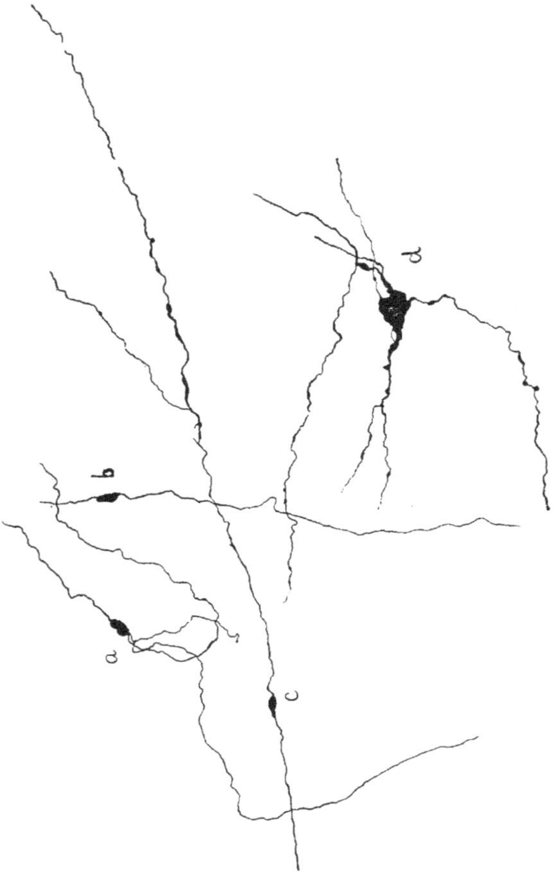

Fig. 30. Peripheral ganglion cells. *a*, *b*, *c*, a group of three in the base of one of the gills. *d*, a cell from the tip of the tongue, on the cephalic surface of the anterior tongue cartilage. Magnification of *a*, *b*, *c*, 200 diameters; of *d*, 350 diameters.

one end plate is formed a fiber continues to form others. I have not seen his second paper (1905) with figures and do not know whether he has seen the simple motor endings in the myotome described in this paper.

2. The distribution of the branchial nerves to the two demibranchs adjacent in each case to the arch in which the nerve runs. Both sensory and motor fibers take part in this arrangement. This means that in the lamprey each nerve supplies one whole gill, while in gnathostomes each nerve divides into pre- and post-

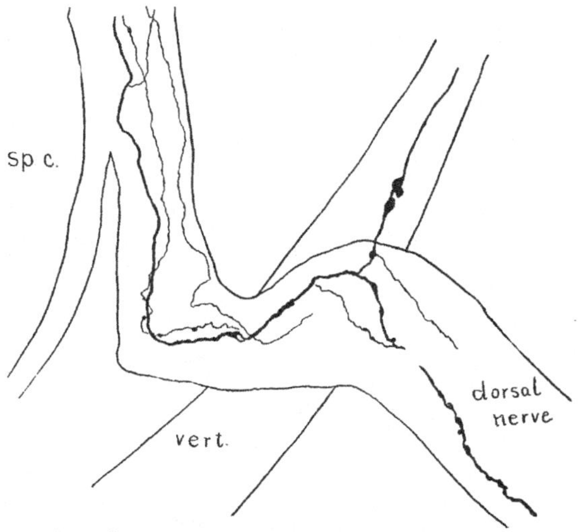

FIG. 31. A medium and two fine fibers in a dorsal nerve root for comparison with motor fibers, etc., *sp.c.*, spinal cord; *vert.*, wall of spinal canal. Magnification, 375 diameters.

trematic rami and supplies the two demibranchs bounding a gill slit. This change is possibly brought about by a downgrowth of the visceral ramus into the next anterior gill arch.

3. The large size of the sympathetic trunk and its evident importance in the branchial region; the distribution of sympathetic fibers in the ventral branchial region by way of the VII nerve; and the endings of sympathetic fibers in the walls of blood vessels. The bodies of most of the sympathetic neurones must be situated in the facial ganglion or within the brain. From this condition in

cyclostomes one would be inclined to expect that the sympathetic ganglion related to the facial nerve in some fishes would be found in process of separation from the facial ganglion proper, as COLE thought was the case in Gadus (see discussion in HERRICK 1900).

4. The existence of peripheral ganglion cells in considerable numbers and in the most diverse regions of the head. RETZIUS (1890) has described subcutaneous ganglion cells in Myxine but, judging from his figures, they do not resemble the cells described here. RETZIUS cites LANGERHANS (1873) as describing subcutaneous ganglion cells in Petromyzon but I have not seen his paper.

5. The slightly differentiated condition of the maxillary and mandibular nerves. The motor portion of the mandibular ramus is clearly differentiated and supplies the muscles of the tongue, which corresponds to the lower jaw. It is possible that the motor trunk is accompanied by some sensory fibers for the covering of the tongue, corresponding to the floor of the mouth in gnathostomes, but they have not been found. The postorbital branches to the skin may be assigned to the mandibular but they are far removed from its motor portion. On the other hand, the maxillaris is largely motor. The muscles supplied by it, however, probably have no counterpart in gnathostomes. If that is so, then it is scarcely right to say that the motor fibers in the maxillaris represent part of the mandibular trunk of gnathostomes. It is simpler and truer to say that the cyclostomes present a peculiar condition of the maxillaris due to the presence of the muscles of the buccal funnel. Whether this be primitive or aberrant, it is the most primitive condition which we know in craniates and in it the trigeminus is indeed peculiar. It is the only nerve in cyclostomes which appears to possess pre- and posttrematic rami. The posttrematic ramus, however, is specialized motor and is lacking in the cutaneous (?) and the branchial sensory and motor components present in the branchial nerves. The absence of the second and third is presumably due to the disappearance of the hyomandibular gill sac. What is commonly called the pre-trematic ramus is a great mixed nerve spreading fan-like to the skin and musculature of the whole anterior part of the head in front of the first gill sac, except the proper muscles of the tongue and velum. This condition in cyclostomes throws a serious doubt on the propriety of calling the maxillaris a pretrematic ramus in any

vertebrate. It is rather a peculiarly large dorsal ramus, with motor components only in cyclostomes. It never acquires the visceral sensory components distributed to the lining of the alimentary canal which characterize rami of the branchial nerves, while the true pretrematic rami never acquire the cutaneous components which characterize the maxillaris. Moreover, no pretrematic rami are developed in the other nerves in petromyzonts. The trigeminus in the lamprey is a branchial nerve like the glossopharyngeus, but supplying the specialized muscles of its arch, wanting in the branches to the next posterior gill sac which has disappeared, and possessed of an extremely large dorsal ramus which supplies muscles as well as skin of the buccal region. It is of course held that the ophthalmicus is a separate nerve.

6. The sensory innervation of the velum comes entirely from the maxillaris. This would imply that the velum is covered by ectoderm, which may be true, or that the velar nerve is a communis component in the maxillaris. The facts that the fibers of the velar nerve are fine and its ganglion cells small and that its ganglion forms the cephalic portion of the gasserian ganglion not mingled with the cells of undoubted cutaneous fibers, lend color to the supposition that this nerve may be communis. The central connections of the fibers were not impregnated.

7. The facial nerve innervates one special muscle related to the tongue, the hyo-hyoideus anterior of FÜRBRINGER. I would suggest that one function of this muscle is to raise the tongue against the floor of the water tube (velar orifice) and so with the m. velo-hyomandibularis internus, which enwraps the œsophagus dorsally, to close both the water tube and the œsophagus during the sucking movements of the buccal funnel.

8. The presence of fibers ending in the intermuscular septa which may serve the muscle sense.

PAPERS CITED.

FÜRBRINGER, P.

1875. Untersuchungen zur vergleichenden Anatomie der Muskulatur des Kopfskelets der Cyclostomen. *Jenaische Zeitschr.*, Bd. 9.

FUSARI.

1901. Presentation de preparations microscopiques demontrant les terminaisons nerveuses dans les muscles striées, dans l'epiderme et dans l'epithelium de la cavite buccale de l'Ammocœtes branchialis. *C. R. Assoc. Anat. 3me. sess., Lyon.*

1905. Contributo allo studio delle terminazioni nervose nei muscoli striati di Ammocœtes branchialis. *Atti Acad. Sc. Torino*, vol. 40.

HERRICK, C. JUDSON.

1900. A contribution upon the cranial nerves of the codfish. *Journ. Comp. Neur.*, vol. 10.

JOHNSTON.
1902. The brain of Petryomyzon. *Journ. Comp. Neur.*, vol. 12.
1905. The cranial nerve components of Petromyzon. *Morph. Jahrb.*, Bd. 34.

KOLTZOFF.
1902. Entwickleungsgeschichte des Kopfes von Petromyzon planeri. *Bull. Soc. Imp. d. Natural d. Moscou*, Année 1901, 1902.

LANGERHANS.
1873. Untersuchungen über Petromyzon planeri. *Freiburg.*

NEAL.
1897. The development of the hypoglossus musculature in Petromyzon and Squalus acanthias. *Anat. Anz.* Bd. 13.

RETZIUS.
1890. Ueber die Ganglionzellen der Cerebrospinalganglien und über subcutane Ganglienzellen bei Myxine glutinosa. *Biol. Untersuch.*, I, 1890.

ON THE SIGNIFICANCE OF THE CALIBER OF THE PARTS OF THE NEURONE IN VERTEBRATES.[1]

BY

J. B. JOHNSTON.

University of Minnesota.

The reflections published here were aroused by the study of the peripheral nerves of the brook lamprey (see p. 569 of this *Journal*). The main facts observed in that study which bear on the present subject were:

1. The great difference in the caliber of fibers in nerve roots or the peripheral nerve trunks. The largest motor fibers are about 70 to 80 times as thick as the finest fibers in the dorsal nerve roots.

2. The great increase in thickness of the motor fibers between their cells of origin and the muscles which they innervate. This increase is greatest in the spinal nerves which supply the myotomes. These fibers begin as cones of origin averaging about 2μ in thickness, decrease to less than 1μ and increase again before leaving the spinal cord to about 3μ. They eventually reach a diameter of 20 to 24μ as they enter the muscles. In one case an increase in thickness from 9 to 19μ was noticed in a distance of .3 mm.; in another case an increase from 4.4 to 8.8μ in the same distance. A similar condition is seen in all the motor nerves, but those to the branchial muscles show the least increase in caliber.

3. The large motor fibers in the lamprey are equal in diameter to the largest medullated fibers in man. The fibers in the lamprey are non-medullated and the species studied is a slender fish not more than 150 mm. in length.

4. The ratio between the number of motor fibers and the number of muscle fibers to be supplied. In the spinal nerves each motor fiber must supply about one hundred muscle fibers. In the cranial nerves supplying specialized muscles the ratio seems to be nearly as large.

[1] Neurological Studies, University of Minnesota, no. 3.

5. The volume of the neuraxone in the case of these motor neurones greatly exceeds that of the cell-body and dendrites combined. The dendrites of the motor neurones (as of other central neurones in the lamprey) are slender smooth fibers, unlike those in the brain of higher forms. The bodies of the motor cells in the spinal cord are usually not greater in diameter than the largest motor fibers at the point where they enter the muscles. In the case of the largest motor neurones measured, I estimate the volume of the cell body and dendrites as equal to that of a cylinder 15μ in diameter and .2 mm. in length, while the motor fibers are often 20μ thick and 2 mm. long. In mammals and man also the volume of the neuraxone often exceeds that of the rest of the neurone, but in this case the greater part of the volume is made up of a conducting fiber. For example, in the case of the motor neurones for the muscles of the lower leg and foot, the volume of the neurone is found chiefly in the fiber connecting the cell-body in the lumbar cord with the motor end plates in the muscles. This fiber is fairly uniform in diameter throughout its length. In the lamprey the portion of the fiber intervening between the cell-body and the first end-branch is a small part of its length and the greatest part of the volume of the neurone lies within the muscle where the end branches are being given off. The conducting portion of the fiber grows as it proceeds, from a slender to a very thick fiber: it is a cone whose apex is at the cell of origin and whose base is in the muscle.

These facts seem to the writer to have a bearing on some important problems of neurology.

The nature of the nerve impulse.

Is the nerve impulse something which is created in any specific part of a neurone and thence merely *conducted* to its destination? This is implied in the ganglionic theory of the neurone, which regards the cell-body as analogous to a battery which discharges currents along the axone. Aside from the valid objections already known, the facts quoted above are inconsistent with such a view. We can scarcely conceive of a cell-body creating an impulse which it discharges into a conductor as thick as itself and fifty times as long. The thick axone must have some other significance than that of a mere conductor.

The same idea that the function of the axone is to *conduct* is implied in the various efforts to discover a special conducting substance or structure in the neurone. Since APÁTHY's description of neurofibrillæ they have been made to do duty as the conducting substance. Here further difficulties arise in the lamprey. If neurofibrillæ in a motor fiber are to supply a hundred muscle fibers, they must be crowded into the motor fiber at its smallest part. If the impulse is merely conducted by these, the impulse itself must arise in the cell-body or dendrites and be measured in strength by the potentiality of that part of the neurone. In the lamprey the impulse starting from the cell is carried through a fiber less than one micron thick and then distributed through a fiber 20 to 24μ thick, and through a hundred branches to as many muscle fibers. Each end-branch of the motor axone is thicker than the axone itself within the spinal cord. I can not help thinking that under these conditions the impulse, if merely conducted, would become so subdivided and attenuated in strength that it would not be effective. All its force would be spent on the first few end branches or would be dissipated in the great expansion of the nerve fiber. On the anatomical side, I can not see how the neurofibrillæ necessary for a hundred motor endings, and for which presumably the large fibers are essential, could be packed into the slender proximal portion of the fiber and still be functionally efficient. Yet APÁTHY, BETHE and others would have us believe that the neurofibrillæ are free and independent throughout their course in the neurone. If the motor impulse were merely conducted by neurofibrillæ, we should expect the cell-body to be large, the motor fiber to be thick at its proximal part, and the combined thickness of the branches to be little if any greater than that of the conducting fiber.

On the other hand, if the neurofibrillæ branch or form a network and if the neuroplasm plays an essential part in conduction, the existence of a slender portion in the axone will not be an obstacle to the passage of an impulse, as will appear below. This is not the place to enter into a discussion of the structural relations of the neurofibrillæ but it may be noted in passing that several authors have described the neurofibrillæ in the cell-body and in the axone as branching, as connected by cross-fibrillæ or as forming an irregular network.

How is the strength of the impulse delivered by a neurone

adapted to the function of that neurone? Is it graded in proportion to the stimulus received? In general, no; and it is probable that this would by no means always result in adaptation or in successful reflex action. Subliminal stimuli do not cause discharge of the neurone unless summated. Adequate stimuli produce responses of normal strength. Increased strength of stimulus in reflexes produces increased response only or chiefly by involving a larger number of neurones. Reference is made here to SHERRINGTON's discussion of the subject of grading of intensity in chapter iii of his book on the Integrative Action of the Nervous System. In general, a neurone in normal and unfatigued condition is able to perform about a certain amount of work and that normal response it gives to any stimulus within a considerable range of intensity.

Is the strength of impulse determined by the size or structure of the cell-body or dendrites? It has usually been thought that some relation exists between the size of a neurone and its activity or the number of tissue elements with which it is related. The nature of this relationship has been very obscure. The writer is inclined to think that the size of the cell-body and dendrites has to do with the nutritive and the receptive functions of the neurone rather than with its effective functions. The volume of the cell-body and dendrites and the ramifications of the latter certainly are important factors in the nutrition of the neurone. Also, those neurones whose axones are long often have large cell-bodies, e.g., motor neurones generally. This is doubtless because the nutrition of the whole neurone is, as is known, in some way dependent upon the nucleus of the cell. On the other hand, it is somewhat commonly true that those neurones which have large and richly branched dendrites may receive impulses from a large number of fibers and perhaps from fibers from various sources. Examples of this are, motor neurones, Purkinje neurones in the cerebellum, pyramidal cells of the cerebral cortex, etc. Examples of neurones whose dendrites are obviously adapted to the reception of impulses from a small number or a definite kind of fibers are: granule cells in the cerebellum, neurones in the tectum opticum related to the optic tract fibers, mitral cells of the olfactory bulb, etc. It is, indeed, difficult to see how the strength of the impulse delivered at the end of the axone could be determined by anything in the cell-body and dendrites without ascribing to the axone the function of mere conduction. The obvious adaptation of the dendrites to the receptive function should lead us to look to the axone for the means of adapting the strength of the impulse to its object.

Can the axone determine the strength of the impulse so as to secure the adaptation to function? The affirmative answer is at once suggested by the facts at hand. In the lamprey the single motor neurone must innervate a hundred muscle fibers. There is nothing about the cell-body or dendrites to suggest special adaptation to this enormous task. The axone, however, increases in diameter until when it enters the muscle it appears quite equal to the work expected of it. If the strength of the impulse is proportional to the caliber of the fiber through which it travels, the effective stimulation of the muscles of the lamprey can readily be understood; otherwise not. Each end branch of a motor fiber is conspicuously thicker than the whole axone at its most slender part in the spinal cord, while the area of the cross section of the thickest part of the motor fiber is more than four hundred times that of its slender portion in the spinal cord. We can not conceive of the motor axone of less than one micron diameter being divided into one hundred branches without increase in diameter. If there were no increase in thickness of nerve fiber and no increase in volume of motor impulse between the spinal cord and the myotomes, I can not understand how effective impulses could be delivered to the muscle fibers. The conclusion reached, then, is that the strength of the impulse is determined by the caliber of the axone and that this is adapted to the work to be performed.

If this statement is correct for the lamprey, is it not substantially true for motor fibers generally? Is not the combined diameter of the end branches of a motor fiber greater than that of its conducting portion? Above all, would not the motor end plate serve especially the function of increasing the strength of the nerve impulse so that it may be effective as it enters the muscle fiber? Miss DUNN (1902) has shown reason to believe that in the leg of the frog there is a conical diminution of nerve fibers toward the periphery. It does not appear, however, whether this diminution takes place in motor or sensory fibers, or both. There is no conflict between her results and the hypothesis here offered, because this hypothesis would be satisfied if a conical enlargement took place in the combined diameter of the end branches, regardless of the thickness of the fiber in its conducting portion.

If the above conclusions are accepted, how must the nature of the nerve impulse be pictured? As a physico-chemical change in the neurone which is propagated from part to part within the neurone, not merely conducted. Such a change must be regarded as an interaction among the substances of the neurone. In such an interaction both the neurofibrillæ and the neuroplasm doubtless play their parts. Under this conception of the nerve impulse, it might travel through a fine fiber and still be delivered in effective strength to a large number of tissue elements, provided only that the caliber of the fiber or the combined caliber of its branches increased sufficiently to multiply the impulse by the number of elements to be supplied. This is exactly what appears to take place in the motor axones in the lamprey. Each motor axone appears to constitute a cone whose apex is at the cell of origin in the spinal cord or brain and whose base is formed by the combined end branches.

The conclusions which this discussion renders probable are that the nerve impulse is a physico-chemical process which consists in an interaction between different substances in the neurone; that it is propagated from part to part of the neurone by the progressive interaction of these substances; that it is not merely conducted by any specific substance in the neurone; that it increases in strength with the increase in caliber of the axone carrying it; and that this last is the chief function of motor end plates and enlargements at the ends of axones wherever such enlargements occur. It may be necessary to add that the practical freedom of the nerve fiber from fatigue and the very slight amount of metabolism connected with the transmission of an impulse must be taken into account in a theory of the nature of the nerve impulse. And it should be said that BETHE, one of the foremost exponents of the important rôle played by neurofibrillæ, recognizes in his fibril acid hypothesis the interaction of fibrillæ and neuroplasm and that in so far this hypothesis conforms to the conclusions reached above.

Significance of the size of the cell-body and dendrites.

If the caliber of the axones is important, may there not be some special significance in the volume of the others parts of the neurone at least in certain cases? Why do the mitral neurones of the olfactory bulb have so large cell-bodies and dendrites? The olfactory fibers are relatively fine. The stimuli which the olfactory cells receive are commonly weak, often subliminal. These fine fibers are met by the largest and coarsest dendrites known in the central nervous system. May the great volume of the mitral neurone provide in some way for the reception of very weak impulses? May it serve to sum up the impulses received at the same time from several fibers—a function that might be called synchronous summation? Whether this particular suggestion shall prove to have any value is of no moment. It seems to the writer that some significance must attach to the great volume of the mitral neurones; and to that of the Purkinje neurones; as also the minute size, coupled with the great number, of the granule cells of the cerebellum. The writer is willing—perhaps more willing than most persons—to grant much to the factors of variation and chance in the determination of the actual form and structure of the elements of the nervous system in a given animal, but here are structures among the most constant and uniform throughout the vertebrate series. The chief feature in that uniformity is that of great volume and this can scarcely be ascribed to accident. The Purkinje cells and the large cells of the vestibular centers seem to be concerned with the preservation of muscular tone. This requires weak rhythmical stimulation and the suggestion presents itself that the large size of these elements may in some way be important for the discharge of rhythmical impulses. The preservation of tone and equilibrium in the lower fishes have been assigned by some to certain neurones of gigantic size, the Müllerian cells and fibers.

A complete consideration of this subject must take into account the caliber of peripheral sensory fibers. Upon this I have little to contribute here. In the lamprey the fineness of visceral sensory fibers which is a fairly constant character in vertebrates reaches an extreme. The cutaneous fibers vary considerably in caliber, while the neuromast fibers are, next to the motor fibers, the thickest in the body. Two significant facts are to be noticed: first, that the fibers supplying the taste organs in the pharynx (or water tube) are coarser than the general visceral fibers, and second, that the fibers of the velar nerve are much finer than the average fibers of the trigeminus going to the skin. The terminal ramifications of the general visceral and cutaneous fibers are very rich, the visceral certainly not less rich than the cutaneous. Since this is true, it can scarcely be said that the caliber of the fibers depends on their functional importance to the organism. The greater diameter of the gustatory fibers as compared with the general visceral may fall under HERRICK's principle of "a correlation between the diameter of the fiber and the functional importance of the fiber, or the physiological importance of its terminal organ, as compared with other organs of the same system" (HERRICK 1902). The case of the velar nerve, however, seems directly opposed to this principle. Of the functional importance of the velum as an organ and of its sensory nerve supply, which is very rich, there can be no doubt. Yet the fibers and their ganglion cells are much smaller than the average of the general cutaneous system, to which, so far as we know, the fibers of the velar nerve belong. Apparently the fibers of the cutaneous system which are characteristically coarse have been reduced in caliber in response to the same influences which have determined the caliber of the visceral sensory fibers. Fibers supplying the internal surfaces of the body are as a rule finer than those supplying the external surface. It seems probable that this is due to some differences in the conditions of stimulation, and this is in agreement with the principle expressed above that the caliber of dendrites is determined by the receptive functions of the neurone. Until we have further experimental data upon the conditions and modes of stimulation of sensory fibers, it would be useless to speculate further concerning the controlling factors here.

The method of specialization.

In the course of the evolution of vertebrates there has been a change from segmental masses of muscle (myotomes) to special muscles. On the functional side the cause of this change is found in the increased number, variety and combinations of muscular movements required of the higher animals by their conditions of life and their habits. Whereas the movements of the adult lamprey consist chiefly in wriggling, sucking, rasping and breathing, we must perform a thousand sorts of movements, massive and minute, complex and clever, in order to earn our daily bread. For all the movements of the lamprey a few large masses of muscle arranged in antagonistic pairs or groups and alternating in their contraction suffice. For us, a great number of special muscles are necessary and each must be capable of a graded series of contractions and of entering into combination with one or another set of muscles to perform this or that particular act. With the specialization of muscles has gone an increase in the number of fibers in the ventral roots. In the frog (BIRGE 1882, HARDESTY 1899) the average number in each ventral root is from six to ten or more times as great as in the lamprey. In man the total number in the ventral roots of one side is 203,700 (INGBERT 1904), or an average for each nerve of 6570. If the last four nerves are left out of account for the sake of a fairer comparison, the number in each of the remaining nerves is 7290. The number is largest in those nerves which supply the largest volume of muscle. After allowance is made for sympathetic fibers in the ventral roots, there is obviously in the higher forms as compared with the lamprey a much larger number of motor fibers and they have a much smaller diameter relative to the body weight. On the other hand, the number of motor fibers in man is much less in proportion to the body weight than in the lamprey and the individual muscle fibers are many times larger. I know of no enumeration or estimate of the number of muscle fibers supplied by one nerve fiber in man or any of the higher animals, although the necessary preliminary observations for the study of this in the frog have been made by Miss DUNN (1900, 1902).

The question of interest on which all this bears is, does specialization of muscles entail a decrease in the number of muscle fibers supplied by each nerve fiber or only a definite distribution of motor fibers to each special muscle with provision for functional isolation in the central nervous system? We know at least that the grading of strength and extent of muscular contraction depends upon bringing into play a larger or smaller number of the fibers of the muscle. It is desirable that the ratio between the number of motor nerve fibers and of muscle fibers supplied should be determined in several animals (frog, mammals, man), and the determination should be made also for different muscles engaged in the several forms of muscular activity. We should expect that in the muscles of the fingers a smaller number of muscle fibers would be supplied by one motor fiber than in muscles concerned in massive movements.

The question of specialization in other organs is of course subject to study along similar lines.

PAPERS CITED.

BIRGE.

1882. Die Zahl der Nervenfasern und der motorischen Ganglienzellen im Rückenmark des Frosches. *Arch. f.* [*Anat. u.*] *Physiol.*

DUNN.

1900. The number and size of the nerve fibers innervating the skin and muscles of the thigh in the frog (Rana virescens brachycephala, Cope). *Jour. Comp. Neurol.*, vol. 10.

1902. On the number and on the relation between diameter and distribution of the nerve fibers innervating the leg of the frog (Rana virescens brachycephala, Cope). *Journ. Comp. Neurol.*, vol. 12.

HARDESTY.

1899. The number and arrangement of the fibers forming the spinal nerves of the frog (Rana virescens). *Jour. Comp. Neurol.*, vol. 9.

HERRICK.

1902. A note on the significance of the size of nerve fibers in fishes. *Jour. Comp. Neurol.*, vol. 12.

INGBERT.

1904. An enumeration of the medullated nerve fibers in the ventral roots of the spinal nerves of man. *Jour. Comp. Neurol. and Psychol.*, vol. 14.

ABERRANT ROOTS AND BRANCHES OF THE ABDUCENT AND HYPOGLOSSAL NERVES.

BY

JOHN LEWIS BREMER, M.D.

(Demonstrator of Histology, Harvard Medical School.)

WITH NINE FIGURES.

My attention was called to the roots of the abducens by a recent paper by ELZE (1907.1) on a 7.0 mm. human embryo, in which he describes the nerve as having two roots, one arising in the usual way from the ventral zone of the medulla, between the origins of the trigeminal and facial nerves; the other, also from the ventral zone of the medulla, but further caudal, about opposite the origin of the glossopharyngeal nerve, and separated from the first root by some branches of the basilar artery. The caudal root runs forward, quite near to the floor of the medulla, and joins the anterior root to make the abducent nerve. This then is merely a caudal prolongation of the origin of the abducens, though showing the tendency of these roots to a segmental arrangement, one root between each two branches of the basilar artery, which are also segmental.

On examining the human embryos in the Harvard Embryological Collection, I find that this is by no means an uncommon occurrence, and may be easily explained by imagining the group of cells in the ventral part of the medulla which gives rise to these fibers to be longer than usual, so that the caudal fibers find it easier to emerge from the brain separately, and run forward outside the brain to join their fellows, instead of passing through the brain wall. A similar origin of the abducens is shown in fig. 6, taken from a chick embryo of 25.0 mm.; in the chick this double origin is almost constant. In one of the human embryos and in one of the pig embryos studied, this process was carried so far that there were two distinct abducent nerves on one side, one from the anterior root and one from the posterior or caudal root, having the same direction but not joining into one nerve. With further growth these separate nerves would probably have merged.

But in the sagittal sections studied, another point was brought out strikingly, confirmed often by the less striking pictures seen in other planes of section. If we examine a model of the brain and the nerves arising from it in an embryo, where the conditions are simpler than, but essentially the same as, in the adult, we notice that there is a gap between the roots of the abducent nerve and those of the hypoglossal nerve, where no ventral roots exist. This is shown in HIS' models of the human brain, in the plates of the brain of a 12 mm. pig embryo described by LEWIS (1903.1), and is figured by many others. In the human embryos studied the striking feature is that this gap is frequently filled, as it were, by smaller roots, emerging in the same line as the hypoglossal roots, and completing a row of nerve bundles, more or less segmentally arranged, continuing the ventral nerve roots of the cord as far forward as the abducens. Moreover, in some cases these roots point toward the forebrain, as though to join the abducens (as in ELZE's embryo) while in other cases roots situated as far forward as these, and even fibers emerging with the abducent roots, turn caudally and run as though to reach the hypoglossal nerve (fig. 1). (Throughout the figures aberrant roots or branches are indicated by *a*, *b*, or *c*.)

In two cases among the human embryos, and in a few more among the embryos of pig and rabbit, the roots combine both directions, so that the abducens receives a caudal root, but from the loop of this a branch runs as though to join the hypoglossal nerve (fig. 2). The roots which run as though to join the hypoglossal nerve, in older embryos in which the cartilage is being laid down for the base of the skull, may even make foramina for themselves, in line with, but more anterior than, the anterior condyloid foramen through which the hypoglossal nerve bundles run. This is shown in fig. 6 from a chick embryo, in which the roots and foramina are similar, but more clearly shown. MCMURRICH (1905, p. 192) states as confirmation of the existence of four fused vertebræ in the occipital bone, first described by FRORIEP (1886.1) that "during the cartilaginous stage of the skull the anterior condyloid foramina are divided into three portions by two cartilaginous partitions which separate the three roots of the hypoglossal nerve." The foramina for ventral roots of which I speak, arising near the

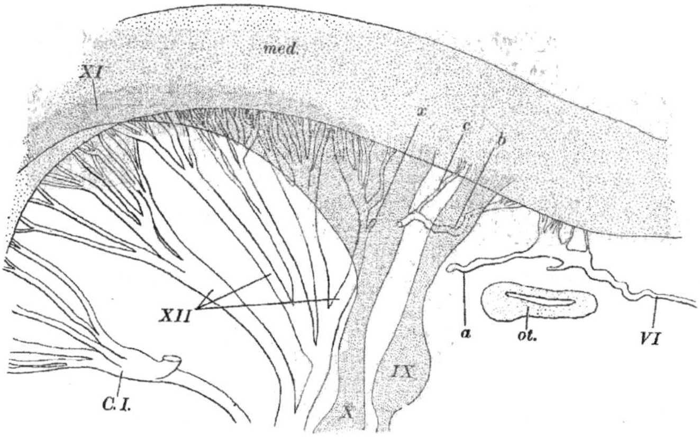

FIG. 1. Graphic reconstruction of human embryo, 10 mm., right side. (H. E. C. series 736, sections 165 to 200.) × 50. *VI, IX, X, XI,* cranial nerves; *XII,* roots of hypoglossal nerve; *CI,* first cervical nerve; *med.,* floor of medulla; *ot.* otocyst; *a,* aberrant branch of abducens; *b,* aberrant root from mesial part of medulla, running laterally and joining *c,* a separate strand of the glossopharyngeal nerve; *x,* degenerating hypoglossal root.

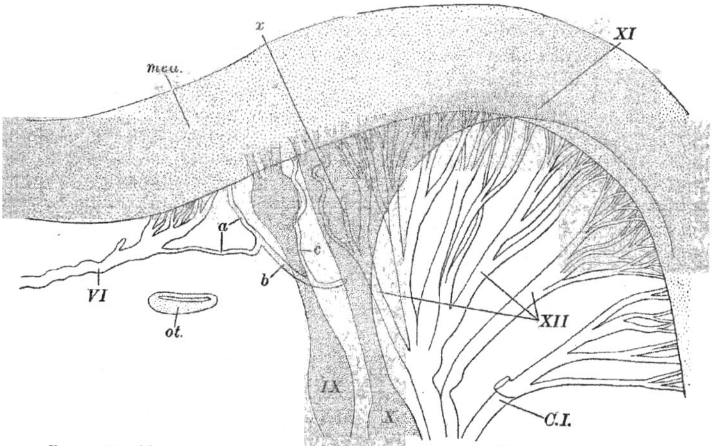

FIG. 2. Graphic reconstruction of same embryo as fig. 1, left side. (H. E. C., series 736, sections 90 to 125.) × 50. *a,* loop of posterior root of abducens; *b,* aberrant branch from this loop, joining *c,* aberrant root from mesial part of medulla; all run ventrally. For other lettering, seen fig. 1.

abducens, are quite distinct from the subdivided anterior condyloid foramen, and may mark the fusion of more anterior vertebræ in the occipital bone. Such a foramen, though existing only for a short time, might cause a weak spot in the forming bone near to the median line, which might serve as a path for a chordoma, a tumor of the notochord which arises (as shown by WILLIAMS[1]) not infrequently in this region.

In one embryo (fig. 3) a root anterior to the hypoglossal roots runs to the jugular foramen, and ends just posterior to the glossopharyngeal nerve; and in other younger embryos the more anterior of these aberrant roots run as though to pass behind the glossopharyngeal nerve, instead of behind the vagus (fig. 1).

Another set of aberrant roots was found in these embryos. All those mentioned heretofore, after emerging from the medulla on either side of the median line, run ventrally, either forward to join the abducens, or caudally to join the hypoglossal nerve, or at least in the general direction of the hypoglossal roots. The fibers of this other set, emerging also from the ventral part of the medulla between the abducent and hypoglossal roots, take a lateral course, at right angles to the hypoglossal roots, or even turn dorsally. Such fibers are not uncommon, to judge by the embryos in the Harvard Embryological Collection. They may arise separately, or as branches of the roots of usual nerves; if they are long enough we can follow them to the mesenchyma at the side and back of the head. They may be looked upon as of two groups: those more caudal, arising with, or just anterior to the roots of the hypoglossal nerve, and those still more anterior, arising with or just posterior to the abducens. The fibers of the first group run to the back of the neck by passing behind the vagus and accessory nerves, while those of the second group pass in front of the vagus, behind the glossopharyngeal nerve. In one embryo (fig. 5) as was the case with the ventrally running roots, fibers from the same root diverge, one branch running in front of, one behind, the conjoined vagus and accessory. In other cases the lateral fiber is made by the junction of several roots, even including a branch of the abducens.

One of these fiber bundles with a dorsal course may be seen in fig. 4, a sagittal view of a human embryo of 11.5 mm. which shows

[1] WILLIAMS, L. W.: Paper in press.

the bundle arising by three ventral roots of its own, and joined by a posterior branch of the abducens, running up beside the medulla to end in the mesenchyma of the top of the head. But those fibers which have a lateral course are shown best in a view of the under surface of the medulla, fig. 5. On the right side of the drawing

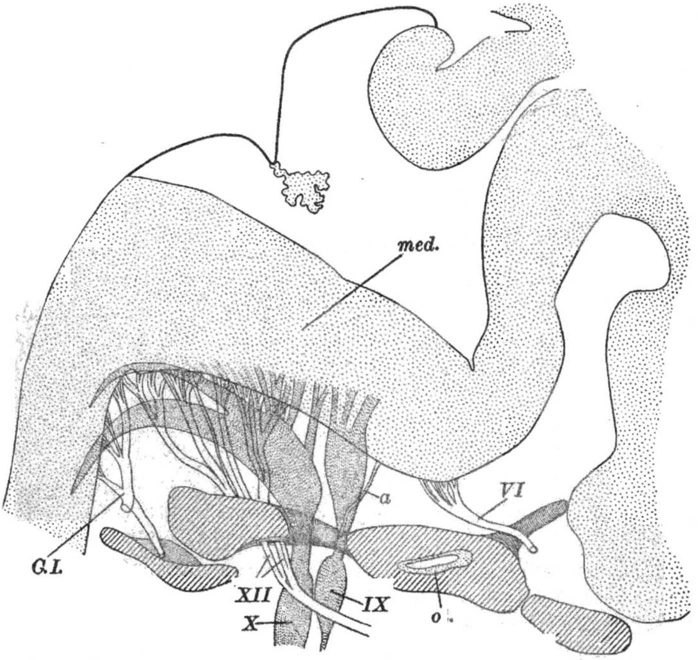

FIG. 3. Graphic reconstruction of human embryo 22.0 mm. (H. E. C. series 851, sections 250 to 320). × 14. The cartilaginous base of the skull is cut in the outer section, and shows the anterior condyloid foramen, through which the hypoglossal roots pass; the jugular foramen is laid open to show an aberrant ventral root, *a*, passing through with the glossopharyngeal nerve. For other lettering, see fig. 1.

(the left side of the embryo) the otocyst is shown lying close to the rounded side of the medulla. Caudal to this is the glossopharyngeal nerve, arising from the lateral zone of the medulla, and cut off as it is running toward us. Behind this is the vagus, also drawn as though cut off, and joined caudally by the accessory, whose long root is shown sweeping forward in a curve at the side of the medulla. Nearer the median line the roots of the hypoglossal nerve are represented at though cut off quite close to their origin; and median to the otocyst is the abducens. Between the anterior roots of the hypoglossal nerve and those of the abducens are several rootlets, from the caudal group of which arise three nerves, two running behind the vagus-accessory trunk, one pointing to the space between the vagus and glossopharyngeal nerves; while from a root near the abducens, a fiber bundle runs as though to join the anterior nerve just mentioned. All of these fibers have a lateral course.

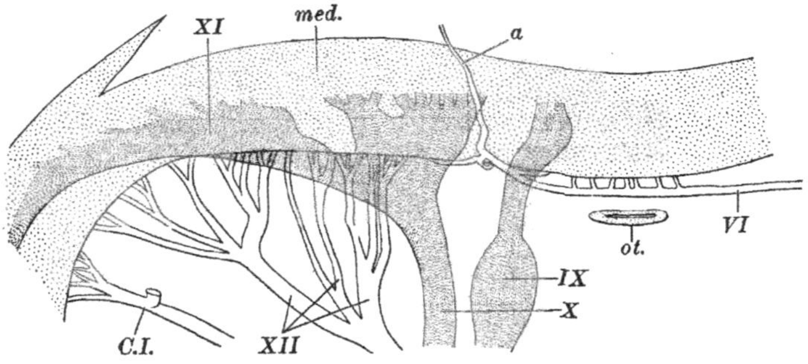

FIG. 4. Graphic reconstruction of human embryo, 11.5 mm. (H. E. C. series 189, sections 136 to 150.) × 50. *a*, dorsally running aberrant nerve, arising by a branch of the abducens and by three ventral roots. For other lettering, see fig. 1.

On the left of the drawing the position of the glossopharyngeal and vagus-accessory trunks is merely indicated while the hypoglossal nerve is represented as having been cut off lower down (i. e., nearer to us) after its various roots have joined to make the main trunk. The anterior root takes a lateral course at first, but after sweeping well outside, turns and joins the trunk lower down. The fibers just anterior to this root, however, persist in their lateral course, and turn caudally to run in the mesenchyma at the side of the head. Anterior to these, other fibers, still running laterally, at right angles to the large nerve trunks, turn forward as two small nerve bundles, pointing in front of the vagus; and from the abducens a posterior branch, joined by fibers from a separate root, also runs laterally toward the same space. In this embryo, then,

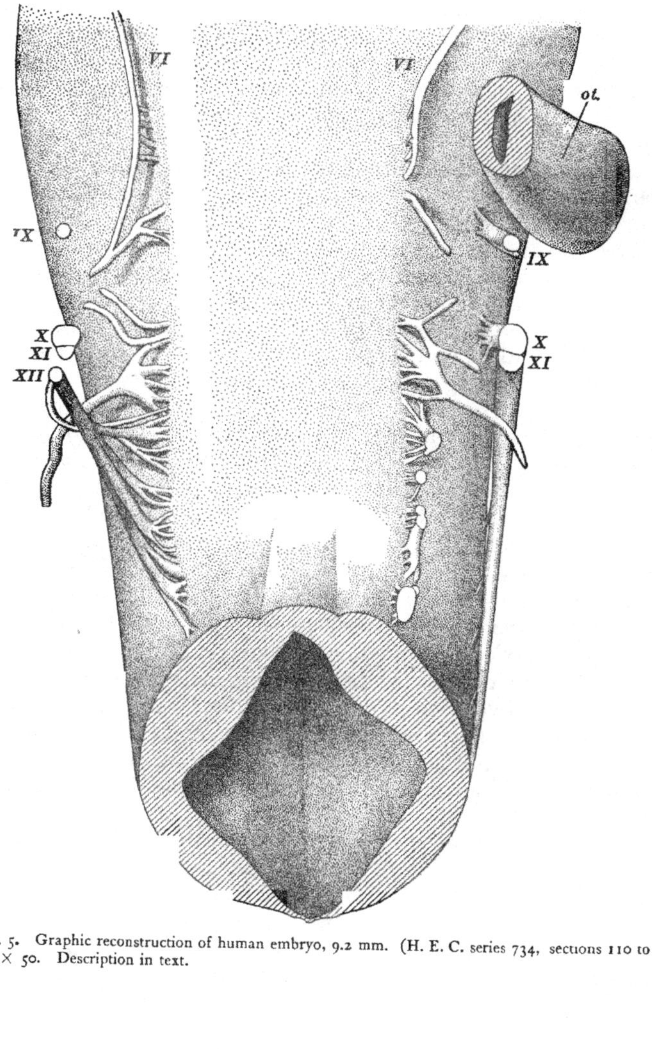

FIG. 5. Graphic reconstruction of human embryo, 9.2 mm. (H. E. C. series 734, sections 110 to 192.) × 50. Description in text.

we see representatives of the two groups mentioned above; for there are fibers with a lateral course which pass behind the vagus-accessory trunk, and others which pass, or at least point, in front of the vagus, behind the glossopharyngeal nerve. In another embryo (fig. 1) a laterally running ventral root not only passes in front of the vagus, but actually joins with a separated bundle of the glossopharyngeal nerve, making a complete ventral root for this nerve.

If we examine the embryos of birds and reptiles, even those of large size, we find conditions which throw light on these aberrant nerves in man and other mammals. In the turtle and the chick there are almost constantly found segmentally arranged roots from the ventral part of the medulla almost as far forward as the abducens, smaller than the main roots of the hypoglossal, of which there are two, with two foramina, on each side (Bronn's *Thierreich*[2]), but still often joining the main roots to form the trunk, and always making foramina for themselves (fig. 6). The most anterior of these small roots often arises opposite the glossopharyngeal nerve, and runs as though to join it, and not the hypoglossal nerve, outside the skull. Moreover, from each of the roots of the hypoglossal nerve, except often the most anterior rudimentary ones, a large branch runs laterally and dorsally, as soon as the root leaves the foramen. Here then we are dealing with branches similar to the dorsal rami of spinal nerves, or at least the motor portion of such rami. These fibers in the spinal nerves innervate the muscles of the back, the dorsal rami of the upper cervical nerves going to the trapezius, which in mammals is innervated chiefly by the accessory nerve; but in birds and reptiles the hypoglossal nerve has dorsal rami running to the muscles that correspond to the trapezius, and, as we might expect in birds and reptiles the accessory nerve is either lacking, or runs as part of the vagus; its place is taken by the dorsal rami of the hypoglossal nerve.

I consider these aberrant fibers which run laterally in the embryos of man and other mammals as vestiges of the dorsal rami of the hypoglossal found in birds and reptiles, whose place is taken in mammals by the accessory nerve; or, if arising nearer the abducens and running in front of the vagus, as vestiges of the dorsal ramus of a ventral root of the glossopharyngeal nerve.

[2] Bd. 6, Abth. iii, 1, p. 149, and Bd. 6, Abth. iv, 1, p. 389.

The discussion of the significance of these aberrant roots seems to me to fall into two parts; first, the question of the homology of the cranial nerves with the spinal nerves; and second, the relation of the different components of these nerves. Most writers agree that the cranial nerves must be serially homologous with the spinal nerves, though with many components lost; and no great difficulty is experienced in adopting the idea that the anatomical nerves, known as the vagus, accessory, and hypoglossal, are really the separated components of several segmental nerves. FRORIEP (1901.2), STREETER (1904.1) and a few others oppose this view,

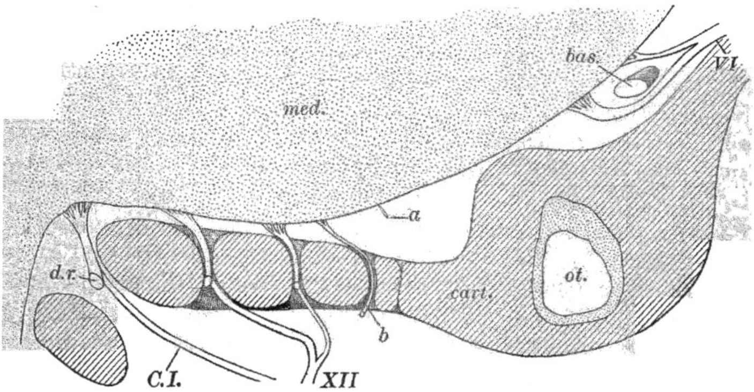

FIG. 6. Graphic reconstruction of chick embryo, 25.0 mm. (H. E. C. series 516, sections 312 to 332.) × 50. *a*, small anterior ventral root, pointing toward anterior foramen; *bas.*, branch of basilar artery; *cart.*, cartilaginous base of skull, pierced by paramedian foramina for the roots of the hypoglossal nerve. The anterior foramen is empty, probably because the nerve root, *a*, has degenerated; *d.r.*, dorsal ramus.

substituting a theory that the cranial nerves represent a secondary system of nerves, connected with the higher, more complex functions of the muscles innervated, characterized by a lateral, instead of a ventral, motor root, and taking the place of the spinal nerves in the brain. STREETER does not mention the nerves anterior to the glossopharyngeal, and so does not explain the abducens, or the other eye-muscle nerves. This secondary system is considered to have grown backward over the spinal nerves, making a wedge-shaped area of cranial system, reaching down the cord as far as the lowest roots of the accessory nerve, overlying an area of the spinal system, which extends as far headward as the anterior roots of the hypoglossal. With the finding of these ventral roots between the hypoglossal nerve and the abducens, it seems to me that we can no longer consider this theory, but must return to the conception of the cranial nerves as modifications of the spinal nerves, but essentially similar to them. The presence of embryonic ventral roots having the same relations to the glossopharyngeal nerve as do the hypoglossal roots to the vagus and accessory complex (i. e., passing just behind the trunk to reach the outer side) further strengthens, if it does not prove the older theory. For, if these ventral roots are vestigial, as I think, we have evidence that the glossopharyngeal nerve originally possessed the same roots as a spinal nerve.

The loss of the ventral roots between the abducens and hypoglossal nerves in the adult should be explained by the absence of muscles derived from the mesodermic segments in this region; and I am led to believe that this reduction in the number of ventral roots is quite extensive, including some of the anterior roots of the hypoglossal nerve. If we examine figs. 1, 2 and 3, the first two of a 10.2 mm. embryo, the last of a 22.0 mm. embryo, we notice that the relative distance between the abducens and hypoglossal roots is greater in the older embryo. This might be accounted for by the growth of the medullary floor between these two points, but, it seems to me, is really due to the loss of some of the more anterior hypoglossal roots. In figs. 1 and 2 at *x* we can see the process going on; and in a reconstruction of a human embryo of 13.6 mm. now being prepared by F. W. THYNG, the same process of degeneration of a hypoglossal root is figured.

We should conceive, therefore, a row of ventral roots from the medulla, in mammals, originally arising as far forward as the abducens, belonging, so to speak, with the vagus-accessory complex, the glossopharyngeal nerve, and possibly with the facial nerve; but soon disappearing because of the failure of the segmental musculature to develop in the head region between the hypoglossal muscles and the eye-muscles. Such a continuous row exists in the young of the cyclostome Bdellostoma (JOHNSTON 1906, p. 190). These vestigial roots are found in sixteen of the twenty-seven human embryos studied, and occur frequently in the embryos of pig, rabbit, sheep, and dog, though not found in the cat and opossum embryos of the Harvard Embryological Collection. They are found almost constantly in the turtle and chick, and in this connection it is interesting to note that FRORIEP (1886.1) states that the occipital bone of the chick is made of five fused vertebræ, instead of four as found in mammals. The relatively large eye of turtle and chick embryos probably is correlated with the common double root of the abducens in these forms. The duration of these vestigial roots is short; they have not been found in human embryos of more than 30.0 mm. in length, at which time the muscles are fairly well laid down; and I have not seen them in the adult, though the possibility of an anomaly of the nerves in this region should not be overlooked. As for the fate of the laterally directed fibers, if they arise with roots of the hypoglossal nerve, they may occassionally be retained to form the small recurrent twigs described in adult anatomy as running to the dura of the anterior condyloid foramen, or to the wall of the jugular vein. These twigs usually appear late, but in fig. 7, taken from a section of a 29.0 mm. human embryo, a small branch from the hypoglossal nerve is seen going to the wall of the jugular vein, and its position, just without the foramen, and its lateral course, suggest that it may be one of the lateral aberrants being converted into one of the recurrent twigs of the adult.

It seems possible that by a more extended study of these vestigial roots we might arrive at a solution of the vexed question of the number of neural segments in the head, at least as far forward as the abducens. I have not seen any traces of similar aberrant roots in the more anterior parts of the head, in conjunction with the trochlear and motor oculi nerves, but, as I have not made an extensive examination of this region, I am not prepared to say that such aberrant roots may not exist.

In regard to the relations of the different components of the cranial nerves, I wish to advance a theory which rests partly on the distribution of, and the course pursued by, these aberrant roots, is supported by many facts long known, but differs in some respects from former theories of the components of the nerves, and seeks to reconcile facts which are not in accord with these theories. JOHNSTON (1906), following GASKELL (1886), divides the nerves into four components, which he calls the somatic efferent or motor, the visceral efferent or motor, the somatic afferent or sensory, and the visceral afferent or sensory. Of the afferent or sensory components I shall say nothing further. The somatic motor fibers arise from cells arranged in groups in the ventral horn of the cord or in the ventral area of the medulla, and the visceral efferent nuclei lie in the lateral horn of the cord and are continued into the medulla as the lateral column of gray matter, subdivided in higher vertebrates into two minor columns, represented by the nucleus ambiguus and the dorsal vagal and glossopharyngeal nuclei. JOHNSTON ascribes to the visceral motor fibers the control of "the smooth muscles in the viscera and elsewhere in the body, the muscles of the heart and blood vessels, cer-

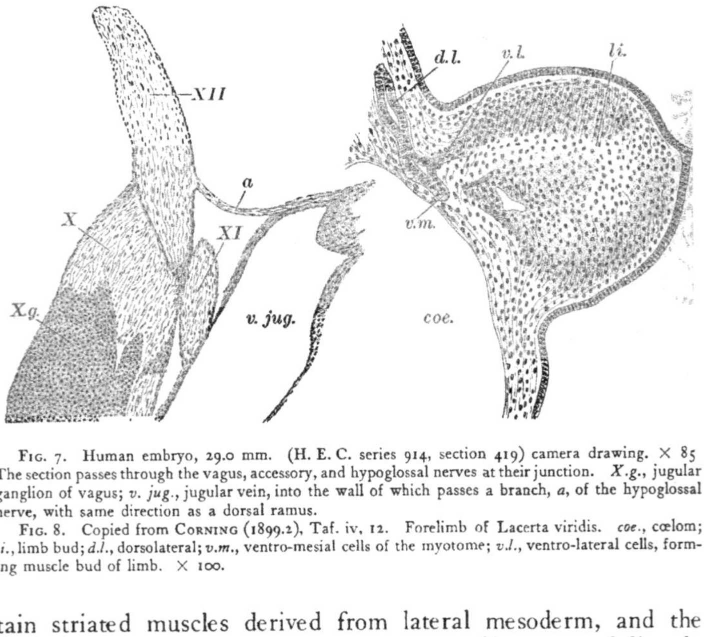

FIG. 7. Human embryo, 29.0 mm. (H. E. C. series 914, section 419) camera drawing. × 85 The section passes through the vagus, accessory, and hypoglossal nerves at their junction. *X.g.*, jugular ganglion of vagus; *v. jug.*, jugular vein, into the wall of which passes a branch, *a*, of the hypoglossal nerve, with same direction as a dorsal ramus.

FIG. 8. Copied from CORNING (1899.2), Taf. iv, 12. Forelimb of Lacerta viridis. *coe.*, cœlom; *l i.*, limb bud; *d.l.*, dorsolateral; *v.m.*, ventro-mesial cells of the myotome; *v.l.*, ventro-lateral cells, forming muscle bud of limb. × 100.

tain striated muscles derived from lateral mesoderm, and the glands of the body;" while the somatic motor fibers control directly "the actions of the typical body muscles; namely, those derived from the dorsal mesoderm or somites." In the cord of vertebrates both of these components run in the ventral root, but in the cord of some lower forms, and in the medulla and pons of vertebrates the somatic motor component alone runs as a ventral root; the visceral motor component becomes a lateral root. This means that the fibers of this component change their point of exit from the line of ventral roots to the postero-lateral groove. In the brain the somatic motor component is represented by the oculomotor, trochlear, abducent, and hypoglossal nerves, whose distribution is to muscles derived from head and occipital somites while the visceral motor component is represented by the motor parts of the other cranial nerves, and, beside innervating the glands, heart, and blood vessels, and the smooth muscles of the intestinal tract, also controls the striated muscles of the head and many in the neck. He explains this peculiar segregation of these special muscles from the other striated muscles of the body by their embryological derivation from lateral mesoderm, instead of from somites, which makes these striated muscles embryologically homologous with the smooth muscles of the intestine and with the heart muscle; and therefore properly supplied by the same nerve component as the smooth muscles.

But JOHNSTON carries this further; the trapezius musculature in all classes of mammals is innervated by the accessory nerve, undoubtedly a lateral root nerve. The striated muscles (except the heart) known to arise from lateral mesoderm, are all originally in connection with the gill pouches, and so may the more justly be homologized with the intestinal muscles, both being in relation with the epithelium of the intestinal tract. In order that the trapezius may be classed in this same category, and therefore justify its innervation, JOHNSTON states that "the only probable explanation is that the shoulder girdle or pectoral arch did not have its origin as a part of the skeleton of a limb, but existed as a branchial arch before the limb was formed;" that the musculature of this girdle belonged to a posterior branchial arch, now lost, and only secondarily became attached to the arm and the trunk.

With this explanation of the innervation of the trapezius musculature I cannot agree. Two facts especially seem to me to militate strongly against it; first, the subdivision of the lateral column of gray matter into a dorsal nucleus of the glossopharyngeal and vagus nerves and a more ventral nucleus ambiguus, a fact left unexplained by JOHNSTON; and second, the innervation of the trapezius muscle by the dorsal rami of the first and second cervical nerves, which often anastomose with the accessory branches. If this muscle is of branchial origin, why should it be innervated by branches of ventral roots?

The theory which I wish to advance is explained in the diagrams, fig. 9. *A* represents a typical hemisection of the spinal cord; the two sensory components of the nerve are represented by fine lines, the visceral motor component by the dotted line, and the somatic motor components by heavy lines. It will be seen that the only difference between this theory and that of JOHNSTON is in the subdivision of the somatic motor component into two parts which I shall call the ventro-mesial and the ventro-lateral components, from the position of their respective nuclei of origin in the ventral horn of the cord; the reason for this subdivision will appear presently.

Let us turn for a moment to the origin of the muscles which these components supply, taking for a theorem that muscles with the same embryological derivation will be innervated by the same components. CORNING (1899.1) in his work on the origin of the muscles in amphibia, gives several figures of the developing myotomes, and he and many others agree that the muscles of the back, except those overlying ones connected with the shoulder and pelvic girdles, develop from the cells in the dorso-lateral portion of the myotome, while the muscles of the sides and ventral part of the trunk develop from the ventral cells of the myotome. I wish to emphasize the lateral position of the dorso-lateral cells of the myotome. VAN BISSELICK (1905.1) in a paper on the innervation of the trunk myotome in Acanthias and Mustelus, finds three distinct divisions of the myotome, a posterior, a lateral, and a ventral division, each supplied by a distinct branch or group of branches of the trunk nerve. The lateral division of the myotome would be in a position to supply cells to form the musculature of the limb. CORNING and many others regard the muscles of the limbs as developing from muscle buds springing from the ventral part of the myotome, and therefore comparable with the ventral body muscles; BARDEEN and LEWIS (1901.1), on the other hand, state that in human embryos the muscles of the limbs arise from mesenchymal cells *in situ*, which have no connection with the myotomes. Many papers, pro and con have been written, notably by BYRNES (1898.1), FIELD (1894.1), VALENTI (1902.1) and KAESTNER (1893.1), and it seems best to me to agree with KAESTNER in considering that the limb muscles of mammals arise from myotomic buds (as do certainly those of amphibia and other lower forms), but at such an early stage that the separate cells of the myotome are not distinguishable from the mesenchymal cells through which they migrate.

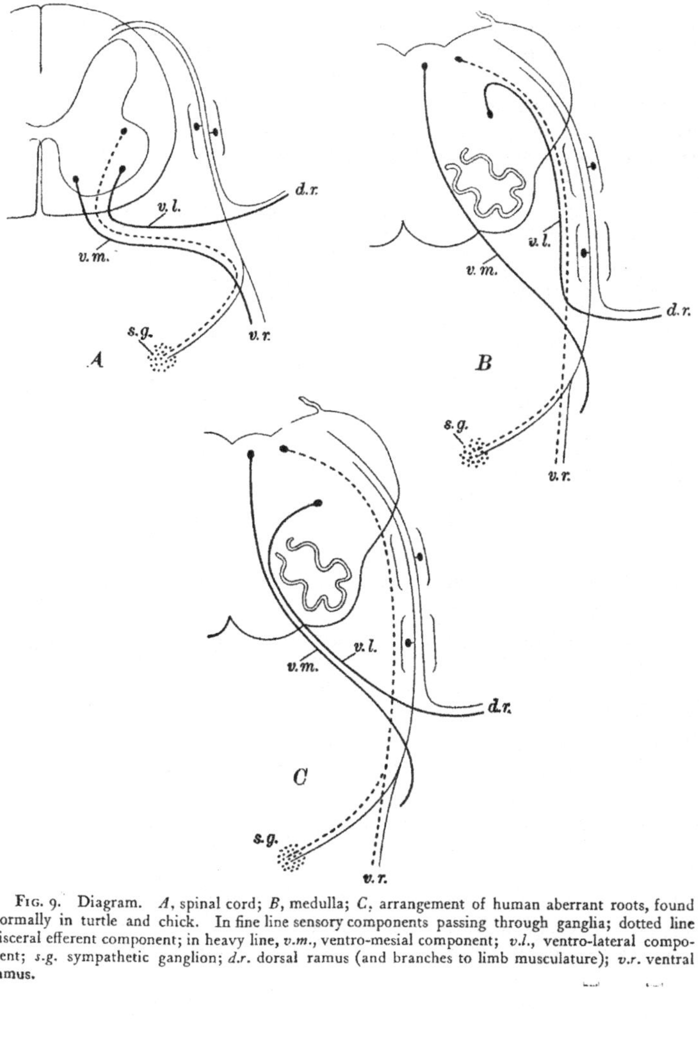

FIG. 9. Diagram. *A*, spinal cord; *B*, medulla; *C*, arrangement of human aberrant roots, found normally in turtle and chick. In fine line sensory components passing through ganglia; dotted line visceral efferent component; in heavy line, *v.m.*, ventro-mesial component; *v.l.*, ventro-lateral component; *s.g.* sympathetic ganglion; *d.r.* dorsal ramus (and branches to limb musculature); *v.r.* ventral ramus.

If we examine CORNING's figures, one of which is reproduced in fig. 8, we can see that the bud which is to form the limb muscles grows distinctly from the lateral part of the myotome, quite separate from the ventral tip from which the body muscles are to arise, so that VAN BISSELICK's three portions are indicated. The cells which are to form the dorsal muscles are also laterally situated, and it seems to me that there is at least as much reason to group the two lateral portions of the myotome, the dorso-lateral and the ventro-lateral cells, and to consider that they give rise to similar muscles, as to group the ventral and the ventro-lateral portions. In other words, I consider the muscles of the limbs as homologous with the muscles of the back and therefore to be supplied by the same nerve components.

JOHNSTON is satisfied to speak of his somatic and visceral motor components as arising from cells in the anterior and lateral horns of the cord, respectively; but the researches, anatomical and physiological, of many workers tend to show that in these broad areas there are several more or less clearly defined groups of cells which have received names, as the ventro-mesial group, the ventro-lateral group, the dorso-mesial group, etc. Experiments have been made to prove that these different groups are the nuclei of origin of certain anatomical nerves, but LAPINSKY (1904.1), in a very elaborate set of tables, shows that this idea has been carried too far; that the fibers to one muscle may arise in several of these cell groups in several segments of the cord, and that these minor groupings are unimportant. A few facts stand out clearly however: there is (1) a median group or groups of cells, extending as a column throughout the cord from the sacral region, and of practically uniform size except for an enlargement at the lower cervical segments; (2) a lateral, or ventro-lateral group also extending throughout the cord, but showing a large increase in size at the lumbar and cervical enlargements; and (3) a more posterior group, described by JOHNSTON as lying between the ventral and dorsal horns, forming the lateral horn; by SANTEE (1907) as "center of the crescent cells," which is not continuous, but present (in the cord) only in the regions of the white rami communicantes of the sympathetic system, and in the region of the roots of the accessory nerve. Both these authors agree that this last group (3) is made of the nerve cells whose axons run to the sympathetic ganglia, and form JOHNSTON's visceral efferent component; with this I agree.

JOHNSTON's somatic efferent component comprises the remaining two columns of cells, whose axons innervate the body muscles. This I have divided into two components, the ventro-mesial and the ventro-lateral, corresponding to the two columns of cells. Recent investigation, summed up by SANTEE, points to the fact that the ventro-mesial cells innervate the ventral trunk muscles, derived from the ventral part of the myotomes, the enlargement of the column at the lower cervical segments being due to the large phrenic nerve which innervates a muscle also derived from the same part of the myotomes. This is the proposed ventro-medial component.

The increase of the ventro-lateral column (2) at the cervical and lumbar enlargements points to the fact that these cells control the muscles of arm and leg, while its continuity throughout the cord points to its control of a smaller body of muscles continuous throughout the trunk, namely, according to this theory—the back muscles. This ventro-lateral component, then, controls the muscles of the back and those of the limbs, which are here considered homologous,[1] and turns laterally from the main nerve trunk either as a dorsal ramus, or as a branch to the limb. This is not without proof, for LAPINSKY's tables agree remarkably well with this conception.

In the medulla oblongata the visceral efferent component takes a new direction, as shown by JOHNSTON, and becomes a lateral root; the ventro-mesial component retains its ventral position, and is represented by the hypoglossal nerve and the eye-muscle nerves; while the ventro-lateral component becomes a lateral root. Two components, then, become lateral, one remains ventral; the nuclei of origin correspond exactly. The nuclei of the hypoglossal nerve and of the eye muscle nerves are median, though no longer ventral, having been forced dorsally by the accumulations of fibers from the motor and sensory decussations; the nucleus of the ventro-lateral component is represented by the nucleus ambiguus; while the nucleus of the visceral efferent component is the dorsal nucleus of the glossopharyngeal, vagus and accessory nerves, and the other motor nuclei in the floor of the IV ventricle. The relative positions of the three columns of gray matter in the cord are maintained, if we consider the opening of the medulla and the changes of fiber tracts.

Let us now see how this theory will account for the cranial nerves; and here we shall consider the vagus-accessory-hypoglossal complex as one nerve, representing the conjoined trunks of several segmental nerves. This complex contains all the components (fig. 9, diag. *B*); the two sensory components run with the vagus to the ganglia found on the root and in a chain extending backward. Sometimes the caudal fibers run to caudal ganglia, FRORIEP'S or hypoglossal ganglia, by way of the hypoglossal trunk which they join where it crosses the vagus. Of the motor components, all three are present; the ventro-mesial efferent runs as the hypoglossal nerve to the muscles of the tongue, derived from the ventro-mesial cells of the myotomes of corresponding segments; the ventro-lateral efferent fibers run in the accessory nerve, and after junction with the other components, as in a spinal nerve, turn dorsally like a dorsal ramus, and innervate one of the muscles of the back, namely the trapezius, joining often with the dorsal rami of the upper cervical nerves; or turn ventrally, like the nerves to the limbs, and innervate the sterno-mastoid, one of the muscles connected with the shoulder girdle, overlying the body muscles, and hence to be classed as a limb muscle. The third motor component, the visceral efferent, is represented by the rest of the accessory nerve and the motor fibers of the vagus, and consists of two varieties of fibers. Running chiefly with the accessory root, but joining the vagus trunk later, are visceral efferent fibers like those in the cord (except for their lateral exit from the medulla) which pass to sympathetic ganglia of the stomach, lungs, heart, etc., and innervate indirectly the involuntary muscles derived from the median part of the cœlom (lateral mesoderm); while the true vagus fibers, also visceral efferents, innervate directly, without the intervention of sympathetic ganglia, the striated muscles of the œsophagus, pharynx and larynx. The vaso-motor and glandular fibers are mingled with both of these varieties, completing the visceral efferent component. All of the components, then, are present.

The difference between this conception of the vago-accessory-hypoglossal complex and that of JOHNSTON is the addition of a somatic motor component, running in the lateral root, and innervating myotomic muscles, "typical body muscles." It places the trapezius muscle and the sterno-mastoid on a par with similar muscles of the lower limb; it explains the junction of the accessory fibers with the dorsal rami of the cervical nerves. Moreover, it is borne out by the conditions found in birds and reptiles, where, as noted above, the hypoglossal nerve roots have true dorsal rami, running to the muscles corresponding to the trapezius, and where the accessory nerve is very small (containing only the visceral efferent fibers). The probability that this is a correct theory is further strengthened by the finding of the laterally running aberrant roots, described above, which could be easily explained as fibers from cells in the nucleus ambiguus, belonging with the ventro-lateral component, which run with the ventral root and then turn dorsally, as in the cord, instead of with the lateral root. (Fig. 9, diag. *C*).

TABLE.

(The sensory components have not been listed)

NERVE.	MOTOR COMPONENTS.
III	ventro-mesial.
IV	ventro-mesial and visceral efferent (?).
V	visceral efferent.
VI	ventro-mesial (probably belongs with V).
VII	visceral efferent.
IX	visceral efferent. (aberrant embryonic ventro-mesial and ventro-lateral components.)
X, XI, XII	ventro-mesial (XII and aberrant). ventro-lateral (X and XI, and aberrant). visceral efferent (X and XI).

It is not necessary to trace the components of the other cranial nerves separately; the accompanying table gives them in outline. The ventro-mesial efferent component is present, with a slight gap (and that this gap is potentially absent is shown by Bdellostoma, and by the ventral roots I have found in embryos) as far forward as the oculomotor nerve. The ventro-lateral efferent component is present as far forward as the nucleus ambiguus extends. Here a word of explanation is necessary. Fibers from the nucleus ambiguus run in the glossopharyngeal and facial nerves, and I expect that it will be found that many of the outer muscles innervated by these nerves are really not derived from lateral mesoderm, like the pharyngeal muscles, but are myotomic muscles, comparable with limb muscles. This would account for the anastomoses between the infra-mandibular branch of the facial nerve, for instance, and branches of the cervical nerves; and for the presence of the laterally running aberrant roots which pass in front of the vagus.

The visceral efferent component is present in the lateral motor roots of the cranial nerves as far forward as the trigeminal, and innervates the muscles derived from the lateral mesoderm the pharyngeal, laryngeal, oral and facial muscles: it also supplies the vaso-motor and excito-glandular fibers to the sympathetic ganglia of the head. This component appears again in the oculomotor nerve, supplying the ciliary ganglion with vaso-motor fibers, and probably innervating the smooth muscles of the eye; but this latter statement is mere conjecture, as the origin of these muscles is not understood. Here the component has resumed its original course taken in the cord, and issues by a ventral root; this is accounted for by the fact that at this level the sides of the brain are no longer opened, as in the medulla, so that the ventral root is the shorter course. Either the EDINGER-WESTPHAL nucleus, which has been supposed to be the center for pupillary reflexes (though this is denied by TSUCHIDA (1906.1), or the nucleus of DARKSCHEWITSCH, which is described as a lateral group of cells, probably represents the dorso-lateral nucleus of this visceral efferent component.

CONCLUSIONS.

In a large percentage of human embryos there are found fibers continuing the line of ventral roots between the hypoglossal nerve and the abducens. Some of these have a ventral course, like the hypoglossal roots, others have a lateral course, like the dorsal rami of spinal nerves. Of each kind some pass behind the vagus and accessory trunk, some between the vagus and the glossopharyngeal nerves. Those in front of the vagus probably represent a ventral root of the glossopharyngeal nerve.

It is suggested that there are three motor components in a typical nerve, arising in three groups of cells in the ventral horn, or in corresponding parts of the medulla, pons, and mid-brain: (1) the ventro-mesial, running to the ventral body muscles: (2) the ventro-lateral, innervating the muscles of the back and of the limbs, which are here considered homologous muscles; and (3) the dorso-lateral, or visceral efferent, supplying the vaso-motor and excito-glandular fibers, and also innervating the muscles derived from lateral mesoderm: i.e., the smooth muscles of the body and the striated muscles of the face, pharynx and larynx.

The ventral mesial components always leave the cord or brain by ventral roots; the ventro-lateral and dorso-lateral components leave also by the ventral root in the cord and mid-brain, but, probably owing to the opening of the medulla, change their point of exit in the hind brain and become lateral roots.

This theory is used to explain the distribution of the cranial nerves.

LIST OF REFERENCES.

BARDEEN, C. R. and LEWIS, W. H.
1901.1. Development of the limbs, body wall and back in man. *Amer. Journ. Anat.*, vol. i.

VAN BISSELICK, J. W.
1905.1. Note on the innervation of the trunk myotone. *Koninkl. Akad. van Wetensch., Amsterdam.*

BYRNES, ESTHER F.
1898.1. On the regeneration of limbs in frogs, etc. *Anat. Anzeiger*, Bd. 15, no. 7.

CORNING, H. K.
1899.1. Ueber die Entwickelung der Kopf- und Extremitäten-muskulatur bei Reptilien. *Morph. Jahrb.*, xxviii, I.

ELZE, C.
1907.1. Beschreibung eines menschlichen Embryo von zirca 7 mm. *Anat. Hefte*, Heft 106

FIELD, H. H.
1894.1. L'embryogenee des Membres. *Rev. Scientif.*, sec. 4, iv.

FRORIEP, A.
1886.1. Zur Entwickleungsgeschichte der Wirbelsäule, etc. *His' Archiv.*
1901.2. Ueber die Ganglienleisten des Kopfs und des Rumpfes und ihre Kreutzung in der Occipitalregion. *Arch. f. Anat. und Physiol., Anat. Abth.*

GASKELL, W. H.
1886. On the structure, distribution and function of the nerves which innervate the visceral and vascular systems. *Journ. of Physiol.*, vol. 8.

JOHNSTON, J. B.
1906. The nervous system of vertebrates. *Philadelphia.*

KAESTNER, S.
1893.1. Die Entwickelung der Extremitäten- und Bauchmusculatur bei den aneuren Amphibien. *Arch. f. Anat. und Physiol., Anat. Abth.*

LAPINSKY, M.
1904.1. Ueber die Lokalisationen motorischer Functionen im Rückenmark. *Deutsch. Zeitschr. f. Nervenheilkunde*, Bd. 26.

LEWIS, F. T.
1903.1. The gross anatomy of a 12 mm. pig. *Amer. Journ. Anat.*, ii, vol. 2.

MCMURRICH, J. P.
1905. The development of the human body. *Philadelphia.*

SANTEE, H. E.
1907. Anatomy of the brain and spinal cord. *Philadelphia.*

STREETER, G. L.
1904.1. The development of the cranial and spinal nerves in the occipital region of the human embryo. *Am. Journ. Anat.*, vol. 4.

TSUCHIDA,
1906.1. *Arbeiten a. d. Hirnanatom. Institut in Zürich*, Heft ii.

VALENTI, G.
1902.1. Sopra le prime fasi di sviluppo della muscolatura degli arti. II (Axolotl).

THE COMMISSURES AND THE NEUROCORD CELLS OF THE BRAIN OF CEREBRATULUS LACTEUS.

BY

CAROLINE BURLING THOMPSON.

WITH THIRTEEN FIGURES.

I. INTRODUCTION.

The material for this paper was obtained partly at Sea Isle City, N. J., during a collecting trip under the direction of Dr. E. G. CONKLIN and Dr. T. H. MONTGOMERY, JR., and partly near Wood's Hole, Mass. To both Dr. CONKLIN and Dr. MONTGOMERY I am indebted for the suggestion that the nervous system of this nemertean should be worked over. The material was sectioned and mounted in the Biological Laboratory of the University of Pennsylvania, and was then laid aside until the present time, but it is a pleasure to recall the kindness and the many helpful suggestions which I received in Dr. CONKLIN's laboratory.

2. METHODS.

Two very large worms were dug up in a small inlet running back from the sea above Sea Isle City. One was fixed in KOSTANECKI's Fluid, the other in 95 per cent alcohol. The former did not give satisfactory results, owing to the gerat shrinkage both of the body as a whole and of the individual tissues, but the worm fixed in 95 per cent alcohol was well extended, and all the tissues were in excellent condition. FLEMMING's Fluid was tried for some much smaller worms found near Wood's Hole, Mass., but the shrinkage of the fibrous parts of the brain was so great that the sections were useless for the study of commissures.

The observations recorded in this paper are based upon the study of the large, well extended worm fixed in 95 per cent alcohol. Only the head has been studied. This was cut in sections about 5μ thick, and stained on the slide with EHRLICH'S hæmatoxylin, followed by ammonia alcohol, and aqueous eosin.

3. THE COMMISSURES.

a. Historical review.—The brains of different species of Cerebratulus have been described by a number of observers, notably HUBRECHT (1887), BÜRGER (1895), COE (1895), and MONTGOMERY (1897). All describe a dorsal commissure, and, immediately posterior to it, a broad stout ventral commissure, the first ventral commissure, but there is great difference of opinion in regard to the number and the character of any additional posterior commissures.

HUBRECHT (1887, pl. xiv, fig. 5) figures that part of the brain of Cerebratulus parkeri which is posterior to the dorsal and the first ventral commissure. There are three commissures (*c.o.*) between the vagus, or œsophageal, nerves (*v.a.*), and the dotted lines (*c.tr.w.*) represent what HUBRECHT terms the ventral metameric connectives of the brain. In the text, p. 77, he states, "*the lower brain lobes are united by the ventral commissures*, separated by a very short distance, till close up to the massive ventral commissure that has been hitherto regarded as the only ventral commissure between the brain lobes. The thin commissures just described are, however, not directly connected with the fibrous core of the brain lobes, which is on the contrary directly continued into the massive inferior commissure, but they seem to derive their fibers from the outer cellular coating of these lobes. They pass underneath the two vagus stems, where these spring from the lower brain lobes, and where these are in their turn in front of the mouth united by transverse commissures." The brain of Drepanophorus lankesteri (pl. ix, fig. 10) has a series of ladder commissures between the lateral nerves.

BÜRGER (1895, taf. 10, fig. 8) figures the brain of Cerebratulus marginatus, in which only the dorsal and the broad ventral commissure are shown. In the text, however, he describes three commissures between the œsophageal nerves; the first of these is slender and thin, the second is also slender, but the third is thick and surrounded by ganglion cells. On p. 320, BÜRGER states "An der nämlichen Stelle" (Abgangstelle der Schlundnerven) "befindet sich die erste Durchbrechung der senkrechten Querwand der Gehirnkapsel, durch welche ein Faseraustausch zwischen den Faserkernen der beiden ventralen Ganglien stattfindet." This "Faseraustausch" is figured on taf. 25, fig. 5, and evidently represents a delicate additional commissure, although BÜRGER does not describe it as such.

MONTGOMERY (1897) describes in the brain of Cerebratulus lacteus a second commissure between the ventral brain lobes, posterior to the broad first ventral commissure, and of much smaller size, and also three between the œsophageal nerves, making a total of five ventral commissures. In Lineus sp. MONTGOMERY finds, posterior to the great first ventral commissure, a second and a third slender commissure between the ventral lobes, and four between the œsophageal nerves, giving a total of seven ventral commissures for this closely related genus.

COE (1895, pl. x, fig. 8) figures the dorsal, the first ventral, and a single œsophageal commissure in the brain of Cerebratulus lacteus.

b. The commissures of Cerebratulus lacteus.—My observations agree with those of the above mentioned workers in the general features of the brain anatomy, but differ from them in the number and the character of the ventral commissures, which must be taken to include those between the ventral brain lobes, whether originating in the cellular sheath or in the fibrous core, and those between the œsophageal nerves.

Fig. 1 is a reconstruction from camera drawings of successive sections, and represents the brain as seen in horizontal optical section. The dorsal brain lobes, *D L*, are formed by the union of numerous small branches which originate in the tip of the head, on each side of the rhynchodeum. The arched dorsal commissure *D C*, unites the two dorsal lobes, and gives off from its median anterior surface a delicate nerve that runs forward to the tip of the head. A similar delicate nerve, the median dorsal nerve, *m d n*, arises on the median posterior surface of the dorsal commissure and runs backward. Just posterior to the dorsal commissure the ventral brain lobes, *V L*, are differentiated from the dorsal lobes, and almost immediately unite in the broad stout band known as the first ventral commissure, V_1; they then extend backward, and with a decrease in size become the lateral cords of the body. The two proboscis nerves, *pr n*, originate on the anterior surface of the first ventral commissure and pass forward and upward into the proboscis at its attachment. From the lateral surfaces of both dorsal and ventral lobes, nerves, not shown in this figure, are given off at irregular intervals, and are not paired with those of the opposite side. The dorsal lobes end posteriorly in the cerebral sense organs, not shown in this figure, which in their turn terminate just in front of the anteriorend of the mouth. The œsophageal nerves, *E N*, arise within the fibrous core of the ventral lobes as follows: A small portion on the medial surface of each brain lobe is constricted off from the rest by a delicate septum, the nerve sheath. These separated portions are the two œsophageal nerves, which, for a short distance, lie within the fibrous sheath of the brain, but farther back pierce the sheath and assume a more medial position.

Behind the broad first ventral commissure comes a series of commissures, *2* to *14*, giving a ladder-like appearance to the brain. Closer investigation reveals thirteen of these commissures, some of which, coming only from the cellular sheath of the brain, may represent the metameric commissures described by HUBRECHT, but others of which, having their roots in the fibrous core of the brain, are commissures that have not been previously described. According to their origin the commissures are of three different kinds: (1) those running from ventral lobe to ventral lobe, whether from the fibrous core or from the cellular sheath, *brain commissures;* (2) those running from ventral lobe to ventral lobe and traversing the substance of the œsophageal nerves, *brain-œsophageal commissures;* (3) those running between the œsophageal nerves only, *œsophageal commissures.*

As many of the commissures are figured here for the first time, they will be described with considerable detail. Their clearness and distinctness are evidently due to the size and the extension of the very favorable material. The following table is a summary of the position, thickness and character of the commissures.

TABLE OF COMMISSURES.

NAME OF COMMISSURE.	ORIGIN OF COMMISSURE.		No. of sections behind preceding commissure.	No. of sections thick.	Dorso-ventral measurement.
		Region in brain.			
dorsal	brain	fibrous core		10	70μ
1st ventral	brain	fibrous core.	15	18	160μ
2d ventral	brain	cellular sheath.	33	2	11μ
3d ventral	brain	cellular sheath and fibrous core	7	3	24μ
4th ventral	brain	cellular sheath	8	3	42μ
5th ventral	brain	fibrous core	4	1	50μ
6th ventral	œsophageal		1	3	53μ
7th ventral	brain	fibrous core, l. root; cellular sheath, r. root,	4	2	42
8th ventral	œsophageal		2	2	50μ
9th ventral	brain	cellular sheath and fibrous core	1	4	52μ
10th ventral	brain-œsophageal	cellular sheath and fibrous core	4	4	42μ
11th ventral	brain-œsophageal	cellular sheath	2	2	21μ
12th ventral	brain-œsophageal	1. root, cellular sheath	4	2	42μ
13th ventral	brain-œsophageal	fibrous core	0	4	42μ
14th ventral	brain-œsophageal	fibrous core	1	10	85μ

Dorsal Commissure.—The dorsal commissure, fig. 1, *D C*, curves forward and also upward, encircling the proboscis at its point of attachment, but the latter curve is not represented in fig. 1, which is approximately in one horizontal plane. From its most anterior point to its posterior ending in the dorsal lobes the dorsal commissure is present in ten sections, and the dorso-ventral measurement is 70μ. Upon the surface of the commissure ganglion cells of type I are very abundant, and a few are found within, between the fibers.

First ventral commissure.—The first ventral commissure, fig. 1, V_1, is the stoutest commissure in the brain, measuring 160μ dorso-ventrally, and eighteen sections in thickness. Its surface is closely invested with a layer of ganglion cells of type I, and great clusters of cells of types II and III are present in the outer part of the cellular sheath, especially on the ventral side.

Second ventral commissure.—The second ventral commissure, fig. 1, *2*, is situated thirty-three sections posterior to the first. The thickness is two sections, and the dorso-ventral measurement is

11μ. This is a *brain* commissure, as the roots come from the cellular sheath of the brain lobes, and it is clear and distinct, though delicate. As the central part is at a more ventral level than the roots, a wide V is formed, which gives the commissure a very distinctive appearance. A few cells of type I are scattered along the surface, but there is no continuous layer. This commissure is intermediate in position to the first two pairs of neurocord cells, and evidently corresponds to the second ventral commissure described by MONTGOMERY (1897) for Cerebratulus and Lineus.

Third ventral commissure.—The third ventral commissure, fig. 1, *3*, is seven sections posterior to the second, and extends through three sections. It is a well defined *brain* commissure, as the roots may be traced into the cellular sheath of the brain lobes. On the right side, fig. 3, two roots are clearly distinguishable, passing toward the brain, one dorsal and one ventral to the right œsophageal nerve, the latter entering the fibrous core. On the left side, only one root is seen, ventral to the œsophageal nerve. Like the second, this commissure also forms a broad V. The central part lies at a more ventral level and is considerably broader than the roots, measuring 24μ dorso-ventrally in the broadest part. The anterior border is thickly beset with ganglion cells of type I. This commissure is here described for the first time in Cerebratulus but seems to correspond, except in its distance from the second, with the third ventral commissure described by MONTGOMERY in Lineus sp.

Fourth ventral commissure.—The fourth ventral commissure, fig. 1, *4*, is situated eight sections posterior to the third. It is three sections thick and has a dorso-ventral measurement of 42μ. The fibers of this commissure are derived from the cellular sheaths of the brain lobes and run from side to side in nearly a straight line at the level of the ventral surface of the œsophageal nerves. A few cells of type I are scattered along the surface of the commissure.

Fifth ventral commissure.—The fifth ventral commissure, fig. 5, comes from the fibrous core of the brain, and lies four sections behind the fourth. The central mass is one section thick and has a dorso-ventral measurement of 50μ. This is the only commissure about which there is any doubt; the roots are clear and distinct, and extend through several sections, fig. 5, *5*, the right root measuring in width 21μ, the left slightly less, and there is a short central mass in one section, fig. 6, *5c*, but the connections between this central part and the roots are not distinguishable. At first it seemed possible that the roots might be merely entering fibers from large cells in the cellular sheath, and this view was supported by the presence of a group of cells of type III, just medial to each of the brain lobes. A further study of succeeding sections showed, however, that the fiber bundles, or roots, may be traced outside the cellular sheath and slightly beyond the medial side of the œsophageal nerves. The question then arises, whether the central part may not be merely the anterior part of the sixth commissure, which begins in the following section. My final conclusion, based upon the careful study of successive sections with the immersion lens, is that the fiber bundles in question are either the roots of a separate, very delicate brain commissure, the fifth, which immediately adjoins the sixth, or the roots of a compound brain-œsophageal commissure, the fifth and the sixth, the anterior fibers of which come from the brain, the posterior fibers from the œsophageal nerves. Since further study of the sections makes the former view slightly more probable, the fifth commissure is represented, fig. 1, as separate from the sixth.

Sixth ventral commissure.—The sixth ventral commissure, fig. 1, *6*, is found one section posterior to the central part of the fifth, and is the first *œsophageal* commissure, running only between the œsophageal nerves. This commissure is three sections thick; in the first two sections the dorso-ventral measurement is 53μ, but is much less in the last section. It is a very clear and well defined commissure, and the passing out of the fibers from the œsophageal nerves is distinctly seen in several sections, as the nerve sheaths are wide open on their medio-ventral surfaces owing to the breadth of the bands of fibers. As the left root originates a few sections posterior to the right, the entire commissure is not in the same frontal plane.

Seventh ventral commissure.—The seventh ventral commissure, fig. 1, *7*, is found four sections posterior to the central part of the sixth; its central mass is present in two sections, with a dorso-ventral measurement of 42μ. The left root is a sharply defined, rather broad band of fibers, and may be traced beneath the left œsophageal nerve through the cellular sheath into the fibrous core of the brain. The right root is delicate and rather indistinct and can be traced only as far as the cellular sheath of the right ventral lobe. Some of the more ventral fibers of the left root seem also to originate in the cellular sheath of the left ventral lobe, hence this commissure may be designated as a brain commissure, derived on the left side from both fibrous core and cellular sheath, but on the right side from the cellular sheath only.

Eighth ventral commissure.—The eighth ventral, fig. 1, *8*, is the second commissure between the œsophageal nerves. It lies two sections behind the seventh; the central part is two sections thick, with a dorso-ventral measurement of 50µ. Owing to the opening of the roots into the œsophageal nerves three or four sections posterior to the central part, the commissure has the form of a broad curve.

Ninth ventral commissure.—The ninth ventral commissure, fig. 1, *9*, is only one section posterior to the eighth. The central part is present in four sections, and measures dorso-ventrally 52µ. Its most median portion is considerably thicker in an antero-posterior direction than the lateral parts near the nerves. Like the eighth, this commissure forms a broad curve, owing to the posterior position of the roots. The fibers pass beneath the œsophageal nerves and may be traced in part into the fibrous core, in part into the cellular sheath of the ventral brain lobes. There is an intimate relation between the fibers of the roots of this commissure and those of the tenth, which will be described below.

Tenth ventral commissure.—The central part of the tenth commissure, fig. 1, *10*, lies four sections posterior to that of the ninth; it is present in four sections, and has a dorso-ventral measurement of 42µ. The anterior surface of the commissure is a straight line, the posterior, a curve, owing to the greater antero-posterior dimension of the median part. The roots come from the fibrous core of the brain lobes, and their fibers then run into and through the œsophageal nerves, making a commissure that may be termed *brain-œsophageal*. The relation between the fibers of the roots of commissures nine and ten will now be described. Leaving the central part of the ninth commissure, the fibers run outward beneath the œsophageal nerves and then slant upward toward the brain lobes, making an oblique fiber band along the lateral surface of each œsophageal nerve. In the following sections the band becomes broader, and the more dorsal fibers, which are from the roots of the tenth commissure, enter the œsophageal nerves on their lateral, and pass out again from their medial surfaces to form the central, œsophageal, part of the tenth commissure.

Eleventh ventral commissure.—The central part of the eleventh commissure, fig. 1, *11*, and figs. 7 to 8, *11*, is situated two sections posterior to the tenth; it is two sections thick, and has a dorso-ventral measurement of 21μ. This is the second brain-œsophageal commissure. The fibers of the central part, figs. 7 to 12, pass above the œsophageal nerves, and are reinforced, especially on the right side, by fibers that issue from the dorsal surface of the nerves. The left root is slender and delicate but may be traced almost to the cellular sheath of the left brain lobe. The right root is stout and runs obliquely backward, through eight sections into the cellular sheath of the right brain lobe, figs. 7 to 12.

Twelfth ventral commissure.—The central part of the twelfth commissure, fig. 1, *12*, lies four sections posterior to the eleventh; it is present in two sections, and measures dorso-ventrally 42μ. This commissure, figs. 9 to 10, may be traced from the cellular sheath of the left brain lobe across to and into the right œsophageal nerve, but no fibers are distinguishable between the right œsophageal nerve and the right brain lobe. The fibers pass above the left œsophageal nerve, and as the nerve sheath is absent at this point there is probably a mingling of nerve substance. In the next two sections, figs. 11 to 12, a very stout bundle of fibers enters the left œsophageal nerve on its lateral surface, and may represent a second root of the same commissure. Fig. 10 is of interest since it contains parts of three commissures, namely, the eleventh, the twelfth and the thirteenth. The fibers at the more dorsal level, *12*, belong to the twelfth, the ventral ones, *13*, to the thirteenth and the root to the right, *11*, to the eleventh. From this it is seen that the twelfth and the thirteenth commissures are in contact with each other, the ventral surface of the former adjoining the dorsal surface of the latter.

Thirteenth ventral commissure.—The anterior part of the thirteenth ventral commissure is in the same section with the posterior part of the twelfth, but ventral to it, fig. 10. At its widest part the dorso-ventral measurement is 60μ. This commissure is present in four sections, and in the last two sections, figs. 12 to 13, the beginnings of the fibers of the fourteenth commissure are seen ventral to the fibers of the thirteenth. The stout distinct roots are several sections posterior to the central part, fig. 1, and may be traced into the fibrous core of the ventral lobes. The fibers from the brain sweep beneath the œsophageal nerves and then upward, forming a broad central loop between. The sheath is absent from the ventral surfaces of the nerves, and an interchange of fibers takes place.

Fourteenth ventral commissure.—Before the loop of the thirteenth commissure has quite disappeared, figs. 12 to 13, other fibers are seen ventral to it which extend across the space between the œsophageal nerves. These fibers may be traced through the ventral part of the œsophageal nerves, and toward the brain lobes, and represent the slender anterior part of the fourteenth and last ventral commissure, which extends altogether through ten sections. It gradually becomes broader until it measures 85μ dorso-ventrally, and in thickness is the second of the ventral commissures. The œsophageal nerves are no longer distinguishable as separate structures but have become a part of the commissure, which is now very wide from side to side and extends from one ventral brain lobe to the other. The very broad roots arise on the dorsal surface of the commissure, and run upward to the brain lobes, entering each fibrous core as a large bundle of fibers several sections posterior to the termination of the central part of the commissure, fig. 1. The œsophageal nerves reappear, and continue backward to the mouth, which begins fifteen sections farther back. This commissure evidently corresponds with the large, third and last, commissure between the œsophageal nerves described by HUBRECHT, BÜRGER and MONTGOMERY.

4. NEUROCORD CELLS.

a. Historical review.—BÜRGER (1894) was the first to distinguish the fourth type of ganglion cells, the giant cells which he terms neurocord cells. He states (1899, p. 105), "Neurochordzellen fand ich bei allen von mir untersuchten Cerebratulen, ferner bei Langia formosa. Das Gehirn besitzt stets nur ein einziges Paar von Neurochordzellen, welches an der medialen Fläche der ventralen Ganglien dort gelagert ist, wo die Schlundnerven entspringen. Zahlreiche Neurochordzellen befinden sich indessen im Ganglienzellbelag der Seitenstamme * * * " The statements of BÜRGER (1895) in regard to the presence of neurocord cells in the Heteronemerteans may be summarized as follows. In several genera a single pair of neurocord cells is found on the medial sides of the ventral brain lobes in the region of the origin of the œsophageal nerves, and numbers of neurocord cells are irregularly distributed along the lateral cords. In the Metanemerteans, Drepanophorus and Prosadenophorus, BÜRGER found that a single pair of neurocord cells occurs in the brain, but that these cells are entirely absent from the lateral cords.

BÜRGER (1895) p. 320 states "Bei C. marginatus sieht man auf einem Querschnitt, welcher die ventralen Ganglien an der Abgangstelle der Schlundnerven getroffen hat, zwei Ganglienzellen von ungewöhnlicher Grösse einander gegenüber liegen, welche um so mehr auffallen, als in diesem Abschnitt des Gesammthirnes nur die kleineren Formen herrschen * * * " He gives the measurements of neurocord cells in two different genera. In Cerebratulus marginatus the diameter across is 20μ, the length 40μ, in Langia formosa the diameter across is 12μ, the length 40μ.

MONTGOMERY (1897) finds in Cerebratulus lacteus three pairs of neurocord cells in the ventral brain lobes, and, like BÜRGER, a large number at unequal intervals along the lateral cords. MONTGOMERY states that the first pair of cells lies in the same section with the beginning of the œsophageal nerves. The third pair lies six sections behind the first, and the cells that compose the second pair, which are not in the same frontal plane, lie between the first and the third pairs. On pp. 402 to 403 the structure of these cells is described. "The structure of the giant ganglion cells IV of Cerebratulus (figs. 27 to 29, 32) has much resemblance to that of cells III of the same species, though there are certain differences which may usually serve to distinguish them.

"The nucleus (fig. 31, *a-e*) may be nearly spherical but is more frequently spherico-oval. It usually has a proximal position within the cell, close to the cell membrane, is seldom central and never distal in position. In it small masses or granules of chromatin (*chr.*) of adequal size are arranged peripherally on the inner surface of the well-marked nuclear membrane; and these do not form a continuous layer, as is frequently the case in the nucleus of III, but are placed at more or less regular distances apart. * * * A thin mass of chromatin envelops the nucleus (*n*). The latter is never absent, is of large size, and almost always peripherally situated; it has thus the same position in the nucleus as the latter has in the cell * * * The cell (figs. 27 to 29, 31) is unusually of a shortened pyriform shape, occasionally nearly spherical, or again elongated (this is the case with the first pair in the brain) * * * As a rule, though not always, these cells are much larger than III.

"The cytoplasm is, especially distally, coarsely vacuolar, more so than in any other ganglion cell; this gives the cell much the same appearance as a slime-producing gland cell."

COE (1895) does not find cells of the fourth type in Cerebratulus lacteus.

The writer (THOMPSON 1901) has found in the brain of Zygeupolia litoralis one pair of neurocord cells, and a pair also in the brain of Micrura cæca.

b. The neurocord cells of Cerebratulus lacteus.—The present investigation differs in one particular from those of the workers quoted above, namely: in the number of the neurocord cells of the brain. Here for the first time are described six pairs of cells and one unpaired cell that in position, in size, and in structure are undoubtedly neurocord cells.

Structure.—For all thirteen cells the general form of the cell body is broad and pear shaped, and the cytoplasm in most cells stains but slightly and contains large vacuoles, although in a few instances the cytoplasm is dark and densely granular. The nucleus is either spherical or slightly flattened, and is always found at the broad end of the cell. Both chromatin and nucleolus are situated at the margin of the nucleus, the latter closely pressed against the nuclear membrane.

The following table gives the size and position of each neurocord cell.

TABLE OF NEUROCORD CELLS.

	WIDTH.		NO. OF SECTIONS THICK.		NO. OF SECTIONS BEHIND PRECEDING PAIR	
	Left cell.	Right cell.	Left cell.	Right cell.	Left cell.	Right cell.
1st pair........	27μ	30μ	3+	4+		
2d pair........	25μ	21μ	1	1	5	5
3d pair........	23μ	21μ	2	2	4	3
4th pair.......	23μ	26μ	1+	1+	3	3
odd cell.......	·	25μ		2+		4
5th pair.......	26μ	25μ	2	2+	3	4
6th pair.......	21μ	21μ	1	1	10	10

Size and position. First pair.—The first pair of neurocord cells is found in the region in which the œsophageal nerves originate from the fibrous core of the ventral brain lobes, fig. 1, n_1, and it therefore agrees in position with the single pair of BÜRGER, and the first pair of MONTGOMERY. The cells of the first pair, fig. 2, ln_1, rn_1, are not symmetrically placed, nor are they exactly equal in size. The right hand cell begins two sections anterior to the left, and extends through four sections, with traces in a fifth; the left cell is present in three sections, with traces in a fourth. The width of the right cell is 30μ and that of the left 27μ. The right cell is very conspicuous on account of its great size and its unusual position, on the ventral surface of the right ventral brain lobe, slightly laterad of the median line of the lobe, and just on the border of the cellular sheath. That the width of this cell seems, in fig. 2, greater than the length may be accounted for by the plane of the section, which has evidently cut the cell at an angle of about 40° to its long axis. The tubule runs obliquely but directly upward into the fibrous core of the brain lobe. The left cell is situated on the medio-ventral surface of the left ventral brain lobe, within the cellular sheath, and its tubule may be traced directly into the fibrous core.

Second pair.—The cells of the second pair, fig. 1, lie in the same section, five sections behind the ending of the first pair, and are only one section thick. In width the left cell measures 25μ, the right cell 21μ. The position of the cells is on the medio-ventral surface of the œsophageal nerves, near the periphery of the cellular sheath of the ventral lobes, and between the second and third ventral commissures.

Third pair.—The neurocord cells of the third pair are not in the same frontal plane, fig. 1, fig. 3, ln_3. The right cell begins three sections, the left cell four sections behind the second pair. Each cell is two sections thick. Their position is near the third ventral commissure on the median side of the œsophageal nerves, the right cell slightly more dorsal than the left and nearer the dorsal blood vessel. In fig. 3 only the left cell is shown, as the right cell ended in the preceding section. The third pair is one of the smaller pairs of neurocord cells, the right cell measuring in width only 21μ, the left cell 23μ; but, in spite of the size, the structure is particularly clear, so that there is no doubt that these cells are of the fourth type.

Fourth pair and the unpaired right hand cell.—The cells of the fourth pair, fig. 1, fig. 4, rn_4, are found in the same section, three sections behind the ending of the third pair. The cells are one section thick, with merely traces in the next, and measure in width, the right cell 26μ, the left 23μ. The fourth pair lies between the third and fourth ventral commissures, but the position in the section of the two cells is not quite the same: the left cell is on the periphery of the left brain lobe medio-ventral to the left œsophageal nerve, the right cell is medio-dorsal to the right œsophageal nerve and just beneath the dorsal blood vessel. A second large cell, 25μ wide, fig. 4, *uc*, presumably of the fourth type, lies alongside of the right hand cell of this pair, and extends through two sections with traces in a third. The tubule of this second cell, together with that of the right hand cell of the pair, may be traced downward through the right œsophageal nerve into the right brain lobe. The line *D V*, fig. 4 indicates the dorso-ventral axis of the œsophageal nerve.

Fifth pair.—The cells of the fifth pair, fig. 1, are not quite in the same frontal plane, the left cell beginning three sections, the right cell four sections behind the fourth pair. This is the second largest pair of neurocord cells; the left cell is two sections thick, and has a width of 26μ, the right cell is two sections thick, with a trace in a third section, and is 25μ wide. This pair is found just anterior to the fourth ventral commissure, and the position in the section of the two cells is similar, both lying on the medial side of the ventral brain lobes, and medio-ventral to the œsophageal nerves.

Sixth pair.—The cells of the sixth and last pair are in the same section, ten sections behind the ending of the fifth pair. This is the smallest pair, as each cell is present in but one section and has a width of only 21μ. They are situated posterior to the sixth ventral commissure, fig. 1, but their position in the section is asymmetrical, the right cell being on the lateral surface of the right œsophageal nerve, between the nerve and the right brain lobe, the left cell on the medial surface of the left œsophageal nerve. The structure of both cells conforms to that of the ganglion cells of the fourth type.

CONCLUSIONS AND SUMMARY.

The present investigation has shown that the ladder-like brain of Cerebratulus lacteus is, in the number of its commissures and neurocord cells, a more complex structure than was heretofore supposed

If we examine the brain of the Metanemertean, Drepanophorus lankesteri, figured by HUBRECHT (1887, plate ix, fig. 10) we find, posterior to the dorsal and the thick first ventral commissure, a series of thin ventral commissures between the lateral nerves. The commissures occur at fairly regular intervals, and the adjacent ones are occasionally connected by irregular fiber bundles.

In the Turbellaria, in the brain of Planocera graffii, figured by LANG (1884, taf. 31, figs. 3 to 4) posterior to the brain are two stout, and many delicate irregular commissures between the lateral nerves, making an intricate network of fibers, but with a generally ladder-like appearance. Again, in the brain of Cestoplana (taf. 31, fig. 2) there is a continuous crossing and interlacing of fibers between the lateral nerves, as far back as the beginning of the proboscis. The nervous system of Gunda segmentata (LANG 1881, taf. xii, fig. 1) is well known on account of the metameric series of commissures between the œsophageal nerves throughout the length of the body.

The comparison of the brains mentioned above with that of Cerebratulus lacteus as described in this paper leads to the conclusion that the brain of Cerebratulus, though complex, is probably of a less specialized and more primitive type than has been supposed.

The greater number of neurocord cells distributed over a greater part of the brain is probably also a primitive character. It is known that they are irregularly placed along the lateral cords, and, the more primitive the brain, the closer is the resemblance in structure of the lateral cords and brain lobes proper.

The presence of the unpaired neurocord cell adjacent to the fourth pair leads me to believe that the number of neurocord cells in the brain is not fixed but variable, and may differ in every individual. The fact that MONTGOMERY found only three pairs of neurocord cells in this same species is additional evidence.

It is probably also true that the number of the ventral commissures varies somewhat with the individual. The stouter commissures, especially those originating in the fibrous core of the brain, would vary least, but the delicate ones, and those derived from the cellular sheath, would be most capable of variation.

1. Thirteen ventral commissures, posterior to the broad, first ventral commissure, are found in the brain of a large well extended individual of Cerebratulus lacteus

2. Of these, six are *brain commissures*, running from ventral lobe to ventral lobe; two are *œsophageal commissures*, running between the œsophageal nerves; five are *brain-œsophageal commissures*, running from ventral lobe to ventral lobe and through the œsophageal nerves.

3. Of the brain commissures, some originate in the fibrous core, some in the cellular sheath of the brain.

4. Six pairs of neurocord cells and one unpaired neurocord cell are found in the ventral lobes of the brain.

5. The brain, though complex in the number of commissures and neurocord cells, is probably of a primitive type, related to that of the Turbellaria.

6. There is probably individual variation in the number of both commissures and neurocord cells.

Wellesley College.
Wellesley, Mass.

BIBLIOGRAPHY.

Bürger, O.
1894. Studien zu einer Revision der Entwicklungsgeschichte der Nemertinen. *Ber. d. Nat. Ges. Freiburg*, Bd. 8.
1895. Die Nemertinen des Golfes von Neapel. *Fauna u. Flora des Golfes von Neapel*, xxii. Monographie.
1897–99. Nemertini. *Bronn's Klassen u. Ordnungen, Bd.* 4, Suppl.

Coe, W. R.
1895. On the anatomy of a species of Nemertean (Cerebratulus lacteus Verrill) with remarks on certain other species. *Trans. Connect. Acad.*, vol. 9.

Hubrecht, A. A. W.
1887. Report on the Nemertea collected by H. M. S. Challenger during the years 1873–1876. *Challenger Report*, vol. 19.

Lang, A.
1881. Der Bau von Gunda segmentata u. d. Verwandtschaft der Platyhelminthen mit Cœlenteraten u. Hirudineen, *Mitth. aus d. Zool. Station zu Neapel*, Bd. iii, Heft 1–2.
1884. Polycladen. *Fauna u. Flora des Golfes von Neapel*, xi. Monographie.

Montgomery, T. H.
1897. Studies on the elements of the central nervous system of the Heteronemertini, *Journ. Morph.*, vol. 13.

Thompson, C. B.
1901. Zygeupolia litoralis, a new Heteronemertean. *Proc. Acad. Nat. Sci., Philadelphia.*

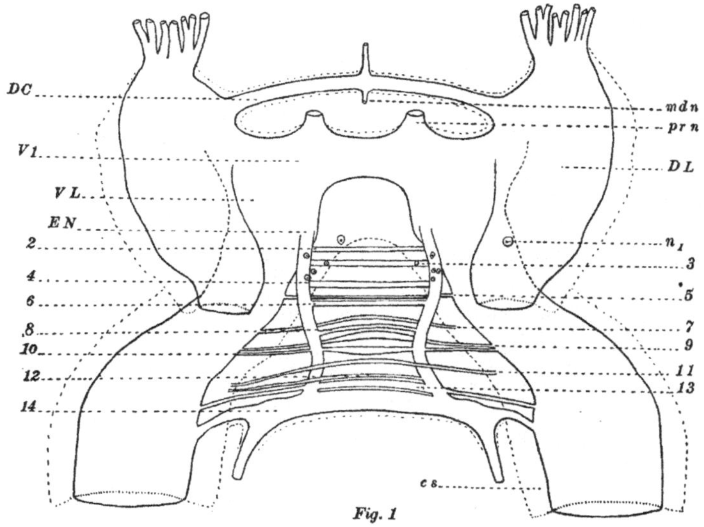

Fig. 1

All figures refer to Cerebratulus lacteus, and are drawn at the level of the stage with the Zeiss camera lucida, and with Zeiss lenses, the combinations of which are given with each figure. The plates have been reduced to about two-thirds of the original size.

FIG. 1. A reconstruction of the brain from a series of transverse sections. Obj. AA, oc. 2, tube length 160 mm. *mdn*, median dorsal nerve; *prn*, proboscis nerve; *DL*, dorsal lobe; *VL*, ventral lobe; n_1, neurocord cell of the first pair; *EN*, œsophageal nerve; *DC*, dorsal commissure; V_1, first ventral commissure; *2–14*, 2d to 14th ventral commissures; *cs*, cellular sheath.

FIG. 2. A transverse section through the ventral lobes in the region of the origin of the œsophageal nerves, and showing the first pair of neurocord cells. Obj. AA, oc. 4, tube length 170 mm. rn_1, right neurocord cell of the first pair; ln_1, left neurocord cell of the first pair; *en*, œsophageal nerve; *fc*, fibrous core; *cs*, cellular sheath.

FIG. 3. A transverse section through the ventral lobes and the third ventral commissure, showing also the left neurocord cell of the third pair. Obj. AA, oc. 4, tube length 170 mm. ln_3 left neurocord cell of the third pair; *3*, third ventral commissure, with two roots on the right side.

FIG. 4. Part of a transverse section, showing the right neurocord cell of the fourth pair, and the unpaired neurocord cell. Obj. homog. immers. $\frac{1}{12}$, oc. 2, tube length 160 mm. rn_4, right neurocord cell of the fourth pair; *uc*, unpaired neurocord cell; *ren*, outline of right œsophageal nerve; *fc*, fibrous core of the right brain lobe; *DV*, dorso-ventral axis of the right œsophageal nerve.

FIGS. 5 and 6. Parts of two consecutive sections through the ventral lobes, showing the fifth ventral commissure. Obj. AA, oc. 4, tube length 160 mm. *5*, fifth ventral commissure: *5c*, central part of fifth ventral commissure.

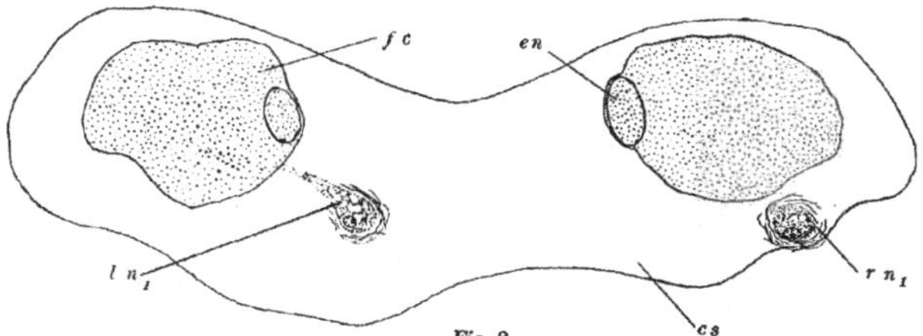

Fig. 2

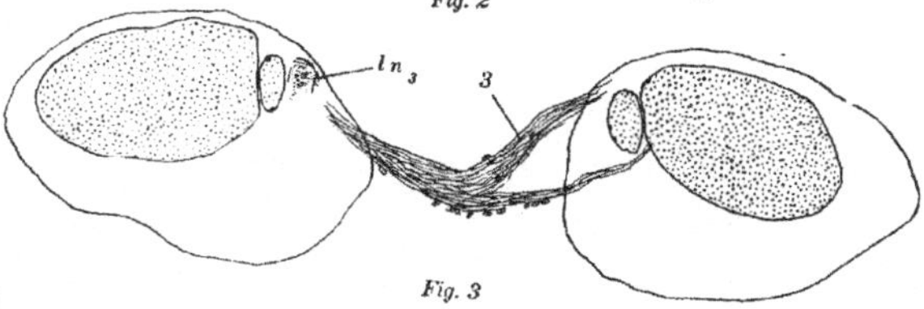

Fig. 3

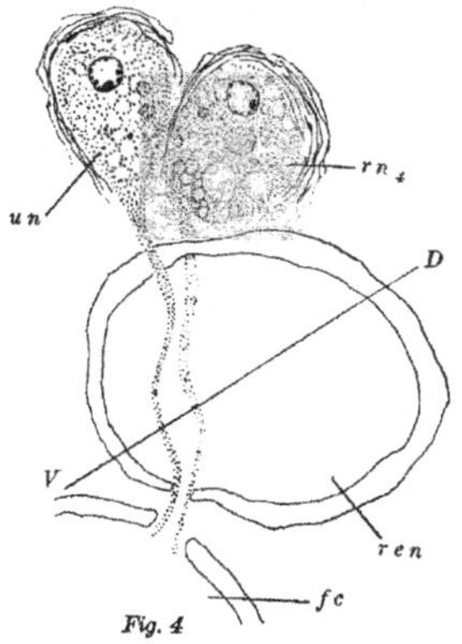

Fig. 4

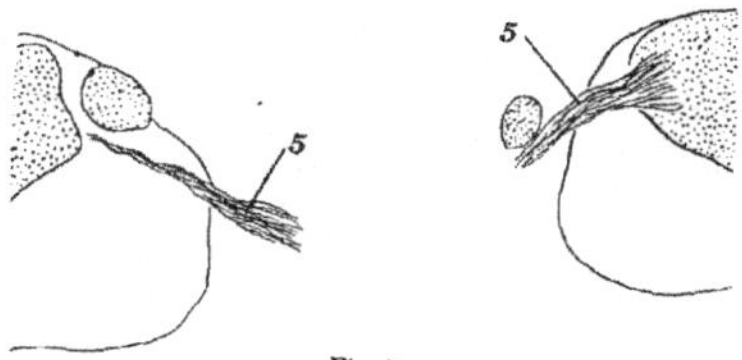

Fig. 5

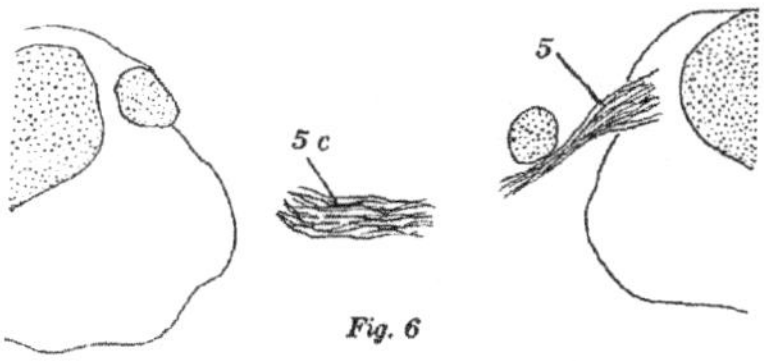

Fig. 6

FIG. 7. Part of a transverse section through the ventral lobes, showing the eleventh ventral commissure. Obj. AA. oc. 4, tube length 160 mm. *cs*, cellular sheath; *fc*, fibrous core; *len*, left œsophageal nerve; *ren*, right œsophageal nerve; *11*, eleventh ventral commissure.

FIG. 8. Part of a transverse section, two sections posterior to that shown in Fig. 7, showing some of the central portion and the right root of the eleventh commissure. Obj. AA, oc. 4, tube length 160 mm.

FIG. 9. Part of a transverse section, two sections posterior to that shown in Fig. 8. Obj. AA, oc. 4, tube length 160 mm. *12*, twelfth ventral commissure; *11*, root of the eleventh ventral commissure.

FIG. 10. Part of a transverse section, one section posterior to that shown in Fig. 9, showing parts of three commissures. Obj. AA, oc. 4, tube length 160 mm. *11*, eleventh ventral commissure; *12*, twelfth ventral commissure; *13*, thirteenth vèntral commissure.

FIG. 11. Part of a transverse section, one section posterior to that shown in Fig. 10. Obj. AA, oc. 4, tube length 160 mm.

FIG. 12. Part of a transverse section, one section posterior to that shown in Fig. 11, showing the broadest part of the thirteenth commissure, the left root of the twelfth, and the beginning of the fourteenth commissure. Obj. AA, oc. 4, tube length 160 mm.

FIG. 13. Part of a transverse section, one section posterior to that shown in Fig. 12, showing the thin posterior part of the thirteenth commissure. Obj. AA, oc. 4, tube length 160 mm. *13*, thirteenth ventral commissure, *14*, fourteenth ventral commissure, *rc*, root of the thirteenth commissure.

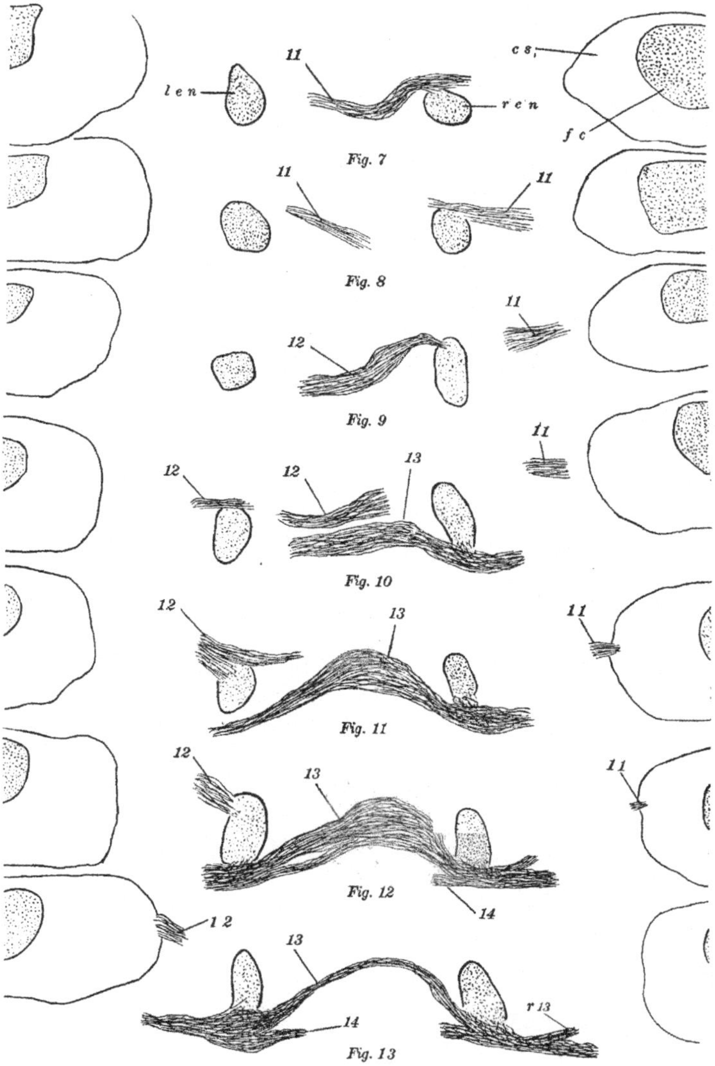

Fig. 7

Fig. 8

Fig. 9

Fig. 10

Fig. 11

Fig. 12

Fig. 13

EDITORIAL.

TWO RECENT TENDENCIES IN CEREBRAL MORPHOLOGY.

Many diverse lines of current biological research are putting emphasis on the functional unity of the living animal body. Normal growth, regulation, coördination of reactions and the manifold phases of adptation all point the same way. The degree of perfection of the integrative function of the nervous system in higher vertebrates is determined not only by the complexity of the central internuncial or associational conduction paths, but quite as much by the extent and character of the differentiation of the receptors and effectors, i.e., the nature of the correspondence (in the Spencerian sense) between the organism and the environment. The central nervous system cannot therefore be studied comparatively to the best advantage by itself, but only in relation with the peripheral nervous system and indeed with the body as a whole.

Recent students of the phylogeny of the vertebrate nervous system, recognizing this principle, have adapted themselves to it in two very different ways. The anatomists of one group have made an especial study of the mechanical factors in cerebral architecture, such as the effects on the brain of ontogenetic or phylogenetic changes in the form of the cranium, position of peripheral organs, vascular supply, arrangements of the non-nervous parts of the brain, etc. The anatomists of the other school have placed more stress upon the conduction paths and have devoted themselves to the exposition of the architectural effects of variations in the physiological importance of the several functional systems of neurones of which the nervous system is composed. Differences in the functional patterns or action systems of animals involve parallel differences in the architecture of the nervous system and the solution of many problems is sought in a comparative study of functional systems of neurones, correlating the variations in anatomical structure with differences in physiological value or behavior.

The first group of anatomists works from the standpoint of developmental mechanics, the nervous masses being considered as shaped more or less passively by surrounding growth forces or by internal pressures and strains due to irregularities in the growth of the masses themselves. The second group lays the emphasis rather upon the functional nervous tissue itself as the determining factor in cerebral morphology.

An excellent illustration of the methods of the mechanical school is furnished by some of the embryological papers of the late Professor HIS, particularly his figures showing very clever mechanical models of the invagination of the neural plate, formation of the neural tube, etc.[1]

As applied to comparative neurology this standpoint received its clearest exposition and most graphic illustration at the hands of Professor RUDOLF BURCKHARDT, whose untimely death last winter interrupted in the midst an exceedingly valuable series of researches. In a personal letter written last January a very few weeks before his death he outlined the motives of his work in these words:

"The principal point—besides the fact that never such large materials have been examined before—is for me that all our views of the central nervous system are still dominated by practical, psychological, physiological, traditions, and that the simple standpoint of vertebrate phylogeny has never been thoroughly kept, as by observing such objects as growing epithelia which are changed by influences of head-formation, and the central need of sensory organs (the latter has been urged most by JOHNSTON). Second, that brain phylogeny must be studied according to phylogeny as it issues from palæontological researches. You will perhaps miss that I do not enter into description of fiber courses, but only into their quantity as a mechanical factor; that I treat the nerve cells as such as of secondary value for the knowledge of brain genesis and the type of the brain, and that on the other side I attribute such a high value to neuroglia. But you will also see that I wanted to regard the brain as a part of the head, and the real head, not the hypothetical of primary metamerism, which for the brain has not much more value than for the skin. There is, in my opinion, a great field for work, particularly for zoölogists,

[1] HIS, WM. Ueber mechanische Grundvorgänge thierischer Formenbildung. *Arch. f. Anat.* [*u. Physiol.*]. 1894.

in applying knowledge from our side to those of the physiologist and pathologist."

These points will be found elaborated and fully illustrated in the extensive series of papers which has come from Professor BURCKHARDT's pen, particularly in the introduction to the last of his papers, on the central nervous system of the selachians as a basis for a phylogeny of the vertebrate brain[2] the first part of which appeared last year. In the letter from which the above extract was taken he wrote that he was at that time engaged upon the final revision of the second and third parts of this paper, and it is to be hoped that the work was sufficiently far advanced at the time of his death as to permit the speedy publication of these two parts.

The most valuable of the concrete results of BURCKHARDT's work so far as published is the demonstration of the conservative character of the non-nervous parts of the brain and the consequent worth of the membranous and ependymal tissues in the study of phylogenetic relationships. This is brought out most graphically in the paper published in 1895, entitled Der Bauplan des Wirbeltiergehirns[3] which is accompanied by a large plate showing diagrammatic sagittal sections of all important types of vertebrate brains with the corresponding regions colored the same way throughout the series. The resemblance of these median and largely membranous parts in the series from Petromyzon to man is very striking.

An important series of neurological researches which in some ways resemble those of BURCKHARDT has been published by Dr. G. STERZI of Padua. The most recent of about a dozen papers relating to the meninges and vascular supply of the central nervous system is the first volume of a comprehensive treatise on the central nervous system of vertebrates.[4] It is announced that the work will be completed in six volumes, of which this first one is devoted to the cyclostomes, and the others are to be upon the fishes, amphibians, reptiles, birds and mammals, respectively. The present volume is divided into two books devoted to the petromyzonts

[2] Das Zentral-Nervensystem der Selachier als Grundlage für eine Phylogenie des Vertebratenhirns. I. Teil. Einleitung und Scymnus lichia. With 5 plates and 64 text-figures. *Nova Acta, Abh. kais. Leop.-Carol. Akad. d. Naturforscher, Halle*, Bd. 73, no. 2, pp. 238–449, 1907.

[3] *Morph. Arbeiten (Schwalbe)*, Bd. 4, no. 2. 1895.

[4] STERZI, G. Il sistema nervoso centrale dei Vertebrati. Ricerche anatomiche ed embriologiche. Vol. 1. Ciclostomi. 732 pp. and 194 figs. *Padua, A. Draghi, Editore.* 1907. Price, Lire 35.

and the myxinoids respectively. The chief topics of consideration in each book are, the morphology of the vertebral canal and cranium, the meninges and the blood vessels and lymphatics, the sheaths of the nerves, and the central nervous system, particularly its hypophysis and membranous parts. In short, the non-nervous parts of the brain receive especial attention, with particular reference to the factors of nutrition, metabolism and mechanical support. All of these subjects are treated from the embryological point of view and the developmental stages are fully figured. The form relations of the brain are studied embryologically and their comparative morphology considered, but the volume contains no other descriptions of internal architecture. Neither the fiber tracts nor the cellular masses are considered.

These researches and many others along the same lines have drawn attention to some very important types of relation between the nervous and the other organs of the body, chiefly mechanical and nutritive. But after all, the principal avenues of relation of the nervous system are the nerves themselves. The sense organs and the organs of response are the immediate instruments of almost all animal activities and the central nervous system reflects every change in peripheral relations. This reflection, however, is not a transient and passive return of the nervous impulse from the receptive to the effective periphery as a light beam rebounds from a mirror; it involves an active process of coördination during the process, and—what is far more important from our present standpoint—a permanent structural change in the coördinating mechanism itself.

The cerebral architecture of every animal species has unquestionably been shaped by its peripheral nervous organs. As animals gradually change their mode of life and different sets of environmental forces impinge on the sensorium, the receptive and effective peripheral organs gradually undergo parallel changes adapted to render the animal more fit to meet the changed environmental conditions. And the central correlation apparatus of these peripheral organs must change its form at the same time or the whole process of the selection of adaptive variations would be abortive. The structure of the central nervous system is in fact very sensitive to changes in environment or mode of life. The eyes of cave animals atrophy; so also do the visual centers of the brain. The acuity of every sense in the action systems of different animals can be measured by the comparative anatomist in terms of the size and structural complexity of the corresponding primary cerebral sensory centers. The habitual type of motor response is no less accurately registered in the permanent organization of the motor cerebral centers. Furthermore, an action system of the rigidly stereotyped sort will be served by a nervous system with the primary reflex centers highly elaborated perhaps, but with the association centers small, while the more plastic types of action system as found in the more intelligent animals are characterized by highly complex association centers and tracts.

Studies in cerebral architecture carried on from the point of view of the analysis of functional systems of neurones, each of which is both a physiological and an anatomical unit, are as far removed as possible from the older descriptive neurology which seemed to aim at mere enumerations of tracts and cell masses with little effective correlation. The best recent work on cerebral architecture aims more or less directly at the analysis of conduction paths and their correlation into definite functional systems.

A great impetus was given to such studies in the comparative field by the analysis of the peripheral nerves into their components and the rearrangement of these components into functional systems, thus facilitating the integration of the more diffuse sensory systems, like the tactile and gustatory, and permitting the study of their central reflex pathways with almost as great precision as the concentrated systems, like the optic. The analysis of the cutaneous nerves of man into their components by the researches of HENRY HEAD and others by a combination of physiological, pathological and anatomical methods promises still more important advances in this direction.

The four-root theory of GASKELL and HIS has been the point of departure, not only for the study of the components of the peripheral nerves, but also for the study of the functional zones of the central nervous system. The progress which has been made in the functional analysis of the brain and the illuminating value of a knowledge of peripheral nerve components in this study (particularly in the medulla oblongata) are illustrated in a striking way by a comparison of the second volume of the sixth edition of

EDINGER's lectures on the Central Nervous System, published in 1904, and the seventh edition, published in 1908.[5]

Dr. EDINGER has, either personally or with the help of other members of his staff, worked over a large part of the field covered by the voluminous literature of comparative neurology of the past decade. He has therefore been able to make this edition of his text-book, like its predecessors, very largely a record of his own observations. This is at the same time a source of great strength and of considerable weakness in his work—of strength because all of the old facts presented come with the added weight of EDINGER's confirmation; of weakness because many equally important facts or theoretical conclusions which do not chance to fall within the scope of the author's personal observation are altogether omitted.

The older literature on the comparative anatomy of the medulla oblongata is a confusing mass of contradictory detail, dominated largely by misleading metameric schemata. The recognition of functional units within the oblongata, each of which stands in relation with a definite component of the peripheral nerves and each of which is integrated in a characteristic manner and has its own special type of secondary reflex pathways—this has made possible a far more simple and comprehensive exposition of the structure of this part of the brain than we have had before. While much remains to be explained in the comparative anatomy of the medulla oblongata, the underlying morphological pattern has been exposed and is found to be surprisingly constant in all vertebrates. This constancy of type grows out of the fact that this part of the brain uniformly serves the simpler vital functions, such as feeding, respiration, etc., whose peripheral mechanisms are broadly similar in vertebrates. Such variations in feeding habits as do occur are, however, accompanied by changes in the details of cerebral structure; as, for example, the development of the enormous vagal lobes of the carp correlated with the peculiar palatal organ of this fish, and the modifications in the sensory termini of the facialis nerve correlated with the peripheral distribution of cutaneous taste buds in Ameiurus and Gadus respectively. Another illustration is furnished by the development of large eyes and optic

[5] EDINGER, L. Vorlesungen über den Bau der Nervösen Zentralorgane des Menschen und der Tiere. Bd. 2. Verlgeichende Anatomie des Gehirns. Seventh Edition. *Leipzig, F. C. W. Vogel.* 1908.

lobes in predacious species which capture moving prey by the sense of sight. The change from aquatic to aërial respiration in the Amphibia involves changes in the medulla oblongata parallel with the atrophy or change of function of the branchial muscles which are as yet imperfectly understood and which can be learned only by a closely correlated examination of the central and peripheral organs of a selected series of forms.

Now returning to Dr. EDINGER's manual, the changes wrought in this edition are quite revolutionary. The first chapter opens with an analysis of the peripheral nerves based on the work of the recent students of nerve components. In later chapters the medulla oblongata is subdivided in accordance with the same criteria along lines which follow in a general way those laid down by the American school of comparative neurologists, though with an entirely new series of illustrative figures and with many differences of interpretation. Most of these differences take the direction of conservatism toward the newer morphological ideas and the result is many cases of confusion, sometimes amounting to actual contradiction, growing out of the imperfect assimilation of the old data and the new morphology. But in the broad view there has been during the past decade a rapid *rapprochement* between the German and American comparative neurologists, particularly in the interpretation of the peripheral nerves and the brain stem, which is very gratifying to those who have painful recollections of the recent (and still all too prevalent) chaos in the morphological interpretaiton of the medulla oblongata and its nerves.

In the midbrain and thalamus there is as yet no such gratifying harmony of interpretation as in the medulla oblongata. In fact, the best course at present open to us in considering the morphology of these regions is a serious application to anatomical study to the end that we may acquire a more precise and comprehensive knowledge of the facts before we attempt to complete our morphological interpretation.

In the chapters of Dr. EDINGER's work devoted to the forebrain we find the most original and the most important of his own contributions to the comparative morphology of the brain. The interpretation of the primordial pallium as a center of correlation for all of the sênse organs of the snout (smell, tactile and somæsthetic sensations of the lips, tongue, etc.) is an exceedingly fruitful suggestion which promises much as a point of departure for the ultimate interpretation of the evolution of the cerebral cortex. This point and many other illustrations of the importance of studying the brain in close functional relation with its peripheral end organs are elaborated in Dr. EDINGER's address printed in the last issue of this *Journal.*[6]

C. J. H.

[6] The relations of comparative anatomy to comparative psychology. This *Journal*, vol. 18, no. 5. 1908.

Spotted something that isn't quite right?

Let us know and we'll fix it for future reprints. And as a thank you for your diligence, we'll send you a free Sothis book of your choice.

publisher@sothis.press

This page intentionally left blank